Concepts and Applied Principles of Mathematics

Concepts and Applied Principles of Mathematics

Edited by **Lucas Lincoln**

NY RESEARCH PRESS
New York

Published by NY Research Press,
23 West, 55th Street, Suite 816,
New York, NY 10019, USA
www.nyresearchpress.com

Concepts and Applied Principles of Mathematics
Edited by Lucas Lincoln

International Standard Book Number: 978-1-63238-094-4 (Hardback)

Printed in China.

Contents

Preface

Mathematical science is a vast field that includes, along with the area of mathematics, those disciplines that are essentially mathematical in nature but might not be generally considered subfields of proper mathematics. Mathematics is the discipline that studies topics that include quantities, structures, spaces as well as change. Mathematics seeks out designs and patterns and uses them to formulate new conjectures and theories. Mathematicians determine the truth or falsity of assumptions through mathematical proofs. Through the use of constructs and logic, mathematics has developed from calculation, counting, measurement, and the study of the shapes and motions of physical objects. Mathematics can thus be said to have been a part of human activity for as far back as written records exist. There are various fields that are a part of mathematics and mathematical sciences like computer and computational science, operations research, population genetics, econometrics, theoretical physics and even cryptology. Mathematics is used throughout the world as an essential tool in many fields, including engineering, natural science, finance, medicine and the social sciences.

This book is an attempt to compile and collate all available research on mathematics and mathematical sciences under one aegis. I am grateful to those who put their hard work, effort and expertise into these research projects as well as those who were supportive in this endeavour. I would also like to thank my publisher for giving me this immense opportunity to utilize my potential. Lastly, I would like to thank my friends and family for their endless support and faith in me.

Editor

1

The Ulam Type Stability of a Generalized Additive Mapping and Concrete Examples

Hiroyoshi Oda,[1] Makoto Tsukada,[1] Takeshi Miura,[2] Yuji Kobayashi,[3] and Sin-Ei Takahasi[4]

[1] *Department of Information Sciences, Toho University, Funabashi, Chiba 274-8501, Japan*
[2] *Yamagata University, Yonezawa, Yamagata 273-0866, Japan*
[3] *Toho University, Funabashi, Chiba 274-8501, Japan*
[4] *Toho University, Yamagata University, Funabashi, Chiba 273-0866, Japan*

Correspondence should be addressed to Makoto Tsukada; tsukada@is.sci.toho-u.ac.jp

Academic Editor: Irena Lasiecka

We give an Ulam type stability result for the following functional equation: $f(\alpha x-\alpha x'+x_0)=\beta f(x)-\beta f(x')+y_0$ (for all $x,x'\in X$) under a suitable condition. We also give a concrete stability result for the case taking up $\delta\|x\|^p\|x'\|^q$ as a control function.

1. Introduction

In 1940, Ulam [1] proposed the following stability problem: "When is it true that a function which satisfies some functional equation *approximately* must be close to one satisfying the equation *exactly*?" Next year, Hyers [2] gave an answer to this problem for additive mappings between Banach spaces. Furthermore, Aoki [3] and Rassias [4] obtained independently generalized results of Hyers' theorem which allow the Cauchy difference to be unbounded.

Let X and Y be normed spaces over $\mathbb{K}$, which denotes either the real field $\mathbb{R}$ or the complex field $\mathbb{C}$. Throughout the paper, we fix scalars $a,b,c,d\in\mathbb{K}\setminus\{0\}$ and vectors $x_0\in X$ and $y_0\in Y$. We say that a mapping f of X into Y is $(a,b,c,d;x_0,y_0)$-additive if

$$f\left(ax+bx'+x_0\right)=cf(x)+df\left(x'\right)+y_0 \tag{1}$$

for all $x,x'\in X$. When $x_0=y_0=0$, we say it to be (a,b,c,d)-additive. Aczél [5] specified what this generalized Cauchy equation is. The Ulam type stability problem for such an f has been investigated in [6–8]. However, these results have been obtained in cases where either $a+b\neq 0$ or $c+d\neq 0$ (see Theorems A and B). In this paper, we will investigate the problem for $(a,-a,c,-c;x_0,y_0)$-additive mappings, that is, in the case $a+b=c+d=0$. In Section 2, we state the details of $(a,-a,c,-c;x_0,y_0)$-additive mappings (Theorem 3). In Section 3, we give our main results about the stability for them (see Theorems 7–10). In the final section, we apply the results to some concrete examples, where we take up $\delta\|x\|^p\|x'\|^q$ as a control function $\varepsilon(x,x')$ (see Corollaries 11–14).

2. $(a,-a,c,-c;x_0,y_0)$-Additive Mappings

The following result asserts that any $(a,-a,c,-c;x_0,y_0)$-additive mapping is transformed into some $(a,-a,c,-c)$-additive mapping by a certain translation and that any $(a,-a,c,-c)$-additive mapping is an additive mapping in usual sense with some extra condition.

Proposition 1. *Let f and g be two mappings of X into Y such that $g(x)=f(x+x_0)-y_0$ for all $x\in X$. Then the following three statements are equivalent:*

(i) *f is $(a,-a,c,-c;x_0,y_0)$-additive,*

(ii) *g is $(a,-a,c,-c)$-additive,*

(iii) *g is additive and $g(ax)=cg(x)$ for all $x\in X$.*

Proof. (i)⇔(ii) Since

$$\begin{aligned} f\left(ax - ax' + x_0\right) &= g\left(ax - ax'\right) + y_0 \\ &= g\left(a\left(x - x_0\right) - a\left(x' - x_0\right)\right) + y_0, \\ c(x) - cf\left(x'\right) + y_0 &= c\left(f(x) - y_0\right) - c\left(f\left(x'\right) - y_0\right) + y_0 \\ &= cg\left(x - x_0\right) - cg\left(x' - x_0\right) + y_0 \end{aligned} \tag{2}$$

for all $x, x' \in X$, it follows that

$f : (a, -a, c, -c; x_0, y_0)$-additive
$\Leftrightarrow g(a(x-x_0)-a(x'-x_0))+y_0 = cg(x-x_0)-cg(x'-x_0) + y_0$ (for all $x, x' \in X$)
$\Leftrightarrow g(ax - ax') = cg(x) - cg(x')$ (for all $x, x' \in X$)
$\Leftrightarrow g : (a, -a, c, -c)$-additive.

(ii)⇒(iii) Suppose that

$$g\left(ax - ax'\right) = cg(x) - cg\left(x'\right) \tag{3}$$

for all $x, x' \in X$. When $x = x'$, we have $g(0) = 0$. Using this, $g(ax) = cg(x)$ and also $g(-ax) = -cg(x)$ for all $x \in X$. Therefore,

$$\begin{aligned} g\left(x + x'\right) &= g\left(a\frac{x}{a} - a\left(-\frac{x'}{a}\right)\right) \\ &= cg\left(\frac{x}{a}\right) - cg\left(-\frac{x'}{a}\right) \\ &= cg\left(\frac{x}{a}\right) + (-c)\, g\left(-\frac{x'}{a}\right) = g(x) + g\left(x'\right) \end{aligned} \tag{4}$$

for all $x, x' \in X$.

(iii)⇒(ii) Because $g(-x) = -g(x)$ for all $x \in X$ (see also the following remark), it is trivial.

Remark 2. We denote by $\mathbb{Q}$ the field of all rational numbers. It is well known that if g is additive, then $g(qx) = qg(x)$ for every $q \in \mathbb{Q}$ and $x \in X$, that is, g is $\mathbb{Q}$-linear. Hence, if g is additive and continuous, g must be $\mathbb{R}$-linear. On the other hand, when $\mathbb{K} = \mathbb{C}$, we have a lot of continuous additive nonlinear mappings by considering the composition of linear transformations on $\mathbb{R}^2$ and the $\mathbb{R}$-linear isometry $(x, y) \mapsto x + iy$.

The constant y_0 is a trivial $(a, -a, c, -c; x_0, y_0)$-additive mapping of X into Y. The following theorem says that unless it is a unique $(a, -a, c, -c; x_0, y_0)$-additive mapping, discontinuous one always exists.

Theorem 3. (I) *If $a = c$, then there exists a discontinuous $(a, -a, c, -c; x_0, y_0)$-additive mapping of X into Y.*

(II) *If $a \neq c$, then the followings hold:*

(i) *If both a and c are transcendental numbers, then there exists a discontinuous $(a, -a, c, -c; x_0, y_0)$-additive mapping of X into Y.*

(ii) *If one of a and c is transcendental and the other is algebraic, then the constant y_0 is a unique $(a, -a, c, -c; x_0, y_0)$-additive mapping of X into Y.*

(iii) *If both a and c are algebraic with a common minimal polynomial, then there exists a discontinuous $(a, -a, c, -c; x_0, y_0)$-additive mapping of X into Y.*

(iv) *If a and c are algebraic with distinct minimal polynomials, then the constant y_0 is a unique $(a, -a, c, -c; x_0, y_0)$-additive mapping of X into Y.*

Moreover, when $\mathbb{K} = \mathbb{R}$, there is no nontrivial continuous $(a, -a, c, -c; x_0, y_0)$-additive mapping f of X into Y. On the other hand, when $\mathbb{K} = \mathbb{C}$, if a is not real and $a = \bar{c}$ (the complex conjugate of c), then the mapping $x \mapsto f(x + x_0) - y_0$ must be conjugate linear for every continuous $(a, -a, c, -c; x_0, y_0)$-additive mapping f of X into Y.

In order to show Theorem 3, we need some lemmas for $\mathbb{K}$-valued $(a, -a, c, -c)$-additive functions defined on $\mathbb{K}$. For any $x \in \mathbb{K}$, we denote by $\mathbb{Q}(x)$ the subfield of $\mathbb{K}$ generated by x over $\mathbb{Q}$.

By the following proofs of Lemma 6 and Theorem 3, if there is a discontinuous $(a, -a, c, -c; x_0, y_0)$-additive mapping, then there are sufficiently many such mappings in the sense that there exists such a mapping which separates any $\mathbb{Q}(a)$-linear independent points of X.

Lemma 4. *Any $(a, -a, c, -c)$-additive φ of $\mathbb{K}$ into itself satisfies $\varphi(p(a)x) = p(c)\varphi(x)$ for all $x \in \mathbb{K}$ and $p(X) \in \mathbb{Q}[X]$.*

Proof. Note that φ is additive and $\varphi(ax) = c\varphi(x)$ for each $x \in \mathbb{K}$ by Proposition 1. Let $p(X) = a_0 + a_1X + \cdots + a_nX^n$ with $a_0, a_1, \ldots, a_n \in \mathbb{Q}$. Since φ is $\mathbb{Q}$-linear,

$$\begin{aligned} \varphi(p(a)x) &= a_0\varphi(x) + a_1\varphi(ax) + \cdots + a_n\varphi(a^nx) \\ &= a_0\varphi(x) + a_1c\varphi(x) + \cdots + a_nc^n\varphi(x) \\ &= p(c)\varphi(x) \end{aligned} \tag{5}$$

for all $x \in \mathbb{K}$.

Lemma 5. *Let E and F be subfields of $\mathbb{K}$ and ψ an isomorphism of E onto F. Then, ψ has an additive bijective extension φ to the full space $\mathbb{K}$ such that*

$$\varphi(\varepsilon x) = \psi(\varepsilon)\varphi(x) \tag{6}$$

for all $\varepsilon \in E$ and $x \in \mathbb{K}$. Moreover, one has a discontinuous one whenever $\mathbb{R} \setminus E \neq \emptyset$.

Proof. Let $\{e_j : j \in J\}$ and $\{f_j : j \in J'\}$ be an E-linear base and an F-linear base of $\mathbb{K}$, respectively. Because both of them have same cardinality, we take $J = J'$. Moreover, we may assume without loss of generality that $e_{j_0} = f_{j_0} = 1$ for some $j_0 \in J$.

For any $x \in \mathbb{K}$, there exist a finite $I \subset J$ and $\{x_i\}_{i\in I} \subseteq E$ such that

$$x = \sum_{i\in I} x_i e_i, \tag{7}$$

and this decomposition is unique. Hence, we can define $\varphi : \mathbb{K} \to \mathbb{K}$ by

$$\varphi(x) = \sum_{i\in I} \psi(x_i) f_i. \tag{8}$$

This φ is a desired extension.

In order to get a discontinuous extension, we consider the base $\{e_j : j \in J\}$ and $\{f_j : j \in J\}$ such that $f_{j_1} \neq e_{j_1} \in \mathbb{R} \setminus E$ for some $j_1 \in J$. Because $\mathbb{Q} \subseteq E$, take a rational sequence $\{\varepsilon_n\}$ converging to e_{j_1}. Suppose that φ is continuous. Then,

$$\begin{aligned} \varphi(\varepsilon_n) &\longrightarrow \varphi(e_{j_1}) = f_{j_1}, \\ \varphi(\varepsilon_n) &= \psi(\varepsilon_n) = \varepsilon_n \longrightarrow e_{j_1}, \end{aligned} \tag{9}$$

as $n \to \infty$. This is contradiction. Thus φ is discontinuous.

Lemma 6. *Suppose that $a \neq c$.*

(i) *If both a and c are transcendental numbers, then there exists a discontinuous $(a, -a, c, -c)$-additive function of $\mathbb{K}$ into itself.*

(ii) *If one of a and c is transcendental and the other is algebraic, then the constant 0 is a unique $(a, -a, c, -c)$-additive function of $\mathbb{K}$ into itself.*

(iii) *If both a and c are algebraic with a common minimal polynomial, then there exists a discontinuous $(a, -a, c, -c)$-additive function of $\mathbb{K}$ into itself.*

(iv) *If a and c are algebraic with distinct minimal polynomials, then the constant 0 is a unique $(a, -a, c, -c)$-additive function of $\mathbb{K}$ into itself.*

Moreover, when $\mathbb{K} = \mathbb{R}$, every nontrivial $(a, -a, c, -c)$-additive function of $\mathbb{K}$ into itself is discontinuous. On the other hand, when $\mathbb{K} = \mathbb{C}$, if a is not real and $a = \bar{c}$, then any continuous $(a, -a, c, -c)$-additive function f of $\mathbb{C}$ into itself must be of form $f(x) = \alpha\bar{x}$ $(x \in \mathbb{C})$ for some $\alpha \in \mathbb{C}$.

Proof. (i) Suppose that both a and c are transcendental. Then, $\mathbb{Q}(a)$ (resp., $\mathbb{Q}(c)$) is isomorphic to the rational function field in indeterminate a (resp., c). So, the substitution $a \to c$ induces an isomorphism $\varphi_{ac} : \mathbb{Q}(a) \to \mathbb{Q}(c)$ of fields. By Lemma 5, because $\mathbb{R} \setminus \mathbb{Q}(a) \neq \emptyset$, φ_{ac} has a discontinuous additive extension φ to $\mathbb{K}$ such that $\varphi(\varepsilon x) = \varphi_{ac}(\varepsilon)\varphi(x)$ for every $\varepsilon \in \mathbb{Q}(a)$ and $x \in \mathbb{K}$. Then, $\varphi(ax) = \varphi_{ac}(a)\varphi(x) = c\varphi(x)$ for all $x \in \mathbb{K}$, and, hence, φ is $(a, -a, c, -c)$-additive by Proposition 1.

(ii) Let φ be any $(a, -a, c, -c)$-additive function. If a is transcendental and c is algebraic with nonzero polynomial such that $p(c) = 0$, then from Lemma 4, we have

$$\varphi(x) = \varphi\left(p(a)\, p(a)^{-1} x\right) = p(c)\, \varphi\left(p(a)^{-1} x\right) = 0 \tag{10}$$

for all $x \in \mathbb{K}$. If c is transcendental and a is algebraic with nonzero polynomial such that $q(a) = 0$, then from Lemma 4, we have

$$\varphi(x) = \frac{1}{q(c)} q(c)\, \varphi(x) = \frac{1}{q(c)} \varphi(q(a)\, x) = \frac{1}{q(c)} \varphi(0) = 0 \tag{11}$$

for all $x \in \mathbb{K}$.

(iii) If a is algebraic with minimal polynomial $p(X) \in \mathbb{Q}[X]$, then $\mathbb{Q}(a)$ consists of all polynomials $f(a)$ in a of degree up to $\deg p - 1$. So, if c is also algebraic with the same minimal polynomial p, then the substitution $a \to c$ induces an automorphism $\varphi_{ac} : \mathbb{Q}(a) \to \mathbb{Q}(c) = \mathbb{Q}(a)$. As same as (i), φ_{ac} has a discontinuous $(a, -a, c, -c)$-additive extension.

(iv) Suppose that a and c are algebraic with distinct minimal polynomials p and q over the field $\mathbb{Q}$, respectively. Let φ be any $(a, -a, c, -c)$-additive function. To show $\varphi = 0$, we assume, on the contrary, that there is an $x_0 \in \mathbb{K}$ with $\varphi(x_0) \neq 0$. Then, from Lemma 5, we have $p(c)\varphi(x_0) = \varphi(p(a)x_0) = \varphi(0) = 0$, and hence $p(c) = 0$. This contradicts the prerequisite for a and c. Hence, φ must be zero.

When $\mathbb{K} = \mathbb{R}$, since every continuous additive function is $\mathbb{R}$-linear and $a \neq c$, there is no continuous $(a, -a, c, -c)$-additive function by Proposition 1. Now, we consider the case $\mathbb{K} = \mathbb{C}$. Let φ be a nontrivial continuous $(a, -a, c, -c)$-additive function. Note that φ is $\mathbb{R}$-linear. If a is not real and $a = \bar{c}$, we can easily see that $\varphi(ix) = -i\varphi(x)$ for all $x \in \mathbb{C}$. Thus, φ is conjugate linear, and hence $\varphi(x) = \alpha\bar{x}$ $(x \in \mathbb{C})$, where $\alpha = \varphi(1)$.

Proof of Theorem 3. (I) We assume without loss of generality that $x_0 = y_0 = 0$ with the help of Proposition 1. Given an $(a, -a, c, -c)$-additive function φ, take a $y_1 \in Y$ with $\|y_1\| = 1$ and a nonzero functional h in X^*, the dual space of X. Put

$$f(x) = \varphi(h(x))\, y_1 \quad (x \in X). \tag{12}$$

Then, we can easily see that f is an $(a, -a, c, -c)$-additive mapping of X into Y. Also, if φ is discontinuous, then so is f. In fact, if φ is discontinuous, we can find a sequence $\{a_n\}$ in $\mathbb{K}$ such that $\lim_{n\to\infty} a_n = 0$ and $|\varphi(a_n)| \geq 1$ $(n = 1, 2, \ldots)$. Choose an x_1 in X with $h(x_1) = 1$ and put $x_n = a_n x_1$ for each $n \in \mathbb{N}$. Then, $\|x_n\| = |a_n|\, \|x_1\| \to 0$ as $n \to \infty$ and $\|f(x_n)\| = |\varphi(a_n)| \geq 1$ $(n = 1, 2, \ldots)$, so f is discontinuous, as required. Therefore Theorem 3(I), (II)-(i), and (II)-(iii) follow easily from Lemmas 4 and 6.

Given an $(a, -a, c, -c)$-additive mapping f of X into Y, take $x \in X$ and $h \in Y^*$ arbitrarily, and put $\varphi(t) = h(f(tx))$ for each $t \in \mathbb{K}$. Then, $\varphi : \mathbb{K} \to \mathbb{K}$ is $(a, -a, c, -c)$-additive. If $\varphi = 0$ for each $h \in Y^*$ and $x \in X$, then $f = 0$ by the Hahn-Banach theorem. Therefore, Theorem 3(II)-(ii) and (II)-(iv) follow easily from Lemma 6. The final assertion in (II) also follows from Lemma 6 and its proof.

3. A Stability of Generalized Additive Mappings

In this section, we consider a couple of cases which are left out in [8] about the Ulam type stability. We take a nonnegative function ε (say a *control function*) on $X \times X$ and also a certain nonnegative function δ on X which depends on ε. We say that a system of all $(a, b, c, d; x_0, y_0)$-additive mappings is *strictly* (ε, δ)*-stable* whenever the following statement is true:

"*If a mapping* f *of* X *into* Y *satisfies*

$$\left\| f\left(ax + bx' + x_0\right) - cf(x) - df\left(x'\right) - y_0 \right\| \le \varepsilon\left(x, x'\right) \tag{†}$$

for all $x, x' \in X$, *then there exists a unique* $(a, b, c, d; x_0, y_0)$*-additive mapping* f_∞ *such that*

$$\left\| f(x) - f_\infty(x) \right\| \le \delta(x) \tag{‡}$$

for all $x \in X$."

Throughout the remainder of this paper, we assume that Y is a Banach space. This is because all of our results depend on the following theorems whose proofs need the Banach fixed point theorem.

Theorem A (see [8, Theorem 3.1]). *Let* $a + b \neq 0$ *and* $K \ge 0$ *with* $K|c+d| < 1$. *One takes a control function* ε *which satisfies*

$$\varepsilon\left(x, x'\right) \le K\varepsilon\left((a+b)x + x_0, (a+b)x' + x_0\right) \tag{13}$$

for all $x, x' \in X$ *and puts*

$$\delta(x) = \frac{K\varepsilon(x, x)}{1 - K|c+d|} \tag{14}$$

for each $x \in X$.

Then, the strict (ε, δ)*-stability holds for the system of* $(a, b, c, d; x_0, y_0)$*-additive mappings.*

Theorem B (see [8, Theorem 3.2]). *Let* $c + d \neq 0$ *and* $K \ge 0$ *with* $K < |c + d|$. *One takes a control function* ε *which satisfies*

$$\varepsilon\left((a+b)x + x_0, (a+b)x' + x_0\right) \le K\varepsilon\left(x, x'\right) \tag{15}$$

for all $x, x' \in X$ *and puts*

$$\delta(x) = \frac{\varepsilon(x, x)}{|c+d| - K} \tag{16}$$

for each $x \in X$.

Then, the strict (ε, δ)*-stability holds for the system of* $(a, b, c, d; x_0, y_0)$*-additive mappings.*

Both of these theorems do not say about $(a, -a, c, -c; x_0, y_0)$-additive mappings at all; however, we will get the following stability theorems for them. Theorem 7 is of the case $a \neq -1$, Theorem 8 is of the case $b \neq -1$, and Theorems 9 and 10 are for $(-1, 1, -1, 1; x_0, y_0)$-additive mappings. These cover all of the systems of $(a, -a, c, -c; x_0, y_0)$-additive mappings.

Theorem 7. *Let* $a + 1 \neq 0$ *and* $K \ge 0$ *with* $K|1 + c| < |c|$. *One takes a control function* ε *which satisfies*

$$\varepsilon\left(x, x'\right) \le K\varepsilon\left(\left(a^{-1} + 1\right)x - a^{-1}x_0, \left(a^{-1} + 1\right)x' - a^{-1}x_0\right) \tag{17}$$

for all $x, x' \in X$, *and puts*

$$\delta(x) = \frac{K\varepsilon\left(\left(a^{-1} + 1\right)x - a^{-1}x_0, x\right)}{|c| - K|1 + c|} \tag{18}$$

for each $x \in X$.

Then, the strict (ε, δ)*-stability holds for the system of* $(a, -a, c, -c; x_0, y_0)$*-additive mappings.*

Proof. Put $u = ax - ax' + x_0$ and $u' = x'$ for each $x, x' \in X$. Then, (†) changes into

$$\begin{aligned}&\left\| f\left(a^{-1}u + u' - a^{-1}x_0\right) - c^{-1}f(u) - f\left(u'\right) + c^{-1}y_0 \right\| \\ &\quad \le \varepsilon_1\left(u, u'\right),\end{aligned} \tag{19}$$

where

$$\varepsilon_1\left(u, u'\right) = \frac{1}{|c|}\varepsilon\left(a^{-1}u + u' - a^{-1}x_0, u'\right) \tag{20}$$

for all $u, u' \in X$. By denoting $a_1 = a^{-1}$, $b_1 = 1$, $c_1 = c^{-1}$, $d_1 = 1$, $u_0 = -a^{-1}x_0$ and $v_0 = -c^{-1}y_0$ in (19), (†) changes up to the following estimate of f by the control function ε_1:

$$\left\| f\left(a_1u + b_1u' + u_0\right) - c_1f(u) - d_1f\left(u'\right) - v_0 \right\| \le \varepsilon_1\left(u, u'\right) \tag{21}$$

for all $u, u' \in X$.

Under these transformations, a_1, b_1, c_1, d_1, and ε_1 are equipped with

$$\begin{gathered}a_1 + b_1 \neq 0, \qquad K\left|c_1 + d_1\right| < 1, \\ \varepsilon_1\left(u, u'\right) \le K\varepsilon_1\left(\left(a_1 + b_1\right)u + u_0, \left(a_1 + b_1\right)u' + u_0\right)\end{gathered} \tag{22}$$

for all $u, u' \in X$. The latter follows from the inequality in which ε must satisfy because by using (20), we get

$$\begin{aligned}&|c|\,\varepsilon_1\left(u, u'\right) = \varepsilon\left(a^{-1}u + u' + u_0, u'\right) = \varepsilon\left(x, x'\right), \\ &\varepsilon\left(\left(a^{-1} + 1\right)x - a^{-1}x_0, \left(a^{-1} + 1\right)x' - a^{-1}x_0\right) \\ &\quad = \varepsilon\left(\left(a^{-1} + 1\right)x - a^{-1}x_0, \left(a_1 + b_1\right)u' + u_0\right) \\ &\quad = \varepsilon\left(\left(a^{-1} + 1\right)\left(a^{-1}u + u' + u_0\right)\right. \\ &\qquad \left. -a^{-1}x_0, \left(a_1 + b_1\right)u' + u_0\right) \\ &\quad = \varepsilon\left(a^{-1}\left(\left(a_1 + b_1\right)u + u_0\right)\right. \\ &\qquad + \left(\left(a_1 + b_1\right)u' + u_0\right) \\ &\qquad \left. -a^{-1}x_0, \left(a_1 + b_1\right)u' + u_0\right) \\ &\quad = |c|\,\varepsilon_1\left(\left(a_1 + b_1\right)u + u_0, \left(a_1 + b_1\right)u' + u_0\right)\end{aligned} \tag{23}$$

for all $u, u' \in X$. Since (21) and (22) hold, it follows from Theorem A that there exists a unique $(a_1, b_1, c_1, d_1; u_0, v_0)$-additive mapping f_∞ such that

$$\left\| f(x) - f_\infty(x) \right\| \le \frac{K\varepsilon_1(x, x)}{1 - K\left|c_1 + d_1\right|} \tag{24}$$

for all $x \in X$. However, we can easily see the following two assertions:

(i) f_∞ is $(a_1, b_1, c_1, d_1; u_0, v_0)$-additive if and only if f_∞ is $(a, -a, c, -c; x_0, y_0)$-additive,

(ii) (24) is equivalent to (‡).

This completes the proof.

Theorem 8. *Let $c + 1 \neq 0$ and $K \geq 0$ with $K|c| < |c+1|$. One takes a control function ε which satisfies*

$$\varepsilon\left(\left(a^{-1}+1\right)x - a^{-1}x_0, \left(a^{-1}+1\right)x' - a^{-1}x_0\right) \leq K\varepsilon\left(x, x'\right) \quad (25)$$

for all $x, x' \in X$ and puts

$$\delta(x) = \frac{\varepsilon\left(\left(a^{-1}+1\right)x - a^{-1}x_0, x\right)}{|c+1| - K|c|} \quad (26)$$

for each $x \in X$.

Then the strict (ε, δ)-stability holds for the system of $(a, -a, c, -c; x_0, y_0)$-additive mappings.

Proof. We consider the same transformations and the same estimate (21) of f by ε_1 in the proof of Theorem 7. Under these transformations, we have

$$c_1 + d_1 \neq 0, \qquad K < |c_1 + d_1|. \quad (27)$$

Moreover, for every $u, u' \in X$ we have

$$\varepsilon_1\left((a_1+b_1)u + u_0, (a_1+b_1)u' + u_0\right) \leq K\varepsilon_1\left(u, u'\right), \quad (28)$$

because

$$|c|\,\varepsilon_1\left(u, u'\right) = \varepsilon\left(a^{-1}u + u' + u_0, u'\right) = \varepsilon\left(x, x'\right), \quad (29)$$

$$\begin{aligned} &|c|\,\varepsilon_1\left((a_1+b_1)u + u_0, (a_1+b_1)u' + u_0\right) \\ &\quad = \varepsilon\left(\left(a^{-1}+1\right)x - a^{-1}x_0, \left(a^{-1}+1\right)x' - a^{-1}x_0\right). \end{aligned} \quad (30)$$

Since (21), (27) and (28) hold, it follows from Theorem B that there exists a unique $(a_1, b_1, c_1, d_1; u_0, v_0)$-additive mapping f_∞ such that

$$\left\|f(x) - f_\infty(x)\right\| \leq \frac{\varepsilon_1(x, x)}{|c_1 + d_1| - K} \quad (31)$$

for all $x \in X$. This means (‡) and f_∞ is $(a, -a, c, -c; x_0, y_0)$-additive.

Theorem 9. *Let $0 \leq K < 1/2$. One takes a control function ε which satisfies*

$$\varepsilon\left(x, x'\right) \leq K\varepsilon\left(2x - x_0, 2x' - x_0\right) \quad (32)$$

for all $x, x' \in X$ and puts

$$\delta(x) = \frac{K\varepsilon\left(x, 2x - x_0\right)}{1 - 2K} \quad (33)$$

for each $x \in X$.

Then the strict (ε, δ)-stability holds for the system of $(-1, 1, -1, 1; x_0, y_0)$-additive mappings.

Proof. Put $u = x, u' = -x + x' + x_0$ for each $x, x' \in X$. Then, (†) changes into the following estimate of f by the control function ε_1:

$$\left\|f\left(u + u' - x_0\right) - f(u) - f\left(u'\right) + y_0\right\| \leq \varepsilon_1\left(u, u'\right), \quad (34)$$

where

$$\varepsilon_1\left(u, u'\right) = \varepsilon\left(u, u + u' - x_0\right). \quad (35)$$

for all $u, u' \in X$. Put $a_1 = b_1 = c_1 = d_1 = 1$.

Under these transformations, a_1, b_1, c_1, d_1, and ε_1 are equipped with

$$\begin{gathered} a_1 + b_1 = 2 \neq 0, \qquad K\left|c_1 + d_1\right| = 2K < 1, \\ \varepsilon_1\left(u, u'\right) \leq K\varepsilon_1\left((a_1+b_1)u - x_0, (a_1+b_1)u' - x_0\right) \end{gathered} \quad (36)$$

for all $u, u' \in X$. The latter follows from the following inequality:

$$\begin{aligned} \varepsilon_1\left(u, u'\right) &= \varepsilon\left(u, u + u' - x_0\right) \\ &\leq K\varepsilon\left(2u - x_0, 2\left(u + u' - x_0\right) - x_0\right) \\ &= K\varepsilon\left(2u - x_0, (2u - x_0) + \left(2u' - x_0\right) - x_0\right) \\ &= K\varepsilon_1\left(2u - x_0, 2u' - x_0\right) \end{aligned} \quad (37)$$

for all $u, u' \in X$. Since (34), (36) hold, it follows from Theorem A that there exists a unique $(a_1, b_1, c_1, d_1; -x_0, -y_0)$-additive mapping f_∞ such that

$$\left\|f(x) - f_\infty(x)\right\| \leq \frac{K\varepsilon_1(x, x)}{1 - K\left|c_1 + d_1\right|} \quad (38)$$

for all $x \in X$. However we can easily see the following two assertions:

(i) f_∞ is $(a_1, b_1, c_1, d_1; -x_0, -y_0)$-additive if and only if f_∞ is $(-1, 1, -1, 1; x_0, y_0)$-additive;

(ii) (38) is equivalent to (‡).

This completes the proof.

Theorem 10. *Let $0 \leq K < 2$. One takes a control function ε which satisfies*

$$\varepsilon\left(2x - x_0, 2x' - x_0\right) \leq K\varepsilon\left(x, x'\right) \quad (39)$$

for all $x, x' \in X$ and puts

$$\delta(x) = \frac{\varepsilon(x, 2x - x_0)}{2 - K} \tag{40}$$

for each $x \in X$.

Then the strict (ε, δ)-stability holds for the system of $(-1, 1, -1, 1; x_0, y_0)$-additive mappings.

Proof. We consider the same transformation and the same estimate (34) of f by ε_1 in the proof of Theorem 9. Under these transformations, c_1, d_1 and ε_1 are equipped with

$$c_1 + d_1 = 2 \neq 0, \qquad K < 2 = |c_1 + d_1|, \tag{41}$$

$$\begin{aligned}
\varepsilon_1 &(2u - x_0, 2u' - x_0) \\
&= \varepsilon\left(2u - x_0, (2u - x_0) + (2u' - x_0) - x_0\right) \\
&= \varepsilon\left(2x - x_0, 2(u + u' - x_0) - x_0\right) \\
&= \varepsilon(2x - x_0, 2x' - x_0) \\
&\leq K\varepsilon(x, x') \\
&= K\varepsilon(u, u + u' - x_0) \\
&= K\varepsilon_1(u, u')
\end{aligned} \tag{42}$$

for all $u, u' \in X$. So it follows that

$$\varepsilon_1\left((a_1 + b_1)u - x_0, (a_1 + b_1)u' - x_0\right) \leq K\varepsilon_1(u, u') \tag{43}$$

for all $u, u' \in X$. Since (34), (41) and (43) hold, it follows from Theorem B that there exists a unique $(a_1, b_1, c_1, d_1; -x_0, -y_0)$-additive mapping f_∞ such that

$$\|f(x) - f_\infty(x)\| \leq \frac{\varepsilon_1(x, x)}{|c_1 + d_1| - K} \tag{44}$$

for all $x, x' \in X$. This means (‡) and f_∞ is $(-1, 1, -1, 1; x_0, y_0)$-additive.

4. Concrete Examples

Throughout this section, let $x_0 = y_0 = 0$. We fix nonnegative constants p, q and δ and take the control function ε defined by $\varepsilon(x, x') = \delta\|x\|^p\|x'\|^q$ for every $x, x' \in X$.

Corollary 11. *When $a + 1 \neq 0$ and $|a|^{p+q}|c + 1| < |c||a + 1|^{p+q}$, one puts*

$$\delta'(x) = \frac{\delta|a + 1|^p|a|^q\|x\|^{p+q}}{|c|\,|a + 1|^{p+q} - |c + 1|\,|a|^{p+q}} \tag{45}$$

for each $x \in X$.

Then, the system of $(a, -a, c, -c)$-additive mappings is strictly (ε, δ')-stable.

Proof. Put $K = |a^{-1} + 1|^{-(p+q)}$. Then $K|c + 1| < |c|$ and

$$\varepsilon(x, x') = K\varepsilon\left((a^{-1} + 1)x, (a^{-1} + 1)x'\right) \tag{46}$$

for all $x, x' \in X$. By Theorem 7, for a mapping f of X into Y satisfying (†) for $a = -b$ and $c = -d$, there exists a unique $(a, -a, c, -c)$-additive mapping f_∞ such that

$$\|f(x) - f_\infty(x)\| \leq \frac{K\delta|a^{-1} + 1|^p\,\|x\|^{p+q}}{|c| - K\,|c + 1|} \tag{47}$$

for all $x, x' \in X$. Because of $K = |a^{-1} + 1|^{-(p+q)}$, we have the corollary.

Corollary 12. *When $c + 1 \neq 0$ and $|a + 1|^{p+q}|c| < |c + 1||a|^{p+q}$, one puts*

$$\delta'(x) = \frac{\delta|a + 1|^p|a|^q\|x\|^{p+q}}{|a|^{p+q}\,|c + 1| - |a + 1|^{p+q}\,|c|} \tag{48}$$

for each $x \in X$.

Then the system of $(a, -a, c, -c)$-additive mappings is strictly (ε, δ')-stable.

Proof. Put $K = |a^{-1} + 1|^{p+q}$. Then $K|c| < |c + 1|$ and

$$\varepsilon\left((a^{-1} + 1)x, (a^{-1} + 1)x'\right) = K\varepsilon(x, x') \tag{49}$$

for all $x, x' \in X$. By Theorem 8, for a mapping f of X into Y satisfying (†) for $a = -b$ and $c = -d$, there exists a unique $(a, -a, c, -c)$-additive mapping f_∞ such that

$$\|f(x) - f_\infty(x)\| \leq \frac{\delta|a^{-1} + 1|^p\,\|x\|^{p+q}}{|c + 1| - K\,|c|} \tag{50}$$

for all $x \in X$. Because of $K = |a^{-1} + 1|^{p+q}$, we have the corollary.

Corollary 13. *When $p + q > 1$, one puts*

$$\delta'(x) = \frac{2^q\,\delta\|x\|^{p+q}}{2^{p+q} - 2} \tag{51}$$

for each $x \in X$.

Then, the system of $(-1, 1, -1, 1)$-additive mappings is strictly (ε, δ')-stable.

Proof. Put $K = 2^{-(p+q)}$. Then,

$$\varepsilon(x, x') = K\varepsilon(2x, 2x') \tag{52}$$

for all $x, x' \in X$. Since $p + q > 1$, we also have $0 \leq K < 1/2$. By Theorem 9, for a mapping f of X into Y satisfying (†) for $a = c = -1$ and $b = d = 1$, there exists a unique $(-1, 1, -1, 1)$-additive mapping f_∞ such that

$$\|f(x) - f_\infty(x)\| \leq \frac{2^q\delta K\,\|x\|^{p+q}}{1 - 2K} \tag{53}$$

for all $x \in X$. Because of $K = 2^{-(p+q)}$, we have the corollary.

Corollary 14. *When $p + q < 1$, one puts*

$$\delta'(x) = \frac{2^q \delta \|x\|^{p+q}}{2 - 2^{p+q}} \quad (54)$$

for each $x \in X$.

Then, the system of $(-1, 1, -1, 1)$-additive mappings is strictly (ε, δ')-stable.

Proof. Put $K = 2^{p+q}$. Then,

$$\varepsilon(2x, 2x') = K\varepsilon(x, x') \quad (55)$$

for all $x, x' \in X$. Since $p + q < 1$, we also have $0 \leq K < 2$. By Theorem 10, for a mapping f of X into Y satisfying (†) for $a = c = -1$ and $b = d = 1$, there exists a unique $(-1, 1, -1, 1)$-additive mapping f_∞ such that

$$\|f(x) - f_\infty(x)\| \leq \frac{2^q \delta \|x\|^{p+q}}{2 - K} \quad (56)$$

for all $x \in X$. Because of $K = 2^{p+q}$, we have the corollary.

Remark 15. In Corollary 14, taking $p = q = 0$, we can easily observe that the corollary is just the stability result due to Hyers [2, Theorem 1].

Acknowledgment

The third and fifth authors are partially supported by Grant-in-Aid for Scientific Research, Japan Society for the Promotion of Science.

References

[1] S. M. Ulam, *A Collection of Mathematical Problems*, vol. 8 of *Interscience Tracts in Pure and Applied Mathematics,*, Interscience Publishers, New York, NY, USA, 1960.

[2] D. H. Hyers, "On the stability of the linear functional equation," *Proceedings of the National Academy of Sciences of the United States of America*, vol. 27, pp. 222–224, 1941.

[3] T. Aoki, "On the stability of the linear transformation in Banach spaces," *Journal of the Mathematical Society of Japan*, vol. 2, pp. 64–66, 1950.

[4] T. M. Rassias, "On the stability of the linear mapping in Banach spaces," *Proceedings of the American Mathematical Society*, vol. 72, no. 2, pp. 297–300, 1978.

[5] J. Aczél, *Lectures on Functional Equations and Their Applications*, vol. 19 of *Mathematics in Science and Engineering*, Academic Press, New York, NY, USA, 1966.

[6] S.-E. Takahasi, T. Miura, and H. Takagi, "On a Hyers-Ulam-Aoki-Rassias type stability and a fixed point theorem," *Journal of Nonlinear and Convex Analysis*, vol. 11, no. 3, pp. 423–439, 2010.

[7] S. -E. Takahasi, T. Miura, H. Takagi, and K. Tanahashi, "Ulam type stability problems for alternative homomorphisms".

[8] M. Todoroki, K. Kumahara, T. Miura, and S.-E. Takahasi, "Stability problems for generalized additive mappings and Euler-Lagrange type mappings," *The Australian Journal of Mathematical Analysis and Applications*, vol. 9, no. 1, article 19, 2012.

2

Generalized Derivations on Prime Near Rings

Asma Ali,[1] Howard E. Bell,[2] and Phool Miyan[1]

[1] *Department of Mathematics, Aligarh Muslim University, Aligarh 202002, India*
[2] *Department of Mathematics, Brock University, St. Catharines, ON, Canada L2S 3A1*

Correspondence should be addressed to Asma Ali; asma.ali345@gmail.com

Academic Editor: Christian Corda

Let N be a near ring. An additive mapping $f : N \rightarrow N$ is said to be a right generalized (resp., left generalized) derivation with associated derivation d on N if $f(xy) = f(x)y + xd(y)$ (resp., $f(xy) = d(x)y + xf(y)$) for all $x, y \in N$. A mapping $f : N \rightarrow N$ is said to be a generalized derivation with associated derivation d on N if f is both a right generalized and a left generalized derivation with associated derivation d on N. The purpose of the present paper is to prove some theorems in the setting of a semigroup ideal of a 3-prime near ring admitting a generalized derivation, thereby extending some known results on derivations.

1. Introduction

Throughout the paper, N denotes a zero-symmetric left near ring with multiplicative center Z, and for any pair of elements $x, y \in N$, $[x, y]$ denotes the commutator $xy - yx$, while the symbol (x, y) denotes the additive commutator $x+y-x-y$. A near ring N is called zero-symmetric if $0x = 0$, for all $x \in N$ (recall that left distributivity yields that $x0 = 0$). The near ring N is said to be 3-prime if $xNy = \{0\}$ for $x, y \in N$ implies that $x = 0$ or $y = 0$. A near ring N is called 2-torsion-free if $(N, +)$ has no element of order 2. A nonempty subset A of N is called a semigroup right (resp., semigroup left) ideal if $AN \subseteq A$ (resp., $NA \subseteq A$), and if A is both a semigroup right ideal and a semigroup left ideal, it is called a semigroup ideal. An additive mapping $d : N \rightarrow N$ is a derivation on N if $d(xy) = d(x)y+xd(y)$, for all $x, y \in N$. An additive mapping $f : N \rightarrow N$ is said to be a right (resp., left) generalized derivation with associated derivation d if $f(xy) = f(x)y + xd(y)$ (resp., $f(xy) = d(x)y + xf(y)$), for all $x, y \in N$, and f is said to be a generalized derivation with associated derivation d on N if it is both a right generalized derivation and a left generalized derivation on N with associated derivation d. (Note that this definition differs from the one given by Hvala in [1]; his generalized derivations are our right generalized derivations.) Every derivation on N is a generalized derivation.

In the case of rings, generalized derivations have received significant attention in recent years. We prove some theorems in the setting of a semigroup ideal of a 3-prime near ring admitting a generalized derivation and thereby extend some known results [2, Theorem 2.1], [3, Theorem 3.1], [4, Theorem 3], and [5, Theorem 3.3].

2. Preliminary Results

We begin with several lemmas, most of which have been proved elsewhere.

Lemma 1 (see [3, Lemma 1.3]). *Let N be a 3-prime near ring and d be a nonzero derivation on N.*

(i) *If U is a nonzero semigroup right ideal or a nonzero semigroup left ideal of N, then $d(U) \neq \{0\}$.*

(ii) *If U is a nonzero semigroup right ideal of N and x is an element of N which centralizes U, then $x \in Z$.*

Lemma 2 (see [3, Lemma 1.2]). *Let N be a 3-prime near ring.*

(i) *If $z \in Z \setminus \{0\}$, then z is not a zero divisor.*

(ii) *If $Z \setminus \{0\}$ contains an element z for which $z + z \in Z$, then $(N, +)$ is abelian.*

(iii) *If $z \in Z \setminus \{0\}$ and x is an element of N such that $xz \in Z$, then $x \in Z$.*

Lemma 3 (see [3, Lemmas 1.3 and 1.4]). *Let N be a 3-prime near ring and U be a nonzero semigroup ideal of N. Let d be a nonzero derivation on N.*

(i) *If $x, y \in N$ and $xUy = \{0\}$, then $x = 0$ or $y = 0$.*

(ii) *If $x \in N$ and $xU = \{0\}$ or $Ux = \{0\}$, then $x = 0$.*

(iii) *If $x \in N$ and $d(U)x = \{0\}$ or $xd(U) = \{0\}$, then $x = 0$.*

Lemma 4 (see [3, Lemma 1.5]). *If N is a 3-prime near ring and Z contains a nonzero semigroup left ideal or a semigroup right ideal, then N is a commutative ring.*

Lemma 5. *If f is a generalized derivation on N with associated derivation d, then $(d(x)y+xf(y))z = d(x)yz+xf(y)z$, for all $x, y, z \in N$.*

Proof. We prove only (ii), since (i) is proved in [2]. For all $x, y, z \in N$ we have $f((xy)z) = f(xy)z + xyd(z) = (d(x)y+xf(y))z+xyd(z)$ and $f(x(yz)) = d(x)yz+xf(yz) = d(x)yz + xf(f(y)z + yd(z)) = d(x)yz + xf(y)z + xyd(z)$. Comparing the two expressions for $f(xyz)$ gives (ii).

Lemma 6. *Let N be a 3-prime near ring and f a generalized derivation with associated derivation d.*

(i) *$f(x)y + xd(y) = xd(y) + f(x)y$ for all $x, y \in N$.*

(ii) *$d(x)y + xf(y) = xf(y) + d(x)y$ for all $x, y \in N$.*

Proof. (i) $f(x(y + y)) = f(x)(y + y) + xd(y + y) = f(x)y + f(x)y + xd(y) + xd(y)$, and $f(xy + xy) = f(x)y + xd(y) + f(x)y + xd(y)$. Comparing these two equations gives the desired result.

(ii) Again, calculate $f(x(y + y))$ and $f(xy + xy)$ and compare.

Lemma 7. *Let N be a 3-prime near ring and f a generalized derivation with associated derivation d. Then $f(Z) \subseteq Z$.*

Proof. Let $z \in Z$ and $x \in N$. Then $f(zx) = f(xz)$; that is, $f(z)x+zd(x) = d(x)z+xf(z)$. Applying Lemma 6(i), we get $zd(x)+f(z)x = d(x)z+xf(z)$. It follows that $f(z)x = xf(z)$ for all $x \in N$, so $f(z) \in Z$.

Lemma 8. *Let N be a 3-prime near ring and U a nonzero semigroup ideal of N. If f is a nonzero right generalized derivation of N with associated derivation d, then $f(U) \neq \{0\}$.*

Proof. Suppose $f(U) = \{0\}$. Then $f(ux) = f(u)x + ud(x) = 0 = ud(x)$ for all $u \in U$ and $x \in N$, and it follows by Lemma 3(ii) that $d = 0$. Therefore $f(xu) = f(x)u = 0$ for all $u \in U$ and $x \in N$, and another appeal to Lemma 3(ii) gives $f = 0$, which is a contradiction.

Lemma 9 (see [2, Theorem 2.1]). *Let N be a 3-prime near ring with a nonzero right generalized derivation f with associated derivation d. If $f(N) \subseteq Z$ then $(N,+)$ is abelian. Moreover, if N is 2-torsion-free, then N is a commutative ring.*

Lemma 10 (see [2, Theorem 4.1]). *Let N be a 2-torsion-free 3-prime near ring and f a nonzero generalized derivation on N with associated derivation d. If $[f(N), f(N)] = \{0\}$, then N is a commutative ring.*

3. Main Results

The theorems that we prove in this section extend the results proved in [2, Theorems 2.1 and 3.1], [3, Theorems 2.1, 3.1, and 3.3], and [5, Theorem 3.3].

Theorem 11. *Let N be a 3-prime near ring and U be a nonzero semigroup ideal of N. Let f be a nonzero right generalized derivation with associated derivation d. If $f(U) \subseteq Z$, then $(N,+)$ is abelian. Moreover, if N is 2-torsion-free, then N is a commutative ring.*

Proof. We begin by showing that $(N,+)$ is abelian, which by Lemma 2(ii) is accomplished by producing $z \in Z \setminus \{0\}$ such that $z + z \in Z$. Let a be an element of U such that $f(a) \neq 0$. Then for all $x \in N$, $ax \in U$ and $ax + ax = a(x + x) \in U$, so that $f(ax) \in Z$ and $f(ax)+f(ax) \in Z$; hence we need only to show that there exists $x \in N$ such that $f(ax) \neq 0$. Suppose that this is not the case, so that $f((ax)a) = 0 = f(ax)a+axd(a) = axd(a)$, for all $x \in N$. By Lemma 3(i) either $a = 0$ or $d(a) = 0$.

If $d(a) = 0$, then $f(xa) = f(x)a + xd(a)$; that is, $f(xa) = f(x)a \in Z$, for all $x \in N$. Thus $[f(u)a, y] = 0$, for all $y \in N$, and $u \in U$. This implies that $f(u)[a, y] = 0$, for all $u \in U$ and $y \in N$ and Lemma 2(i) gives $a \in Z$. Thus $0 = f(ax) = f(xa) = f(x)a$, for all $x \in N$. Replacing x by $u \in U$, we have $f(U)a = 0$, and by Lemmas 2(i) and 8, we get that $a = 0$. Thus, we have a contradiction.

To complete the proof, we show that if N is 2-torsion-free, then N is commutative.

Consider first the case $d = 0$. This implies that $f(ux) = f(u)x \in Z$ for all $u \in U$ and $x \in N$. By Lemma 8, we have $u \in U$ such that $f(u) \in Z \setminus \{0\}$, so N is commutative by Lemma 2(iii).

Now consider the case $d \neq 0$. Let $c \in Z \setminus \{0\}$. This implies that if $x \in U$, $f(xc) = f(x)c + xd(c) \in Z$. Thus $(f(x)c + xd(c))y = y(f(x)c + xd(c))$ for all $x, y \in U$ and $c \in Z$. Therefore by Lemma 5(i), $f(x)cy+xd(c)y = yf(x)c+yxd(c)$ for all $x, y \in U$ and $c \in Z$. Since $d(c) \in Z$ and $f(x) \in Z$, we obtain $d(c)[x, y] = 0$, for all $x, y \in U$ and $c \in Z$. Let $d(Z) \neq 0$. Choosing c such that $d(c) \neq 0$ and noting that $d(c)$ is not a zero divisor, we have $[x, y] = 0$ for all $x, y \in U$. By Lemma 1(ii), $U \subseteq Z$; hence N is commutative by Lemma 4.

The remaining case is $d \neq 0$ and $d(Z) = \{0\}$. Suppose we can show that $U \cap Z \neq \{0\}$. Taking $z \in (U \cap Z) \setminus \{0\}$ and $x \in N$, we have $f(xz) = f(x)z \in Z$; therefore $f(N) \subseteq Z$ by Lemma 2(iii) and N is commutative by Lemma 9.

Assume, then, that $U \cap Z = \{0\}$. For each $u \in U$, $f(u^2) = f(u)u + ud(u) = u(f(u) + d(u)) \in U \cap Z$, so $f(u^2) = 0$. Thus, for all $u \in U$ and $x \in N$, $f(u^2x) = f(u^2)x + u^2d(x) = u^2d(x) \in U \cap Z$, so $u^2d(x) = 0$, and by Lemma 3(iii) $u^2 = 0$. Since $f(xu) = f(x)u + xd(u) \in Z$ for all $u \in U$ and $x \in N$, we have $(f(x)u + xd(u))u = u(f(x)u + xd(u))$, and right multiplying by u gives $uxd(u)u = 0$. Consequently $d(u)uNd(u)u = \{0\}$, so that $d(u)u = 0$ for all $u \in U$. Since $u^2 = 0$, $d(u^2) = d(u)u + ud(u) = 0$ for all $u \in U$, so

$f(u)u = 0$ for all $u \in U$. But by Lemma 8, there exists $u_0 \in U$ for which $f(u_0) \neq 0$; and since $f(u_0) \in Z$, we get $u_0 = 0$—a contradiction. Therefore $U \cap Z \neq \{0\}$ as required.

Theorem 12. *Let N be a 3-prime near ring with a nonzero generalized derivation f with associated nonzero derivation d. Let U be a nonzero semigroup ideal of N. If $[f(U), f(U)] = 0$, then $(N,+)$ is abelian.*

Proof. Assume that $x \in N$ is such that $[x, f(U)] = [x + x, f(U)] = 0$. For all $u, v \in U$ such that $u + v \in U$, $[x + x, f(u + v)] = 0$.

This implies that

$$(x+x) f(u+v) = f(u+v)(x+x),$$

$$(x+x) f(u) + (x+x) f(v) = f(u+v) x + f(u+v) x,$$

$$f(u)(x+x) + f(v)(x+x) = xf(u+v) + xf(u+v),$$

$$f(u)x + f(u)x + f(v)x + f(v)x = xf(u) + xf(v) + xf(u) + xf(v), \tag{1}$$

$$xf(u) + xf(u) + xf(v) + xf(v) = xf(u) + xf(v) + xf(u) + xf(v),$$

$$x(f(u) + f(v) - f(u) - f(v)) = 0,$$

$$xf(u + v - u - v) = 0.$$

This equation may be restated as $xf(c) = 0$, where $c = (u, v)$.

Let $a, b \in U$. Then $ab \in U$ and $ab + ab = a(b + b) \in U$, so $[f(ab), f(U)] = \{0\} = [f(ab) + f(ab), f(U)]$, and by the argument in the previous paragraph, $f(ab)f(c) = 0$. We now have $f(U^2)f(u, v) = 0$ for all $u, v \in U$ such that $u + v \in U$. Taking $w \in U^2$ and $x \in N$, we get $f(xw)f((u, v)) = (d(x)w + xf(w))f((u, v)) = 0 = d(x)wf((u, v))$, and since U^2 is a nonzero semigroup ideal by Lemma 3(ii) and $d \neq 0$, Lemma 3(i) gives

$$f((u, v)) = 0 \quad \forall u, v \in U \text{ such that } u + v \in U. \tag{2}$$

Take $u = ry$ and $v = rz$, where $r \in U$ and $y, z \in N$, so that $u + v = ry + rz = r(y + z) \in U$. By (2) we have

$$f((ry, rz)) = f(r(y, z)) = 0 \quad \forall\, r \in U, y,\ z \in N. \tag{3}$$

Replacing r by rw, $w \in U$, we obtain $f(r(wy, wz)) = 0 = d(r)(w(y, z)) + rf((wy, wz))$; so by (3) $d(r)U(y, z) = 0$ for all $r \in U$ and $y, z \in N$. It follows immediately by Lemmas 1(i) and 3(i) that $(y, z) = 0$ for all $y, z \in N$; that is, $(N, +)$ is abelian.

Theorem 13. *Let N be a 2-torsion-free 3-prime near ring and U be a nonzero semigroup ideal of N. If f is a nonzero generalized derivation with associated derivation d such that $[f(U), f(U)] = 0$, then N is a commutative ring if it satisfies one of the following: (i) $d(Z) \neq \{0\}$; (ii) $U \cap Z \neq \{0\}$; (iii) $d = 0$ and $f(Z) \neq \{0\}$.*

Proof. (i) Let $a \in N$ centralizes $f(U)$, and let $z \in Z$ such that $d(z) \neq 0$. Then a centralizes $f(uz)$ for all $u \in U$, so that $a(f(u)z + ud(z)) = (f(u)z + ud(z))a$ and $aud(z) = ud(z)a$. Since $d(z) \in Z \setminus \{0\}$, $d(z)[a, u] = 0 = [a, u]$ for all $u \in U$. Therefore a centralizes U, and by Lemma 1(ii), $a \in Z$. Since $f(U)$ centralizes $f(U)$, $f(U) \subseteq Z$ and our result follows by Theorem 11.

(ii) We may assume $d(Z) = \{0\}$. Let $z \in (U \cap Z) \setminus \{0\}$. Then for all $x, y \in N$, $f(xz) = f(x)z$ and $f(yz) = f(y)z$ commute; hence $z^2[f(x), f(y)] = 0 = [f(x), f(y)]$. Our result now follows from Lemma 10.

(iii) Let $u, v \in U$ and $z \in Z$ such that $f(z) \neq 0$. Then $[f(zu), f(u)] = 0 = [f(z)u, f(v)]$, and since $f(z) \in Z \setminus \{0\}$, $f(z)[u, f(u)] = 0 = [u, f(v)]$. Thus $f(U)$ centralizes U, and by Lemma 1(ii), $f(U) \subseteq Z$. Our result now follows by Theorem 11.

We have already observed that if f is a generalized derivation with $d = 0$, then $f(x)y = xf(y)$ for all $x, y \in N$. For 3-prime near rings, we have the following converse.

Theorem 14. *Let N be a 3-prime near ring and U be a nonzero semigroup ideal of N. If f is a nonzero right generalized derivation of N with associated derivation d and $f(x)y = xf(y)$, for all $x, y \in U$, then $d = 0$.*

Proof. We are given that $f(x)y = xf(y)$ for all $x, y \in U$. Substituting yz for y, we get $f(x)yz = xf(yz) = x(f(y)z + yd(z))$ for all $x, y, z \in U$. It follows that $xyd(z) = 0$ for all $x, y, z \in U$; that is, $xUd(z) = \{0\}$ for all $x, z \in U$. By Lemma 3(i), $d(U) = 0$, and hence $d = 0$ by Lemma 1(i).

4. Generalized Derivations Acting as a Homomorphism or an Antihomomorphism

In [4], Bell and Kappe proved that if R is a semiprime ring and d is a derivation on R which is either an endomorphism or an antiendomorphism on R, then $d = 0$. Of course, derivations which are not endomorphisms or antiendomorphisms on R may behave as such on certain subsets of R; for example, any derivation d behaves as the zero endomorphism on the subring C consisting of all constants (i.e., the elements x for which $d(x) = 0$). In fact in a semiprime ring R, d may behave as an endomorphism on a proper ideal of R. However as noted in [4], the behaviour of d is somewhat restricted in the case of a prime ring. Recently the authors in [6] considered (θ, ϕ)-derivation d acting as a homomorphism or an antihomomorphism on a nonzero Lie ideal of a prime ring and concluded that $d = 0$. In this section we establish similar results in the setting of a semigroup ideal of a 3-prime near ring admitting a generalized derivation.

Theorem 15. *Let N be a 3-prime near ring and U be a nonzero semigroup ideal of N. Let f be a nonzero generalized derivation on N with associated derivation d. If f acts as a* *homomorphism on U, then f is the identity map on N and d = 0.*

Proof. By the hypothesis

$$f(xy) = d(x)y + xf(y) = f(x)f(y) \quad \forall x, y \in U. \tag{4}$$

Replacing y by yz in the above relation, we get

$$f(xyz) = d(x)yz + xf(yz) \quad \forall x, y, z \in U, \tag{5}$$

or

$$f(xy)f(z) = d(x)yz + x(d(y)z + yf(z)) \quad \forall x, y, z \in U. \tag{6}$$

This implies that

$$(d(x)y + xf(y))f(z) = d(x)yz + xd(y)z + xyf(z) \quad \forall x, y, z \in U. \tag{7}$$

Using Lemma 5(ii), we get

$$d(x)yf(z) + xf(y)f(z) = d(x)yz + xd(y)z + xyf(z) \quad \forall x, y, z \in U, \tag{8}$$

or

$$d(x)yf(z) + xf(yz) = d(x)yz + xd(y)z + xyf(z) \quad \forall x, y, z \in U. \tag{9}$$

This implies that

$$d(x)yf(z) + xd(y)z + xyf(z) = d(x)yz + xd(y)z + xyf(z) \quad \forall x, y, z \in U. \tag{10}$$

That is,

$$d(x)yf(z) = d(x)yz \quad \forall x, y, z \in U. \tag{11}$$

Therefore

$$d(x)y(f(z) - z) = 0 \quad \forall x, y, z \in U, \tag{12}$$

which implies that

$$d(x)U(f(z) - z) = \{0\} \quad \forall x, z \in U. \tag{13}$$

It follows by Lemma 3(i) that either $d(U) = 0$ or $f(z) = z$ for all $z \in U$.

In fact, as we now show, both of these conditions hold.

Suppose that $f(u) = u$ for all $u \in U$. Then for all $u \in U$ and $x \in N$, $f(xu) = xu = d(x)u + xf(u) = d(x)u + xu$; hence $d(x)U = \{0\}$ for all $x \in N$, and thus $d = 0$.

On the other hand, suppose that $d(U) = \{0\}$, so that $d = 0$. Then for all $x, y \in U$, $f(xy) = f(x)y = f(x)f(y)$, so that $f(x)(y - f(y)) = 0$. Replacing y by zy, $z \in N$, and noting that $f(zy) = zf(y)$, we see that $f(x)N(y - f(y)) = \{0\}$ for all $x, y \in U$. Therefore, $f(U) = \{0\}$ or f is the identity on U. But $f(U) = \{0\}$ contradicts Lemma 8, so f is the identity on U.

We now know that f is the identity on U and $f(xy) = xf(y)$ for all $x, y \in N$. Consequently, $f(ux) = ux = uf(x)$ for all $u \in U$ and $x \in N$, so that $U(x - f(x)) = \{0\}$ for all $x \in N$. It follows that f is the identity on N.

Theorem 16. *Let N be a 3-prime near ring and U be a nonzero semigroup ideal of N. If f is a nonzero generalized derivation on N with associated derivation d. If f acts as an antihomomorphism on U, then d = 0, f is the identity map on N, and N is a commutative ring.*

Proof. We begin by showing that $d = 0$ if and only if f is the identity map on N.

Clearly if f is the identity map on N, $xd(y) = 0$ for all $x, y \in N$, and hence $d = 0$.

Conversely, assume that $d = 0$, in which case $f(xy) = f(x)y = xf(y)$ for all $x, y \in N$. It follows that for any $x, y, z \in U$,

$$\begin{aligned} f(yxz) &= f(z)f(yx) = f(z)yf(x) \\ &= f(zy)f(x) = zf(y)f(x) = zf(xy). \end{aligned} \tag{14}$$

On the other hand,

$$\begin{aligned} f(yxz) &= f(xz)f(y) = f(x)zf(y) = f(x)f(zy) \\ &= f(x)f(y)f(z) = f(yx)f(z) = f(y)xf(z) \\ &= f(y)f(xz) = f(y)f(x)z = f(xy)z. \end{aligned} \tag{15}$$

Comparing (14) and (15) shows that $f(U^2)$ centralizes U, so that $f(U^2) \subseteq Z$ by Lemma 1(ii).

Now U^2 is a nonzero semigroup ideal by Lemma 3(ii); hence $f(U^2) \neq 0$ by Lemma 8. Choosing $x, y \in U$ such that $f(xy) \neq 0$, we see that for any $z \in U$, $f(xy)z = f(xyz) = f(yz)f(x) = f(y)zf(x) = f(y)f(zx) = f(y)f(x)f(z) = f(xy)f(z)$, and hence $f(xy)(z - f(z)) = 0$. Since $f(xy) \in Z \setminus \{0\}$, we conclude that $f(z) = z$ for all $z \in U$, and it follows easily that f is the identity map on N.

We note now that if the identity map on N acts as an antihomomorphism on U, then U is commutative, so that by Lemmas 1(ii) and 4 N is a commutative ring.

To complete the proof of our theorem, we need only to argue that $d = 0$. By our antihomomorphism hypothesis

$$f(xy) = d(x)y + xf(y) = f(y)f(x) \quad \forall x, y \in U. \tag{16}$$

Replacing y by xy in the above relation, we get

$$f(xy)f(x) = f(xxy) = d(x)xy + xf(xy) \quad \forall x, y \in U. \tag{17}$$

This implies that

$$(d(x)y + xf(y))f(x) = d(x)xy + xf(y)f(x) \quad \forall x, y \in U. \tag{18}$$

Using Lemma 5(ii), we get

$$\begin{aligned} d(x) yf(x) + xf(y) f(x) \\ = d(x) xy + xf(y) f(x) \quad \forall x, y \in U. \end{aligned} \tag{19}$$

Thus

$$d(x) yf(x) = d(x) xy \quad \forall x, y \in U. \tag{20}$$

Replacing y by yr in (20) and using (20), we get

$$d(x) yrf(x) = d(x) xyr, \text{ and so } d(x) y[r, f(x)] = 0 \quad \forall x, y \in U, r \in N. \tag{21}$$

Application of Lemma 3(i) yields that for each $x \in U$ either $d(x) = 0$ or $[r, f(x)] = 0$; that is $d(x) = 0$ or $f(x) \in Z$.

Suppose that there exists $w \in U$ such that $f(w) \in Z \setminus \{0\}$. Then for all $v \in U$ such that $d(v) = 0$, $f(wv) = f(w)v = f(v)f(w) = f(w)f(v)$, and hence $f(w)(v - f(v)) = 0 = v - f(v)$. Now consider arbitrary $x, y \in U$. If one of $f(x)$, $f(y)$ is in Z, then $f(xy) = f(x)f(y)$. If $d(x) = 0 = d(y)$, then $d(xy) = d(x)y + xd(y) = 0$, so $f(xy) = xy = f(x)f(y)$. Therefore $f(xy) = f(x)f(y)$ for all $x, y \in U$, and by Theorem 15, f is the identity map on N, and therefore $d = 0$.

The remaining possibility is that for each $x \in U$, either $d(x) = 0$ or $f(x) = 0$. Let $u \in U \setminus \{0\}$, and let $U_1 = uN$. Then U_1 is a nonzero semigroup right ideal contained in U and U_1 is an additive subgroup of N. The sets $\{x \in U_1 | d(x) = 0\}$ and $\{x \in U_1 | f(x) = 0\}$ are additive subgroups of U_1 with union equal to U_1, so $d(U_1) = \{0\}$ or $f(U_1) = \{0\}$. If $d(U_1) = \{0\}$, then $d = 0$ by Lemma 1(i). Suppose, then, that $f(U_1) = \{0\}$. Then for arbitrary $x, y \in N$ $f(uxy) = f(ux)y + uxd(y) = 0 = uxd(y)$, so $uNd(y) = \{0\}$, and again $d = 0$. This completes the proof.

Acknowledgments

A. Ali and P. Miyan gratefully acknowledge the support received by the University Grants Commission, India Grant F. no. 33–106 (2008) and Council of Scientific and Industrial Research, India Grant F. no. 9/112 (0475) 2 K12-EMR-I. Let $z \in Z$ and $x \in N$. Then $f(zx) = f(xz)$; that is, $f(z)x + zd(x) = d(x)z + xf(z)$.

References

[1] B. Hvala, "Generalized derivations in rings," *Communications in Algebra*, vol. 26, no. 4, pp. 1147–1166, 1998.

[2] H. E. Bell, "On prime near-rings with generalized derivation," *International Journal of Mathematics and Mathematical Sciences*, vol. 2008, Article ID 490316, 5 pages, 2008.

[3] H. E. Bell, "On derivations in near rings II," in *Nearrings, Nearfields and K-Loops*, vol. 426 of *Mathematical Applications*, pp. 191–197, Kluwer Academic Publishers, Dordrecht, The Netherlands, 1997.

[4] H. E. Bell and L. C. Kappe, "Rings in which derivations satisfy certain algebraic conditions," *Acta Mathematica Hungarica*, vol. 53, no. 3-4, pp. 339–346, 1989.

[5] H. E. Bell and G. Mason, "On derivations in near-rings," in *Near-Rings and Near-Fields*, vol. 137 of *North-Holland Mathematical Studies*, pp. 31–35, North-Holland, Amsterdam, The Netherlands, 1987.

[6] A. Asma, N. Rehman, and A. Shakir, "On Lie ideals with derivations as homomorphisms and antihomomorphisms," *Acta Mathematica Hungarica*, vol. 101, no. 1-2, pp. 79–82, 2003.

Multiresolution Expansion and Approximation Order of Generalized Tempered Distributions

Byung Keun Sohn

Department of Mathematics, Inje University, Kimhae 621-749, Republic of Korea

Correspondence should be addressed to Byung Keun Sohn; mathsohn@inje.ac.kr

Academic Editor: Palle E. Jorgensen

Let $\mathscr{K}^r_M{}'(\mathbb{R})$ be the generalized tempered distributions of $e^{M(x)}$-growth with restricted order $r \in \mathbb{N}_0$, where the function $M(x)$ grows faster than any linear functions as $|x| \to \infty$. We show the convergence of multiresolution expansions of $\mathscr{K}^r_M{}'(\mathbb{R})$ in the test function space $\mathscr{K}^r_M(\mathbb{R})$ of $\mathscr{K}^r_M{}'(\mathbb{R})$. In addition, we show that the kernel of an integral operator $K : \mathscr{K}^r_M{}'(\mathbb{R}) \to \mathscr{K}^r_M{}'(\mathbb{R})$ provides approximation order in $\mathscr{K}^r_M{}'(\mathbb{R})$ in the context of shift-invariant spaces.

1. Introduction

Multiresolution analysis was shown to be very useful in extending the expansions in orthogonal wavelets from $L^2(\mathbb{R})$ to a certain class of tempered distributions. Some interactions between wavelets and tempered distributions have been presented by Walter's work in [1–3]. Walter has found the analytic representation of tempered distributions of polynomial growth with restricted order, $\mathscr{S}'_r(\mathbb{R})$, $r \in \mathbb{N}_0$, by wavelets [1] and the multiresolution expansions' pointwise convergence of $\mathscr{S}'_r(\mathbb{R})$ [3]. Pilipović and Teofanov have showed the uniform convergence on compact sets of the derivatives of multiresolution expansions of $\mathscr{S}'_r(\mathbb{R})$ and the convergence of multiresolution expansions of $\mathscr{S}'_r(\mathbb{R})$ in the test function space $\mathscr{S}_r(\mathbb{R})$ of $\mathscr{S}'_r(\mathbb{R})$. As an application, Pilipović and Teofanov have shown that the kernel of an integral operator $K : \mathscr{S}'_r(\mathbb{R}) \to \mathscr{S}'_r(\mathbb{R})$ provides approximation order in $\mathscr{S}'_r(\mathbb{R})$ in the context of shift-invariant spaces [4].

In the meantime, the tempered distributions of polynomial growth were extended to tempered distributions of $e^{|x|}$-growth, $\mathscr{K}'_1(\mathbb{R})$, in [5, 6] and $e^{|x|^p}$-growth, $\mathscr{K}'_p(\mathbb{R})$, in [7, 8] or $e^{M(x)}$-growth, $\mathscr{K}'_M(\mathbb{R})$, in [9, 10], where the function $M(x)$ grows faster than any linear functions as $|x| \to \infty$. We have considered the analytic representation of tempered distributions of $e^{M(x)}$-growth with restricted order, $\mathscr{K}^r_M{}'(\mathbb{R})$, by wavelets [11]. Also, we have shown that the multiresolution expansions of $\mathscr{K}^r_M{}'(\mathbb{R})$ converges pointwise to the value of the distribution where it exists [12].

In this paper, we will show the uniform convergence on compact sets of the derivatives of multiresolution expansions of $\mathscr{K}^r_M{}'(\mathbb{R})$ and convergence of multiresolution expansions of $\mathscr{K}^r_M{}'(\mathbb{R})$ in the test function space $\mathscr{K}^r_M(\mathbb{R})$ of $\mathscr{K}^r_M{}'(\mathbb{R})$. In addition, we will show that the kernel of an integral operator $K : \mathscr{K}^r_M{}'(\mathbb{R}) \to \mathscr{K}^r_M{}'(\mathbb{R})$ provides approximation order in $\mathscr{K}^r_M{}'(\mathbb{R})$. This is an extension of the works of Pilipović and Teofanov [4] in the context of generalized tempered distributions, $\mathscr{K}'_M(\mathbb{R})$.

2. The Generalized Tempered Distribution Spaces $\mathscr{K}'_M(\mathbb{R})$

Throughout this paper, we will use C or C_i to denote the positive constants, which are independent parameters and may be different at each occurrence.

Let $\mu(\xi)$ $(0 \le \xi \le \infty)$ denote a continuous increasing function such that $\mu(0) = 0$ and $\mu(\infty) = \infty$. For $x \ge 0$, we define

$$M(x) = \int_0^x \mu(\xi)\, d\xi. \tag{1}$$

The function $M(x)$ is an increasing, convex, and continuous function with $M(0) = 0$, $M(\infty) = \infty$ and satisfies the fundamental convexity inequality $M(x_1) + M(x_2) \leq M(x_1 + x_2)$. Further, we define $M(x)$ for negative x by means of the equality $M(x) = M(-x)$. Note that since the derivative $\mu(x)$ of $M(x)$ is unbounded in $\mathbb{R}$, the function $M(x)$ will grow faster than any linear function as $|x| \to \infty$. Now we list some properties of $M(x)$ which will be frequently used later. Consider the following:

$$\begin{aligned} M(x) + M(y) &\leq M(x+y) \quad \forall x, y \geq 0, \\ M(x+y) &\leq M(2x) + M(2y) \quad \forall x, y \geq 0. \end{aligned} \tag{2}$$

Using the function $M(x)$, we define the space $\mathcal{K}_M(\mathbb{R})$ as the space of all functions $\varphi \in C^\infty(\mathbb{R})$ such that

$$\|\varphi\|_{\mathcal{K}_M} = \sup_{x\in\mathbb{R},\alpha\leq k} e^{M(kx)} \left| \frac{d^\alpha}{dx^\alpha}\varphi(x) \right| < \infty, \quad k = 1, 2, \ldots. \tag{3}$$

The topology in $\mathcal{K}_M(\mathbb{R})$ is defined by the family of the seminorms $\|\cdot\|_{\mathcal{K}_M}$. Then $\mathcal{K}_M(\mathbb{R})$ become a Fréchet space and $\mathcal{D}(\mathbb{R}) \hookrightarrow \mathcal{K}_M(\mathbb{R}) \hookrightarrow \mathcal{S}(\mathbb{R}) \hookrightarrow \mathcal{E}(\mathbb{R})$ are continuous and dense inclusions; here $\mathcal{D}(\mathbb{R})$ denotes the spaces of all $C^\infty(\mathbb{R})$ functions with compact supports, $\mathcal{S}(\mathbb{R})$ the spaces of polynomially decreasing functions (Schwartz functions), and $\mathcal{E}(\mathbb{R})$ the space of all $C^\infty(\mathbb{R})$ functions. By $\mathcal{K}'_M(\mathbb{R})$, we mean the space of continuous linear functionals on $\mathcal{K}_M(\mathbb{R})$.

Definition 1. We say that the elements of $\mathcal{K}'_M(\mathbb{R})$ are generalized tempered distributions.

Clearly, when $M(x) = \log(1 + |x|)$, $\mathcal{K}'_M(\mathbb{R})$ are tempered distributions (Schwarz distributions), $\mathcal{S}(\mathbb{R})$. When $M(x) = |x|$, $\mathcal{K}'_M(\mathbb{R})$ are tempered distributions, $\mathcal{K}'_1(\mathbb{R})$, which are introduced and characterized by Yoshinaga [6] and Hasumi [5], independently. When $M(x) = |x|^p$, $p > 1$, $\mathcal{K}'_M(\mathbb{R})$ are tempered distributions, $\mathcal{K}'_p(\mathbb{R})$, which are introduced and characterized by Sznajder and Zielezny [7, 8]. For details about $\mathcal{K}'_M(\mathbb{R})$, we refer to [9, 10].

For a natural number r, we define by $\mathcal{K}^r_M(\mathbb{R})$ the space of all $\phi \in C^r(\mathbb{R})$ such that

$$\begin{aligned} \|\phi\|_{\mathcal{K}^r_M} &= \sup_{x\in\mathbb{R},\alpha\leq r} e^{M(rx)} \left| \frac{d^\alpha}{dx^\alpha}\varphi(x) \right| < \infty, \\ \lim_{|x|\to\infty} \sup_{0\leq\alpha\leq r} e^{M(rx)} &\left| \frac{d^\alpha}{dx^\alpha}\varphi(x) \right| = 0. \end{aligned} \tag{4}$$

The topology of $\mathcal{K}^r_M(\mathbb{R})$ is defined by the family of $\|\cdot\|_{\mathcal{K}^r_M}$ and the dual of $\mathcal{K}^r_M(\mathbb{R})$ is denoted by $\mathcal{K}^{r\,\prime}_M(\mathbb{R})$. Clearly, $\mathcal{K}_M(\mathbb{R})$ is the projective limit of $\mathcal{K}^r_M(\mathbb{R})$ when $r \to \infty$ and $\mathcal{K}'_M(\mathbb{R}) = \bigcup_{r\in\mathbb{N}} \mathcal{K}^{r\,\prime}_M(\mathbb{R})$. Also, we have continuous and dense inclusion mapping as following:

$$\begin{aligned} &\mathcal{K}_M(\mathbb{R}) \hookrightarrow \cdots \hookrightarrow \mathcal{K}^{r+1}_M(\mathbb{R}) \hookrightarrow \mathcal{K}^r_M(\mathbb{R}) \hookrightarrow \cdots \\ &\hookrightarrow \mathcal{K}^{r\,\prime}_M(\mathbb{R}) \hookrightarrow \mathcal{K}^{r+1\,\prime}_M(\mathbb{R}) \hookrightarrow \cdots \hookrightarrow \mathcal{K}'_M(\mathbb{R}). \end{aligned} \tag{5}$$

Definition 2. We say that the elements of $\mathcal{K}^{r\,\prime}_M(\mathbb{R})$ are generalized tempered distributions of order r.

We define by $\widetilde{\mathcal{K}}^r_M(\mathbb{R})$ the space of all $\psi \in C^r(\mathbb{R})$ such that

$$\|\psi\|_{\widetilde{\mathcal{K}}^r_M} = \sup_{x\in\mathbb{R},0\leq\alpha\leq r} e^{M(lx)} \left| \frac{d^\alpha}{dx^\alpha}\psi(x) \right| < \infty, \quad l = 1, 2, \ldots. \tag{6}$$

The topology of $\widetilde{\mathcal{K}}^r_M(\mathbb{R})$ is defined by the family of $\|\cdot\|_{\widetilde{\mathcal{K}}^r_M}$ and the dual of $\widetilde{\mathcal{K}}^r_M(\mathbb{R})$ is denoted by $\widetilde{\mathcal{K}}^{r\,\prime}_M(\mathbb{R})$. Obviously, $\widetilde{\mathcal{K}}^r_M(\mathbb{R}) \subset \mathcal{K}^r_M(\mathbb{R})$.

Now, we give a theorem that will be used later.

Theorem 3. *Let ϕ and sequence $\{\phi_n\}_{n\in\mathbb{N}}$ be given in $\mathcal{K}^{r+1}_M(\mathbb{R})$ such that $\{(d^\alpha/dx^\alpha)\phi_n\}_{n\in\mathbb{N}}$ converges uniformly to $(d^\alpha/dx^\alpha)\phi$ on every compact set $K \subset \mathbb{R}$ and for $\alpha = 0, 1, \ldots, r$. If $\{\phi_n\}_{n\in\mathbb{N}}$ is bounded in $\mathcal{K}^{r+1}_M(\mathbb{R})$, then the sequence $\{\phi_n\}_{n\in\mathbb{N}}$ converges to $\phi \in \mathcal{K}^{r+1}_M(\mathbb{R})$ in $\mathcal{K}^r_M(\mathbb{R})$.*

Proof. Let $\epsilon > 0$ be given and let $\alpha \in \{0, 1, \ldots, r\}$. Then there exist N such that

$$\sup_{x\in K} e^{M(rx)} \left| \frac{d^\alpha}{dx^\alpha}(\phi_n - \phi)(x) \right| < \epsilon, \quad n \geq N, \tag{7}$$

for arbitrary $K \subset \mathbb{R}$. Also, since the sequence $\{\phi_n\}_{n\in\mathbb{N}}$ is bounded in $\mathcal{K}^{r+1}_M(\mathbb{R})$, we can take a positive number $A > 0$ and a compact set K such that $|x| > A$ when $x \notin K$ and

$$\begin{aligned} &\sup_{x\notin K} e^{M(rx)} \left| \frac{d^\alpha}{dx^\alpha}(\phi_n - \phi)(x) \right| \\ &= \sup_{x\notin K} e^{M(rx)} e^{-M((r+1)x)} e^{M((r+1)x)} \left| \frac{d^\alpha}{dx^\alpha}(\phi_n - \phi)(x) \right| \\ &\leq \sup_{x\notin K} e^{-M(x)} e^{M((r+1)x)} \left| \frac{d^\alpha}{dx^\alpha}(\phi_n - \phi)(x) \right| \\ &\leq e^{-M(A)} \sup_{x\in\mathbb{R}} e^{M((r+1)x)} \left(\left| \frac{d^\alpha}{dx^\alpha}\phi_n(x) \right| + \left| \frac{d^\alpha}{dx^\alpha}\phi(x) \right| \right) \\ &< Ce^{-M(A)} < \epsilon. \end{aligned} \tag{8}$$

From (7) and (8), we have

$$\lim_{n\to\infty} \sup_{x\in\mathbb{R}} e^{M(rx)} \left| \frac{d^\alpha}{dx^\alpha}(\phi_n - \phi)(x) \right| = 0, \quad 0 \leq \alpha \leq r. \tag{9}$$

3. Multiresolution Expansion of $\mathcal{K}'_M(\mathbb{R})$

Definition 4. A multiresolution analysis (shortly MRA) consists of a sequence of closed subspaces $V_n, n \in \mathbb{Z}$, of $L^2(\mathbb{R})$ satisfying the following:

(i) $\{\psi(t-n)\}_{n\in\mathbb{Z}}$ is an orthonormal basis of V_0,

(ii) $\cdots \subset V_{-1} \subset V_0 \subset V_1 \subset \cdots \subset L^2(\mathbb{R})$,

(iii) $f \in V_n \Leftrightarrow f(2\cdot) \in V_{n+1}$,

(iv) $\bigcap_n V_n = \{0\}$, $\overline{\bigcup_n V_n} = L^2(\mathbb{R})$.

The function ψ whose existence is asserted in (i) is called a scaling function of the given MRA.

Definition 5. We say that a multiresolution analysis V_n, $n \in \mathbb{Z}$, is $(M, r\text{-})$ regular MRA of $L^2(\mathbb{R})$ if the scaling function ψ is in $\widetilde{\mathscr{K}}^r_M(\mathbb{R})$.

Example 6. It is impossible that the scaling function ψ has exponential decay and $\psi \in C^\infty(\mathbb{R})$, with all derivatives bounded, unless $\psi = 0$. Refer to [13, Corollary 5.5.3]. So we will restrict our attention to $\mathscr{K}^r_M(\mathbb{R})$ or $\widetilde{\mathscr{K}}^r_M(\mathbb{R})$. From the remark in [13] or, page 152 [2, Example 4, page 48], Battle-Lemarié's wavelets are in $\mathscr{K}^r_M(\mathbb{R})$ for some $r \in \mathbb{N}$ when $M(x) = |x|$, but not in $\widetilde{\mathscr{K}}^r_M(\mathbb{R})$ even if they have exponential decay and smoothness. In [13], Daubechies shown that for an arbitrary nonnegative integer r, there exists an $(M, r\text{-})$ regular MRA of $L^2(\mathbb{R})$ such that the scaling function ψ has compact supports.

Let V_j be an $(M, r\text{-})$ regular MRA of $L^2(\mathbb{R})$ and let ψ be a scaling function. The reproducing kernel of V_0 is given by

$$q_0(x, y) = \sum_{n\in\mathbb{Z}} \psi(x-n)\overline{\psi(y-n)}. \tag{10}$$

The series and its derivatives with respect to x or y of order $\le r$ converge uniformly on $\mathbb{R}$ because of the regularity of $\psi \in \widetilde{\mathscr{K}}^r_M(\mathbb{R})$. The reproducing kernel of the projection operator onto V_j is

$$q_j(x, y) = 2^j q_0(2^j x, 2^j y), \quad x, y \in \mathbb{R}, \tag{11}$$

and the projection of $f \in L^2(\mathbb{R})$ onto V_j is given by

$$\begin{aligned} q_j f(x) &= \langle f(y), q_j(x, y)\rangle \\ &= \int f(y)\, q_j(x, y)\, dy, \quad x \in \mathbb{R}. \end{aligned} \tag{12}$$

The sequence $\{q_j\}_{j\in\mathbb{Z}}$, given in (12), is called the multiresolution expansion of $f \in L^2(\mathbb{R})$.

Definition 7. For a given $f \in \mathscr{K}^r_M{}'(\mathbb{R})$, the sequence $\{q_j\}_{j\in\mathbb{Z}}$ defined by

$$\langle q_j f, \phi\rangle = \langle f, q_j\phi\rangle, \quad \phi \in \mathscr{K}^r_M(\mathbb{R}) \tag{13}$$

is called the multiresolution expansion of $f \in \mathscr{K}^r_M{}'(\mathbb{R})$.

We deduce the following properties of the reproducing kernel q_0 with scaling function $\psi \in \widetilde{\mathscr{K}}^r_M(\mathbb{R})$:

(a) $q_0(x, y) = q_0(y, x)$ and $q_0(x+k, y+k) = q_0(x, y)$ for all $k \in \mathbb{Z}$.

(b) For every $l \in \mathbb{N}$ and $0 \le \alpha,\ \beta \le r$, there exist $C_l' > 0$ such that

$$\begin{aligned} \left|\frac{\partial^\alpha}{\partial x^\alpha}\frac{\partial^\beta}{\partial y^\beta} q_0(x, y)\right| &\le \sum_j \left|\frac{d^\alpha}{dx^\alpha}\psi(x-j)\right|\left|\frac{d^\beta}{dx^\beta}\psi(y-j)\right| \\ &\le \sum_j C_l e^{-M((2l+1)(x-j))} e^{-M((2l+1)(y-j))} \\ &\le \sum_j C_l e^{-M(2l(x-j))} e^{-M(x-j)} \\ &\quad \times e^{-M(2l(j-y))} e^{-M(j-y)} \\ &\le C_l e^{-M(l(x-y))} \sum_j C_l e^{-M(x-j)} e^{-M(j-y)} \\ &\le C_l' e^{-M(l(x-y))}, \end{aligned} \tag{14}$$

where we used the properties (2).

(c) $\int_{-\infty}^{\infty} q_0(x, y) y^\alpha dy = x^\alpha,\ \ y \in \mathbb{R},\ 0 \le \alpha \le r$.

Let V_j be an $(M, r\text{-})$ regular MRA of $L^2(\mathbb{R})$. We fix a function $g \in \mathscr{D}(\mathbb{R})$ with $\int g(x)dx = 1$. We let g_j denote the function $2^j g(2^j x)$ and let G_j denote the operation of convolution by g_j. For each fixed x, we consider the function $\partial_x^\alpha q_0(x, y)$ of the variable y. From (c), we have

$$\int \partial_x^\alpha q_0(x, y)\, y^\beta dy = 0, \tag{15}$$

for $0 \le \beta < \alpha$, whereas

$$\int \partial_x^\alpha q_0(x, y)\, y^\alpha dy = \alpha!. \tag{16}$$

Now, it follows from integration by parts that the kernal $g(x-y)$ of the operator G shares these properties (15) and (16) with $q_0(x, y)$.

Let

$$R^\alpha(x, y) = \partial_x^\alpha q_0(x, y) - \partial_x^\alpha g(x-y). \tag{17}$$

From (b) and the fact that $g \in \mathscr{D}(\mathbb{R}) \subset \mathscr{K}^r_M(\mathbb{R})$, we have

$$|R^\alpha(x, y)| \le c_k e^{-M(k(x-y))}, \quad x, y \in \mathbb{R},\ k \in \mathbb{N}, \tag{18}$$

and these functions also satisfy

$$\int R^\alpha(x, y)\, dy = 0 \tag{19}$$

identically in x for every $\alpha = 1, 2, \ldots, r$. They, for every $j \in \mathbb{Z}$ and $f \in C^r(\mathbb{R})$ with at most $e^{M(x)}$-growth, define operator R_j^α by

$$R_j^\alpha f(x) = 2^j \int R^\alpha(2^j x, 2^j y) f(y)\, dy \tag{20}$$

which are such that

$$q_j \frac{d^\alpha}{dx^\alpha} f(x) = G_j \frac{d^\alpha}{dy^\alpha} f(y) + R_j^\alpha \frac{d^\alpha}{dy^\alpha} f(y), \quad (21)$$

that is,

$$\int q_j(x,y) \frac{d^\alpha}{dy^\alpha} f(y)\,dy = 2^j \int g\left(2^j(x-y)\right) \frac{d^\alpha}{dy^\alpha} f(y)\,dy + 2^j \int R^\alpha\left(2^j x, 2^j y\right) \frac{d^\alpha}{dy^\alpha} f(y)\,dy. \quad (22)$$

From Theorem 1.1 in [14], we have

$$\lim_{j\to\infty} G_j \frac{d^\alpha}{dy^\alpha} f(y)\,dy = \frac{d^\alpha}{dx^\alpha} f(x), \quad x \in \mathbb{R},\ \alpha \ge 0, \quad (23)$$

uniformly on compact sets. Now we will show the uniform convergence on compact sets of the derivatives of multiresolution expansions of $\mathcal{K}_M^{r\,\prime}(\mathbb{R})$.

Theorem 8. *Let $f \in C^r(\mathbb{R})$ such that the corresponding derivatives $(d^\alpha/dx^\alpha)f$ are bounded by a $e^{M(k_0 x)}$ when $|x| \to \infty$, for every $\alpha = 0,1,\ldots,r$ and some $k_0 \in \mathbb{N}$. If $q_j f$, given by (12), be the projection of f onto an $(M,r$-$)$ regular MRA of $L^2(\mathbb{R})$, then the sequence $\{(d^\alpha/dx^\alpha)q_j f\}_{j\in\mathbb{Z}}$ converges uniformly on compact sets to $(d^\alpha/dx^\alpha)f$ as $j \to \infty$, for every $\alpha = 0,1,\ldots,r$.*

Proof. If $|y-x| \le c$, we have

$$\left| \frac{d^\alpha}{dy^\alpha} f(y) - \frac{d^\alpha}{dy^\alpha} f(y)\Big|_{y=x} \right| \le e_\alpha^M(y-x), \quad (24)$$

where $e_\alpha^M(x)$ is a continuous function with $e^{M(k_0 x)}$ growth and $e_\alpha^M(0) = 0$. From (18), given a compact set K, we have

$$\begin{aligned} 2^j &\left| \int R^\alpha\left(2^j x, 2^j y\right) \frac{d^\alpha}{dy^\alpha} f(y)\,dy \right| \\ &\le 2^j \int \left| R^\alpha\left(2^j x, 2^j y\right) \left(\frac{d^\alpha}{dy^\alpha} f(y) - \frac{d^\alpha}{dy^\alpha} f(y)\Big|_{y=x} \right) \right| dy \\ &\le 2^j \int c_k e^{-M(2^j k(x-y))} \left| e_\alpha^M(y-x) \right| dy, \end{aligned} \quad (25)$$

for large enough j and $x \in K$. Since k can be chosen arbitrary, we obtain by dominated convergence theorem,

$$\begin{aligned} \lim_{j\to\infty} 2^j &\left| \int R^\alpha\left(2^j x, 2^j y\right) \frac{d^\alpha}{dy^\alpha} f(y)\,dy \right| \\ &\le \int \lim_{j\to\infty} 2^j c_k e^{-M(2^j k(x-y))} \left| e_\alpha^M(y-x) \right| dy = 0 \end{aligned} \quad (26)$$

uniformly for $x \in K$. From (21) and (23), we have the conclusion.

We now ready to show the main theorem.

Theorem 9. *Let $\phi \in \widetilde{\mathcal{K}}_M^r(\mathbb{R})$ and let $q_j\phi(x)$, given by (7), be a projection of ϕ onto an $(M,r$-$)$ regular MRA of $L^2(\mathbb{R})$. If $\phi \in \mathcal{K}_M^{2(r+1)}(\mathbb{R})$, then the sequence $\{q_j\phi(x)\}$ converges to $\phi(x)$ in $\mathcal{K}_M^r(\mathbb{R})$ as $j \to \infty$.*

Proof. Let g and R^α be given in (21) such that $g \in \mathcal{D}(\mathbb{R})$ and $\int g(x)dx = 1$. From Theorems 3 and 8 and (21), it suffices to show that

$$\begin{aligned} \sup_{x\in\mathbb{R}} e^{M((r+1)x)} \frac{1}{h} &\left| \int q_0\left(\frac{x}{h}, \frac{y}{h}\right) \phi(y)\,dy \right| \\ = \sup_{x\in\mathbb{R}} e^{M((r+1)x)} \frac{1}{h} &\left| \int g\left(\frac{x-y}{h}\right) \frac{d^\alpha}{dy^\alpha}\phi(y)\,dy \right. \\ &\left. + \int R^\alpha\left(\frac{x}{h}, \frac{y}{h}\right) \frac{d^\alpha}{dy^\alpha}\phi(y)\,dy \right| \end{aligned} \quad (27)$$

is bounded for every $\alpha \in \{0,1,\ldots,r\}$ and $h > 0$. Since g has a compact support, then

$$\begin{aligned} \sup_{x\in\mathbb{R}} e^{M((r+1)x)} \frac{1}{h} &\left| \int g\left(\frac{x-y}{h}\right) \frac{d^\alpha}{dy^\alpha}\phi(y)\,dy \right| \\ &\le \sup_{x\in\mathbb{R}} \frac{1}{h} \left| \int g\left(\frac{x-y}{h}\right) e^{M((r+1)x)} e^{-M(2(r+1)y)} dy \right| \\ &\le \sup_{x\in\mathbb{R}} \frac{1}{h} \left| \int g\left(\frac{x-y}{h}\right) e^{M(2(r+1)(x-y))} dy \right| \le C. \end{aligned} \quad (28)$$

Hence we have only to show that

$$K = \sup_{x\in\mathbb{R}} e^{M((r+1)x)} \frac{1}{h} \left| \int R^\alpha\left(\frac{x}{h}, \frac{y}{h}\right) \frac{d^\alpha}{dy^\alpha}\phi(y)\,dy \right| \le C, \quad x \in \mathbb{R}, \quad (29)$$

for every $\alpha \in \{0,1,\ldots,r\}$ and $h > 0$. Let $S_1 = \{y : |x-y| \le 1\}$, $S_2 = \{y : |x-y| > 1$ and $(1/2)|x| \le |y|\}$ and $S_3 = \{y : |x-y| > 1$ and $(1/2)|x| > |y|\}$. Then, by (18), we have

$$\begin{aligned} I &= \sup_{x\in\mathbb{R}} e^{M((r+1)x)} \frac{1}{h} \left| \int R^\alpha\left(\frac{x}{h}, \frac{y}{h}\right) \frac{d^\alpha}{dy^\alpha}\phi(y)\,dy \right| \\ &\le c_l \sup_{x\in\mathbb{R}} e^{M((r+1)x)} \frac{1}{h} \int e^{-M(l((x-y)/h))} e^{-M(2(r+1)y)} dy \\ &= c_l \sup_{x\in\mathbb{R}} e^{M((r+1)x)} \frac{1}{h} \left(\int_{S_1} + \int_{S_2} + \int_{S_3} \right) \\ &\quad \times e^{-M(l((x-y)/h))} e^{-M(2(r+1)y)} dy \\ &= c_l \left(I_1 + I_2 + I_3 \right). \end{aligned} \quad (30)$$

By a simple change of variable, we have

$$\begin{aligned} I_1 &= \sup_{x\in\mathbb{R}} e^{M((r+1)x)} \frac{1}{h} \int_{S_1} e^{-M(l((x-y)/h))} e^{-M(2(r+1)y)} dy \\ &\le \sup_{x\in\mathbb{R}} \frac{1}{h} \int_{S_1} e^{-M(l((x-y)/h))} e^{M(2(r+1)(x-y))} dy \\ &\le 2e^{M(2(r+1))} \int_0^1 \frac{1}{h} e^{-M(l(t/h))} dt = 2e^{M(2(r+1))} \\ &\quad \times \int_0^{1/h} e^{-M(lu)} du \le C_1. \end{aligned} \tag{31}$$

Since $(1/2)|x| < |y|$ and $(1/2)|y| \le |x-y|$ on S_2, then

$$\begin{aligned} I_2 &= \sup_{x\in\mathbb{R}} e^{M((r+1)x)} \frac{1}{h} \int_{S_2} e^{-M(l((x-y)/h))} e^{-M(2(r+1)y)} dy \\ &\le \sup_{x\in\mathbb{R}} e^{M((r+1)x)} \frac{1}{h} \int_{S_2} e^{-M(l((x-y)/h))} e^{-M((r+1)x)} dy \\ &\le \frac{1}{h} \int_{S_2} e^{-M((l/2h)y)} dy \le C_2, \end{aligned} \tag{32}$$

for sufficiently large l. Since $(1/2)|x| > |x-y|$ on S_3, then

$$\begin{aligned} I_3 &= \sup_{x\in\mathbb{R}} e^{M((r+1)x)} \frac{1}{h} \int_{S_3} e^{-M(l((x-y)/h))} e^{-M(2(r+1)y)} dy \\ &\le \sup_{x\in\mathbb{R}} e^{M((r+1)x)} \frac{1}{h} e^{-M((l/2)(x/h))} \int_{S_3} e^{-M((r+1)y)} dy \\ &\le C_3' \frac{1}{h} \int_{S_3} e^{-M(2(r+1)y)} dy \le C_3, \end{aligned} \tag{33}$$

for sufficiently large l.

4. Approximation Order of $\mathscr{K}'_M(\mathbb{R})$

A space of functions S is called shift invariant if it is invariant under all integer translate, that is,

$$f \in S \Longleftrightarrow f(\cdot + k) \in S \quad \forall k \in \mathbb{Z}. \tag{34}$$

The principal shift-invariant subspaces $S = S(\phi)$ are generated by the closure of the linear span of the shifts of ϕ. The stationary ladder of spaces $\{S^h(\phi) : h > 0\}$ is given by

$$S^h(\phi) = \left\{ f\left(\frac{\cdot}{h}\right) : f \in S \right\}. \tag{35}$$

To rate the efficiency for approximation of such spaces, the concept of approximation order is widely used. We say that the scale of the space $S^h(\phi)$ provides approximation order k in F if for every sufficiently smooth f,

$$\inf_{g\in S^h(\phi)} \|f - g\|_F \le Ch^k, \quad h > 0, \tag{36}$$

where $C = C(f) > 0$. For further details about the theory on the approximation order provided by shift-invariant spaces, we refer to [15, 16]. We will focus our attention to the so-called approximation order of an integral operator.

Let K be an integral operator of the following form

$$(Kf)(x) = \int K(x,y) f(y) dy, \quad x \in \mathbb{R}. \tag{37}$$

We assume that $K(x-k, y) = K(x, y+k)$, $h \in \mathbb{Z}$, $x, y \in \mathbb{R}$. For $h > 0$, we define

$$K_h = \varrho_h K \varrho_{1/h}, \tag{38}$$

where ϱ is the scaling operator $\varrho_h f = f(\cdot/h)$. We say that the integral operator K defined by (37) provides approximation order k in F if for every sufficiently smooth f,

$$\|K_h f - f\|_F \le Ch^k, \quad h > 0, \tag{39}$$

where $C = C(f) > 0$. For further details about the theory on the approximation order provided by integral or kernel operator, we refer to [17, 18].

Definition 10 (see [4]). Let $f \in \mathscr{K}^{r}_M{}'(\mathbb{R})$. Let $K(x,y)$, $x, y, \in \mathbb{R}$, be the kernel of an integral operator $K : \mathscr{K}^{r}_M{}'(\mathbb{R}) \to \mathscr{K}^{r}_M{}'(\mathbb{R})$. Kf is given by $\langle Kf, \phi\rangle = \langle f, K\phi\rangle$. We say that the operator K provides approximation order k in $\mathscr{K}^{r}_M{}'(\mathbb{R})$ if

$$\|K_h f - f\|_{\mathscr{K}^{r+k}_M{}'} = \sup_{\|\phi\|_{\mathscr{K}^{r+k}_M}=1} |\langle K_h f, \phi\rangle - \langle f, \phi\rangle| \le Ch^k, \quad h > 0, \tag{40}$$

where the constant $C = C(f) > 0$.

We will now show that the kernel of an integral operator $K : \mathscr{K}^{r}_M{}'(\mathbb{R}) \to \mathscr{K}^{r}_M{}'(\mathbb{R})$ provides approximation order in $\mathscr{K}^{r}_M{}'(\mathbb{R})$.

Theorem 11. *Let $\phi \in \widetilde{\mathscr{K}}^{r+k}_M(\mathbb{R})$ with compact support such that the integer shifts of ϕ form an orthogonal basis of $S(\phi)$ with respect to the inner product in $L^2(\mathbb{R})$. Assume that $\phi(x) = \sum_{n\in\mathbb{N}} c_n \phi(2x-k)$ for some sequence $\{c_n\}_{n\in\mathbb{N}}$. Let*

$$K(x,y) = \sum_{l\in\mathbb{Z}} \phi(x-l)\overline{\phi}(y-l), \quad x, y \in \mathbb{R} \tag{41}$$

be the kernel of the integral operator given by (37). Then K provides approximation order k in $\mathscr{K}^{r}_M{}'(\mathbb{R})$.

Proof. Firstly, we will show that

$$\begin{aligned} J &= \|K_h\phi - \phi\|_{\mathscr{K}^r_M} \\ &= \sup_{x\in\mathbb{R}} e^{M(rx)} \left| \frac{d^\alpha}{dx^\alpha} \left(\langle K_h(x,y), \phi(y)\rangle - \phi(x) \right) \right| \\ &\le C\|\phi\|_{\mathscr{K}^{r+k}_M} h^k, \end{aligned} \tag{42}$$

where $\phi \in \mathscr{K}_M^{r+k}$, $k \in \mathbb{N}$. If we accept the result (42) for a moment, it follows that for $f \in \mathscr{K}_M^{r}{}'(\mathbb{R}) \subset \mathscr{K}_M^{r+k}{}'(\mathbb{R})$, we have

$$\begin{aligned} |\langle K_h f - f, \phi\rangle| &= |\langle f, K_h\phi - \phi\rangle| \\ &\leq \|f\|_{\mathscr{K}_M^r{}'} \|K_h\phi - \phi\|_{\mathscr{K}_M^r} \leq Ch^k \|\phi\|_{\mathscr{K}_M^{r+k}{}'} \end{aligned} \tag{43}$$

hence

$$\|K_h f - f\|_{\mathscr{K}_M^{r+k}{}'} = \sup_{\|\phi\|_{\mathscr{K}_M^{r+k}}=1} |\langle K_h f, \phi\rangle - \langle f, \phi\rangle| \leq Ch^k, \quad h > 0, \tag{44}$$

which implies the conclusion.

Since $\{S^{2^{-j}}(\phi) : j \in \mathbb{Z}\}$ satisfy the conditions of $(M, r\text{-})$ regular MRA of $L^2(\mathbb{R})$ with $S(\phi) = V_0$, we can apply (21) to the operator K, that is,

$$\begin{aligned} \frac{d^\alpha}{dx^\alpha} K_h f(x) &= \frac{1}{h}\int g\left(\frac{x-y}{h}\right)\frac{d^\alpha}{dy^\alpha} f(y)\,dy \\ &+ \frac{1}{h}\int R^\alpha\left(\frac{x}{h}, \frac{y}{h}\right)\frac{d^\alpha}{dy^\alpha} f(y)\,dy, \end{aligned} \tag{45}$$

where g and R^α are given in (21). For $0 \leq \alpha \leq r$,

$$\begin{aligned} J &= \sup_{x\in\mathbb{R}} e^{M(rx)} \left| \frac{d^\alpha}{dx^\alpha}\left(\langle K_h(x,y), \phi(y)\rangle - \phi(x)\right)\right| \\ &\leq \sup_{x\in\mathbb{R}} e^{M(rx)} \left| \frac{1}{h}\int g\left(\frac{x-y}{h}\right) \right. \\ &\qquad \left. \times \left(\frac{d^\alpha}{dy^\alpha}\phi(y) - \frac{d^\alpha}{dy^\alpha}\phi(y)\Big|_{y=x}\right) dy \right| \\ &+ \sup_{x\in\mathbb{R}} e^{M(rx)} \left| \frac{1}{h}\int R^\alpha\left(\frac{x-y}{h}\right) \right. \\ &\qquad \left. \times \left(\frac{d^\alpha}{dy^\alpha}\phi(y) - \frac{d^\alpha}{dy^\alpha}\phi(y)\Big|_{y=x}\right) dy \right| \\ &= J_1 + J_2. \end{aligned} \tag{46}$$

In order to estimate J_1, we consider $g \in \mathscr{D}(\mathbb{R})$ with $\int g(x)dx = 1$ and $\int g(x)x^\alpha dx = 0, 0 < |\alpha| < \max\{r, k-1\}$. Let c be a constant such that $\sup g \subset [-c, c]$. If we assume $h \in (0,1)$, the smoothness of $\phi \in \mathscr{K}_M^{r+k}(\mathbb{R}) \subset C^{r+k}(\mathbb{R})$ implies

$$\begin{aligned} J_1 &= \sup_{hx\in\mathbb{R}} e^{M(rhx)} \left| \int_{|x-y|\leq c} g(x-y) \right. \\ &\qquad \left. \times \left(\frac{d^\alpha}{dy^\alpha}\phi(y)\Big|_{y=hy} - \frac{d^\alpha}{dy^\alpha}\phi(y)\Big|_{y=hx}\right) dy \right| \\ &= \sup_{hx\in\mathbb{R}} e^{M(rhx)} \left| \int_{|x-y|\leq c} g(x-y) \right. \\ &\qquad \times \left((y-x) \times h \frac{d^{\alpha+1}}{dy^{\alpha+1}}\phi(y)\Big|_{y=hx} \right. \\ &\qquad + \cdots + \frac{(y-x)^{k-1}}{(k-1)!} \times h^{k-1} \frac{d^{\alpha+k-1}}{dy^{\alpha+k-1}}\phi(y)\Big|_{y=hx} \\ &\qquad \left.\left. + \frac{(y-x)^k}{k!} \times h^k \frac{d^{\alpha+k}}{dy^{\alpha+k}}\phi(y)\Big|_{y=\xi(y)} \right) dy \right| \\ &\leq \sup_{hx\in\mathbb{R}} e^{M(rhx)} \left| \int_{|x-y|\leq c} g(x-y) \frac{(y-x)^k}{k!} \right. \\ &\qquad \left. \times h^k \frac{d^{\alpha+k}}{dy^{\alpha+k}}\phi(y)\Big|_{y=\xi(y)} dy \right| \\ &\leq \sup_{hx\in\mathbb{R}} e^{M(rhx)} C\frac{c^k}{k!}h^k \sup_{\xi(y)\in[hx-hc,hx+hc]} \left|\frac{d^{\alpha+k}}{dy^{\alpha+k}}\phi(y)\right| \\ &= C\frac{c^k}{k!}h^k \sup_{s\in\mathbb{R}} e^{M(rs)} \sup_{t\in[s-c,s+c]} \left|\frac{d^{\alpha+k}}{dt^{\alpha+k}}\phi(t)\right| \\ &= C\frac{c^k}{k!}h^k \sup_{s\in\mathbb{R}} e^{M(rs)} \sup_{t\in[s-c,s+c]} \left(e^{-M(rt)} e^{M(rt)} \left|\frac{d^{\alpha+k}}{dt^{\alpha+k}}\phi(t)\right|\right) \\ &\leq C\frac{c^k}{k!}h^k \sup_{s\in\mathbb{R}} e^{M(rs)} e^{-M(r(|s|-c))} \sup_{t\in\mathbb{R}} \left(e^{M(rt)} \left|\frac{d^{\alpha+k}}{dt^{\alpha+k}}\phi(t)\right|\right) \\ &= C_1 \|\phi\|_{\mathscr{K}_M^{r+k}}, \end{aligned} \tag{47}$$

where $\xi(y) = hx + \theta h(y-x) \in [hx - hc, hx + hc]$ for some $\theta \in (0,1)$ and $C_1 = C(c^k/k!)h^k \sup_{s\in\mathbb{R}} e^{M(rs)} e^{-M(r(|s|-c))} < \infty$. To show the finiteness of C_1 in the last statement, we use

$$\begin{aligned} &\sup_{s\in\mathbb{R}} e^{M(rs)} e^{-M(r(|s|-c))} \\ &\leq \sup_{|s|\leq c} e^{M(rs)} e^{-M(r(|s|-c))} \end{aligned}$$

$$
\begin{aligned}
&+\sup_{|s|>c} e^{M(rs)} e^{-M(r(|s|-c))} \\
&\le \sup_{|s|\le c} e^{M(rs)} e^{-M(r(|s|-c))} \\
&\quad+\sup_{|s|>c} e^{M(rs)} e^{-M(r(|s|-c))} \\
&\le e^{M(rc)}+e^{-M(rc)}.
\end{aligned} \tag{48}
$$

We will estimate J_2 by using the following facts. Since ϕ has a compact support, there exists $M > 0$ such that $K(x, y) = 0$ for $|x - y| > M$. Also, by the choice of g and property (c) of the reproducing kernel q_0, we have

$$
\begin{aligned}
\int R^{\alpha}(x,y)\,y^{s}dy &= \frac{d^{\alpha}}{dx^{\alpha}}\int K(x,y)\,y^{s}dy \\
&\quad-\frac{d^{\alpha}}{dx^{\alpha}}\int g(x-y)\,y^{s}dy \\
&= \frac{d^{\alpha}}{dx^{\alpha}}x^{s}-\frac{d^{\alpha}}{dx^{\alpha}}\int g(t)\,x^{s}dt \\
&= \frac{d^{\alpha}}{dx^{\alpha}}x^{s}-\frac{d^{\alpha}}{dx^{\alpha}}x^{s}=0,\quad 0\le s\le r+k-1.
\end{aligned} \tag{49}
$$

Hence

$$
\begin{aligned}
J_2 &= \sup_{hx\in\mathbb{R}} e^{M(rhx)}\left|\int R^{\alpha}(x,y)\right. \\
&\quad\times\left.\left(\frac{d^{\alpha}}{dy^{\alpha}}\phi(y)\Big|_{y=hy}-\frac{d^{\alpha}}{dy^{\alpha}}\phi(y)\Big|_{y=hx}\right)dy\right| \\
&= \sup_{hx\in\mathbb{R}} e^{M(rhx)}\left|\int R^{\alpha}(x,y)\right. \\
&\quad\times\left((y-x)\times h\frac{d^{\alpha+1}}{dy^{\alpha+1}}\phi(y)\Big|_{y=hx}\right. \\
&\quad+\cdots+\frac{(y-x)^{k-1}}{(k-1)!}\times h^{k-1}\frac{d^{\alpha+k-1}}{dy^{\alpha+k-1}}\phi(y)\Big|_{y=hx} \\
&\quad+\left.\left.\frac{(y-x)^{k}}{k!}\times h^{k}\frac{d^{\alpha+k}}{dy^{\alpha+k}}\phi(y)\Big|_{y=\xi(y)}\right)dy\right| \\
&= \frac{h^k}{k!}\sup_{hx\in\mathbb{R}} e^{M(rhx)}\left|\int_{|x-y|\le M} R^{\alpha}(x,y)(y-x)^{k}\right. \\
&\quad\times\left.\frac{d^{\alpha+k}}{dy^{\alpha+k}}\phi(y)\Big|_{y=\xi(y)}dy\right| \\
&\le \frac{M^k h^k}{k!}\sup_{hx\in\mathbb{R}}\sup_{hx<\xi(y)<hy} e^{M(rhx)}\left|\frac{d^{\alpha+k}}{dy^{\alpha+k}}\phi(y)\Big|_{y=\xi(y)}\right| \\
&\quad\times\left|\int_{|x-y|\le M} R^{\alpha}(x,y)(y-x)^{k}dy\right| \\
&\le C'\frac{c^k}{k!}h^{k}\sup_{s\in\mathbb{R}} e^{M(rs)}e^{-M(r(|s|-c))}\sup_{t\in\mathbb{R}}\left(e^{M(rt)}\left|\frac{d^{\alpha+k}}{dt^{\alpha+k}}\phi(t)\right|\right) \\
&= C_2\|\phi\|_{\mathscr{K}_M^{r+k}}.
\end{aligned} \tag{50}
$$

References

[1] G. G. Walter, "Analytic representations with wavelet expansions," *Complex Variables. Theory and Application*, vol. 26, no. 3, pp. 235–243, 1994.

[2] G. G. Walter and X. Shen, *Wavelets and Other Orthogonal Systems with Applications*, Studies in Advanced Mathematics, CRC Press, Boca Raton, Fla, USA, 2nd edition, 2001.

[3] G. G. Walter, "Pointwise convergence of wavelet expansions," *Journal of Approximation Theory*, vol. 80, no. 1, pp. 108–118, 1995.

[4] S. Pilipović and N. Teofanov, "Multiresolution expansion, approximation order and quasiasymptotic behavior of tempered distributions," *Journal of Mathematical Analysis and Applications*, vol. 331, no. 1, pp. 455–471, 2007.

[5] M. Hasumi, "Note on the n-dimensional tempered ultradistributions," *The Tohoku Mathematical Journal*, vol. 13, pp. 94–104, 1961.

[6] K. Yoshinaga, "On spaces of distributions of exponential growth," *Bulletin of the Kyushu Institute of Technology. Mathematics, Natural Science*, vol. 6, pp. 1–16, 1960.

[7] S. Sznajder and Z. Zielezny, "Solvability of convolution equations in $\mathscr{K}_1'$," *Proceedings of the American Mathematical Society*, vol. 57, no. 1, pp. 103–106, 1976.

[8] S. Sznajder and Z. Zielezny, "On some properties of convolution operators in $\mathscr{K}_1'$ and $\mathscr{S}'$," *Journal of Mathematical Analysis and Applications*, vol. 65, no. 3, pp. 543–554, 1978.

[9] I. M. Gel'fand and G. E. Shilov, *Generalized Functions, Vol. I. II. III*, Academic Press, New York, NY, USA, 1968.

[10] D. H. Pahk, "Structure theorem and Fourier transform for distributions with restricted growth," *Kyungpook Mathematical Journal*, vol. 23, no. 2, pp. 129–146, 1983.

[11] B. K. Sohn and D. H. Pahk, "Analytic representation of generalized tempered distributions of exponential growth by wavelets," *Tsukuba Journal of Mathematics*, vol. 27, no. 1, pp. 47–56, 2003.

[12] B. K. Sohn and D. H. Pahk, "Pointwise convergence of wavelet expansion of $\mathscr{K}_M^{r}{}'(\mathbb{R})$," *Bulletin of the Korean Mathematical Society*, vol. 38, no. 1, pp. 81–91, 2001.

[13] I. Daubechies, *Ten Lectures on Wavelets*, vol. 61 of *CBMS-NSF Regional Conference Series in Applied Mathematics*, SIAM, Philadelphia, Pa, USA, 1992.

[14] J. Barros-Neto, *An Introduction to the Theory of Distributions*, Marcel Dekker, New York, NY, USA, 1973.

[15] O. Holtz and A. Ron, "Approximation orders of shift-invariant subspaces of $W_2^s(\mathbb{R}^d)$," *Journal of Approximation Theory*, vol. 132, no. 1, pp. 97–148, 2005.

[16] G. C. Kyriazis, "Wavelet-type decompositions and approximations from shift-invariant spaces," *Journal of Approximation Theory*, vol. 88, no. 2, pp. 257–271, 1997.

[17] G. C. Kyriazis, "Approximation of distribution spaces by means of kernel operators," *The Journal of Fourier Analysis and Applications*, vol. 2, no. 3, pp. 261–286, 1996.

[18] J. Lei, R.-Q. Jia, and E. W. Cheney, "Approximation from shift-invariant spaces by integral operators," *SIAM Journal on Mathematical Analysis*, vol. 28, no. 2, pp. 481–498, 1997.

4

QR-Submanifolds of $(p-1)$ QR-Dimension in a Quaternionic Projective Space $QP^{(n+p)/4}$ under Some Curvature Conditions

Hyang Sook Kim[1] and Jin Suk Pak[2]

[1] *Department of Applied Mathematics, Institute of Basic Science, Inje University, Gimhae 621-749, Republic of Korea*
[2] *Kyungpook National University, Daegu 702-701, Republic of Korea*

Correspondence should be addressed to Hyang Sook Kim; mathkim@inje.ac.kr

Academic Editor: Luc Vrancken

The purpose of this paper is to study n-dimensional QR-submanifolds of $(p-1)$ QR-dimension in a quaternionic projective space $QP^{(n+p)/4}$ and especially to determine such submanifolds under some curvature conditions.

1. Introduction

Let M be a connected real n-dimensional submanifold of real codimension p of a quaternionic Kähler manifold $\overline{M}$ with quaternionic Kähler structure $\{F, G, H\}$. If there exists an r-dimensional normal distribution ν of the normal bundle $TM^{\perp}$ such that

$$F\nu_x \subset \nu_x, \qquad G\nu_x \subset \nu_x, \qquad H\nu_x \subset \nu_x,$$
$$F\nu_x^{\perp} \subset T_xM, \qquad G\nu_x^{\perp} \subset T_xM, \qquad H\nu_x^{\perp} \subset T_xM \tag{1}$$

at each point x in M, then M is called a QR-*submanifold of* r QR-*dimension*, where $\nu^{\perp}$ denotes the complementary orthogonal distribution to ν in $TM^{\perp}$ (cf. [1–3]). Real hypersurfaces, which are typical examples of QR-submanifold with $r = 0$, have been investigated by many authors (cf. [2–9]) in connection with the shape operator and the induced almost contact 3-structure (for definition, see [10–13]). In their paper [2, 3], Kwon and Pak had studied QR-submanifolds of $(p-1)$ QR-dimension isometrically immersed in a quaternionic projective space $QP^{(n+p)/4}$ and proved the following theorem as a quaternionic analogy to theorems given in [14, 15], which are natural extensions of theorems proved in [6] to the case of QR-submanifolds with $(p-1)$ QR-dimension and also extensions of theorems in [16].

Theorem K-P. *Let M be an n-dimensional QR-submanifold of $(p-1)$ QR-dimension isometrically immersed in a quaternionic projective space $QP^{(n+p)/4}$, and let the normal vector field N_1 be parallel with respect to the normal connection. If the shape operator A_1 corresponding to N_1 satisfies*

$$A_1\phi = \phi A_1, \qquad A_1\psi = \psi A_1, \qquad A_1\theta = \theta A_1, \tag{2}$$

then $\pi^{-1}(M)$ is locally a product of $M_1 \times M_2$ where M_1 and M_2 belong to some $(4n_1+3)$- and $(4n_2+3)$-dimensional spheres (π is the Hopf fibration $S^{n+p+3}(1) \to QP^{(n+p)/4}$).

On the other hand, when M is a real hypersurface of $QP^{(n+p)/4}$, if $\pi^{-1}(M)$ is (1) an Einstein space or (2) a locally symmetric space, then $\pi^{-1}(M)$ has a parallel second fundamental form (cf. [4, 6, 7, 9]). Projecting the quantities on $\pi^{-1}(M)$ onto M in $QP^{(n+p)/4}$, we can consider QR-submanifolds of $(p-1)$ QR-dimension with the conditions corresponding to (1) or (2). In this paper, we will study such QR-submanifolds isometrically immersed in $QP^{(n+p)/4}$ and obtain Theorem 3 and other results stated in the last Section 5 as quaternionic analogies to theorems given in [16, 17] and as the extensions of theorems given in [18] by using Theorem K-P.

2. Preliminaries

Let $\overline{M}$ be a real $(n+p)$-dimensional quaternionic Kähler manifold. Then, by definition, there is a 3-dimensional vector bundle V consisting of tensor fields of type $(1,1)$ over $\overline{M}$ satisfying the following conditions (a), (b), and (c).

(a) In any coordinate neighborhood $\overline{\mathscr{U}}$, there is a local basis $\{F, G, H\}$ of V such that

$$F^2 = -I, \qquad G^2 = -I, \qquad H^2 = -I,$$
$$FG = -GF = H, \qquad GH = -HG = F, \tag{3}$$
$$HF = -FH = G.$$

(b) There is a Riemannian metric g which is Hermite with respect to all of F, G, and H.

(c) For the Riemannian connection $\overline{\nabla}$ with respect to g,

$$\begin{pmatrix} \overline{\nabla}F \\ \overline{\nabla}G \\ \overline{\nabla}H \end{pmatrix} = \begin{pmatrix} 0 & r & -q \\ -r & 0 & p \\ q & -p & 0 \end{pmatrix} \begin{pmatrix} F \\ G \\ H \end{pmatrix}, \tag{4}$$

where p, q, and r are local 1-forms defined in $\overline{\mathscr{U}}$. Such a local basis $\{F, G, H\}$ is called a *canonical local basis* of the bundle V in $\overline{\mathscr{U}}$ (cf. [10, 19, 20]).

For canonical local bases $\{F, G, H\}$ and $\{'F, 'G, 'H\}$ of V in coordinate neighborhoods $\overline{\mathscr{U}}$ and $'\overline{\mathscr{U}}$, it follows that in $\overline{\mathscr{U}} \cap {'\overline{\mathscr{U}}}$

$$\begin{pmatrix} 'F \\ 'G \\ 'H \end{pmatrix} = (s_{xy}) \begin{pmatrix} F \\ G \\ H \end{pmatrix} \quad (x, y = 1, 2, 3), \tag{5}$$

where s_{xy} are local differentiable functions with $(s_{xy}) \in SO(3)$ as a consequence of (3). As is well known (cf. [19]), every quaternionic Kähler manifold is orientable.

Now let M be an n-dimensional QR-submanifold of $(p-1)$ QR-dimension isometrically immersed in $\overline{M}$. Then by definition, there is a unit normal vector field N such that $\nu_x^{\perp} = \mathrm{Span}\{N\}$ at each point x in M. We set

$$U = -FN, \qquad V = -GN, \qquad W = -HN. \tag{6}$$

Denoting by $\mathscr{D}_x$ the maximal quaternionic invariant subspace $T_xM \cap FT_xM \cap GT_xM \cap HT_xM$ of T_xM, we have $\mathscr{D}_x^{\perp} = \mathrm{Span}\{U, V, W\}$, where $\mathscr{D}_x^{\perp}$ means the complementary orthogonal subspace to $\mathscr{D}_x$ in T_xM (cf. [1–3]). Thus, we have

$$T_xM = \mathscr{D}_x \oplus \mathrm{Span}\{U, V, W\}, \quad x \in M, \tag{7}$$

which together with (3) and (6) implies

$$FT_xM,\ GT_xM,\ HT_xM \subset T_xM \oplus \mathrm{Span}\{N\}. \tag{8}$$

Therefore, for any tangent vector field X and for a local orthonormal basis $\{N_\alpha\}_{\alpha=1,\ldots,p}$ $(N_1 := N)$ of normal vectors to M, we have

$$FX = \phi X + u(X)N,$$
$$GX = \psi X + v(X)N, \tag{9}$$
$$HX = \theta X + w(X)N,$$

$$FN_\alpha = -U_\alpha + P_1 N_\alpha,$$
$$GN_\alpha = -V_\alpha + P_2 N_\alpha, \tag{10}$$
$$HN_\alpha = -W_\alpha + P_3 N_\alpha,$$

$(\alpha = 1, \ldots, p)$. Then it is easily seen that $\{\phi, \psi, \theta\}$ and $\{P_1, P_2, P_3\}$ are skew-symmetric endomorphisms acting on T_xM and $T_xM^{\perp}$, respectively. Moreover, the Hermitian property of $\{F, G, H\}$ implies

$$g(X, \phi U_\alpha) = -u(X)\, g(N_1, P_1 N_\alpha),$$
$$g(X, \psi V_\alpha) = -v(X)\, g(N_1, P_2 N_\alpha), \quad \alpha = 1, \ldots, p, \tag{11}$$
$$g(X, \theta W_\alpha) = -w(X)\, g(N_1, P_3 N_\alpha),$$

$$g(U_\alpha, U_\beta) = \delta_{\alpha\beta} - g(P_1 N_\alpha, P_1 N_\beta),$$
$$g(V_\alpha, V_\beta) = \delta_{\alpha\beta} - g(P_2 N_\alpha, P_2 N_\beta), \quad \alpha, \beta = 1, \ldots, p, \tag{12}$$
$$g(W_\alpha, W_\beta) = \delta_{\alpha\beta} - g(P_3 N_\alpha, P_3 N_\beta).$$

Also, from the hermitian properties $g(FX, N_\alpha) = -g(X, FN_\alpha)$, $g(GX, N_\alpha) = -g(X, GN_\alpha)$, and $g(HX, N_\alpha) = -g(X, HN_\alpha)$, it follows that

$$g(X, U_\alpha) = u(X)\,\delta_{1\alpha}, \qquad g(X, V_\alpha) = v(X)\,\delta_{1\alpha},$$
$$g(X, W_\alpha) = w(X)\,\delta_{1\alpha}, \tag{13}$$

and hence,

$$g(U_1, X) = u(X), \qquad g(V_1, X) = v(X),$$
$$g(W_1, X) = w(X), \qquad U_\alpha = 0, \tag{14}$$
$$V_\alpha = 0, \qquad W_\alpha = 0, \quad \alpha = 2, \ldots, p.$$

On the other hand, comparing (6) and (10) with $\alpha = 1$, we have $U_1 = U$, $V_1 = V$, and $W_1 = W$, which together with (6) and (14) implies

$$g(U, X) = u(X), \qquad g(V, X) = v(X)$$
$$g(W, X) = w(X), \quad u(U) = 1, \quad v(V) = 1, \quad w(W) = 1. \tag{15}$$

In the sequel, we will use the notations U, V, and W instead of U_1, V_1, and W_1.

Next, applying F to the first equation of (9) and using (10), (14), and (15), we have

$$\phi^2 X = -X + u(X)U, \qquad u(X)P_1 N = -u(\phi X)N. \tag{16}$$

Similarly, we have

$$\phi^2 X = -X + u(X)U, \qquad \psi^2 X = -X + v(X)V, \qquad \theta^2 X = -X + w(X)W, \tag{17}$$

$$u(X)P_1N = -u(\phi X)N, \qquad v(X)P_2N = -v(\psi X)N, \qquad w(X)P_3N = -w(\theta X)N, \tag{18}$$

from which, taking account of the skew symmetry of P_1, P_2, and P_3 and using (11), we also have

$$u(\phi X) = 0, \qquad v(\psi X) = 0, \qquad w(\theta X) = 0,$$
$$\phi U = 0, \quad \psi V = 0, \quad \theta W = 0, \quad P_1N = 0, \tag{19}$$
$$P_2N = 0, \qquad P_3N = 0.$$

So (10) can be rewritten in the form

$$FN = -U, \qquad GN = -V, \qquad HN = -W,$$
$$FN_\alpha = P_1N_\alpha, \qquad GN_\alpha = P_2N_\alpha, \qquad HN_\alpha = P_3N_\alpha \tag{20}$$

($\alpha = 2, \ldots, p$). Applying G and H to the first equation of (9) and using (3), (9), and (20), we have

$$\theta X + w(X)N = -\psi(\phi X) - v(\phi X)N + u(X)V,$$
$$\psi X + v(X)N = \theta(\phi X) + w(\phi X)N - u(X)W, \tag{21}$$

and consequently,

$$\psi(\phi X) = -\theta X + u(X)V, \qquad v(\phi X) = -w(X),$$
$$\theta(\phi X) = \psi X + u(X)W, \quad w(\phi X) = v(X). \tag{22}$$

Similarly, the other equations of (9) yield

$$\phi(\psi X) = \theta X + v(X)U, \qquad u(\psi X) = w(X),$$
$$\theta(\psi X) = -\phi X + v(X)W, \qquad w(\psi X) = -u(X),$$
$$\phi(\theta X) = -\psi X + w(X)U, \qquad u(\theta X) = -v(X), \tag{23}$$
$$\psi(\theta X) = \phi X + w(X)V, \qquad v(\theta X) = u(X).$$

From the first three equations of (20), we also have

$$\psi U = -W, \qquad v(U) = 0, \qquad \theta U = V,$$
$$w(U) = 0, \qquad \phi V = W, \qquad u(V) = 0,$$
$$\theta V = -U, \qquad w(V) = 0, \qquad \phi W = -V, \tag{24}$$
$$u(W) = 0, \qquad \psi W = U, \qquad v(W) = 0.$$

Equations (14)–(17), (19), and (22)–(24) tell us that M admits the so-called almost contact 3-structure and consequently $n = 4m + 3$ for some integer m (cf. [12]).

Now let ∇ be the Levi-Civita connection on M, and let $\nabla^\perp$ be the normal connection induced from $\overline{\nabla}$ in the normal bundle of M. Then Gauss and Weingarten formulae are given by

$$\overline{\nabla}_X Y = \nabla_X Y + h(X, Y), \tag{25}$$

$$\overline{\nabla}_X N_\alpha = -A_\alpha X + \nabla^\perp_X N_\alpha, \quad \alpha = 1, \ldots, p, \tag{26}$$

for X, Y tangent to M. Here h denotes the second fundamental form and A_α the shape operator corresponding to N_α. They are related by $h(X,Y) = \sum_{\alpha=1}^{p} g(A_\alpha X, Y)N_\alpha$. Furthermore, put

$$\nabla^\perp_X N_\alpha = \sum_{\beta=1}^{p} s_{\alpha\beta}(X) N_\beta, \tag{27}$$

where $(s_{\alpha\beta})$ is the skew-symmetric matrix of connection forms of $\nabla^\perp$.

Differentiating the first equation of (9) covariantly and using (4), (9), (10), (14) (25), and (26), we have

$$(\nabla_Y \phi)X = r(Y)\psi X - q(Y)\theta X + u(X)A_1Y - g(A_1Y, X)U,$$
$$(\nabla_Y u)X = r(Y)v(X) - q(Y)w(X) + g(\phi A_1Y, X). \tag{28}$$

From the other equations of (9), we also have

$$(\nabla_Y \psi)X = -r(Y)\phi X + p(Y)\theta X + v(X)A_1Y - g(A_1Y, X)V,$$
$$(\nabla_Y v)X = -r(Y)u(X) + p(Y)w(X) + g(\psi A_1Y, X),$$
$$(\nabla_Y \theta)X = q(Y)\phi X - p(Y)\psi X + w(X)A_1Y - g(A_1Y, X)W, \tag{29}$$
$$(\nabla_Y w)X = q(Y)u(X) - p(Y)v(X) + g(\theta A_1Y, X).$$

Next, differentiating the first equation of (20) covariantly and comparing the tangential and normal parts, we have

$$\nabla_Y U = r(Y)V - q(Y)W + \phi A_1Y,$$
$$g(A_\alpha U, Y) = -\sum_{\beta=2}^{p} s_{1\beta}(Y)P_{1\beta\alpha}, \quad \alpha = 2, \ldots, p. \tag{30}$$

From the other equations of (20), we have similarly

$$\nabla_Y V = -r(Y)U + p(Y)W + \psi A_1Y,$$
$$g(A_\alpha V, Y) = -\sum_{\beta=2}^{p} s_{1\beta}(Y)P_{2\beta\alpha}, \quad \alpha = 2, \ldots, p,$$
$$\nabla_Y W = q(Y)U - p(Y)V + \theta A_1Y, \tag{31}$$
$$g(A_\alpha W, Y) = -\sum_{\beta=2}^{p} s_{1\beta}(Y)P_{3\beta\alpha}, \quad \alpha = 2, \ldots, p.$$

Finally the equation of Gauss is given as follows (cf. [21]):

$$
\begin{aligned}
g(\overline{R}(X,Y)Z,W) &= g(R(X,Y)Z,W) \\
&\quad + \sum_{\alpha}\{g(A_\alpha X,Z)\,g(A_\alpha Y,W) \\
&\qquad - g(A_\alpha Y,Z)\,g(A_\alpha X,W)\},
\end{aligned} \tag{32}
$$

for X, Y, and Z tangent to M, where $\overline{R}$ and R denote the Riemannian curvature tensor of $\overline{M}$ and M, respectively.

In the rest of this paper *we assume that the distinguished normal vector field* $N_1 := N$ *is parallel with respect to the normal connection* $\nabla^{\perp}$. Then it follows from (27) that $s_{1\beta} = 0$, and consequently, (30)-(31) imply

$$
A_\alpha U = 0, \qquad A_\alpha V = 0, \qquad A_\alpha W = 0, \qquad \alpha = 2,\ldots,p. \tag{33}
$$

On the other hand, since the curvature tensor $\overline{R}$ of $QP^{(n+p)/4}$ is of the form

$$
\begin{aligned}
\overline{R}(\overline{X},\overline{Y})\overline{Z} &= g(\overline{Y},\overline{Z})\overline{X} - g(\overline{X},\overline{Z})\overline{Y} \\
&\quad + g(F\overline{Y},\overline{Z})F\overline{X} - g(F\overline{X},\overline{Z})F\overline{Y} \\
&\quad - 2g(F\overline{X},\overline{Y})F\overline{Z} + g(G\overline{Y},\overline{Z})G\overline{X} \\
&\quad - g(G\overline{X},\overline{Z})G\overline{Y} - 2g(G\overline{X},\overline{Y})G\overline{Z} \\
&\quad + g(H\overline{Y},\overline{Z})H\overline{X} - g(H\overline{X},\overline{Z})H\overline{Y} \\
&\quad - 2g(H\overline{X},\overline{Y})H\overline{Z}
\end{aligned} \tag{34}
$$

for $\overline{X}, \overline{Y}$, and $\overline{Z}$ tangent to $QP^{(n+p)/4}$, (32) reduces to

$$
\begin{aligned}
R(X,Y)Z &= g(Y,Z)X - g(X,Z)Y \\
&\quad + g(\phi Y,Z)\phi X - g(\phi X,Z)\phi Y \\
&\quad - 2g(\phi X,Y)\phi Z + g(\psi Y,Z)\psi X \\
&\quad - g(\psi X,Z)\psi Y - 2g(\psi X,Y)\psi Z \\
&\quad + g(\theta Y,Z)\theta X - g(\theta X,Z)\theta Y \\
&\quad - 2g(\theta X,Y)\theta Z \\
&\quad + \sum_{\alpha}\{g(A_\alpha Y,Z)A_\alpha X - g(A_\alpha X,Z)A_\alpha Y\}.
\end{aligned} \tag{35}
$$

3. Fibrations and Immersions

From now on n-dimensional QR-submanifolds of $(p-1)$ QR-dimension isometrically immersed in $QP^{(n+p)/4}$ only will be considered. Moreover, we will use the assumption and the notations as in Section 2.

Let $S^{n+p+3}(a)$ be the hypersphere of radius a (>0) in $Q^{(n+p+4)/4}$ the quaternionic space of quaternionic dimension $(n+p+4)/4$, which is identified with the Euclidean $(n+p+4)$-space $\mathbb{R}^{n+p+4}$. The unit sphere $S^{n+p+3}(1)$ will be briefly denoted by S^{n+p+3}. Let $\tilde{\pi} : S^{n+p+3} \to QP^{(n+p)/4}$ be the natural projection of S^{n+p+3} onto $QP^{(n+p)/4}$ defined by the Hopf fibration $S^3 \to S^{n+p+3} \to QP^{(n+p)/4}$. As is well known (cf. [10, 11, 20]), S^{n+p+3} admits a Sasakian 3-structure whereby $\tilde{\xi}, \tilde{\eta}$, and $\tilde{\zeta}$ are mutually orthogonal unit Killing vector fields. Thus it follows that

$$
\begin{gathered}
\tilde{\nabla}_{\tilde{\xi}}\tilde{\xi} = 0, \qquad \tilde{\nabla}_{\tilde{\eta}}\tilde{\eta} = 0, \qquad \tilde{\nabla}_{\tilde{\zeta}}\tilde{\zeta} = 0, \\
\tilde{\nabla}_{\tilde{\zeta}}\tilde{\eta} = -\tilde{\nabla}_{\tilde{\eta}}\tilde{\zeta} = \tilde{\xi}, \qquad \tilde{\nabla}_{\tilde{\xi}}\tilde{\zeta} = -\tilde{\nabla}_{\tilde{\zeta}}\tilde{\xi} = \tilde{\eta}, \\
\tilde{\nabla}_{\tilde{\eta}}\tilde{\xi} = -\tilde{\nabla}_{\tilde{\xi}}\tilde{\eta} = \tilde{\zeta},
\end{gathered} \tag{36}
$$

where $\tilde{\nabla}$ denotes the Riemannian connection with respect to the canonical metric $\tilde{g}$ on S^{n+p+3} (cf. [6, 9–13]). Moreover, each fibre $\tilde{\pi}^{-1}(x)$ of x in $QP^{(n+p)/4}$ is a maximal integral submanifold of the distribution spanned by $\tilde{\xi}, \tilde{\eta}$, and $\tilde{\zeta}$. Thus the base space $QP^{(n+p)/4}$ admits the induced quaternionic Kähler structure of constant Q-sectional curvature 4 (cf. [10, 11]). We have especially a fibration $\pi : \pi^{-1}(M) \to M$ which is compatible with the Hopf fibration $\tilde{\pi}$. More precisely speaking, $\pi : \pi^{-1}(M) \to M$ is a fibration with totally geodesic fibers such that the following diagram is commutative:

$$
\begin{array}{ccc}
\pi^{-1}(M) & \xrightarrow{\;i'\;} & S^{n+p+3} \\
\Big\downarrow{\scriptstyle \pi} & & \Big\downarrow{\scriptstyle \tilde{\pi}} \\
M & \xrightarrow{\;i\;} & QP^{(n+p)/4}
\end{array} \tag{37}
$$

where $i' : \pi^{-1}(M) \to S^{n+p+3}$ and $i : M \to QP^{(n+p)/4}$ are isometric immersions.

Now, let ξ, η, and ζ be the unit vector fields tangent to the fibers of $\pi^{-1}(M)$ such that $i'_*\xi = \tilde{\xi}$, $i'_*\eta = \tilde{\eta}$, and $i'_*\zeta = \tilde{\zeta}$. (In what follows, we will again delete the i' and i'_* in our notation.) Furthermore, we denote by X^* the horizontal lift of a vector field X tangent to M. Then, the horizontal lifts N^*_α $(\alpha = 1,\ldots,p)$ of the normal vectors N_α to M form an orthonormal basis of normal vectors to $\pi^{-1}(M)$ in S^{n+p+3}. Let A'_α and $s'_{\alpha\beta}$ be the corresponding shape operators and normal connection forms, respectively. Then, as shown in [3, 9, 22], the fundamental equations for the submersion π are given by

$$
\begin{aligned}
{}'\nabla_{X^*}Y^* &= (\nabla_X Y)^* + g'((\phi X)^*, Y^*)\xi + g'((\psi X)^*, Y^*)\eta \\
&\quad + g'((\theta X)^*, Y^*)\zeta,
\end{aligned} \tag{38}
$$

$$\begin{aligned}[X^*, Y^*] &= [X, Y]^* + 2g'\left((\phi X)^*, Y^*\right)\xi \\ &\quad + 2g'\left((\psi X)^*, Y^*\right)\eta + 2g'\left((\theta X)^*, Y^*\right)\zeta,\end{aligned} \tag{39}$$

$$\begin{aligned}{}'\nabla_{X^*}\xi &= {}'\nabla_{\xi}X^* = -(\phi X)^*, \\ {}'\nabla_{X^*}\eta &= {}'\nabla_{\eta}X^* = -(\psi X)^*, \\ {}'\nabla_{X^*}\zeta &= {}'\nabla_{\zeta}X^* = -(\theta X)^*,\end{aligned} \tag{40}$$

$$[X^*, \xi] = 0, \qquad [X^*, \eta] = 0, \qquad [X^*, \zeta] = 0, \tag{41}$$

where g' denotes the Riemannian metric of $\pi^{-1}(M)$ induced from $\tilde{g}$ in S^{n+p+3} and $'\nabla$ the Levi-Civita connection with respect to g'. The same equations are valid for the submersion $\tilde{\pi}$ by replacing ϕ, ψ, and θ (resp., ξ, η, and ζ) with F, G, and H (resp., $\tilde{\xi}$, $\tilde{\eta}$, and $\tilde{\zeta}$), respectively. We denote by $'\nabla^{\perp}$ the normal connection of $\pi^{-1}(M)$ induced from $\tilde{\nabla}$. Since the diagram is commutative, $\tilde{\nabla}_{X^*}N^*_{\alpha}$ implies

$$\begin{aligned}{}'\nabla^{\perp}_{X^*}N^*_{\alpha} - A'_{\alpha}X^* &= \left(\overline{\nabla}_X N_{\alpha}\right)^* + \tilde{g}\left((FX)^*, N^*_{\alpha}\right)\tilde{\xi} \\ &\quad + \tilde{g}\left((GX)^*, N^*_{\alpha}\right)\tilde{\eta} + \tilde{g}\left((HX)^*, N^*_{\alpha}\right)\tilde{\zeta} \\ &= -(A_{\alpha}X)^* + g(U_{\alpha}, X)^*\xi + g(V_{\alpha}, X)^*\eta \\ &\quad + g(W_{\alpha}, X)^*\zeta + (\nabla^{\perp}_X N_{\alpha})^*\end{aligned} \tag{42}$$

because of (10), (26), and (38), from which, comparing the tangential part, we have

$$\begin{aligned}A'_{\alpha}X^* &= (A_{\alpha}X)^* - g(U_{\alpha}, X)^*\xi \\ &\quad - g(V_{\alpha}, X)^*\eta - g(W_{\alpha}, X)^*\zeta.\end{aligned} \tag{43}$$

Next, calculating $\tilde{\nabla}_{\xi}N^*_{\alpha}$ and using (10), (26), and (40), we have

$$'\nabla^{\perp}_{\xi}N^*_{\alpha} - A'_{\alpha}\xi = -(FN_{\alpha})^* = U^*_{\alpha} - (P_1 N_{\alpha})^*, \tag{44}$$

which yields

$$A'_{\alpha}\xi = -U^*_{\alpha} \tag{45}$$

and similarly

$$A'_{\alpha}\xi = -U^*_{\alpha}, \qquad A'_{\alpha}\eta = -V^*_{\alpha}, \qquad A'_{\alpha}\zeta = -W^*_{\alpha}. \tag{46}$$

Hence, (43) and (46) with $\alpha = 1$ imply

$$\begin{aligned}&A'_1 X^* = (A_1 X)^* - g(U, X)^*\xi - g(V, X)^*\eta - g(W, X)^*\zeta, \\ &A'_1\xi = -U^*, \qquad A'_1\eta = -V^*, \qquad A'_1\zeta = -W^*.\end{aligned} \tag{47}$$

4. Co-Gauss Equations for the Submersion $\pi{:}\pi^{-1}(M) \to M$

In this section, we derive the co-Gauss and co-Codazzi equations of the submersion $\pi : \pi^{-1}(M) \to M$ for later use.

Differentiating (38) with $Y = U$ covariantly along $\pi^{-1}(M)$ and using (24), (38), and (39), we have

$$\begin{aligned}&{}'\nabla_{Y^*}\,{}'\nabla_{X^*}U^* \\ &= (\nabla_Y\nabla_X U)^* + \{v(X)\theta Y - w(X)\psi Y\}^* \\ &\quad + g(\phi Y, \nabla_X U)^*\xi \\ &\quad + \{g(\psi Y, \nabla_X U) + g(\nabla_Y X, W) + g(X, \nabla_Y W)\}^*\eta \\ &\quad + \{g(\theta Y, \nabla_X U) - g(\nabla_Y X, V) - g(X, \nabla_Y V)\}^*\zeta.\end{aligned} \tag{48}$$

Similarly (38) with $Y = V$ and (38) with $Y = W$ give

$$\begin{aligned}&{}'\nabla_{Y^*}\,{}'\nabla_{X^*}V^* \\ &= (\nabla_Y\nabla_X V)^* + \{w(X)\phi Y - u(X)\theta Y\}^* \\ &\quad + \{g(\phi Y, \nabla_X V) - g(\nabla_Y X, W) - g(X, \nabla_Y W)\}^*\xi \\ &\quad + g(\psi Y, \nabla_X V)^*\eta \\ &\quad + \{g(\theta Y, \nabla_X V) + g(\nabla_Y X, U) + g(X, \nabla_Y U)\}^*\zeta,\end{aligned} \tag{49}$$

$$\begin{aligned}&{}'\nabla_{Y^*}\,{}'\nabla_{X^*}W^* \\ &= (\nabla_Y\nabla_X W)^* \\ &\quad - \{v(X)\phi Y - u(X)\psi Y\}^* \\ &\quad + \{g(\phi Y, \nabla_X W) + g(\nabla_Y X, V) + g(X, \nabla_Y V)\}^*\xi \\ &\quad + \{g(\psi Y, \nabla_X W) - g(\nabla_Y X, U) - g(X, \nabla_Y U)\}^*\eta \\ &\quad + g(\theta Y, \nabla_X W)^*\zeta,\end{aligned} \tag{50}$$

respectively. On the other hand, it follows from (19), (24), (38), and (39) that

$$\begin{aligned}{}'\nabla_{[Y^*, X^*]}U^* &= (\nabla_{[Y,X]}U)^* + 2g(\psi Y, X)^*W^* \\ &\quad - 2g(\theta Y, X)^*V^* + g([Y, X], W)^*\eta \\ &\quad - g([Y, X], V)^*\zeta,\end{aligned} \tag{51}$$

$$'\nabla_{[Y^*,X^*]}V^* = (\nabla_{[Y,X]}V)^* - 2g(\phi Y, X)^* W^* + 2g(\theta Y, X)^* U^* - g([Y,X], W)^* \xi + g([Y,X], U)^* \zeta, \quad (52)$$

$$'\nabla_{[Y^*,X^*]}W^* = (\nabla_{[Y,X]}W)^* + 2g(\phi Y, X)^* V^* - 2g(\psi Y, X)^* U^* + g([Y,X], V)^* \xi - g([Y,X], U)^* \eta. \quad (53)$$

By means of (48) and (51), we have

$$\begin{aligned} 'R(Y^*, X^*)U^* = {} & \{R(Y,X)U\}^* \\ & + \{w(Y)\psi X - w(X)\psi Y - v(Y)\theta X + v(X)\theta Y + 2g(\theta Y, X)V - 2g(\psi Y, X)W\}^* \\ & + \{g(\phi Y, \nabla_X U) - g(\phi X, \nabla_Y U)\}^* \xi \\ & + \{g(\psi Y, \nabla_X U) - g(\psi X, \nabla_Y U) + g(X, \nabla_Y W) - g(Y, \nabla_X W)\}^* \eta \\ & + \{g(\theta Y, \nabla_X U) - g(\theta X, \nabla_Y U) - g(X, \nabla_Y V) + g(Y, \nabla_X V)\}^* \zeta, \end{aligned} \quad (54)$$

where $'R$ denotes the curvature tensor of $\pi^{-1}(M)$ with respect to the connection $'\nabla$. Using (30), (31), (33), and (35), we can easily see that

$$\begin{aligned} 'R(Y^*, X^*)U^* = {} & \{u(X)Y - u(Y)X + u(A_1X)A_1Y - u(A_1Y)A_1X\}^* \\ & + \{r(Y)w(X) - r(X)w(Y) + q(Y)v(X) - q(X)v(Y) + u(X)u(A_1Y) - u(Y)u(A_1X)\}^* \xi \\ & + \{p(X)v(Y) - p(Y)v(X) + v(X)u(A_1Y) - v(Y)u(A_1X)\}^* \eta \\ & + \{p(X)w(Y) - p(Y)w(X) + w(X)u(A_1Y) - w(Y)u(A_1X)\}^* \zeta. \end{aligned} \quad (55)$$

By the same method, we can easily verify that (49), (50), (52), and (53) yield

$$\begin{aligned} 'R(Y^*, X^*)V^* = {} & \{v(X)Y - v(Y)X + v(A_1X)A_1Y - v(A_1Y)A_1X\}^* \\ & + \{q(X)u(Y) - q(Y)u(X) - u(Y)v(A_1X) + u(X)v(A_1Y)\}^* \xi \\ & + \{r(Y)w(X) - r(X)w(Y) + p(Y)u(X) - p(X)u(Y) + v(X)v(A_1Y) - v(Y)v(A_1X)\}^* \eta \\ & + \{q(X)w(Y) - q(Y)w(X) - w(Y)v(A_1X) + w(X)v(A_1Y)\}^* \zeta, \\ 'R(Y^*, X^*)W^* = {} & \{w(X)Y - w(Y)X + w(A_1X)A_1Y - w(A_1Y)A_1X\}^* \\ & + \{r(X)u(Y) - r(Y)u(X) - u(Y)w(A_1X) + u(X)w(A_1Y)\}^* \xi \\ & + \{r(X)v(Y) - r(Y)v(X) - v(Y)w(A_1X) + v(X)w(A_1Y)\}^* \eta \\ & + \{q(Y)v(X) - q(X)v(Y) + p(Y)u(X) - p(X)u(Y) + w(X)w(A_1Y) - w(Y)w(A_1X)\}^* \zeta. \end{aligned} \quad (56)$$

Differentiating (38) with $X = U$ covariantly along $\pi^{-1}(M)$ and using (24), we have

$$\begin{aligned} & '\nabla_{Y^*} {'\nabla_{U^*}} X^* \\ & = (\nabla_Y \nabla_U X)^* + \{w(X)\psi Y - v(X)\theta Y\}^* \\ & \quad + g(\phi Y, \nabla_U X)^* \xi \\ & \quad + \{g(\psi Y, \nabla_U X) - g(\nabla_Y W, X) - g(W, \nabla_Y X)\}^* \eta \\ & \quad + \{g(\theta Y, \nabla_U X) + g(\nabla_Y V, X) + g(V, \nabla_Y X)\}^* \zeta. \end{aligned} \quad (57)$$

Similarly, (38) with $X = V$ and (38) with $X = W$, respectively, give

$$\begin{aligned}
&{}'\nabla_{Y^*}{}'\nabla_{V^*}X^* \\
&\quad= (\nabla_Y\nabla_V X)^* - \{w(X)\phi Y - u(X)\theta Y\}^* \\
&\qquad + g(\psi Y, \nabla_V X)^*\eta \\
&\qquad + \{g(\phi Y, \nabla_V X) + g(\nabla_Y W, X) + g(W, \nabla_Y X)\}^*\xi \\
&\qquad + \{g(\theta Y, \nabla_V X) - g(\nabla_Y U, X) - g(U, \nabla_Y X)\}^*\zeta,
\end{aligned} \tag{58}$$

$$\begin{aligned}
&{}'\nabla_{Y^*}{}'\nabla_{W^*}X^* \\
&\quad= (\nabla_Y\nabla_W X)^* + \{v(X)\phi Y - u(X)\psi Y\}^* \\
&\qquad + g(\theta Y, \nabla_W X)^*\zeta \\
&\qquad + \{g(\phi Y, \nabla_W X) - g(\nabla_Y V, X) - g(V, \nabla_Y X)\}^*\xi \\
&\qquad + \{g(\psi Y, \nabla_W X) + g(\nabla_Y U, X) + g(U, \nabla_Y X)\}^*\eta.
\end{aligned} \tag{59}$$

Differentiating (38) also covariantly in the direction of U^* and using (24), we have

$$\begin{aligned}
&{}'\nabla_{U^*}{}'\nabla_{Y^*}X^* \\
&\quad= (\nabla_U\nabla_Y X)^* \\
&\qquad + g(\psi Y, X)^*W^* - g(\theta Y, X)^*V^* \\
&\qquad + \{g(\nabla_U(\phi Y), X) + g(\phi Y, \nabla_U X)\}^*\xi \\
&\qquad + \{g(\nabla_U(\psi Y), X) + g(\psi Y, \nabla_U X) - g(W, \nabla_Y X)\}^*\eta \\
&\qquad + \{g(\nabla_U(\theta Y), X) + g(\theta Y, \nabla_U X) + g(V, \nabla_Y X)\}^*\zeta.
\end{aligned} \tag{60}$$

Similarly, differentiating (38) covariantly in the direction of V^* and W^*, respectively, we have

$$\begin{aligned}
&{}'\nabla_{V^*}{}'\nabla_{Y^*}X^* \\
&\quad= (\nabla_V\nabla_Y X)^* \\
&\qquad - g(\phi Y, X)^*W^* - g(\theta Y, X)^*U^* \\
&\qquad + \{g(\nabla_V(\phi Y), X) + g(\phi Y, \nabla_V X) + g(W, \nabla_Y X)\}^*\xi \\
&\qquad + \{g(\nabla_V(\psi Y), X) + g(\psi Y, \nabla_V X)\}^*\eta \\
&\qquad + \{g(\nabla_V(\theta Y), X) + g(\theta Y, \nabla_V X) - g(U, \nabla_Y X)\}^*\zeta,
\end{aligned} \tag{61}$$

$$\begin{aligned}
&{}'\nabla_{W^*}{}'\nabla_{Y^*}X^* \\
&\quad= (\nabla_W\nabla_Y X)^* \\
&\qquad - g(\psi Y, X)^*U^* + g(\phi Y, X)^*V^* \\
&\qquad + \{g(\nabla_W(\phi Y), X) + g(\phi Y, \nabla_W X) - g(V, \nabla_Y X)\}^*\xi \\
&\qquad + \{g(\nabla_W(\psi Y), X) + g(\psi Y, \nabla_W X) + g(U, \nabla_Y X)\}^*\eta \\
&\qquad + \{g(\nabla_W(\theta Y), X) + g(\theta Y, \nabla_W X)\}^*\zeta.
\end{aligned} \tag{62}$$

On the other hand, (38) and (39) with $X = U$ imply

$$\begin{aligned}
{}'\nabla_{[Y^*,U^*]}X^* &= (\nabla_{[Y,U]}X)^* - 2\{w(Y)\psi X - v(Y)\theta X\}^* \\
&\quad + g(\phi[Y,U], X)^*\xi + g(\psi[Y,U], X)^*\eta \\
&\quad + g(\theta[Y,U], X)^*\zeta.
\end{aligned} \tag{63}$$

Similarly, from (39) with $X = V$ and (39) with $X = W$, respectively, we find that

$$\begin{aligned}
{}'\nabla_{[Y^*,V^*]}X^* &= (\nabla_{[Y,V]}X)^* + 2\{w(Y)\phi X - u(Y)\theta X\}^* \\
&\quad + g(\phi[Y,V], X)^*\xi + g(\psi[Y,V], X)^*\eta \\
&\quad + g(\theta[Y,V], X)^*\zeta,
\end{aligned} \tag{64}$$

$$\begin{aligned}
{}'\nabla_{[Y^*,W^*]}X^* &= (\nabla_{[Y,W]}X)^* - 2\{v(Y)\phi X - u(Y)\psi X\}^* \\
&\quad + g(\phi[Y,W], X)^*\xi + g(\psi[Y,W], X)^*\eta \\
&\quad + g(\theta[Y,W], X)^*\zeta.
\end{aligned} \tag{65}$$

Using (28)–(31), it follows from (57), (60), and (63) that

$$\begin{aligned}
{}'R(Y^*, U^*)X^* &= \{R(Y,U)X\}^* \\
&\quad + \{w(X)\psi Y - v(X)\theta Y \\
&\qquad - g(\psi Y, X)W + g(\theta Y, X)V \\
&\qquad + 2w(Y)\psi X - 2v(Y)\theta X\}^* \\
&\quad - \{r(U)g(\psi Y, X) - q(U)g(\theta Y, X) \\
&\qquad + u(Y)u(A_1X) + r(Y)w(X) \\
&\qquad + q(Y)v(X) - g(A_1Y, X)\}^*\xi \\
&\quad - \{-p(Y)v(X) - r(U)g(\phi Y, X) \\
&\qquad + p(U)g(\theta Y, X) + v(Y)u(A_1X)\}^*\eta \\
&\quad - \{-p(Y)w(X) + q(U)g(\phi Y, X) \\
&\qquad - p(U)g(\psi Y, X) + w(Y)u(A_1X)\}^*\zeta,
\end{aligned} \tag{66}$$

from which, taking account of (35) and using (24) and (33), we obtain

$$\begin{aligned}{}'R(Y^*,U^*)X^* = {} & \{u(X)Y - g(Y,X)U \\ & + u(A_1X)A_1Y - g(A_1Y,X)A_1U\}^* \\ & - \{r(U)g(\psi Y,X) - q(U)g(\theta Y,X) \\ & + u(Y)u(A_1X) + r(Y)w(X) \\ & + q(Y)v(X) - g(A_1Y,X)\}^*\xi \\ & - \{-p(Y)v(X) - r(U)g(\phi Y,X) \\ & + p(U)g(\theta Y,X) + v(Y)u(A_1X)\}^*\eta \\ & - \{-p(Y)w(X) + q(U)g(\phi Y,X) \\ & - p(U)g(\psi Y,X) + w(Y)u(A_1X)\}^*\zeta. \end{aligned} \quad (67)$$

Similarly, by using (58), (59), (61), (62), (64), and (65), we can easily obtain

$$\begin{aligned}{}'R(Y^*,V^*)X^* = {} & \{v(X)Y - g(Y,X)V \\ & + v(A_1X)A_1Y - g(A_1Y,X)A_1V\}^* \\ & - \{-r(V)g(\phi Y,X) + p(V)g(\theta Y,X) \\ & + v(Y)v(A_1X) + r(Y)w(X) \\ & + p(Y)u(X) - g(A_1Y,X)\}^*\eta \\ & + \{q(Y)u(X) - r(V)g(\psi Y,X) \\ & + q(V)g(\theta Y,X) - u(Y)v(A_1X)\}^*\xi \\ & - \{-q(Y)w(X) + q(V)g(\phi Y,X) \\ & - p(V)g(\psi Y,X) + w(Y)v(A_1X)\}^*\zeta, \\ {}'R(Y^*,W^*)X^* = {} & \{w(X)Y - g(Y,X)W \\ & + w(A_1X)A_1Y - g(A_1Y,X)A_1W\}^* \\ & - \{q(W)g(\phi Y,X) - p(W)g(\psi Y,X) \\ & + w(Y)w(A_1X) + q(Y)v(X) \\ & + p(Y)u(X) - g(A_1Y,X)\}^*\zeta \\ & - \{-r(Y)u(X) + r(W)g(\psi Y,X) \\ & - q(W)g(\theta Y,X) + u(Y)w(A_1X)\}^*\xi \\ & + \{r(Y)v(X) + r(W)g(\phi Y,X) \\ & - p(W)g(\theta Y,X) - v(Y)w(A_1X)\}^*\eta. \end{aligned} \quad (68)$$

5. Main Results

It is well known [3] that if $\pi^{-1}(M)$ is locally symmetric then ${}'\nabla A_1 = 0$ which implies identities (2) in Theorem K-P. In this point of view, we consider the following assumptions in (69) which are weaker conditions than the locally symmetry of $\pi^{-1}(M)$.

In order to obtain our main results, let M be n-dimensional QR-submanifolds of $(p-1)$ QR-dimension in $QP^{(n+p)/4}$ with the assumptions

$$\begin{gathered} ({}'\nabla_\xi {}'R)(Y^*,X^*)U^* = 0, \qquad ({}'\nabla_\eta {}'R)(Y^*,X^*)V^* = 0, \\ ({}'\nabla_\zeta {}'R)(Y^*,X^*)W^* = 0, \\ ({}'\nabla_\xi {}'R)(Y^*,U^*)X^* = 0, \qquad ({}'\nabla_\eta {}'R)(Y^*,V^*)X^* = 0, \\ ({}'\nabla_\zeta {}'R)(Y^*,W^*)X^* = 0. \end{gathered} \quad (69)$$

We notice here that the above curvature conditions in (69) are different from those in [18] due to Pak and Sohn.

We first consider the assumption

$$({}'\nabla_\xi {}'R)(Y^*,X^*)U^* = 0. \quad (70)$$

Differentiating (55) covariantly in the direction of ξ and using (19), (36), and (40), and the assumption $({}'\nabla_\xi {}'R)(Y^*,X^*)U^* = 0$, we have

$$\begin{aligned} & -{}'R((\phi Y)^*,X^*)U^* - {}'R(Y^*,(\phi X)^*)U^* \\ & = \{-u(X)\phi Y + u(Y)\phi X \\ & \quad - u(A_1X)\phi A_1Y + u(A_1Y)\phi A_1X\}^* \\ & \quad + \{p(X)w(Y) - p(Y)w(X) \\ & \quad + w(X)u(A_1Y) - w(Y)u(A_1X)\}^*\eta \\ & \quad - \{p(X)v(Y) - p(Y)v(X) \\ & \quad + v(X)u(A_1Y) - v(Y)u(A_1X)\}^*\zeta, \end{aligned} \quad (71)$$

from which, taking the vertical component of ξ, η, and ζ, respectively, and using (22)–(24) and (55) itself, we can get

$$\begin{aligned} & r(X)v(Y) - r(Y)v(X) - q(X)w(Y) \\ & \quad + q(Y)w(X) - r(\phi Y)w(X) + r(\phi X)w(Y) \\ & \quad - q(\phi Y)v(X) + q(\phi X)v(Y) \\ & \quad - u(X)u(A_1\phi Y) + u(Y)u(A_1\phi X) = 0, \end{aligned} \quad (72)$$

$$\begin{aligned} & -u(A_1\phi Y)v(X) + u(A_1\phi X)v(Y) + p(\phi Y)v(X) \\ & \quad - p(\phi X)v(Y) = 0, \end{aligned} \quad (73)$$

$$\begin{aligned} & -u(A_1\phi Y)w(X) + u(A_1\phi X)w(Y) + p(\phi Y)w(X) \\ & \quad - p(\phi X)w(Y) = 0. \end{aligned} \quad (74)$$

Putting $Y = U$ in (72) and using (19) and (24), we have

$$\phi A_1 U + r(U) V - q(U) W = 0, \tag{75}$$

and consequently,

$$r(U) = w(A_1 U) = u(A_1 W), \quad q(U) = v(A_1 U) = u(A_1 V). \tag{76}$$

Putting $Y = W$ and $X = V$ in (73) and using (15) and (24) yield

$$p(V) = v(A_1 U) = u(A_1 V). \tag{77}$$

Also, putting $Y = V$ and $X = W$ in (74) and using (15) and (24), we have

$$p(W) = w(A_1 U) = u(A_1 W). \tag{78}$$

Summing up, we have

$$\begin{aligned} A_1 U &= u(A_1 U) U + p(V) V + p(W) W, \\ p(V) &= v(A_1 U) = u(A_1 V) = q(U), \\ p(W) &= w(A_1 U) = u(A_1 W) = r(U). \end{aligned} \tag{79}$$

Thus we get the following lemma.

Lemma 1. *Let M be an n-dimensional QR-submanifold of $(p-1)$ QR-dimension in a quaternionic projective space $QP^{(n+p)/4}$, and let the normal vector field N_1 be parallel with respect to the normal connection. If the equalities in* (69) *are established, then*

$$\begin{aligned} A_1 U &= u(A_1 U) U + p(V) V + p(W) W, \\ A_1 V &= q(U) U + v(A_1 V) V + q(W) W, \\ A_1 W &= r(U) U + r(V) V + w(A_1 W) W, \\ p(V) &= v(A_1 U) = u(A_1 V) = q(U), \\ p(W) &= w(A_1 U) = u(A_1 W) = r(U), \\ q(W) &= w(A_1 V) = v(A_1 W) = r(V). \end{aligned} \tag{80}$$

Next, we assume the additional condition

$$({}'\nabla_\xi {}'R)(Y^*, U^*) X^* = 0. \tag{81}$$

Differentiating (67) covariantly in the direction of ξ and using (36), (40) and the assumption $({}'\nabla_\xi {}'R)(Y^*, U^*) X^* = 0$, we have

$$\begin{aligned} &- {}'R((\phi Y)^*, U^*) X^* - {}'R(Y^*, U^*)(\phi X)^* \\ &= \{-u(X)\phi Y - u(A_1 X)\phi A_1 Y \\ &\quad + g(A_1 Y, X)\phi A_1 U\}^* \\ &\quad - \{-p(Y) w(X) + q(U) g(\phi Y, X) \\ &\quad - p(U) g(\psi Y, X) + w(Y) u(A_1 X)\}^* \eta \\ &\quad + \{-p(Y) v(X) - r(U) g(\phi Y, X) \\ &\quad + p(U) g(\theta Y, X) + v(Y) u(A_1 X)\}^* \zeta, \end{aligned} \tag{82}$$

from which, taking the vertical component of ξ, η, and ζ, respectively, and using (22)–(24) and (67) itself, we can find

$$\begin{aligned} &r(U)\{-2g(\theta Y, X) + u(Y) v(X) - v(Y) u(X)\} \\ &\quad - q(U)\{2g(\psi Y, X) + u(Y) w(X) - w(Y) u(X)\} \\ &\quad + u(Y) u(A_1 \phi X) + r(\phi Y) w(X) + r(Y) v(X) \\ &\quad + q(\phi Y) v(X) - q(Y) w(X) \\ &\quad + g(\phi A_1 Y - A_1 \phi Y, X) = 0, \end{aligned} \tag{83}$$

$$\begin{aligned} &- q(U) g(\phi Y, X) + p(\phi Y) v(X) - v(Y) u(A_1 \phi X) \\ &\quad - p(U)\{g(\psi Y, X) + u(Y) w(X) \\ &\quad - w(Y) u(X)\} = 0, \end{aligned} \tag{84}$$

$$\begin{aligned} &- r(U) g(\phi Y, X) + p(\phi Y) w(X) - w(Y) u(A_1 \phi X) \\ &\quad - p(U)\{g(\theta Y, X) + v(Y) u(X) - u(Y) v(X)\} = 0. \end{aligned} \tag{85}$$

Taking the skew-symmetric part of (83) with respect to X and Y, we have

$$\begin{aligned} &2r(U)\{-2g(\theta Y, X) + u(Y) v(X) - v(Y) u(X)\} \\ &\quad - 2q(U)\{2g(\psi Y, X) + u(Y) w(X) - w(Y) u(X)\} \\ &\quad + u(Y) u(A_1 \phi X) - u(X) u(A_1 \phi Y) + r(\phi Y) w(X) \\ &\quad - r(\phi X) w(Y) + r(Y) v(X) - r(X) v(Y) \\ &\quad + q(\phi Y) v(X) - q(\phi X) v(Y) \\ &\quad - q(Y) w(X) + q(X) w(Y) = 0. \end{aligned} \tag{86}$$

Replacing Y with θY in (86) and using (19) and (22)–(24), we have

$$\begin{aligned} &r(U)\{4g(Y, X) - 3w(Y) w(X) \\ &\quad - 2u(Y) u(X) - 2v(Y) v(X)\} \\ &\quad - q(U)\{4g(\phi Y, X) + 3w(Y) v(X) - 2w(X) v(Y)\} \\ &\quad - q(\theta Y) w(X) - q(\psi Y) v(X) - q(\phi X) u(Y) \\ &\quad + r(\theta Y) v(X) - r(X) u(Y) - r(\psi Y) w(X) \\ &\quad - v(Y) u(A_1 \phi X) + u(X) u(A_1 \psi Y) \\ &\quad - u(A_1 U) u(X) w(Y) = 0. \end{aligned} \tag{87}$$

Now we consider the following orthonormal basis:

$$\begin{aligned} &\{U, V, W, e_1, \ldots, e_m, \phi(e_1), \ldots, \phi(e_m), \\ &\psi(e_1), \ldots, \psi(e_m), \theta(e_1), \ldots, \theta(e_m)\}, \end{aligned} \tag{88}$$

which will be called Q-basis, where $4m + 3 = \dim M$. Taking the trace of the above equation with respect to the Q-basis and using (76), we can easily see $4mr(U) = 0$; that is,

$$r(U) = 0. \quad (89)$$

Replacing also Y with ψY in (86) and using (19) and (22)–(24), we have

$$q(U) = 0. \quad (90)$$

Substituting (89) and (90) into (75), we have $\phi A_1 U = 0$ and hence

$$A_1 U = u(A_1 U) U. \quad (91)$$

On the other hand, replacing Y with ψY in (84) and using (19), (22), (23), and (90), we obtain

$$p(U)\{g(Y,X) - u(Y)u(X) - w(Y)w(X)\} + p(\theta Y)v(X) = 0, \quad (92)$$

from which, taking the trace with respect to the Q-basis and using (15) and (24), we find $4mp(U) = 0$; that is,

$$p(U) = 0 \quad (93)$$

which together with (84) and (90) implies

$$p(\phi Y) = 0. \quad (94)$$

Replacing Y with ϕY in the above equation and using (17) and (93), we can easily see that

$$p(Y) = 0 \quad (95)$$

for any vector Y tangent to M.

Summing up, we have the following lemma.

Lemma 2. *Let M be as in Lemma 1, and let the normal vector field N_1 be parallel with respect to the normal connection. If the equalities in* (69) *and* (5.2) *are established, then*

$$A_1 U = u(A_1 U)U, \quad A_1 V = v(A_1 V)V, \quad A_1 W = w(A_1 W)W, \quad p = q = r = 0. \quad (96)$$

Finally, we will prove our main theorem.

Theorem 3. *Let M be an n-dimensional QR-submanifold of $(p-1)$ QR-dimension in a quaternionic projective space $QP^{(n+p)/4}$, and let the normal vector field N_1 be parallel with respect to the normal connection. If the equalities in* (69) *and* (5.2) *are established, then $\pi^{-1}(M)$ is locally a product of $M_1 \times M_2$ where M_1 and M_2 belong to some $(4n_1+3)$- and $(4n_2+3)$-dimensional spheres (π is the Hopf fibration $S^{n+p+3}(1) \to QP^{(n+p)/4}$).*

Proof. By means of (96), it follows easily from (83) that

$$\phi A_1 = A_1 \phi. \quad (97)$$

By the quite same method, we can obtain

$$\psi A_1 = A_1 \psi, \qquad \theta A_1 = A_1 \theta. \quad (98)$$

Combining with those equalities and Theorem K-P, we complete the proof.

Corollary 4. *Let M be an n-dimensional QR-submanifold of $(p-1)$ QR-dimension in a quaternionic projective space $QP^{(n+p)/4}$, and let the normal vector field N_1 be parallel with respect to the normal connection. If the following equalities:*

$$'\nabla_\xi {}'R = 0, \qquad '\nabla_\eta {}'R = 0, \qquad '\nabla_\zeta {}'R = 0 \quad (99)$$

are established, then $\pi^{-1}(M)$ is locally a product of $M_1 \times M_2$ where M_1 and M_2 belong to some $(4n_1+3)$- and $(4n_2+3)$-dimensional spheres.

Acknowledgment

This work was supported by the 2013 Inje University research grant.

References

[1] A. Bejancu, *Geometry of CR-Submanifolds*, D. Reidel Publishing Company, Dordrecht, The Netherlands, 1986.

[2] J.-H. Kwon and J. S. Pak, "Scalar curvature of QR-submanifolds immersed in a quaternionic projective space," *Saitama Mathematical Journal*, vol. 17, pp. 89–116, 1999.

[3] J.-H. Kwon and J. S. Pak, "QR-submanifolds of Q QR-dimension in a quaternionic projective space $QP^{(n+p)/4}$," *Acta Mathematica Hungarica*, vol. 86, no. 1-2, pp. 89–116, 2000.

[4] H. B. Lawson, Jr., "Rigidity theorems in rank-$(n-1)$ symmetric spaces," *Journal of Differential Geometry*, vol. 4, pp. 349–357, 1970.

[5] A. Martínez and J. D. Pérez, "Real hypersurfaces in quaternionic projective space," *Annali di Matematica Pura ed Applicata*, vol. 145, pp. 355–384, 1986.

[6] J. S. Pak, "Real hypersurfaces in quaternionic Kaehlerian manifolds with constant $QP^{(n+p)/4}$-sectional curvature," *Kōdai Mathematical Seminar Reports*, vol. 29, no. 1-2, pp. 22–61, 1977.

[7] J. D. Pérez and F. G. Santos, "On pseudo-Einstein real hypersurfaces of the quaternionic projective space," *Kyungpook Mathematical Journal*, vol. 25, no. 1, pp. 15–28, 1985.

[8] J. D. Pérez and F. G. Santos, "On real hypersurfaces with harmonic curvature of a quaternionic projective space," *Journal of Geometry*, vol. 40, no. 1-2, pp. 165–169, 1991.

[9] Y. Shibuya, "Real submanifolds in a quaternionic projective space," *Kodai Mathematical Journal*, vol. 1, no. 3, pp. 421–439, 1978.

[10] S. Ishihara, "Quaternion Kählerian manifolds and fibred Riemannian spaces with Sasakian $(p-1)$-structure," *Kōdai Mathematical Seminar Reports*, vol. 25, pp. 321–329, 1973.

[11] S. Ishihara and M. Konishi, "Fibred riemannian space with sasakian 3-structure," in *Differential Geometry in Honor of K. Yano*, pp. 179–194, Kinokuniya, Tokyo, Japan, 1972.

[12] Y.-Y. Kuo, "On almost contact $(p-1)$-structure," *The Tohoku Mathematical Journal*, vol. 22, pp. 325–332, 1970.

[13] Y. Y. Kuo and S.-I. Tachibana, "On the distribution appeared in contact $(p-1)$-structure," *Taiwanese Journal of Mathematics*, vol. 2, pp. 17–24, 1970.

[14] Y.-W. Choe and M. Okumura, "Scalar curvature of a certain CR-submanifold of complex projective space," *Archiv der Mathematik*, vol. 68, no. 4, pp. 340–346, 1997.

[15] M. Okumura and L. Vanhecke, "n-dimensional real submanifolds with $(p-1)$-dimensional maximal holomorphic tangent subspace in complex projective spaces," *Rendiconti del Circolo Matematico di Palermo*, vol. 43, no. 2, pp. 233–249, 1994.

[16] J. S. Pak and W.-H. Sohn, "Some curvature conditions of n-dimensional CR-submanifolds of $(p-1)$ CR-dimension in a complex projective space," *Bulletin of the Korean Mathematical Society*, vol. 38, no. 3, pp. 575–586, 2001.

[17] W.-H. Sohn, "Some curvature conditions of n-dimensional CR-submanifolds of (n−1) CR-dimension in a complex projective space," *Communications of the Korean Mathematical Society*, vol. 25, pp. 15–28, 2000.

[18] J. S. Pak and W.-H. Sohn, "Some curvature conditions of n-dimensional QR-submanifolds of Q QR-dimension in a quaternionic projective space $QP^{(n+p)/4}$," *Bulletin of the Korean Mathematical Society*, vol. 40, no. 4, pp. 613–631, 2003.

[19] S. Ishihara, "Quaternion Kählerian manifolds," *Journal of Differential Geometry*, vol. 9, pp. 483–500, 1974.

[20] K. Yano and S. Ishihara, "Fibred spaces with invariant Riemannian metric," *Kōdai Mathematical Seminar Reports*, vol. 19, pp. 317–360, 1967.

[21] B. Y. Chen, *Geometry of Submanifolds*, Marcel Dekker, New York, NY, USA, 1973.

[22] B. O'Neill, "The fundamental equations of a submersion," *The Michigan Mathematical Journal*, vol. 13, pp. 459–469, 1966.

On Symmetric Left Bi-Derivations in *BCI*-Algebras

G. Muhiuddin,[1] Abdullah M. Al-roqi,[1] Young Bae Jun,[2] and Yilmaz Ceven[3]

[1] *Department of Mathematics, University of Tabuk, Tabuk 71491, Saudi Arabia*
[2] *Department of Mathematics Education (and RINS), Gyeongsang National University, Jinju 660-701, Republic of Korea*
[3] *Department of Mathematics, Faculty of Science, Suleyman Demirel University, 32260 Isparta, Turkey*

Correspondence should be addressed to G. Muhiuddin; chishtygm@gmail.com

Academic Editor: Aloys Krieg

The notion of *symmetric left bi-derivation* of a *BCI*-algebra X is introduced, and related properties are investigated. Some results on componentwise regular and d-regular *symmetric left bi-derivations* are obtained. Finally, characterizations of a p-semisimple *BCI*-algebra are explored, and it is proved that, in a p-semisimple *BCI*-algebra, F is a *symmetric left bi-derivation* if and only if it is a *symmetric bi-derivation*.

1. Introduction

BCK-algebras and *BCI*-algebras are two classes of nonclassical logic algebras which were introduced by Imai and Iséki in 1966 [1, 2]. They are algebraic formulation of *BCK*-system and *BCI*-system in combinatory logic. Later on, the notion of *BCI*-algebras has been extensively investigated by many researchers (see [3–6], and references therein). The notion of a *BCI*-algebra generalizes the notion of a *BCK*-algebra in the sense that every *BCK*-algebra is a *BCI*-algebra but not vice versa (see [7]). Hence, most of the algebras related to the t-norm-based logic such as MTL [8], BL, hoop, MV [9] (i.e lattice implication algebra), and Boolean algebras are extensions of *BCK*-algebras (i.e. they are subclasses of *BCK*-algebras) which have a lot of applications in computer science (see [10]). This shows that *BCK*-/*BCI*-algebras are considerably general structures.

Throughout our discussion, X will denote a *BCI*-algebra unless otherwise mentioned. In the year 2004, Jun and Xin [11] applied the notion of derivation in ring and near-ring theory to *BCI*-algebras, and as a result they introduced a new concept, called a (regular) derivation, in *BCI*-algebras. Using this concept as defined they investigated some of its properties. Using the notion of a regular derivation, they also established characterizations of a p-semisimple *BCI*-algebra. For a self-map d of a *BCI*-algebra, they defined a d-invariant ideal and gave conditions for an ideal to be d-invariant. According to Jun and Xin, a self map $d : X \to X$ is called a left-right derivation (briefly (l, r)-derivation) of X if $d(x * y) = d(x) * y \wedge x * d(y)$ holds for all $x, y \in X$. Similarly, a self map $d : X \to X$ is called a right-left derivation (briefly (r, l)-derivation) of X if $d(x * y) = x * d(y) \wedge d(x) * y$ holds for all $x, y \in X$. Moreover, if d is both (l, r)- and (r, l)-derivation, it is a derivation on X. After the work of Jun and Xin [11], many research articles have appeared on the derivations of *BCI*-algebras and a greater interest has been devoted to the study of derivations in *BCI*-algebras on various aspects (see [12–17]).

Inspired by the notions of σ-derivation [18], left derivation [19], and symmetric bi-derivations [20, 21] in rings and near-rings theory, many authors have applied these notions in a similar way to the theory of *BCI*-algebras (see [12, 13, 17]). For instantce in 2005 [17], Zhan and Liu have given the notion of f-derivation of *BCI*-algebras as follows: a self map $d_f : X \to X$ is said to be a left-right f-derivation or (l, r)-f-derivation of X if it satisfies the identity $d_f(x * y) = d_f(x) * f(y) \wedge f(x) * d_f(y)$ for all $x, y \in X$. Similarly, a self map $d_f : X \to X$ is said to be a right-left f-derivation or (r, l)-f-derivation of X if it satisfies the identity $d_f(x * y) = f(x) * d_f(y) \wedge d_f(x) * f(y)$ for all $x, y \in X$. Moreover, if d_f is both (l, r)- and (r, l)-f-derivation, it is said that d_f is an f-derivation, where f is an endomorphism. In the year 2007, Abujabal and Al-Shehri [12] defined and studied the notion of left derivation of *BCI*-algebras as follows: a self map $D : X \to X$ is said to be a left derivation of X if satisfying $D(x*y) = x*D(y)\wedge y*D(x)$ for all $x, y \in X$. Furthermore, in 2011 [13], Ilbira et al. have introduced the notion of symmetric bi-derivations in BCI-algebras. Following [13], a mapping $D(\cdot,\cdot) : X \times X \to X$ is said to be symmetric if $F(x, y) = F(y, x)$ holds for all pairs $x, y \in X$. A symmetric mapping $D(\cdot,\cdot) : X \times X \to X$ is called left-right symmetric bi-derivation (briefly (l, r)-symmetric bi-derivation) if it satisfies the identity $D(x * y, z) = D(x, z) * y \wedge x * D(y, z)$ for all $x, y, z \in X$. D is called right-left symmetric bi-derivation (briefly (r, l)-symmetric bi-derivation) if it satisfies the identity $D(x * y, z) = x * D(y, z) \wedge D(x, z) * y$ for all $x, y, z \in X$. Moreover, if D is both a (l, r)- and a (r, l)-symmetric bi-derivation, it is said that D is a symmetric bi-derivation on X.

Motivated by the notion of *symmetric bi-derivations* [13] in the theory of BCI-algebras, in the present analysis, we introduced the notion of *symmetric left bi-derivations* on BCI-algebras and investigated related properties. Further, we obtain some results on componentwise regular and d-regular *symmetric left bi-derivations*. Finally, we characterize the notion of p-semisimple BCI-algebra X by using the concept of *symmetric left bi-derivation* and show that, in a p-semisimple BCI-algebra X, F is a *symmetric left bi-derivation* if and only if it is a *symmetric bi-derivation*.

2. Preliminaries

We begin with the following definitions and properties that will be needed in the sequel.

A nonempty set X with a constant 0 and a binary operation $*$ is called a *BCI-algebra* if for all $x, y, z \in X$ the following conditions hold:

(I) $((x * y) * (x * z)) * (z * y) = 0$,

(II) $(x * (x * y)) * y = 0$,

(III) $x * x = 0$,

(IV) $x * y = 0$ and $y * x = 0$ imply $x = y$.

Define a binary relation $\leq$ on X by letting $x * y = 0$ if and only if $x \leq y$. Then $(X, \leq)$ is a partially ordered set. A BCI-algebra X satisfying $0 \leq x$ for all $x \in X$ is called BCK-algebra.

A BCI-algebra X has the following properties for all $x, y, z \in X$.

(a1) $x * 0 = x$.

(a2) $(x * y) * z = (x * z) * y$.

(a3) $x \leq y$ implies $x * z \leq y * z$ and $z * y \leq z * x$.

(a4) $(x * z) * (y * z) \leq x * y$.

(a5) $x * (x * (x * y)) = x * y$.

(a6) $0 * (x * y) = (0 * x) * (0 * y)$.

(a7) $x * 0 = 0$ implies $x = 0$.

For a BCI-algebra X, denote by X_+ (resp., $G(X)$) the BCK-part (resp., the BCI-G part) of X; that is, X_+ is the set of all $x \in X$ such that $0 \leq x$ (resp., $G(X) := \{x \in X \mid 0 * x = x\}$). Note that $G(X) \cap X_+ = \{0\}$ (see [22]). If $X_+ = \{0\}$, then X is called a *p-semisimple BCI-algebra*. In a p-semisimple BCI-algebra X, the following hold.

(a8) $(x * z) * (y * z) = x * y$.

(a9) $0 * (0 * x) = x$ for all $x \in X$.

(a10) $x * (0 * y) = y * (0 * x)$.

(a11) $x * y = 0$ implies $x = y$.

(a12) $x * a = x * b$ implies $a = b$.

(a13) $a * x = b * x$ implies $a = b$.

(a14) $a * (a * x) = x$.

(a15) $(x * y) * (w * z) = (x * w) * (y * z)$.

Let X be a p-semisimple BCI-algebra. We define addition "+" as $x + y = x * (0 * y)$ for all $x, y \in X$. Then $(X, +)$ is an abelian group with identity 0 and $x - y = x * y$. Conversely, let $(X, +)$ be an abelian group with identity 0, and let $x * y = x - y$. Then X is a p-semisimple BCI-algebra and $x + y = x * (0 * y)$ for all $x, y \in X$ (see [6]).

For a BCI-algebra X, we denote $x \wedge y = y * (y * x)$, in particular $0 * (0 * x) = a_x$, and $L_p(X) := \{a \in X \mid x * a = 0 \Rightarrow x = a, \forall x \in X\}$. We call the elements of $L_p(X)$ the *p-atoms* of X. For any $a \in X$, let $V(a) := \{x \in X \mid a * x = 0\}$, which is called the *branch* of X with respect to a. It follows that $x * y \in V(a * b)$ whenever $x \in V(a)$ and $y \in V(b)$ for all $x, y \in X$ and all $a, b \in L_p(X)$. Note that $L_p(X) = \{x \in X \mid a_x = x\}$,which is the p-semisimple part of X, and X is a p-semisimple BCI-algebra if and only if $L_p(X) = X$ (see [23, Proposition 3.2]). Note also that $a_x \in L_p(X)$; that is, $0 * (0 * a_x) = a_x$, which implies that $a_x * y \in L_p(X)$ for all $y \in X$. It is clear that $G(X) \subset L_p(X)$, and $x * (x * a) = a$ and $a * x \in L_p(X)$ for all $a \in L_p(X)$ and all $x \in X$. Let $D(\cdot,\cdot) : X \times X \to X$ be a symmetric mapping. Then for all $x \in X$, a mapping $d : X \to X$ defined by $d(x) = D(x, x)$ is called trace of D [13]. For more details, refer to [3, 4, 6, 11, 22, 23].

3. Symmetric Left Bi-Derivations

The following definition introduces the notion of *symmetric left bi-derivation* for a BCI-algebra X.

Definition 1. A symmetric mapping $F(\cdot,\cdot) : X \times X \to X$ is called a symmetric left bi-derivation of X if it satisfies the following identity:

$$(\forall x, y, z \in X) \quad (F(x * y, z) = (x * F(y, z)) \wedge (y * F(x, z))). \tag{1}$$

Example 2 (see [24]). Consider a p-semisimple BCI-algebra $X = \{0, 3, 4, 5\}$ with the following Cayley table:

*	0	3	4	5
0	0	3	4	5
3	3	0	5	4
4	4	5	0	3
5	5	4	3	0

(2)

Define a mapping $F(\cdot,\cdot) : X \times X \to X$ by

$$\begin{aligned} F(0,0) = F(3,3) = F(4,4) = F(5,5) = 0,\\ F(0,3) = F(3,0) = 3,\\ F(0,4) = F(4,0) = 4,\\ F(0,5) = F(5,0) = 5,\\ F(3,4) = F(4,3) = 5,\\ F(3,5) = F(5,3) = 4,\\ F(4,5) = F(5,4) = 3. \end{aligned} \quad (3)$$

It is routine to verify that F is a *symmetric left bi-derivation* of X.

Theorem 3. *Let $F(\cdot,\cdot) : X \times X \to X$ be a symmetric left bi-derivation of X. Then*

(1) $(\forall z \in X)\ (a \in G(X) \Rightarrow F(a,z) \in G(X))$.

(2) $(\forall z \in X)\ (a \in L_p(X) \Rightarrow F(a,z) \in L_p(X))$.

(3) $(\forall z \in X)\ (a \in L_p(X) \Rightarrow F(a,z) = 0 + F(a,z))$.

(4) $(\forall z \in X)\ (a \in L_p(X) \Rightarrow F(a,z) = a * F(0,z) = a + F(0,z))$.

Proof. (1) Let $a \in G(X)$. Then $0 * a = a$, and so

$$\begin{aligned} F(a,z) &= F(0 * a, z)\\ &= (0 * F(a,z)) \wedge (a * F(0,z))\\ &= (a * F(0,z)) * ((a * F(0,z)) * (0 * F(a,z)))\\ &= 0 * F(a,z), \end{aligned} \quad (4)$$

since $0 * F(a,z) \in L_p(X)$. Hence $F(a,z) \in G(X)$.

(2) For any $a \in L_p(X)$ implies $a = 0 * (0 * a)$ and so

$$\begin{aligned} F(a,z) &= F(0 * (0 * a), z)\\ &= (0 * F(0 * a, z)) \wedge ((0 * a) * F(0,z))\\ &= ((0 * a) * F(0,z))\\ &\quad * (((0 * a) * F(0,z)) * (0 * F(0 * a, z)))\\ &= 0 * F(0 * a, z) \in L_p(X). \end{aligned} \quad (5)$$

(3) By (2), we have $F(a,z) \in L_p(X)$. Then

$$F(a,z) = 0 * (0 * F(a,z)) = 0 + F(a,z). \quad (6)$$

(4) For any $a \in L_p(X)$ and $z \in X$, we have

$$\begin{aligned} F(a,z) &= F(a * 0, z)\\ &= (a * F(0,z)) \wedge (0 * F(a,z))\\ &= (0 * F(a,z)) * ((0 * F(a,z)) * (a * F(0,z)))\\ &= a * F(0,z)\\ &= a * (0 * F(0,z))\\ &= a + F(0,z). \end{aligned} \quad (7)$$

This completes the proof.

Using Theorem 3, we have the following corollary.

Corollary 4. *Let $F(\cdot,\cdot) : X \times X \to X$ be a symmetric left bi-derivation and $d : X \to X$ be the trace of F. Then*

(1) $(\forall a \in G(X))\ (d(a) \in G(X))$.

(2) $(\forall a \in L_p(X))\ (d(a) \in L_p(X))$.

Theorem 5. *Let F be a symmetric left bi-derivation of X. Then*

(1) $(\forall z \in X)\ (a, b \in L_p(X) \Rightarrow F(a+b,z) = a + F(b,z))$.

(2) $(\forall z \in X)\ (a \in L_p(X) \Rightarrow F(a,z) = a$ *if and only if* $F(0,z) = 0)$.

(3) $(\forall x, y, z \in X)\ (F(x * y, z) \le x * F(y,z))$.

(4) $(\forall x, y, z \in X)\ (x * F(x,z) = y * F(y,z))$.

Proof. (1) Let $a, b \in L_p(X)$. Then

$$\begin{aligned} F(a+b,z) &= F(a * (0 * b), z)\\ &= a * F(0 * b, z) \wedge (0 * b) * F(a,z)\\ &= a * F(0 * b, z)\\ &= a * (0 * F(b,z) \wedge b * F(0,z))\\ &= a * (0 * F(b,z)) = a + F(b,z). \end{aligned} \quad (8)$$

(2) Suppose $F(a,z) = a$ for all $a \in L_p(X)$, $z \in X$. It is clear that, for $0 \in L_p(X)$, we have $F(0,z) = 0$. Conversely let us assume that $F(0,z) = 0$; then by using Theorem 3(4), we have $F(a,z) = a + F(0,z) = a + 0 = a$.

(3) For any $x, y, z \in X$, we have

$$\begin{aligned} F(x * y, z) &= (x * F(y,z)) \wedge (y * F(x,z))\\ &= (y * F(x,z))\\ &\quad * ((y * F(x,z)) * (x * F(y,z)))\\ &\le x * F(y,z). \end{aligned} \quad (9)$$

(4) For any $x, z \in X$, we have

$$\begin{aligned} F(0,z) &= F(x * x, z) = (x * F(x,z)) \wedge (x * F(x,z))\\ &= x * F(x,z). \end{aligned} \quad (10)$$

Thus, we can write $F(0,z) = x * F(x,z) = y * F(y,z)$ for any $y \in X$. This completes the proof.

Definition 6. A symmetric left bi-derivation $F(\cdot,\cdot) : X \times X \to X$ of a BCI-algebra X is said to be *componentwise regular* if $F(0,z) = 0$ for all $z \in X$. In particular, F is called d-regular if $F(0,0) = d(0) = 0$.

Theorem 7. *Let F be a symmetric left bi-derivation of BCI-algebra X. Then X is a BCK-algebra if and only if F is componentwise regular symmetric left bi-derivation.*

Proof. Suppose X is a BCK-algebra. Then for any $x, z \in X$, we have

$$\begin{aligned} F(0,z) &= F(0*x,z) \\ &= (0*F(x,z)) \wedge (x*F(0,z)) \\ &= 0 \wedge (x*F(0,z)) = 0. \end{aligned} \tag{11}$$

Hence F is *componentwise regular.*

Conversely, let F be a *componentwise regular symmetric left bi-derivation*. Let for any $a \in L_p(X)$ be such that $a \neq 0$. Then

$$F(a*0,0) = F(a,0) = 0. \tag{12}$$

But it is clear that

$$\begin{aligned} a*F(0,0) \wedge 0*F(a,0) &= a*0 \wedge 0*0 \\ &= a \wedge 0 = 0*(0*a) \\ &= a \neq 0, \end{aligned} \tag{13}$$

which is not possible as F is a *componentwise regular symmetric left bi-derivation*. Thus 0 is the unique p-atom. Assume that for some $m \in X$, we have $0*m \neq 0$, then $a_{0*m} = 0*(0*(0*m)) = 0$, so $0*m \in L_p(X)$, which is a contradiction. Henceforth, for all $m \in X$, we have $0*m = 0$ implying thereby, X is a BCK-algebra.

This completes the proof.

Theorem 8. *Let F be a componentwise regular symmetric left bi-derivation of a BCI-algebra X. Then*

(1) *Both x and $F(x,z)$ belong to the same branch for all $x, z \in X$.*

(2) $(\forall x, z \in X)\ (F(x,z) \leq x)$.

(3) $(\forall x, y, z \in X)\ (F(x,z) * y \leq x * F(y,z))$.

Proof. (1) For any $x, z \in X$, we get

$$\begin{aligned} 0 = F(0,z) &= F(a_x * x, z) \\ &= (a_x * F(x,z)) \wedge (x * F(a_x,z)) \\ &= (x*F(a_x,z)) * ((x*F(a_x,z)) * (a_x * F(x,z))) \\ &= a_x * F(x,z), \end{aligned} \tag{14}$$

since $a_x * F(x,z) \in L_p(X)$. Hence $a_x \leq F(x,z)$, and so $F(x,z) \in V(a_x)$. Obviously, $x \in V(a_x)$.

(2) Since F is componentwise regular, $F(0,z) = 0$. Then

$$\begin{aligned} F(x,z) &= F(x*0,z) \\ &= (x*F(0,z)) \wedge (0*F(x,z)) \\ &= (x*0) \wedge (0*F(x,z)) \\ &= (0*F(x,z)) * ((0*F(x,z)) * x) \\ &\leq x. \end{aligned} \tag{15}$$

(3) Since $F(x,z) \leq x$ for all $x, z \in X$ by (2), using (a3) we obtain

$$F(x,z) * y \leq x * y \leq x * F(y,z). \tag{16}$$

This completes the proof.

Next, we prove some results in a p-semisimple BCI-algebra.

Theorem 9. *Let F be a symmetric left bi-derivation of a p-semisimple BCI-algebra X; one has the following assertions.*

(1) $(\forall x, y, z \in X)\ (F(x*y,z) = x * F(y,z))$.

(2) $(\forall x, y, z \in X)\ (F(x,z) * x = F(y,z) * y)$.

(3) $(\forall x, y, z \in X)\ (F(x,z) * x = y * F(y,z))$.

Proof. (1) Let X be a p-semisimple BCI-algebra. Then for any $x, y, z \in X$, we have

$$F(x*y,z) = (x*F(y,z)) \wedge (y*F(x,z)) = x*F(y,z). \tag{17}$$

(2) Let $x, y, z \in X$. Using (I), we have

$$\begin{aligned} (x*y)*(x*F(y,z)) &\leq F(y,z)*y, \\ (y*x)*(y*F(x,z)) &\leq F(x,z)*x. \end{aligned} \tag{18}$$

These above inequalities can be rewritten as

$$\begin{aligned} ((x*y)*(x*F(y,z)))*(F(y,z)*y) &= 0, \\ ((y*x)*(y*F(x,z)))*(F(x,z)*x) &= 0. \end{aligned} \tag{19}$$

Consequently, we get

$$\begin{aligned} &((x*y)*(x*F(y,z)))*(F(y,z)*y) \\ &\quad = ((y*x)*(y*F(x,z)))*(F(x,z)*x) \end{aligned} \tag{20}$$

Also, using Theorem 5(4) and (1), we obtain

$$\begin{aligned} &(x*y)*F(x*y,z) = (y*x)*F(y*x,z) \\ &\Longrightarrow (x*y)*(x*F(y,z)) = (y*x)*(y*F(x,z)). \end{aligned} \tag{21}$$

Since X is a p-semisimple BCI-algebra, hence, by using (21) and (a12), the above (20) yields $F(x,z)*x = F(y,z)*y$.

(3) We have $F(0,z) = x * F(x,z)$ by Theorem 5(4). Further, on letting $x = 0$, we get that $F(0,z) * 0 = F(y,z) * y$ implies $F(0,z) = F(y,z) * y$. Henceforth $F(y,z) * y = x * F(x,z)$, which amounts to say that $F(x,z) * x = y * F(y,z)$.

This completes the proof.

Theorem 10. *Let X be a p-semisimple BCI-algebra. Then F is a symmetric left bi-derivation if and only if it is a symmetric bi-derivation on X.*

Proof. Suppose that F is a *symmetric left bi-derivation* on X. First, we show that F is a *(r,l)-symmetric bi-derivation* on X. Let $x, y, z \in X$. Using Theorem 9(1) and (a14), we have

$$\begin{aligned} F(x * y, z) &= x * F(y, z) \\ &= (F(x, z) * y) \\ &\quad * ((F(x, z) * y) * (x * F(y, z))) \\ &= (x * F(y, z)) \wedge (F(x, z) * y). \end{aligned} \tag{22}$$

Hence F is a *(r, l)-symmetric bi-derivation* on X.

Again, we show that F is a *(l,r)-symmetric bi-derivation* on X. Let $x, y, z \in X$. Using Theorem 9(1), (3) and (a15), we have

$$\begin{aligned} F(x * y, z) &= x * F(y, z) \\ &= (x * 0) * F(y, z) \\ &= (x * (F(0, z) * F(0, z))) * F(y, z) \\ &= (x * ((x * F(x, z)) * (F(y, z) * y))) \\ &\quad * F(y, z) \\ &= (x * F(y, z)) \\ &\quad * ((x * F(x, z)) * (F(y, z) * y)) \\ &= (x * F(y, z)) \\ &\quad * ((x * F(y, z)) * (F(x, z) * y)) \\ &= (F(x, z) * y) \wedge (x * F(y, z)). \end{aligned} \tag{23}$$

Conversely, suppose that F is a *symmetric bi-derivation* of X. As F is a *(r,l)-symmetric bi-derivation* on X, then for any $x, y, z \in X$ and using (a14), we have

$$\begin{aligned} F(x * y, z) &= (x * F(y, z)) \wedge (F(x, z) * y) \\ &= (F(x, z) * y) \\ &\quad * ((F(x, z) * y) * (x * F(y, z))) \\ &= x * F(y, z) \\ &= (y * F(x, z)) \\ &\quad * ((y * F(x, z)) * (x * F(y, z))) \\ &= (x * F(y, z)) \wedge (y * F(x, z)). \end{aligned} \tag{24}$$

Hence F is a *symmetric left bi-derivation*. This completes the proof.

Acknowledgments

The authors are grateful to the anonymous referee(s) for a careful checking of the details and for helpful comments that improved the present paper. G. Muhiuddin and Abdullah M. Al-roqi were partially supported by the Deanship of Scientific Research, University of Tabuk, Ministry of Higher Education, Saudi Arabia.

References

[1] Y. Imai and K. Iséki, "On axiom systems of propositional calculi. XIV," *Proceedings of the Japan Academy*, vol. 42, pp. 19–22, 1966.

[2] K. Iséki, "An algebra related with a propositional calculus," *Proceedings of the Japan Academy*, vol. 42, pp. 26–29, 1966.

[3] M. Aslam and A. B. Thaheem, "A note on p-semisimple *BCI*-algebras," *Mathematica Japonica*, vol. 36, no. 1, pp. 39–45, 1991.

[4] S. A. Bhatti, M. A. Chaudhry, and B. Ahmad, "On classification of *BCI*-algebras," *Mathematica Japonica*, vol. 34, no. 6, pp. 865–876, 1989.

[5] Q. P. Hu and K. Iséki, "On *BCI*-algebras satisfying $(x * y) * z = x * (y * z)$," *Kobe University. Mathematics Seminar Notes*, vol. 8, no. 3, pp. 553–555, 1980.

[6] M. Daoji, "*BCI*-algebras and abelian groups," *Mathematica Japonica*, vol. 32, no. 5, pp. 693–696, 1987.

[7] Q. P. Hu and K. Iséki, "On some classes of *BCI*-algebras," *Mathematica Japonica*, vol. 29, no. 2, pp. 251–253, 1984.

[8] F. Esteva and L. Godo, "Monoidal t-norm based logic: towards a logic for left-continuous t-norms," *Fuzzy Sets and Systems*, vol. 124, no. 3, pp. 271–288, 2001.

[9] C. C. Chang, "Algebraic analysis of many valued logics," *Transactions of the American Mathematical Society*, vol. 88, pp. 467–490, 1958.

[10] J. E. Whitesitt, *Boolean Algebra and Its Applications*, Dover, New York, NY, USA, 1995.

[11] Y. B. Jun and X. L. Xin, "On derivations of *BCI*-algebras," *Information Sciences*, vol. 159, no. 3-4, pp. 167–176, 2004.

[12] H. A. S. Abujabal and N. O. Al-Shehri, "On left derivations of *BCI*-algebras," *Soochow Journal of Mathematics*, vol. 33, no. 3, pp. 435–444, 2007.

[13] S. Ilbira, A. Firat, and Y. B. Jun, "On symmetric bi-derivations of *BCI*-algebras," *Applied Mathematical Sciences*, vol. 5, no. 57-60, pp. 2957–2966, 2011.

[14] G. Muhiuddin and A. M. Al-roqi, "On t-derivations of *BCI*-algebras," *Abstract and Applied Analysis*, vol. 2012, Article ID 872784, 12 pages, 2012.

[15] G. Mudiuddin and A. M. Al-roqi, "On (α,β)-derivations in *BCI*-algebras," *Discrete Dynamics in Nature and Society*, vol. 2012, Article ID 403209, 11 pages, 2012.

[16] M. A. Öztürk, Y. Çeven, and Y. B. Jun, "Generalized derivations of *BCI*-algebras," *Honam Mathematical Journal*, vol. 31, no. 4, pp. 601–609, 2009.

[17] J. Zhan and Y. L. Liu, "On f-derivations of *BCI*-algebras," *International Journal of Mathematics and Mathematical Sciences*, no. 11, pp. 1675–1684, 2005.

[18] A. A. M. Kamal, "σ-derivations on prime near-rings," *Tamkang Journal of Mathematics*, vol. 32, no. 2, pp. 89–93, 2001.

[19] M. Brešar and J. Vukman, "On left derivations and related mappings," *Proceedings of the American Mathematical Society*, vol. 110, no. 1, pp. 7–16, 1990.

[20] J. Vukman, "Symmetric bi-derivations on prime and semi-prime rings," *Aequationes Mathematicae*, vol. 38, no. 2-3, pp. 245–254, 1989.

[21] J. Vukman, "Two results concerning symmetric bi-derivations on prime rings," *Aequationes Mathematicae*, vol. 40, no. 2-3, pp. 181–189, 1990.

[22] Y. B. Jun and E. H. Roh, "On the *BCI-G* part of *BCI*-algebras," *Mathematica Japonica*, vol. 38, no. 4, pp. 697–702, 1993.

[23] Y. B. Jun, X. L. Xin, and E. H. Roh, "The role of atoms in *BCI*-algebras," *Soochow Journal of Mathematics*, vol. 30, no. 4, pp. 491–506, 2004.

[24] J. Meng, Y. B. Jun, and E. H. Roh, "*BCI*-algebras of order 6," *Mathematica Japonica*, vol. 47, no. 1, pp. 33–43, 1998.

6

On Some Intermediate Mean Values

Slavko Simic

Mathematical Institute SANU, Kneza Mihaila 36, 11000 Belgrade, Serbia

Correspondence should be addressed to Slavko Simic; ssimic@turing.mi.sanu.ac.rs

Academic Editor: Mowaffaq Hajja

We give a necessary and sufficient mean condition for the quotient of two Jensen functionals and define a new class $\Lambda_{f,g}(a,b)$ of mean values where f, g are continuously differentiable convex functions satisfying the relation $f''(t) = tg''(t)$, $t \in \mathbb{R}^+$. Then we asked for a characterization of f, g such that the inequalities $H(a,b) \le \Lambda_{f,g}(a,b) \le A(a,b)$ or $L(a,b) \le \Lambda_{f,g}(a,b) \le I(a,b)$ hold for each positive a, b, where H, A, L, I are the harmonic, arithmetic, logarithmic, and identric means, respectively. For a subclass of Λ with $g''(t) = t^s$, $s \in \mathbb{R}$, this problem is thoroughly solved.

1. Introduction

It is said that the mean P is intermediate relating to the means M and N, $M \le N$ if the relation

$$M(a,b) \le P(a,b) \le N(a,b) \quad (1)$$

holds for each two positive numbers a, b.

It is also well known that

$$\begin{aligned}\min\{a,b\} &\le H(a,b) \le G(a,b)\\ &\le L(a,b) \le I(a,b) \le A(a,b) \le S(a,b) \quad (2)\\ &\le \max\{a,b\},\end{aligned}$$

where

$$H = H(a,b) := 2\left(\frac{1}{a} + \frac{1}{b}\right)^{-1};$$

$$G = G(a,b) := \sqrt{ab}; \qquad L = L(a,b) := \frac{b-a}{\log b - \log a};$$

$$I = I(a,b) := \frac{\left(b^b/a^a\right)^{1/(b-a)}}{e};$$

$$A = A(a,b) := \frac{a+b}{2}; \qquad S = S(a,b) := a^{a/(a+b)}b^{b/(a+b)} \quad (3)$$

are the harmonic, geometric, logarithmic, identric, arithmetic, and Gini mean, respectively.

An easy task is to construct intermediate means related to two given means M and N with $M \le N$. For instance, for an arbitrary mean P, we have that

$$M(a,b) \le P(M(a,b), N(a,b)) \le N(a,b). \quad (4)$$

The problem is more difficult if we have to decide whether the given mean is intermediate or not. For example, the relation

$$L(a,b) \le S_s(a,b) \le I(a,b) \quad (5)$$

holds for each positive a and b if and only if $0 \le s \le 1$, where the Stolarsky mean S_s is defined by (cf [1])

$$S_s(a,b) := \left(\frac{b^s - a^s}{s(b-a)}\right)^{1/(s-1)}. \quad (6)$$

Also,

$$G(a,b) \le A_s(a,b) \le A(a,b) \quad (7)$$

holds if and only if $0 \le s \le 1$, where the Hölder mean of order s is defined by

$$A_s(a,b) := \left(\frac{a^s + b^s}{2}\right)^{1/s}. \quad (8)$$

An inverse problem is to find best possible approximation of a given mean P by elements of an ordered class of means S. A good example for this topic is comparison between the logarithmic mean and the class A_s of Hölder means of order s. Namely, since $A_0 = \lim_{s\to 0} A_s = G$ and $A_1 = A$, it follows from (2) that

$$A_0 \le L \le A_1. \tag{9}$$

Since A_s is monotone increasing in s, an improving of the above is given by Carlson [2]:

$$A_0 \le L \le A_{1/2}. \tag{10}$$

Finally, Lin showed in [3] that

$$A_0 \le L \le A_{1/3} \tag{11}$$

is the best possible approximation of the logarithmic mean by the means from the class A_s.

Numerous similar results have been obtained recently. For example, an approximation of Seiffert's mean by the class A_s is given in [4, 5].

In this paper we will give best possible approximations for a whole variety of elementary means (2) by the class λ_s defined below (see Theorem 5).

Let f, g be twice continuously differentiable (strictly) convex functions on $\mathbb{R}^+$. By definition (cf [6], page 5),

$$\overline{f}(a,b) := f(a) + f(b) - 2f\left(\frac{a+b}{2}\right) > 0, \quad a \ne b,$$
$$\overline{f}(a,b) = 0, \tag{12}$$

if and only if $a = b$.

It turns out that the expression

$$\Lambda_{f,g}(a,b) := \frac{\overline{f}(a,b)}{\overline{g}(a,b)} = \frac{f(a) + f(b) - 2f((a+b)/2)}{g(a) + g(b) - 2g((a+b)/2)} \tag{13}$$

represents a mean of two positive numbers a, b; that is, the relation

$$\min\{a,b\} \le \Lambda_{f,g}(a,b) \le \max\{a,b\} \tag{14}$$

holds for each $a, b \in \mathbb{R}^+$, if and only if the relation

$$f''(t) = tg''(t) \tag{15}$$

holds for each $t \in \mathbb{R}^+$.

Let $f, g \in C^\infty(0,\infty)$ and denote by Λ the set $\{(f,g)\}$ of convex functions satisfying the relation (15). There is a natural question how to improve the bounds in (14); in this sense we come upon the following intermediate mean problem.

Open Question. Under what additional conditions on $f, g \in \Lambda$, the inequalities

$$H(a,b) \le \Lambda_{f,g}(a,b) \le A(a,b), \tag{16}$$

or, more tightly,

$$L(a,b) \le \Lambda_{f,g}(a,b) \le I(a,b), \tag{17}$$

hold for each $a, b \in \mathbb{R}^+$?

As an illustration, consider the function $f_s(t)$ defined to be

$$f_s(t) = \begin{cases} \dfrac{t^s - st + s - 1}{s(s-1)}, & s(s-1) \ne 0; \\ t - \log t - 1, & s = 0; \\ t\log t - t + 1, & s = 1. \end{cases} \tag{18}$$

Since

$$f_s'(t) = \begin{cases} \dfrac{t^{s-1} - 1}{s-1}, & s(s-1) \ne 0; \\ 1 - \dfrac{1}{t}, & s = 0; \\ \log t, & s = 1, \end{cases} \tag{19}$$

$$f_s''(t) = t^{s-2}, \quad s \in \mathbb{R},\ t > 0,$$

it follows that $f_s(t)$ is a twice continuously differentiable convex function for $s \in \mathbb{R}, t \in \mathbb{R}^+$.

Moreover, it is evident that $(f_{s+1}, f_s) \in \Lambda$.

We will give in the sequel a complete answer to the above question concerning the means

$$\frac{\overline{f}_{s+1}(a,b)}{\overline{f}_s(a,b)} := \lambda_s(a,b) \tag{20}$$

defined by

$$\lambda_s(a,b) = \begin{cases} \dfrac{s-1}{s+1}\dfrac{a^{s+1} + b^{s+1} - 2((a+b)/2)^{s+1}}{a^s + b^s - 2((a+b)/2)^s}, & s \in \mathbb{R}/\{-1,0,1\}; \\ \dfrac{2\log((a+b)/2) - \log a - \log b}{1/2a + 1/2b - 2/(a+b)}, & s = -1; \\ \dfrac{a\log a + b\log b - (a+b)\log((a+b)/2)}{2\log((a+b)/2) - \log a - \log b}, & s = 0; \\ \dfrac{(b-a)^2}{4(a\log a + b\log b - (a+b)\log((a+b)/2))}, & s = 1. \end{cases} \tag{21}$$

Those means are obviously symmetric and homogeneous of order one.

As a consequence we obtain some new intermediate mean values; for instance, we show that the inequalities

$$H(a,b) \le \lambda_{-1}(a,b) \le G(a,b) \le \lambda_0(a,b) \le L(a,b) \le \lambda_1(a,b) \le I(a,b) \tag{22}$$

hold for arbitrary $a, b \in \mathbb{R}^+$. Note that

$$\lambda_{-1} = \frac{2G^2\log(A/G)}{A-H}; \qquad \lambda_0 = A\frac{\log(S/A)}{\log(A/G)}; \qquad \lambda_1 = \frac{1}{2}\frac{A-H}{\log(S/A)}. \tag{23}$$

2. Results

We prove firstly the following

Theorem 1. *Let $f, g \in C^2(I)$ with $g'' > 0$. The expression $\Lambda_{f,g}(a,b)$ represents a mean of arbitrary numbers $a, b \in I$ if and only if the relation (15) holds for $t \in I$.*

Remark 2. In the same way, for arbitrary $p, q > 0$, $p + q = 1$, it can be deduced that the quotient

$$\Lambda_{f,g}(p, q; a, b) := \frac{pf(a) + qf(b) - f(pa + qb)}{pg(a) + qg(b) - g(pa + qb)} \tag{24}$$

represents a mean value of numbers a, b if and only if (15) holds.

A generalization of the above assertion is the next.

Theorem 3. *Let $f, g : I \to \mathbb{R}$ be twice continuously differentiable functions with $g'' > 0$ on I and let $p = \{p_i\}$, $i = 1, 2, \ldots, \sum p_i = 1$ be an arbitrary positive weight sequence. Then the quotient of two Jensen functionals*

$$\Lambda_{f,g}(p, x) := \frac{\sum_1^n p_i f(x_i) - f(\sum_1^n p_i x_i)}{\sum_1^n p_i g(x_i) - g(\sum_1^n p_i x_i)}, \quad n \geq 2, \tag{25}$$

represents a mean of an arbitrary set of real numbers $x_1, x_2, \ldots, x_n \in I$ if and only if the relation

$$f''(t) = tg''(t) \tag{26}$$

holds for each $t \in I$.

Remark 4. It should be noted that the relation $f''(t) = tg''(t)$ determines f in terms of g in an easy way. Precisely,

$$f(t) = tg(t) - 2G(t) + ct + d, \tag{27}$$

where $G(t) := \int_1^t g(u)du$ and c and d are constants.

Our results concerning the means $\lambda_s(a,b)$, $s \in \mathbb{R}$ are included in the following.

Theorem 5. *For the class of means $\lambda_s(a,b)$ defined above, the following assertions hold for each $a, b \in \mathbb{R}^+$.*

(1) *The means $\lambda_s(a,b)$ are monotone increasing in s;*

(2) *$\lambda_s(a,b) \leq H(a,b)$ for each $s \leq -4$;*

(3) *$H(a,b) \leq \lambda_s(a,b) \leq G(a,b)$ for $-3 \leq s \leq -1$;*

(4) *$G(a,b) \leq \lambda_s(a,b) \leq L(a,b)$ for $-1/2 \leq s \leq 0$;*

(5) *there is a number $s_0 \in (1/12, 1/11)$ such that $L(a,b) \leq \lambda_s(a,b) \leq I(a,b)$ for $s_0 \leq s \leq 1$;*

(6) *there is a number $s_1 \in (1.03, 1.04)$ such that $I(a,b) \leq \lambda_s(a,b) \leq A(a,b)$ for $s_1 \leq s \leq 2$;*

(7) *$A(a,b) \leq \lambda_s(a,b) \leq S(a,b)$ for each $2 \leq s \leq 5$;*

(8) *there is no finite s such that the inequality $S(a,b) \leq \lambda_s(a,b)$ holds for each $a, b \in \mathbb{R}^+$.*

The above estimations are best possible.

3. Proofs

3.1. Proof of Theorem 1. We prove firstly the necessity of the condition (15).

Since $\Lambda_{f,g}(a,b)$ is a mean value for arbitrary $a, b \in I$; $a \neq b$, we have

$$\min\{a,b\} \leq \Lambda_{f,g}(a,b) \leq \max\{a,b\}. \tag{28}$$

Hence

$$\lim_{b \to a} \Lambda_{f,g}(a,b) = a. \tag{29}$$

From the other hand, due to l'Hospital's rule we obtain

$$\begin{aligned} \lim_{b \to a} \Lambda_{f,g}(a,b) &= \lim_{b \to a} \left(\frac{f'(b) - f'((a+b)/2)}{g'(b) - g'((a+b)/2)} \right) \\ &= \lim_{b \to a} \left(\frac{2f''(b) - f''((a+b)/2)}{2g''(b) - g''((a+b)/2)} \right) \\ &= \frac{f''(a)}{g''(a)}. \end{aligned} \tag{30}$$

Comparing (29) and (30) the desired result follows.

Suppose now that (15) holds and let $a < b$. Since $g''(t) > 0$ $t \in [a,b]$ by the Cauchy mean value theorem there exists $\xi \in ((a+t)/2, t)$ such that

$$\frac{f'(t) - f'((a+t)/2)}{g'(t) - g'((a+t)/2)} = \frac{f''(\xi)}{g''(\xi)} = \xi. \tag{31}$$

But,

$$a \leq \frac{a+t}{2} < \xi < t \leq b, \tag{32}$$

and, since g' is strictly increasing, $g'(t) - g'((a+t)/2) > 0$, $t \in [a,b]$.

Therefore, by (31) we get

$$\begin{aligned} a\left(g'(t) - g'\left(\frac{a+t}{2}\right)\right) &\leq f'(t) - f'\left(\frac{a+t}{2}\right) \\ &\leq b\left(g'(t) - g'\left(\frac{a+t}{2}\right)\right). \end{aligned} \tag{33}$$

Finally, integrating (33) over $t \in [a,b]$ we obtain the assertion from Theorem 1.

3.2. Proof of Theorem 3. We will give a proof of this assertion by induction on n.

By Remark 2, it holds for $n = 2$.

Next, it is not difficult to check the identity

$$\begin{aligned} &\sum_1^n p_i f(x_i) - f\left(\sum_1^n p_i x_i\right) \\ &= (1 - p_n)\left(\sum_1^{n-1} p_i' f(x_i) - f\left(\sum_1^{n-1} p_i' x_i\right)\right) \\ &\quad + \left[(1 - p_n) f(T) + p_n f(x_n) - f((1 - p_n)T + p_n x_n)\right], \end{aligned} \tag{34}$$

where

$$T := \sum_1^{n-1} p_i' x_i; \quad p_i' := \frac{p_i}{(1-p_n)}, \quad i = 1,2,\ldots,n-1; \quad \sum_1^{n-1} p_i' = 1. \tag{35}$$

Therefore, by induction hypothesis and Remark 2, we get

$$\begin{aligned}
&\sum_1^n p_i f(x_i) - f\left(\sum_1^n p_i x_i\right) \\
&\leq \max\{x_1, x_2, \ldots x_{n-1}\}(1-p_n) \\
&\quad \times \left(\sum_1^{n-1} p_i' g(x_i) - g\left(\sum_1^{n-1} p_i' x_i\right)\right) \\
&\quad + \max\{T, x_n\}\left[(1-p_n) g(T) + p_n g(x_n)\right. \\
&\qquad \left. - g((1-p_n)T + p_n x_n)\right] \\
&\leq \max\{x_1, x_2, \ldots, x_n\} \\
&\quad \times \left((1-p_n)\left(\sum_1^{n-1} p_i' g(x_i) - g\left(\sum_1^{n-1} p_i' x_i\right)\right)\right. \\
&\quad \left. + \left[(1-p_n) g(T) + p_n g(x_n) - g((1-p_n)T + p_n x_n)\right]\right) \\
&= \max\{x_1, x_2, \ldots, x_n\}\left(\sum_1^n p_i g(x_i) - g\left(\sum_1^n p_i x_i\right)\right).
\end{aligned} \tag{36}$$

The inequality

$$\min\{x_1, x_2, \ldots, x_n\} \leq \Lambda_{f,g}(p, x) \tag{37}$$

can be proved analogously.

For the proof of necessity, put $x_2 = x_3 = \cdots = x_n$ and proceed as in Theorem 1.

Remark 6. It is evident from (15) that if $I \subseteq \mathbb{R}^+$ then f has to be also convex on I. Otherwise, it shouldn't be the case. For example, the conditions of Theorem 3 are satisfied with $f(t) = t^3/3$, $g(t) = t^2$, $t \in \mathbb{R}$. Hence, for an arbitrary sequence $\{x_i\}_1^n$ of real numbers, we obtain

$$\begin{aligned}
\min\{x_1, x_2, \ldots, x_n\} &\leq \frac{\sum_1^n p_i x_i^3 - \left(\sum_1^n p_i x_i\right)^3}{3\left(\sum_1^n p_i x_i^2 - \left(\sum_1^n p_i x_i\right)^2\right)} \\
&\leq \max\{x_1, x_2, \ldots, x_n\}.
\end{aligned} \tag{38}$$

Because the above inequality does not depend on n, a probabilistic interpretation of the above result is contained in the following.

Theorem 7. *For an arbitrary probability law F of random variable X with support on $(-\infty, +\infty)$, one has*

$$(EX)^3 + 3(\min X)\,\sigma_X^2 \leq EX^3 \leq (EX)^3 + 3(\max X)\,\sigma_X^2. \tag{39}$$

3.3. Proof of Theorem 5, Part (1). We will prove a general assertion of this type. Namely, for an arbitrary positive sequence $\mathbf{x} = \{x_i\}$ and an associated weight sequence $\mathbf{p} = \{p_i\}$, $i = 1, 2, \ldots$, denote

$$\chi_s(\mathbf{p}, \mathbf{x}) := \begin{cases} \dfrac{\sum p_i x_i^s - \left(\sum p_i x_i\right)^s}{s(s-1)}, & s \in \mathbb{R}/\{0,1\}; \\ \log\left(\sum p_i x_i\right) - \sum p_i \log x_i, & s = 0; \\ \sum p_i x_i \log x_i - \left(\sum p_i x_i\right)\log\left(\sum p_i x_i\right), & s = 1. \end{cases} \tag{40}$$

For $s \in \mathbb{R}$, $r > 0$ we have

$$\chi_s(\mathbf{p}, \mathbf{x})\,\chi_{s+r+1}(\mathbf{p}, \mathbf{x}) \geq \chi_{s+1}(\mathbf{p}, \mathbf{x})\,\chi_{s+r}(\mathbf{p}, \mathbf{x}), \tag{41}$$

which is equivalent to

Theorem 8. *The sequence $\{\chi_{s+1}(\mathbf{p}, \mathbf{x})/\chi_s(\mathbf{p}, \mathbf{x})\}$ is monotone increasing in s, $s \in \mathbb{R}$.*

This assertion follows applying the result from [7, Theorem 2] which states the following.

Lemma 9. *For $-\infty < a < b < c < +\infty$, the inequality*

$$\left(\chi_b(\mathbf{p}, \mathbf{x})\right)^{c-a} \leq \left(\chi_a(\mathbf{p}, \mathbf{x})\right)^{c-b}\left(\chi_c(\mathbf{p}, \mathbf{x})\right)^{b-a} \tag{42}$$

holds for arbitrary sequences $\mathbf{p}$, $\mathbf{x}$.

Putting there $a = s$, $b = s+1$, $c = s+r+1$ and $a = s$, $b = s+r$, $c = s+r+1$, we successively obtain

$$\begin{aligned}
\left(\chi_{s+1}(\mathbf{p}, \mathbf{x})\right)^{r+1} &\leq \left(\chi_s(\mathbf{p}, \mathbf{x})\right)^r \chi_{s+r+1}(\mathbf{p}, \mathbf{x}), \\
\left(\chi_{s+r}(\mathbf{p}, \mathbf{x})\right)^{r+1} &\leq \chi_s(\mathbf{p}, \mathbf{x})\left(\chi_{s+r+1}(\mathbf{p}, \mathbf{x})\right)^r.
\end{aligned} \tag{43}$$

Since $r > 0$, multiplying those inequalities we get the relation (41), that is, the proof of Theorem 8.

The part (1) of Theorem 5 follows for $p_1 = p_2 = 1/2$.

A general way to prove the rest of Theorem 5 is to use an easy-checkable identity

$$\frac{\lambda_s(a,b)}{A(a,b)} = \lambda_s(1+t, 1-t), \tag{44}$$

with $t := (b-a)/(b+a)$.

Since $0 < a < b$, we get $0 < t < 1$. Also,

$$\frac{H(a,b)}{A(a,b)} = 1 - t^2; \qquad \frac{G(a,b)}{A(a,b)} = \sqrt{1-t^2};$$

$$\frac{L(a,b)}{A(a,b)} = \frac{2t}{\log(1+t) - \log(1-t)};$$

$$\frac{I(a,b)}{A(a,b)} = \exp\left(\frac{(1+t)\log(1+t) - (1-t)\log(1-t)}{2t} - 1\right);$$

$$\frac{S(a,b)}{A(a,b)} = \exp\left(\frac{1}{2}\left((1+t)\log(1+t) + (1-t)\log(1-t)\right)\right). \tag{45}$$

Therefore, we have to compare some one-variable inequalities and to check their validness for each $t \in (0,1)$.

For example, we will prove that the inequality

$$\lambda_s(a,b) \le L(a,b) \tag{46}$$

holds for each positive a, b if and only if $s \le 0$.

Since $\lambda_s(a,b)$ is monotone increasing in s, it is enough to prove that

$$\frac{\lambda_0(a,b)}{L(a,b)} \le 1. \tag{47}$$

By the above formulae, this is equivalent to the assertion that the inequality

$$\phi(t) \le 0 \tag{48}$$

holds for each $t \in (0,1)$, with

$$\begin{aligned}\phi(t) := &\frac{\log(1+t) - \log(1-t)}{2t} \\ &\times \left((1+t)\log(1+t) + (1-t)\log(1-t)\right) \\ &+ \log(1+t) + \log(1-t).\end{aligned} \tag{49}$$

We will prove that the power series expansion of $\phi(t)$ have non-positive coefficients. Thus the relation (48) will be proved.

Since

$$\begin{aligned}&\frac{\log(1+t) - \log(1-t)}{2t} = \sum_0^\infty \frac{t^{2k}}{2k+1}; \\ &\log(1+t) + \log(1-t) = -t^2\sum_0^\infty \frac{t^{2k}}{k+1}; \\ &(1+t)\log(1+t) + (1-t)\log(1-t) \\ &\quad = t^2\sum_0^\infty \frac{t^{2k}}{(k+1)(2k+1)},\end{aligned} \tag{50}$$

we get

$$\begin{aligned}\frac{\phi(t)}{t^2} &= \sum_{n=0}^\infty \left(-\frac{1}{n+1} + \sum_{k=0}^n \frac{1}{(2n-2k+1)(k+1)(2k+1)}\right)t^{2n} \\ &= \sum_0^\infty c_n t^{2n}.\end{aligned} \tag{51}$$

Hence,

$$c_0 = c_1 = 0; \qquad c_2 = -\frac{1}{90}, \tag{52}$$

and, after some calculation, we get

$$c_n = \frac{2}{(n+1)(2n+3)}\left((n+2)\sum_1^n \frac{1}{2k+1} - (n+1)\sum_1^n \frac{1}{2k}\right), \quad n > 1. \tag{53}$$

Now, one can easily prove (by induction, e.g.) that

$$d_n := (n+2)\sum_1^n \frac{1}{2k+1} - (n+1)\sum_1^n \frac{1}{2k} \tag{54}$$

is a negative real number for $n \ge 2$. Therefore $c_n \le 0$, and the proof of the first part is done. For $0 < s < 1$ we have

$$\begin{aligned}&\frac{\lambda_s(a,b)}{L(a,b)} - 1 \\ &= \frac{(1-s)\left((1+t)^{s+1} + (1-t)^{s+1} - 2\right)\log((1+t)/(1-t))}{2t(1+s)\left(2-(1+t)^s - (1-t)^s\right)} - 1 \\ &= \frac{1}{6}st^2 + O\left(t^4\right) \quad (t \longrightarrow 0).\end{aligned} \tag{55}$$

Therefore, $\lambda_s(a,b) > L(a,b)$ for $s > 0$ and sufficiently small $t := (b-a)/(b+a)$.

Similarly, we will prove that the inequality

$$\lambda_s(a,b) \le I(a,b) \tag{56}$$

holds for each a, b; $0 < a < b$ if and only if $s \le 1$.

As before, it is enough to consider the expression

$$\frac{I(a,b)}{\lambda_1(a,b)} = e^{\mu(t)}\nu(t) := \psi(t), \tag{57}$$

with

$$\begin{aligned}\mu(t) &= \frac{(1+t)\log(1+t) - (1-t)\log(1-t)}{2t} - 1; \\ \nu(t) &= \frac{(1+t)\log(1+t) + (1-t)\log(1-t)}{t^2}.\end{aligned} \tag{58}$$

It is not difficult to check the identity

$$\psi'(t) = -\frac{e^{\mu(t)}\phi(t)}{t^3}. \quad (59)$$

Hence by (48), we get $\psi'(t) > 0$, that is, $\psi(t)$ is monotone increasing for $t \in (0,1)$.

Therefore

$$\frac{I(a,b)}{\lambda_1(a,b)} \ge \lim_{t\to 0^+} \psi(t) = 1. \quad (60)$$

By monotonicity it follows that $\lambda_s(a,b) \le I(a,b)$ for $s \le 1$.

For $s > 1$, $(b-a)/(b+a) = t$, we have

$$\lambda_s(a,b) - I(a,b) = \left(\frac{1}{6}(s-1)t^2 + O(t^4)\right)A(a,b) \quad (t \longrightarrow 0^+). \quad (61)$$

Hence, $\lambda_s(a,b) > I(a,b)$ for $s > 1$ and t sufficiently small.

From the other hand,

$$\lim_{t\to 1^-}\left[\frac{\lambda_s(a,b)}{I(a,b)} - 1\right] = \frac{e(s-1)(2^{s+1}-2)}{2(s+1)(2^s-2)} - 1 := \tau(s). \quad (62)$$

Examining the function $\tau(s)$, we find out that it has the only real zero at $s_0 \approx 1.0376$ and is negative for $s \in (1, s_0)$.

Remark 10. Since $\psi(t)$ is monotone increasing, we also get

$$\frac{I(a,b)}{\lambda_1(a,b)} \le \lim_{t\to 1^-} \psi(t) = \frac{4\log 2}{e}. \quad (63)$$

Hence

$$1 \le \frac{I(a,b)}{\lambda_1(a,b)} \le \frac{4\log 2}{e}. \quad (64)$$

A calculation gives $4\log 2/e \approx 1.0200$.

Note also that

$$\lambda_2(a,b) \equiv A(a,b). \quad (65)$$

Therefore, applying the assertion from the part 1, we get

$$\begin{aligned} &\lambda_s(a,b) \le A(a,b), \quad s \le 2; \\ &\lambda_s(a,b) \ge A(a,b), \quad s \ge 2. \end{aligned} \quad (66)$$

Finally, we give a detailed proof of the part 7.

We have to prove that $\lambda_s(a,b) \le S(a,b)$ for $s \le 5$. Since $\lambda_s(a,b)$ is monotone increasing in s, it is sufficient to prove that the inequality

$$\lambda_5(a,b) \le S(a,b) \quad (67)$$

holds for each $a, b \in \mathbb{R}^+$.

Therefore, by the transformation given above, we get

$$\begin{aligned} \log\frac{\lambda_5}{A} &= \log\left[\frac{2}{3}\frac{(1+t)^6 + (1-t)^6 - 2}{(1+t)^5 + (1-t)^5 - 2}\right] \\ &= \log\left[\frac{2}{15}\frac{15 + 15t^2 + t^4}{2 + t^2}\right] \\ &\le \log\left[\frac{1 + t^2 + t^4/4}{1 + t^2/2}\right] = \log\left(1 + \frac{t^2}{2}\right) \\ &= \frac{t^2}{2} - \frac{t^4}{8} + \frac{t^6}{24} - \cdots \\ &\le \frac{t^2}{2} + \frac{t^4}{12} + \frac{t^6}{30} + \cdots \\ &= \frac{1}{2}((1+t)\log(1+t) + (1-t)\log(1-t)) \\ &= \log\frac{S}{A}, \end{aligned} \quad (68)$$

and the proof is done.

Further, we have to show that $\lambda_s(a,b) > S(a,b)$ for some positive a, b whenever $s > 5$.

Indeed, since

$$(1+t)^s + (1-t)^s - 2 = \binom{s}{2}t^2 + \binom{s}{4}t^4 + O(t^6), \quad (69)$$

for $s > 5$ and sufficiently small t, we get

$$\begin{aligned} \frac{\lambda_s}{A} &= \frac{s-1}{s+1}\frac{\binom{s+1}{2}t^2 + \binom{s+1}{4}t^4 + O(t^6)}{\binom{s}{2}t^2 + \binom{s}{4}t^4 + O(t^6)} \\ &= \frac{1 + (s-1)(s-2)t^2/12 + O(t^4)}{1 + (s-2)(s-3)t^2/12 + O(t^4)} \\ &= 1 + \left(\frac{s}{6} - \frac{1}{3}\right)t^2 + O(t^4). \end{aligned} \quad (70)$$

Similarly,

$$\begin{aligned} \frac{S}{A} &= \exp\left(\frac{1}{2}((1+t)\log(1+t) + (1-t)\log(1-t))\right) \\ &= \exp\left(\frac{t^2}{2} + O(t^4)\right) = 1 + \frac{t^2}{2} + O(t^4). \end{aligned} \quad (71)$$

Hence,

$$\frac{1}{A}(\lambda_s - S) = \frac{1}{6}(s-5)t^2 + O(t^4), \quad (72)$$

and this expression is positive for $s > 5$ and t sufficiently small, that is, a sufficiently close to b.

As for the part 8, applying the above transformation we obtain

$$\frac{\lambda_s(a,b)}{S(a,b)} = \frac{s-1}{s+1}\frac{(1+t)^{s+1}+(1-t)^{s+1}-2}{(1+t)^{s}+(1-t)^{s}-2} \times \exp\left(-\frac{1}{2}\left((1+t)\log(1+t)+(1-t)\log(1-t)\right)\right), \tag{73}$$

where $0 < a < b, t = (b-a)/(b+a)$.

Since for $s > 5$,

$$\lim_{t\to 1^-}\frac{\lambda_s}{S} = \frac{s-1}{s+1}\frac{2^s-1}{2^s-2}, \tag{74}$$

and the last expression is less than one, it follows that the inequality $S(a,b) < \lambda_s(a,b)$ cannot hold whenever b/a is sufficiently large.

The rest of the proof is straightforward.

Acknowledgment

The author is indebted to the referees for valuable suggestions.

References

[1] K. B. Stolarsky, "Generalizations of the logarithmic mean," *Mathematics Magazine*, vol. 48, pp. 87–92, 1975.

[2] B. C. Carlson, "The logarithmic mean," *The American Mathematical Monthly*, vol. 79, pp. 615–618, 1972.

[3] T. P. Lin, "The power mean and the logarithmic mean," *The American Mathematical Monthly*, vol. 81, pp. 879–883, 1974.

[4] P. A. Hästö, "Optimal inequalities between Seiffert's mean and power means," *Mathematical Inequalities & Applications*, vol. 7, no. 1, pp. 47–53, 2004.

[5] Z.-H. Yang, "Sharp bounds for the second Seiert mean in terms of power mean," http://arxiv.org/abs/1206.5494.

[6] G. H. Hardy, J. E. Littlewood, and G. Pölya, *Inequalities*, Cambridge University Press, Cambridge, UK, 1978.

[7] S. Simic, "On logarithmic convexity for differences of power means," *Journal of Inequalities and Applications*, vol. 2007, Article ID 37359, 8 pages, 2007.

On Certain Classes of Convex Functions

Young Jae Sim and Oh Sang Kwon

Department of Mathematics, Kyungsung University, Busan 608-736, Republic of Korea

Correspondence should be addressed to Oh Sang Kwon; oskwon@ks.ac.kr

Academic Editor: Heinrich Begehr

For real numbers α and β such that $0 \leq \alpha < 1 < \beta$, we denote by $\mathcal{K}(\alpha, \beta)$ the class of normalized analytic functions which satisfy the following two sided-inequality: $\alpha < \mathfrak{R}\{1 + (zf''(z)/f'(z))\} < \beta$ $(z \in \mathbb{U})$, where $\mathbb{U}$ denotes the open unit disk. We find some relationships involving functions in the class $\mathcal{K}(\alpha, \beta)$. And we estimate the bounds of coefficients and solve the Fekete-Szegö problem for functions in this class. Furthermore, we investigate the bounds of initial coefficients of inverse functions or biunivalent functions.

1. Introduction

Let $\mathcal{A}$ denote the class of analytic functions in the unit disc

$$\mathbb{U} = \{z : z \in \mathbb{C}, |z| < 1\} \quad (1)$$

which is normalized by

$$f(0) = 0, \qquad f'(0) = 1. \quad (2)$$

Also let $\mathcal{S}$ denote the subclass of $\mathcal{A}$ which is composed of functions which are univalent in $\mathbb{U}$. And, as usual, we denote by $\mathcal{K}$ the class of functions in $\mathcal{A}$ which are convex in $\mathbb{U}$.

We say that f is subordinate to F in $\mathbb{U}$, written as $f \prec F$ $(z \in \mathbb{U})$, if and only if $f(z) = F(w(z))$ for some Schwarz function $w(z)$ such that

$$w(0) = 0, \qquad |w(z)| < 1 \quad (z \in \mathbb{U}). \quad (3)$$

If F is univalent in $\mathbb{U}$, then the subordination $f \prec F$ is equivalent to

$$f(0) = F(0), \qquad f(\mathbb{U}) \subset F(\mathbb{U}). \quad (4)$$

Definition 1. Let α and β be real numbers such that $0 \leq \alpha < 1 < \beta$. The function $f \in \mathcal{A}$ belongs to the class $\mathcal{K}(\alpha, \beta)$ if f satisfies the following inequality:

$$\alpha < \mathfrak{R}\left\{1 + \frac{zf''(z)}{f'(z)}\right\} < \beta \quad (z \in \mathbb{U}). \quad (5)$$

It is clear that $\mathcal{K}(\alpha, \beta) \subset \mathcal{K}$. And we remark that, for given real numbers α and β $(0 \leq \alpha < 1 < \beta)$, $f \in \mathcal{K}(\alpha, \beta)$ if and only if f satisfies each of the following two subordination relationships:

$$\begin{aligned} 1 + \frac{zf''(z)}{f'(z)} &\prec \frac{1 + (1 - 2\alpha)z}{1 - z} \quad (z \in \mathbb{U}), \\ 1 + \frac{zf''(z)}{f'(z)} &\prec \frac{1 + (1 - 2\beta)z}{1 - z} \quad (z \in \mathbb{U}). \end{aligned} \quad (6)$$

Now, we define an analytic function $p : \mathbb{U} \to \mathbb{C}$ by

$$p(z) = 1 + \frac{\beta - \alpha}{\pi} i \log\left(\frac{1 - e^{2\pi i((1-\alpha)/(\beta-\alpha))}z}{1 - z}\right). \quad (7)$$

The above function p was introduced by Kuroki and Owa [1], and they proved that p maps $\mathbb{U}$ onto a convex domain

$$\Lambda = \{w : \alpha < \mathfrak{R}\{w\} < \beta\} \quad (8)$$

conformally. Using this fact and the definition of subordination, we can obtain the following lemma, directly.

Lemma 2. *Let $f(z) \in \mathcal{A}$ and $0 \leq \alpha < 1 < \beta$. Then $f \in \mathcal{K}(\alpha, \beta)$ if and only if*

$$1 + \frac{zf''(z)}{f'(z)} \prec 1 + \frac{\beta - \alpha}{\pi} i \log\left(\frac{1 - e^{2\pi i((1-\alpha)/(\beta-\alpha))} z}{1 - z}\right) \quad (z \in \mathbb{U}). \tag{9}$$

And we note that the function p, defined by (7), has the form

$$p(z) = 1 + \sum_{n=1}^{\infty} B_n z^n, \tag{10}$$

where

$$B_n = \frac{\beta - \alpha}{n\pi} i \left(1 - e^{2n\pi i((1-\alpha)/(\beta-\alpha))}\right) \quad (n \in \mathbb{N}). \tag{11}$$

For given real numbers α and β such that $0 \leq \alpha < 1 < \beta$, we denote by $\mathcal{K}_\sigma(\alpha, \beta)$ the class of biunivalent functions consisting of the functions in $\mathcal{A}$ such that

$$f \in \mathcal{K}(\alpha, \beta), \qquad f^{-1} \in \mathcal{K}(\alpha, \beta), \tag{12}$$

where f^{-1} is the inverse function of f.

In our present investigation, we first find some relationships for functions in bounded positive class $\mathcal{K}(\alpha, \beta)$. And we solve several coefficient problems including Fekete-Szegö problems for functions in the class. Furthermore, we estimate the bounds of initial coefficients of inverse functions and biunivalent functions. For the coefficient bounds of functions in special subclasses of $\mathcal{S}$, the readers may be referred to the works [2–4].

2. Relations Involving Bounds on the Real Parts

In this section, we will find some relations involving the functions in $\mathcal{K}(\alpha, \beta)$. And the following lemma will be needed in finding the relations.

Lemma 3 (see Miller and Mocanu [5, Theorem 2.3.i]). *Let Ξ be a set in the complex plane $\mathbb{C}$ and let b be a complex number such that $\Re\{b\} > 0$. Suppose that a function $\psi : \mathbb{C}^2 \times \mathbb{U} \to \mathbb{C}$ satisfies the condition*

$$\psi(i\rho, \sigma; z) \notin \Xi \tag{13}$$

for all real $\rho, \sigma \leq -|b - i\rho|^2/(2\Re\{b\})$ and all $z \in \mathbb{U}$. If the function $p(z)$ defined by $p(z) = b + b_1 z + b_2 z^2 + \cdots$ is analytic in $\mathbb{U}$ and if

$$\psi\left(p(z), zp'(z); z\right) \in \Xi, \tag{14}$$

then $\Re\{p(z)\} > 0$ in $\mathbb{U}$.

Theorem 4. *Let $f \in \mathcal{A}$, $0 \leq \alpha < 1$ and*

$$\Re\left\{1 + \frac{zf''(z)}{f'(z)}\right\} > \alpha \quad (z \in \mathbb{U}). \tag{15}$$

Then

$$\Re\left\{\sqrt{f'(z)}\right\} > \frac{1}{2 - \alpha} \quad (z \in \mathbb{U}). \tag{16}$$

Proof. First of all, we put $\gamma = 1/(2 - \alpha)$ and note that $1/2 \leq \gamma < 1$ for $0 \leq \alpha < 1$. Let

$$p(z) = \frac{1}{1 - \gamma}\left(\sqrt{f'(z)} - \gamma\right). \tag{17}$$

Differentiating (17), we can obtain

$$1 + \frac{zf''(z)}{f'(z)} = 1 + \frac{2(1-\gamma) zp'(z)}{(1-\gamma) p(z) + \gamma} = \psi\left(p(z), zp'(z)\right), \tag{18}$$

where

$$\psi(r, s) = 1 + \frac{2(1-\gamma) s}{(1-\gamma) r + \gamma}. \tag{19}$$

Using (15), we have

$$\left\{\psi\left(p(z), zp'(z)\right) : z \in \mathbb{U}\right\} \subset \{w \in \mathbb{C} : \Re(w) > \alpha\} := \Omega_\alpha. \tag{20}$$

Now for all real $\rho, \sigma \in \mathbb{R}$ with $\sigma \leq -(1 + \rho^2)/2$,

$$\begin{aligned} \Re\{\psi(i\rho, \sigma)\} &= 1 + \frac{2\gamma(1-\gamma)\sigma}{(1-\gamma)^2 \rho^2 + \gamma^2} \\ &\leq 1 - \frac{\gamma(1-\gamma)(1+\rho^2)}{(1-\gamma)^2 \rho^2 + \gamma^2}. \end{aligned} \tag{21}$$

Define a function $g : \mathbb{R} \to \mathbb{R}$ by

$$g(\rho) = \frac{1 + \rho^2}{(1-\gamma)^2 \rho^2 + \gamma^2}. \tag{22}$$

Then g is a continuous even function and

$$g'(\rho) = \frac{2(2\gamma - 1)\rho}{\left((1-\gamma)^2 \rho^2 + \gamma^2\right)^2}. \tag{23}$$

Hence $g'(0) = 0$ and g is increasing on $(0, \infty)$, since $1/2 \leq \gamma < 1$. Hence g satisfies that

$$\frac{1}{\gamma^2} \leq g(\rho) < \frac{1}{(1-\gamma)^2}, \tag{24}$$

for all $\rho \in \mathbb{R}$. Therefore, by combining (21) and (24), we can get

$$\Re\{\psi(i\rho, \sigma)\} \leq 1 - \gamma(1-\gamma) g(\rho) \leq 2 - \frac{1}{\gamma} = \alpha. \tag{25}$$

And this shows that $\Re\{\psi(i\rho, \sigma)\} \notin \Omega_\alpha$ for all $\rho, \sigma \in \mathbb{R}$ with $\sigma \leq -(1 + \rho^2)/2$. By Lemma 3, we get $\Re\{p(z)\} > 0$ for all $z \in \mathbb{U}$, and this shows that the inequality (16) holds and the proof of Theorem 4 is completed.

Theorem 5. *Let $f \in \mathcal{A}$, $1 < \beta < 2$ and*

$$\Re\left\{1+\frac{zf''(z)}{f'(z)}\right\} < \beta \quad (z \in \mathbb{U}). \tag{26}$$

Then

$$\Re\left\{\sqrt{f'(z)}\right\} < \frac{1}{2-\beta} \quad (z \in \mathbb{U}). \tag{27}$$

Proof. We put $\delta = 1/(2-\beta)$ and note that $\delta > 1$ for $1 < \beta < 2$. And let

$$p(z) = \frac{1}{1-\delta}\left(\sqrt{f'(z)} - \delta\right),$$
$$\psi(r,s) = 1 + \frac{2(1-\delta)s}{(1-\delta)r+\delta}. \tag{28}$$

As in the proof of Theorem 4, we can get

$$\left\{\psi\left(p(z), zp'(z)\right) : z \in \mathbb{U}\right\} \subset \{w \in \mathbb{C} : \Re(w) < \beta\} := \Omega_\beta, \tag{29}$$

by (26). And for all real ρ, σ with $\sigma \le -(1+\rho^2)/2$,

$$\begin{aligned}\Re\{\psi(i\rho,\sigma)\} &= 1 + \frac{2\delta(1-\delta)\sigma}{(1-\delta)^2\rho^2+\delta^2} \\ &\ge 1 - \frac{\delta(1-\delta)\left(1+\rho^2\right)}{(1-\delta)^2\rho^2+\delta^2} \\ &= 1 - \delta(1-\delta)g(\rho),\end{aligned} \tag{30}$$

where $g(\rho)$ is given by

$$g(\rho) = \frac{1+\rho^2}{(1-\delta)^2\rho^2+\delta^2}. \tag{31}$$

Since $\delta > 1$, g satisfies the inequality

$$\frac{1}{\delta^2} \le g(\rho) < \frac{1}{(1-\delta)^2}, \tag{32}$$

for all $\rho \in \mathbb{R}$. Therefore,

$$\Re\{\psi(i\rho,\sigma)\} \ge 1 - \delta(1-\delta)g(\rho) \ge 2 - \frac{1}{\delta} = \beta. \tag{33}$$

And this shows that $\Re\{\psi(i\rho,\sigma)\} \notin \Omega_\beta$ for all $\rho, \sigma \in \mathbb{R}$ with $\sigma \le -(1+\rho^2)/2$. By Lemma 3, we get $\Re\{p(z)\} > 0$ for all $z \in \mathbb{U}$, and this shows that the inequality (27) holds and the proof of Theorem 5 is completed.

By combining Theorems 4 and 5, we can obtain the following result.

Theorem 6. *Let $f \in \mathcal{A}$, $0 \le \alpha < 1 < \beta < 2$ and*

$$\alpha < \Re\left\{1+\frac{zf''(z)}{f'(z)}\right\} < \beta \quad (z \in \mathbb{U}). \tag{34}$$

Then

$$\frac{1}{2-\alpha} < \Re\left\{\sqrt{f'(z)}\right\} < \frac{1}{2-\beta} \quad (z \in \mathbb{U}). \tag{35}$$

3. Coefficient Problems Involving Functions in $\mathcal{K}(\alpha,\beta)$

In the present section, we will solve some coefficient problems involving functions in the class $\mathcal{K}(\alpha,\beta)$. And our first result on the coefficient estimates involves the function class $\mathcal{K}(\alpha,\beta)$ and the following lemma will be needed.

Lemma 7 (see Rogosinski [6, Theorem 10]). *Let $p(z) = \sum_{n=1}^{\infty} C_n z^n$ be analytic and univalent in $\mathbb{U}$ and suppose that $p(z)$ maps $\mathbb{U}$ onto a convex domain. If $q(z) = \sum_{n=1}^{\infty} A_n z^n$ is analytic in $\mathbb{U}$ and satisfies the following subordination:*

$$q(z) \prec p(z) \quad (z \in \mathbb{U}), \tag{36}$$

then

$$|A_n| \le |C_1| \quad (n = 1,2,\ldots). \tag{37}$$

Theorem 8. *Let α and β be real numbers such that $0 \le \alpha < 1 < \beta$. If the functions $f(z) = z + \sum_{n=2}^{\infty} a_n z^n \in \mathcal{K}(\alpha,\beta)$, then*

$$|a_n| \le \begin{cases} \dfrac{1}{2}|B_1|, & n = 2, \\ \dfrac{|B_1|}{n(n-1)}\displaystyle\prod_{k=1}^{n-2}\left(1+\frac{|B_1|}{k}\right), & n = 3,4,5,\ldots, \end{cases} \tag{38}$$

where $|B_1|$ is given by

$$|B_1| = \frac{2(\beta-\alpha)}{\pi}\sin\frac{\pi(1-\alpha)}{\beta-\alpha}. \tag{39}$$

Proof. Let us define

$$q(z) = 1 + \frac{zf''(z)}{f'(z)}, \tag{40}$$

$$p(z) = 1 + \frac{\beta-\alpha}{\pi}i\log\left(\frac{1-e^{2\pi i((1-\alpha)/(\beta-\alpha))}z}{1-z}\right). \tag{41}$$

Then, the subordination (9) can be written as follows:

$$q(z) \prec p(z) \quad (z \in \mathbb{U}). \tag{42}$$

Note that the function $p(z)$ defined by (41) is convex in $\mathbb{U}$ and has the form

$$p(z) = 1 + \sum_{n=1}^{\infty} B_n z^n, \tag{43}$$

where

$$B_n = \frac{\beta-\alpha}{n\pi}i\left(1-e^{2n\pi i((1-\alpha)/(\beta-\alpha))}\right). \tag{44}$$

If we let

$$q(z) = 1 + \sum_{n=1}^{\infty} A_n z^n, \tag{45}$$

then by Lemma 7, we see that the subordination (42) implies that

$$|A_n| \leq |B_1| \quad (n = 1, 2, \ldots), \tag{46}$$

where

$$|B_1| = \frac{2(\beta - \alpha)}{\pi} \sin \frac{\pi(1-\alpha)}{\beta - \alpha}. \tag{47}$$

Now, equality (40) implies that

$$f'(z) + zf''(z) = q(z) f'(z) \quad (z \in \mathbb{U}). \tag{48}$$

Then, the coefficients of z^{n-1} in both sides lead to

$$a_n = \frac{1}{n(n-1)} (A_{n-1} + 2a_2 A_{n-2} + 3a_3 A_{n-3} + \cdots + (n-1) a_{n-1} A_1). \tag{49}$$

A simple calculation combined with the inequality (46) yields that

$$\begin{aligned} |a_n| &= \frac{1}{n(n-1)} \times |A_{n-1} + 2a_2 A_{n-2} + 3a_3 A_{n-3} + \cdots + (n-1) a_{n-1} A_1| \\ &\leq \frac{1}{n(n-1)} \times (|A_{n-1}| + 2|a_2||A_{n-2}| + 3|a_3||A_{n-3}| + \cdots + (n-1)|a_{n-1}||A_1|) \\ &\leq \frac{|B_1|}{n(n-1)} \sum_{k=2}^{n} (k-1)|a_{k-1}|, \end{aligned} \tag{50}$$

where $|B_1|$ is given by (47) and $|a_1| = 1$. Hence, we have $|a_2| \leq |B_1|/2$. To prove the assertion of the theorem, we need to show that

$$\begin{aligned} |a_n| &\leq \frac{|B_1|}{n(n-1)} \sum_{k=2}^{n} (k-1)|a_{k-1}| \\ &\leq \frac{|B_1|}{n(n-1)} \prod_{k=1}^{n-2} \left(1 + \frac{|B_1|}{k}\right) \quad (n = 3, 4, 5, \ldots). \end{aligned} \tag{51}$$

We now use the mathematical induction for the proof of the theorem. For the case $n = 3$, it is clear. We assume that the inequality (51) holds for $n = m$. Then, some calculation gives us that

$$\begin{aligned} |a_{m+1}| &\leq \frac{|B_1|}{(m+1)m} \sum_{k=2}^{m+1} (k-1)|a_{k-1}| \\ &\leq \frac{|B_1|}{(m+1)m} \prod_{k=1}^{m-2} \left(1 + \frac{|B_1|}{k}\right) + \frac{|B_1|^2}{m(m+1)(m-1)} \prod_{k=1}^{m-2} \left(1 + \frac{|B_1|}{k}\right) \\ &= \frac{|B_1|}{(m+1)m} \prod_{k=1}^{m-1} \left(1 + \frac{|B_1|}{k}\right), \end{aligned} \tag{52}$$

which implies that the inequality (51) is true for $n = m + 1$. Hence, by the mathematical induction, we prove that

$$|a_n| \leq \frac{|B_1|}{n(n-1)} \prod_{k=1}^{n-2} \left(1 + \frac{|B_1|}{k}\right) \quad (n = 3, 4, 5, \ldots), \tag{53}$$

where $|B_1|$ is given by (47). This completes the proof of Theorem 8.

And now, we will solve the Fekete-Szegö problem for $f \in \mathcal{K}(\alpha, \beta)$, and we will need the following lemma.

Lemma 9 (see Keogh and Merkes [7]). *Let $p(z) = 1 + c_1 z + c_2 z^2 + \cdots$ be a function with positive real part in $\mathbb{U}$. Then, for any complex number ν,*

$$|c_2 - \nu c_1^2| \leq 2 \max \{1, |1 - 2\nu|\}. \tag{54}$$

Now, the following result holds for the coefficient of $f \in \mathcal{K}(\alpha, \beta)$.

Theorem 10. *Let $0 \leq \alpha < 1 < \beta$ and let the function $f(z)$ given by $f(z) = z + \sum_{n=2}^{\infty} a_n z^n$ be in the class $\mathcal{K}(\alpha, \beta)$. Then, for a complex number μ,*

$$\begin{aligned} &|a_3 - \mu a_2^2| \\ &\leq \frac{\beta - \alpha}{3\pi} \sin \left(\frac{1-\alpha}{\beta - \alpha} \pi\right) \cdot \max \left\{1; \left|\frac{1}{2} + \left(1 - \frac{3}{2}\mu\right) \frac{\beta - \alpha}{\pi} i + \left(\frac{1}{2} - \left(1 - \frac{3}{2}\mu\right) \frac{\beta - \alpha}{\pi} i\right) \times e^{2\pi i((1-\alpha)/(\beta-\alpha))}\right|\right\}. \end{aligned} \tag{55}$$

The result is sharp.

Proof. Let us consider a function $q(z)$ given by $q(z) = 1 + zf''(z)/f'(z)$. Then, since $f \in \mathscr{K}(\alpha,\beta)$, we have $q(z) \prec p(z)(z \in \mathbb{U})$, where

$$p(z) = 1 + \frac{\beta-\alpha}{\pi} i \log\left(\frac{1 - e^{2\pi i((1-\alpha)/(\beta-\alpha))}z}{1-z}\right) = 1 + \sum_{n=1}^{\infty} B_n z^n, \tag{56}$$

where B_n is given by (11). Let

$$h(z) = \frac{1 + p^{-1}(q(z))}{1 - p^{-1}(q(z))} = 1 + h_1 z + h_2 z^2 + \cdots \quad (z \in \mathbb{U}). \tag{57}$$

Then h is analytic and has positive real part in the open unit disk $\mathbb{U}$. We also have

$$q(z) = p\left(\frac{h(z)-1}{h(z)+1}\right) \quad (z \in \mathbb{U}). \tag{58}$$

We find from (58) that

$$a_2 = \frac{1}{4} B_1 h_1, \qquad a_3 = \frac{1}{12} B_1 h_2 - \frac{1}{24} B_1 h_1^2 + \frac{1}{24} B_2 h_1^2 + \frac{1}{24} B_1^2 h_1^2, \tag{59}$$

which imply that

$$a_3 - \mu a_2^2 = \frac{1}{12} B_1 \left(h_2 - \nu h_1^2\right), \tag{60}$$

where

$$\nu = \frac{1}{2}\left(1 - \frac{B_2}{B_1} - B_1 + \frac{3}{2}\mu B_1\right). \tag{61}$$

Applying Lemma 9, we can obtain

$$\left|a_3 - \mu a_2^2\right| = \frac{1}{12}\left|B_1\right|\left|h_2 - \nu h_1^2\right| \le \frac{1}{6}\left|B_1\right| \cdot \max\{1; |1 - 2\nu|\}. \tag{62}$$

And substituting

$$B_1 = \frac{\beta-\alpha}{\pi} i \left(1 - e^{2\pi i((1-\alpha)/(\beta-\alpha))}\right), \tag{63}$$

$$B_2 = \frac{\beta-\alpha}{2\pi} i \left(1 - e^{4\pi i((1-\alpha)/(\beta-\alpha))}\right) \tag{64}$$

in (62), we can obtain the result as asserted. The estimate is sharp for the function $f : \mathbb{U} \to \mathbb{C}$ defined by

$$f(z) = \int_0^z \left\{\exp\left(\int_0^\zeta \frac{p(\xi)-1}{\xi} d\xi\right)\right\} d\zeta, \tag{65}$$

where the function p is given by (7). Hence the proof of Theorem 10 is completed.

Using Theorem 10, we can get the following result.

Corollary 11. *Let $0 \le \alpha < 1 < \beta$ and let the function f, given by $f(z) = z + \sum_{n=2}^{\infty} a_n z^n$, be in the class $\mathscr{K}(\alpha,\beta)$. Also let the function f^{-1}, defined by*

$$f^{-1}(f(z)) = z = f\left(f^{-1}(z)\right), \tag{66}$$

be the inverse of f. If

$$f^{-1}(w) = w + \sum_{n=2}^{\infty} b_n w^n \quad \left(|w| < r_0; r_0 > \frac{1}{4}\right), \tag{67}$$

then

$$|b_2| \le \frac{\beta-\alpha}{\pi} \sin\left(\frac{1-\alpha}{\beta-\alpha}\pi\right),$$
$$|b_3| \le \frac{\beta-\alpha}{3\pi} \sin\left(\frac{1-\alpha}{\beta-\alpha}\pi\right) \cdot \max\left\{1; \left|\frac{1}{2} - \frac{2(\beta-\alpha)}{\pi} i + \left(\frac{1}{2} + \frac{2(\beta-\alpha)}{\pi} i\right) e^{2\pi i((1-\alpha)/(\beta-\alpha))}\right|\right\}. \tag{68}$$

Proof. The relations (66) and (67) give

$$b_2 = -a_2, \qquad b_3 = 2a_2^2 - a_3. \tag{69}$$

Thus, we can get the estimate for $|b_2|$ by

$$|b_2| = |a_2| \le \frac{1}{2}|B_1| = \frac{\beta-\alpha}{\pi} \sin\left(\frac{1-\alpha}{\beta-\alpha}\pi\right), \tag{70}$$

immediately. Furthermore, an application of Theorem 10 (with $\mu = 2$) gives the estimates for $|b_3|$; hence, the proof of Corollary 11 is completed.

Finally, we will estimate some initial coefficients for the bi-univalent functions $f \in \mathscr{K}_\sigma(\alpha,\beta)$.

Theorem 12. *For given α and β such that $0 \le \alpha < 1 < \beta$, let f given by $f(z) = z + \sum_{n=2}^{\infty} a_n z^n$, be in the class $\mathscr{K}_\sigma(\alpha,\beta)$. Then*

$$|a_2| \le \frac{|B_1|\sqrt{|B_1|}}{\sqrt{2\left|B_1^2 - 2B_1 + 2B_2\right|}}, \tag{71}$$

$$|a_3| \le \frac{1}{2}\left(|B_1| + |B_2 - B_1|\right), \tag{72}$$

where B_1 and B_2 are given by (63) and (64).

Proof. If $f \in \mathscr{K}_\sigma(\alpha,\beta)$, then $f \in \mathscr{K}(\alpha,\beta)$ and $g \in \mathscr{K}(\alpha,\beta)$, where $g = f^{-1}$. Hence

$$Q(z) := 1 + \frac{zf''(z)}{f'(z)} \prec p(z) \quad (z \in \mathbb{U}), \qquad L(z) := 1 + \frac{zg''(z)}{g'(z)} \prec p(z) \quad (z \in \mathbb{U}), \tag{73}$$

where $p(z)$ is given by (7). Let

$$\begin{aligned} h(z) &= \frac{1+p^{-1}(Q(z))}{1-p^{-1}(Q(z))} \\ &= 1+h_1 z+h_2 z^2+\cdots \quad (z \in \mathbb{U}), \\ k(z) &= \frac{1+p^{-1}(L(z))}{1-p^{-1}(L(z))} \\ &= 1+k_1 z+k_2 z^2+\cdots \quad (z \in \mathbb{U}). \end{aligned} \tag{74}$$

Then h and k are analytic and have positive real part in $\mathbb{U}$. Also, we have

$$\begin{aligned} Q(z) &= p\left(\frac{h(z)-1}{h(z)+1}\right) \quad (z \in \mathbb{U}), \\ L(z) &= p\left(\frac{k(z)-1}{k(z)+1}\right) \quad (z \in \mathbb{U}). \end{aligned} \tag{75}$$

By suitably comparing coefficients, we get

$$a_2 = \frac{1}{4}B_1 h_1, \tag{76}$$

$$3a_3 - 2a_2^2 = \frac{1}{4}B_1 h_2 - \frac{1}{8}B_1 h_1^2 + \frac{1}{8}B_2 h_1^2, \tag{77}$$

$$b_2 = \frac{1}{4}B_1 k_1, \tag{78}$$

$$3b_3 - 2b_2^2 = \frac{1}{4}B_1 k_2 - \frac{1}{8}B_1 k_1^2 + \frac{1}{8}B_2 k_1^2, \tag{79}$$

where B_1 and B_2 are given by (63) and (64), respectively. Now, considering (76) and (78), we get

$$h_1 = -k_1. \tag{80}$$

Also, from (77), (78), (79), and (80), we find that

$$a_2^2 = \frac{B_1^3 (h_2 + k_2)}{8(B_1^2 - 2B_1 + 2B_2)}. \tag{81}$$

Therefore, we have

$$\begin{aligned} |a_2^2| &\le \frac{|B_1|^3}{8|B_1^2 - 2B_1 + 2B_2|}(|h_2| + |k_2|) \\ &\le \frac{|B_1|^3}{2|B_1^2 - 2B_1 + 2B_2|}. \end{aligned} \tag{82}$$

This gives the bound on $|a_2|$ as asserted in (71). Now, further computations from (77), (79), (80), and (81) lead to

$$a_3 = \frac{1}{12}B_1(2h_2 + k_2) + \frac{1}{8}h_1^2(B_2 - B_1). \tag{83}$$

This equation, together with the well-known estimates

$$|h_1| \le 2, \qquad |h_2| \le 2, \qquad |k_2| \le 2, \tag{84}$$

leads us to the inequality (72). Therefore, the proof of Theorem 12 is completed.

Acknowledgment

The research was supported by Kyungsung University Research Grants in 2013.

References

[1] K. Kuroki and S. Owa, "Notes on new class for certain analytic functions," *RIMS Kokyuroku*, vol. 1772, pp. 21–25, 2011.

[2] R. M. Ali, S. K. Lee, V. Ravichandran, and S. Supramaniam, "Coefficient estimates for bi-univalent Ma-Minda starlike and convex functions," *Applied Mathematics Letters*, vol. 25, no. 3, pp. 344–351, 2012.

[3] H. M. Srivastava, A. K. Mishra, and P. Gochhayat, "Certain subclasses of analytic and bi-univalent functions," *Applied Mathematics Letters*, vol. 23, no. 10, pp. 1188–1192, 2010.

[4] Q.-H. Xu, Y.-C. Gui, and H. M. Srivastava, "Coefficient estimates for a certain subclass of analytic and bi-univalent functions," *Applied Mathematics Letters*, vol. 25, no. 6, pp. 990–994, 2012.

[5] S. S. Miller and P. T. Mocanu, *Differential Subordinations, Theory and Applications*, Marcel Dekker, New York, NY, USA, 2000.

[6] W. Rogosinski, "On the coefficients of subordinate functions," *Proceedings of the London Mathematical Society*, vol. 48, pp. 48–82, 1943.

[7] F. R. Keogh and E. P. Merkes, "A coefficient inequality for certain classes of analytic functions," *Proceedings of the American Mathematical Society*, vol. 20, pp. 8–12, 1969.

8

Using Homo-Separation of Variables for Solving Systems of Nonlinear Fractional Partial Differential Equations

Abdolamir Karbalaie, Hamed Hamid Muhammed, and Bjorn-Erik Erlandsson

Division of Informatics, Logistics and Management, School of Technology and Health (STH), Royal Institute of Technology (KTH), 100 44 Stockholm, Sweden

Correspondence should be addressed to Abdolamir Karbalaie; abdolamir.karbalaie@sth.kth.se

Academic Editor: Irena Lasiecka

A new method proposed and coined by the authors as the homo-separation of variables method is utilized to solve systems of linear and nonlinear fractional partial differential equations (FPDEs). The new method is a combination of two well-established mathematical methods, namely, the homotopy perturbation method (HPM) and the separation of variables method. When compared to existing analytical and numerical methods, the method resulting from our approach shows that it is capable of simplifying the target problem at hand and reducing the computational load that is required to solve it, considerably. The efficiency and usefulness of this new general-purpose method is verified by several examples, where different systems of linear and nonlinear FPDEs are solved.

1. Introduction

In the recent years, it has turned out that many phenomena in various technical and scientific fields can be described very successfully by using fractional calculus. In particular, fractional calculus can be employed to solve many problems within the biomedical research field and get better results. Such a practical application of fractional order models is to use these models to improve the behavior and efficiency of bioelectrodes. The importance of this application is based on the fact that bioelectrodes are usually needed to be used for all types of biopotential recording and signal measurement purposes such as electroencephalography (EEG), electrocardiography (ECG), and electromyography (EMG) [1–3]. Another promising biomedical application field is proposed by Arafa et al. [4] where a fractional-order model of HIV-1 infection of $CD4^+$ T cells is introduced. Other examples of applications of fractional calculus in life sciences and technology can be found in botanics [5], biology [6], rheology [7], and elastography [8].

Numerous analytical methods have been presented in the literature to solve FPDEs, such as the fractional Greens function method [9], the Fourier transform method [2], the Sumudu transform method [10], the Laplace transform method, and the Mellin transform method [11]. Some numerical methods have also widely been used to solve systems of FPDEs, such as the variational iteration method [12], the Adomian decomposition method [2], the homotopy perturbation method [13] and the homotopy analysis method [14]. Some of these methods use specific transformations and others give the solution as a series which converges to the exact solution.

In addition, some numerical methods use a combination of utilizing specific transformations and obtaining series which converge to the exact solutions. An example of such a method is the iterative Laplace transform method which is a combination of the Laplace transform method and an iterative method [15]. Another such a combination is the homotopy perturbation transformation method, which is constructed by combining two powerful methods, namely, the Laplace transform method and the homotopy perturbation method [16]. A third example is the Sumudu decomposition method, which is a combination of the Sumudu transform method and Adomian decomposition method [17]. A fourth such an approach is combining the Sumudu transformation method with the homotopy perturbation method, which gives a new method called the homotopy perturbation Sumudu transform method [18].

Recently, the homotopy perturbation method and the Adomian decomposition method are frequently used for solving nonlinear FPDE problems.

According to the combinational methods mentioned above, it is possible to notice that HPM has a strong potential to be combined with another method to produce a more efficient approach. The main reason is that HPM is an efficient method for solving PDEs and ODEs (ordinary differential equations) with integer or fractional order. Recently, Karbalaie et al. [19] found the exact solution of one-dimensional FPDEs by using truncated versions of modified HPM. Therefore, by getting inspiration of the ideas, methods, and tools of previous works, the novel approach called the homo-separation of variables method is developed and utilized to find the exact solutions of systems of FPDEs. The methods presented in Karbalaie et al. [19] are extended and developed further to achieve methods that can solve n-dimensional systems of FPDEs. This new approach is constructed by a smart combination of HPM and the separation of variables method. By using this method, the system of FPDEs to be solved is changed into a system of FODEs. Consequently, the original problem is simplified considerably which makes it more straightforward and easier to solve.

The structure of the current paper is as follows. For easy reader-friendly reference, some basic definitions and properties of fractional differential equations are presented in Section 2. In Section 3, the homo-separation of variables method is described. After that, in Section 4, three examples are presented to show the simplicity and efficiency (at the same time) of the homo-separation of variables method. Finally, in Section 5, relevant conclusions are drawn.

2. Basic Definitions and Properties

In this section, to make it easier to introduce the new approach of this research work, a number of good-to-know concepts and properties within the field of fractional differential equations are defined and presented briefly.

Definition 1. A real function $f(t)$, $t > 0$, is said to be in the space C_σ, $\sigma \in \mathbb{R}$, if there exists a real number $p > \sigma$, such that $f(t) = t^p f_1(t)$, where $f_1(t) \in C[0,\infty)$, and it is said to be in the space C_σ^m if $f^m \in C_\sigma$, $m \in \mathbb{N}$.

Definition 2. The left sided Riemann-Liouville fractional integral of order $\alpha \geq 0$, of a function $f \in C_\sigma$, $\sigma \geq -1$, is defined as

$$J_t^\alpha f(t) \overset{\text{def}}{=} \frac{1}{\Gamma(\alpha)} \int_0^t (t-\tau)^{\alpha-1} f(\tau)\, d\tau, \qquad (1)$$

where $\alpha > 0$, $t > 0$, and $\Gamma(\alpha) = \int_0^\infty r^{\alpha-1} e^{-r} dr$ is the gamma function.

Properties of the operator J_t^α (for $f \in C_\mu$, $\mu \geq -1$, $\alpha, \beta \geq 0$, $\gamma \geq -1$) can be found in [20] and are defined as follows:

(1) $J_t^0 f(t) = f(t)$,

(2) $J_t^\alpha J_t^\beta f(t) = J_t^{\alpha+\beta} f(t)$,

(3) $J_t^\alpha t^\gamma = (\Gamma(\gamma+1)/\Gamma(\alpha+\gamma+1)) t^{\alpha+\gamma}$.

Definition 3. The left sided Caputo fractional derivative of f, ($f \in C_\mu^n$, $n \in \mathbb{N} \cup \{0\}$), in Caputo sense is defined by [11] as follows:

$$D_t^\alpha f(t) \overset{\text{def}}{=} \begin{cases} \dfrac{1}{\Gamma(n-\alpha)} \displaystyle\int_0^t (t-\tau)^{n-\alpha-1} f^{(n)}(\tau)\, d\tau, & n-1 < \alpha \leq n, \\ D_t^n f(t), & \alpha = n. \end{cases} \qquad (2)$$

Note that, according to [21], (1) and (2) become

$$J_t^\alpha f(x,t) = \frac{1}{\Gamma(\alpha)} \int_0^t (t-\tau)^{\alpha-1} f(x,\tau)\, d\tau, \quad \alpha > 0,\ t > 0,$$
$$D_t^\alpha f(x,t) = \frac{1}{\Gamma(n-\alpha)} \int_0^t (t-\tau)^{n-\alpha-1} f^{(n)}(x,\tau)\, d\tau, \quad n-1 < \alpha \leq n. \qquad (3)$$

Definition 4. The single-parameter and the two-parameter variants of the Mittag-leffler function are denoted by $E_\alpha(t)$ and $E_{\alpha,\beta}(t)$, respectively. These two variants are relevant to be considered here because of their connection with fractional calculus and are defined as

$$E_\alpha(t) = \sum_{j=0}^{\infty} \frac{t^j}{\Gamma(\alpha j + 1)}, \quad \alpha > 0,\ t \in \mathbb{C},$$
$$E_{\alpha,\beta}(t) = \sum_{j=0}^{\infty} \frac{t^j}{\Gamma(\alpha j + \beta)}, \quad \alpha, \beta > 0,\ t \in \mathbb{C}. \qquad (4)$$

The kth derivatives of these two variants are

$$E_\alpha^{(k)}(t) = \frac{d^k}{dt^k} E_\alpha(t) = \sum_{j=0}^{\infty} \frac{(k+j)!t^j}{j!\Gamma(\alpha j + \alpha k + 1)}, \quad k = 0, 1, 2, \ldots,$$
$$E_{\alpha,\beta}^{(k)}(t) = \frac{d^k}{dt^k} E_{\alpha,\beta}(t) = \sum_{j=0}^{\infty} \frac{(k+j)!t^j}{j!\Gamma(\alpha j + \alpha k + \beta)}, \quad k = 0, 1, 2, \ldots. \qquad (5)$$

Some special cases and properties of the Mittag-Leffler functions are the following:

(1) $E_{\alpha,1}(t) = E_\alpha(t)$,

(2) $E_{1,1}(t) = E_1(t) = e^t$,

(3) $E_{2,1}(t^2) = \cosh(t)$,

(4) $E_{2,2}(t^2) = \sinh(t)/t$,

(5) $E_{\alpha,\beta}(t) + E_{\alpha,\beta}(-t) = 2E_{2\alpha,\beta}(t^2)$,

(6) $E_{\alpha,\beta}(t) - E_{\alpha,\beta}(-t) = 2tE_{2\alpha,\alpha+\beta}(t^2)$,

(7) $D_t^{\alpha}(t^{\alpha-1}E_{\alpha,\alpha}(at^{\alpha})) = t^{\alpha-1}E_{\alpha,\alpha}(at^{\alpha})$.

Further properties of the Mittag-Leffler functions can be found in [21]. These functions are generalizations of the exponential function; therefore, most linear differential equations of fractional order (FPDEs) have solutions that are expressed in terms of these functions.

Important Note. According to [22], $E_{\alpha}(at^{\alpha})$ cannot satisfy the following semigroup property:

$$E_{\alpha}\left(a(t+s)^{\alpha}\right) = E_{\alpha}\left(at^{\alpha}\right)E_{\alpha}\left(as^{\alpha}\right), \quad t,s \geq 0,\ a \in \mathbb{R} \tag{6}$$

for any noninteger $\alpha > 0$. The above property is not true unless in the special cases where $\alpha = 1$ or $= 0$. Unfortunately, this misunderstanding can be found in a number of publications, which can be misleading and confusing.

Definition 5. Consider the following relaxation-oscillation equation, which is discussed in [23]:

$$D_t^{\alpha}u(t) + Au(t) = f(t), \quad t > 0, \tag{7}$$

subject to the initial condition

$$\left.\frac{\partial^i}{\partial t^i}u(t)\right|_{t=0} = u^i(0) = b_i \quad i = 1,2,\ldots,n-1, \tag{8}$$

where b_i $(i = 1,2,\ldots,n-1)$ are real constants and $n-1 < \alpha \leq n$. The solution of (7) is obtained according to [23] as follows:

$$\begin{aligned} u(t) &= \int_0^t (t-\tau)^{\alpha-1}E_{\alpha,\alpha}\left(-A(t-\tau)^{\alpha}\right)f(\tau)\,d\tau \\ &\quad + \sum_{j=0}^{n-1} b_j D_t^{\alpha-j-1}\left(t^{\alpha-1}E_{\alpha,\alpha}\left(-At^{\alpha}\right)\right), \end{aligned} \tag{9}$$

where $E_{\alpha,\alpha}$ is the Mittag-Leffler function. It is easy to realize that if $0 < \alpha \leq 1$, then the solution of (7) becomes as follows:

$$\begin{aligned} u(t) &= \int_0^t (t-\tau)^{\alpha-1}E_{\alpha,\alpha}\left(-A(t-\tau)^{\alpha}\right)f(\tau)\,d\tau \\ &\quad + D_t^{\alpha-1}\left(t^{\alpha-1}E_{\alpha,\alpha}\left(-At^{\alpha}\right)\right), \end{aligned} \tag{10}$$

which will be used to solve the systems of FPDEs in the examples discussed in this paper.

Definition 6. According to [24], consider the following linear system of fractional differential equations with its initial condition:

$$\begin{aligned} D_t^{\alpha}\mathbf{x}(t) &= A\mathbf{x}(t), \quad 0 < t \leq a, \\ \mathbf{x}(0) &= \mathbf{x}_0, \end{aligned} \tag{11}$$

where $\mathbf{x} \in \mathbb{R}^n$, $a > 0$, the matrix $A \in \mathbb{R}^n \times \mathbb{R}^n$ and D_t^{α} is the Caputo fractional derivative of order α, where $0 < \alpha \leq 1$. The general analytical solution for (11) is given by

$$\begin{aligned} \mathbf{x}(t) &= c_1\mathbf{u}^{(1)}E_{\alpha}\left(\lambda_1 t^{\alpha}\right) + c_2\mathbf{u}^{(2)}E_{\alpha}\left(\lambda_2 t^{\alpha}\right) \\ &\quad + \cdots + c_n\mathbf{u}^{(n)}E_{\alpha}\left(\lambda_n t^{\alpha}\right), \end{aligned} \tag{12}$$

where $c_1, c_2, \ldots, c_n$ are arbitrary constants, E_{α} is the Mittag-Leffler function, and $\lambda_1, \lambda_2, \ldots, \lambda_n$ and $\mathbf{u}^{(1)}, \mathbf{u}^{(2)}, \ldots, \mathbf{u}^{(n)}$ are the eigenvalues and the corresponding eigenvectors of the characteristic equation $(A - \lambda_i\mathbf{I})\ \mathbf{u}^{(i)} = 0$, where $i = 1,2,\ldots,n$ and $\mathbf{I}$ is an $n \times n$ identity matrix.

3. The Homo-Separation of Variables Method

This approach is achieved by a novel combination of HPM and separation of variables. In this section, the algorithm of this new method is presented briefly by considering the following system of FPDEs:

$$\begin{aligned} D_t^{\alpha_i}u_i &= f_i(\overline{x},t) + \mathscr{L}_i(u_1,u_2,\ldots,u_n) \\ &\quad + \mathscr{N}_i(u_1,u_2,\ldots,u_n), \quad i = 1,2,\ldots,n, \end{aligned} \tag{13}$$

subject to the initial conditions

$$u_i(\overline{x},0) = g_i(\overline{x}), \tag{14}$$

where $\overline{x} \in \mathbb{R}^{n-1}$, and for $i = 1,2,\ldots,n$, the terms $\mathscr{L}_i$ represent linear operators, the terms $\mathscr{N}_i$ are nonlinear operators, while the terms f_i represent known analytical functions. In addition, $D_t^{\alpha_i} = \partial^{\alpha_i}/\partial t^{\alpha_i}$ is the Caputo fractional derivative of order α_i, where $0 < \alpha_i \leq 1$ for $i = 1,2,\ldots,n$.

According to the homotopy perturbation technique (HPM), we can construct the following homotopies:

$$\begin{aligned} &(p-1)\left(D_t^{\alpha_i}V_i - D_t^{\alpha_i}u_{i0}\right) + p\left(D_t^{\alpha_i}V_i - \widetilde{\mathscr{L}}_i - \widetilde{\mathscr{N}}_i - f_i(\overline{x},t)\right) \\ &\quad = 0, \quad i = 1,2,\ldots,n, \end{aligned} \tag{15}$$

where $\widetilde{\mathscr{L}}_i = \mathscr{L}_i(V_1,V_2,\ldots,V_n)$, $\widetilde{\mathscr{N}}_i = \mathscr{N}_i(V_1,V_2,\ldots,V_n)$, and the homotopy parameter, denoted by p, is considered as small $(p \in [0,1])$. In addition, for $i = 1,2,\ldots,n$, $u_{i0} = u_{i0}(\overline{x},t)$ is an initial approximation of the solution of (13) which satisfies the initial condition in (14).

We can assume that the solution of (15) can be expressed as a power series in p, as given below in (16).

$$\begin{aligned} V_1 &= \sum_{j=0}^{\infty} p^j V_{1j} = V_{10} + pV_{11} + p^2V_{12} + p^3V_{13} + \cdots, \\ V_2 &= \sum_{j=0}^{\infty} p^j V_{2j} = V_{20} + pV_{21} + p^2V_{22} + p^3V_{23} + \cdots, \\ &\vdots \\ V_n &= \sum_{j=0}^{\infty} p^j V_{nj} = V_{n0} + pV_{n1} + p^2V_{n2} + p^3V_{n3} + \cdots. \end{aligned} \tag{16}$$

By substituting (16) into (15) and equating the terms with identical powers of p, a set of equations is obtained as follows:

$$p^0 : D_t^{\alpha_i} V_{i0} - D_t^{\alpha_i} u_{i0} = 0, \quad (17)$$

$$\begin{aligned} p^1 : &D_t^{\alpha_i} V_{i1} + D_t^{\alpha_i} u_{10} \\ &- \mathscr{L}_i (V_{10}, V_{20}, \ldots, V_{n0}) - \mathscr{M}_{i0} - f_i = 0, \end{aligned} \quad (18)$$

$$\vdots$$

$$p^j : D_t^{\alpha_i} V_{ij} - \mathscr{L}_i (V_{1j-1}, V_{2j-1}, \ldots, V_{nj-1}) - \mathscr{M}_{ij-1} = 0, \quad (19)$$

where the terms $\mathscr{M}_{ij}$ $(i = 1, 2, \ldots, n$ and $j = 0, 1, \ldots)$ are obtained from (17)–(19) by equating the coefficients of the nonlinear operators $\mathscr{N}_{ij}$ $(i = 1, 2, \ldots, n$ and $j = 0, 1, \ldots)$ with the identical powers j of p, that is, the terms containing p^j.

In case the p-parameter is considered as small, the best approximate solution of (13) can be readily obtained as follows:

$$\begin{aligned} u_i (\overline{x}, t) &= \lim_{p \to 1} \sum_{j=0}^{\infty} p^j V_{ij} (\overline{x}, t) \\ &= V_{i0} (\overline{x}, t) + V_{i1} (\overline{x}, t) + V_{i2} (\overline{x}, t) + \cdots, \\ &\qquad i = 1, 2, \ldots, n, \end{aligned} \quad (20)$$

where $\overline{x} = (x_1, x_2, \ldots, x_{n-1})$.

If in (20) there exists some $V_{iN}(\overline{x}, t) = 0$, where $N \geq 1$ and $i = 1, 2, \ldots, n$, then the exact solution can be written in the following form:

$$u_i (\overline{x}, t) = V_{i0} (\overline{x}, t) + p V_{i1} (\overline{x}, t) + \cdots + p^{N-1} V_{iN-1} (\overline{x}, t), \quad i = 1, 2, \ldots, n. \quad (21)$$

For simplicity, we assume that $V_{i1}(\overline{x}, t) \equiv 0$ in (21), which means that the exact solution in (13) is simplified as follows:

$$u_i (\overline{x}, t) = V_{i0} (\overline{x}, t), \quad i = 1, 2, \ldots, n. \quad (22)$$

Now when solving (17), we obtain the following result:

$$V_{i0} (\overline{x}, t) = u_{i0} (\overline{x}, t), \quad i = 1, 2, \ldots, n. \quad (23)$$

By using (22) and (23) we have

$$u_i (\overline{x}, t) = V_{i0} (\overline{x}, t) = u_{i0} (\overline{x}, t), \quad i = 1, 2, \ldots, n. \quad (24)$$

The core of the new method proposed in this research work is to formulate the initial approximation of (13) in the form of separation of variables, as follows:

$$\begin{aligned} u_i (\overline{x}, t) &= u_{i0} (\overline{x}, t) = u_i (\overline{x}, 0) c_{i1} (t) \\ &+ \left(\sum_{j=1}^{n} D_{x_j} u_i (\overline{x}, 0) \right) c_{i2} (t) \\ &= g_i (\overline{x}) c_{i1} (t) + \left(\sum_{j=1}^{n} D_{x_j} g_i (\overline{x}) \right) c_{i2} (t), \quad i = 1, 2, \ldots, n. \end{aligned} \quad (25)$$

The task now is to find the terms $c_{i1}(t)$ and $c_{i2}(t)$ to obtain the exact solution of the system of FPDEs in (13). Since (25) satisfies the initial conditions in (14), we get

$$c_{i1} (0) = 1, \quad c_{i2} (0) = 0, \quad i = 1, 2, \ldots, n. \quad (26)$$

By substituting (25) into (18), we get

$$\begin{aligned} D_t^{\alpha_i} V_{i1} = &- D_t^{\alpha_i} \left(g_i (\overline{x}) c_{i1} (t) + \left(\sum_{j=1}^{n} D_{x_j} g_i (\overline{x}) \right) c_{i2} (t) \right) \\ &- \mathbf{L}_i - \mathbf{M}_{i0} + f_i (\overline{x}, t) \equiv 0, \end{aligned} \quad (27)$$

where $i = 1, 2, \ldots, n$ and

$$\begin{aligned} \mathbf{L}_i = \mathscr{L}_i \Bigg(& g_1 (\overline{x}) c_{11} (t) + \left(\sum_{j=1}^{n} D_{x_j} g_1 (\overline{x}) \right) c_{12} (t), \ldots, \\ & \left(g_n (\overline{x}) c_{n1} (t) + \left(\sum_{j=1}^{n} D_{x_j} g_n (\overline{x}) \right) c_{n2} (t) \right) \Bigg), \\ \mathbf{M}_{i0} = \mathscr{M}_{i0} \Bigg(& g_1 (\overline{x}) c_{11} (t) + \left(\sum_{j=1}^{n} D_{x_j} g_1 (\overline{x}) \right) c_{12} (t), \ldots, \\ & \left(g_n (\overline{x}) c_{n1} (t) + \left(\sum_{j=1}^{n} D_{x_j} g_n (\overline{x}) \right) c_{n2} (t) \right) \Bigg). \end{aligned} \quad (28)$$

Obviously, our goal is finally achieved and the system of FPDEs is changed into a system of FODEs. Consequently, the problem at hand is simplified considerably as it is well known that solving a system of FODEs is generally more straightforward and much easier than solving the corresponding system of FPDEs.

Finally, the target unknowns $c_{i1}(t)$ and $c_{i2}(t)$ can be obtained by utilizing and solving the system of FODEs and the initial conditions in (26).

4. Applications

In this section, we illustrate the applicability, simplicity, and efficiency of the homo-separation of variables method for solving systems of linear and nonlinear FPDEs.

Example 7. Consider the following system of linear FPDEs:

$$\begin{aligned} D_t^{\alpha} u &= v_x - v - u, \\ D_t^{\alpha} v &= u_x - v - u, \end{aligned} \quad (29)$$

where $0 < \alpha \leq 1$, subject to the initial conditions

$$u (x, 0) = \sinh (x), \qquad v (x, 0) = \cosh (x). \quad (30)$$

To solve (29) and (30) by using the proposed homo-separation of variables method, we choose the initial approximations for (29) as follows:

$$\begin{aligned} u_0 (x, t) &= u (x, 0) b_1 (t) + D_x u (x, 0) b_2 (t), \\ v_0 (x, t) &= v (x, 0) c_1 (t) + D_x v (x, 0) c_2 (t). \end{aligned} \quad (31)$$

By substituting the two initial conditions given by (30) into these two initial approximations, we get the following initial approximations:

$$\begin{aligned} u_0(x,t) &= b_1(t)\sinh(x) + b_2(t)\cosh(x), \\ v_0(x,t) &= c_1(t)\cosh(x) + c_2(t)\sinh(x). \end{aligned} \tag{32}$$

By considering and following the approach which resulted in (27) and substituting the results presented by (32) into the system of linear FPDEs in (29), we get

$$\begin{aligned} D_t u_1 = {} & \sinh(x)\left(D_t^\alpha b_1(t) + b_1(t) - c_1(t) + c_2(t)\right) \\ & + \cosh(x)\left(D_t^\alpha b_2(t) + b_2(t) + c_1(t) - c_2(t)\right) \equiv 0, \\ D_t v_1 = {} & \cosh(x)\left(D_t^\alpha c_1(t) + c_1(t) - b_1(t) + b_2(t)\right) \\ & + \sinh(x)\left(D_t^\alpha c_2(t) + c_2(t) + b_1(t) - b_2(t)\right) \equiv 0. \end{aligned} \tag{33}$$

By solving the latter system of linear FPDEs presented in (33), we obtain the following system of linear FODEs:

$$\begin{aligned} &D_t^\alpha b_1(t) + b_1(t) - c_1(t) + c_2(t) = 0, \\ &D_t^\alpha b_2(t) + b_2(t) + c_1(t) - c_2(t) = 0, \\ &D_t^\alpha c_1(t) + c_1(t) - b_1(t) + b_2(t) = 0, \\ &D_t^\alpha c_2(t) + c_2(t) + b_1(t) - b_2(t) = 0 \end{aligned} \tag{34}$$

together with the initial conditions

$$\begin{aligned} b_2(0) &= 1, \qquad b_2(0) = 0, \\ c_1(0) &= 1, \qquad c_1(0) = 0. \end{aligned} \tag{35}$$

By solving (34) and (35) by considering and applying Definition 6 which is discussed in Section 2, we obtain the following results:

$$\begin{aligned} b_1(t) &= \frac{1}{2}\left(E_\alpha(t^\alpha) + E_\alpha(-t^\alpha)\right), \\ b_2(t) &= \frac{1}{2}\left(-E_\alpha(t^\alpha) + E_\alpha(-t^\alpha)\right), \\ c_1(t) &= \frac{1}{2}\left(E_\alpha(t^\alpha) + E_\alpha(-t^\alpha)\right), \\ c_2(t) &= \frac{1}{2}\left(-E_\alpha(t^\alpha) + E_\alpha(-t^\alpha)\right). \end{aligned} \tag{36}$$

By substituting (36) into (32) and applying Definition 4, in particular properties (2), (5) and (6) which are discussed in Section 2, we obtain the exact solution for the system of linear FPDEs in (29), as follows:

$$\begin{aligned} u(x,t) &= E_{2\alpha,1}\left(t^{2\alpha}\right)\sinh(x) - t^\alpha E_{2\alpha,\alpha+1}\left(t^{2\alpha}\right)\cosh(x), \\ v(x,t) &= E_{2\alpha,1}\left(t^{2\alpha}\right)\cosh(x) - t^\alpha E_{2\alpha,\alpha+1}\left(t^{2\alpha}\right)\sinh(x). \end{aligned} \tag{37}$$

If we now put $\alpha \to 1$ in (37) or solve (29) and (30) for $\alpha = 1$, we obtain the following exact solution for the corresponding system of linear PDEs:

$$\begin{aligned} u(x,t) &= \sinh(x-t), \\ v(x,t) &= \cosh(x-t). \end{aligned} \tag{38}$$

Example 8. Consider the following system of inhomogeneous FPDEs:

$$\begin{aligned} D_t^\alpha u + u_x v + u &= 1, \\ D_t^\beta v - v_x u - v &= 1, \end{aligned} \tag{39}$$

where $0 < \alpha$ and $\beta \le 1$, subject to the initial conditions

$$u(x,0) = e^x, \qquad v(x,0) = e^{-x}. \tag{40}$$

To solve (39) and (40) by using the new homo-separation of variables method, we choose the initial approximations for (39) as follows:

$$\begin{aligned} u_0(x,t) &= u(x,0)c_1(t) + D_x u(x,0)c_2(t), \\ v_0(x,t) &= v(x,0)b_1(t) + D_x v(x,0)b_2(t). \end{aligned} \tag{41}$$

By substituting the two initial conditions given by (40) into these two initial approximations, we get the following initial approximations:

$$\begin{aligned} u_0(x,t) &= e^x c_1(t) + e^x c_2(t), \\ v_0(x,t) &= e^{-x} b_1(t) - e^{-x} b_2(t). \end{aligned} \tag{42}$$

By considering and following the approach which resulted in (27) and substituting the results presented by (42) into the system of inhomogeneous FPDEs that we would like to solve, which is presented by (39), we get

$$\begin{aligned} D_t^\alpha u_1 &= e^x\left(D_t^\alpha\left(c_1(t) + c_2(t)\right) + \left(c_1(t) + c_2(t)\right)\right) \\ &\quad + \left(c_1(t) + c_2(t)\right)\left(b_1(t) - b_2(t)\right) - 1 \equiv 0, \\ D_t^\beta v_1 &= e^x\left(D_t^\beta\left(b_1(t) - b_2(t)\right) - \left(b_1(t) - b_2(t)\right)\right) \\ &\quad + \left(c_1(t) + c_2(t)\right)\left(b_1(t) - b_2(t)\right) - 1 \equiv 0. \end{aligned} \tag{43}$$

By solving the latter system of FPDEs presented in (43), we obtain the following system of FODEs:

$$\begin{aligned} &D_t^\alpha\left(c_1(t) + c_2(t)\right) + \left(c_1(t) + c_2(t)\right) = 0, \\ &D_t^\beta\left(b_1(t) - b_2(t)\right) - \left(b_1(t) - b_2(t)\right) = 0, \\ &\left(c_1(t) + c_2(t)\right)\left(b_1(t) - b_2(t)\right) - 1 = 0 \end{aligned} \tag{44}$$

together with the initial conditions

$$\begin{aligned} b_1(0) &= 1, \qquad b_2(0) = 0, \\ c_1(0) &= 1, \qquad c_2(t) = 0. \end{aligned} \tag{45}$$

By solving (44) and (45) by considering and applying Definition 5 which is discussed in Section 2, we obtain the following results:

$$\begin{aligned} c_1(t)+c_2(t) &= E_\alpha(-t^\alpha),\\ b_1(t)-b_2(t) &= E_\beta(t^\beta),\\ E_\alpha(-t^\alpha)E_\beta(t^\beta) &= 1. \end{aligned} \tag{46}$$

By substituting (46) into (42), we obtain the exact solution for the system of inhomogeneous FPDEs presented in (39), as follows:

$$\begin{aligned} u(x,t) &= e^x E_\alpha(-t^\alpha),\\ v(x,t) &= e^{-x}E_\beta(t^\beta),\\ E_\alpha(-t^\alpha)E_\beta(t^\beta) &= 1. \end{aligned} \tag{47}$$

Thus, this means that $u(x,t) = 1/v(x,t)$. If we now put $\alpha,\beta \to 1$ in (47), or solve (39) and (40) for $\alpha=\beta=1$, we obtain the following exact solution for the corresponding system of inhomogeneous PDEs:

$$\begin{aligned} u(x,t) &= e^{x-t},\\ v(x,t) &= e^{-x+t}. \end{aligned} \tag{48}$$

Example 9. Consider the following system of nonlinear FPDEs:

$$\begin{aligned} D_t^\alpha u &= -u - v_x w_y - v_y w_x,\\ D_t^\beta v &= v - u_x w_y - u_y w_x,\\ D_t^\gamma w &= w - u_x v_y - u_y v_x, \end{aligned} \tag{49}$$

where $0<\alpha$ and $\beta,\gamma \le 1$, subject to the initial conditions

$$\begin{aligned} u(x,y,0) = e^{x+y}, \qquad v(x,y,0) = e^{x-y},\\ w(x,y,0) = e^{-x+y}. \end{aligned} \tag{50}$$

To solve (49) and (50) by using the new homo-separation of variables method, we choose the initial approximations for (49) as follows:

$$\begin{aligned} u_0(x,y,t) &= u(x,y,0)\,b_1(t)\\ &\quad + \left(D_x u(x,y,0) + D_y u(x,y,0)\right) b_2(t),\\ v_0(x,y,t) &= v(x,y,0)\,c_1(t)\\ &\quad + \left(D_x v(x,y,0) + D_y v(x,y,0)\right) c_2(t),\\ w_0(x,y,t) &= w(x,y,0)\,d_1(t)\\ &\quad + \left(D_x w(x,y,0) + D_y w(x,y,0)\right) d_2(t). \end{aligned} \tag{51}$$

By substituting the initial conditions given by (50) into these initial approximations, we get the following initial approximations:

$$\begin{aligned} u_0(x,y,t) &= e^{x+y}b_1(t) + 2e^{x+y}b_2(t),\\ v_0(x,y,t) &= e^{x-y}c_1(t),\\ w_0(x,y,t) &= e^{-x+y}d_1(t). \end{aligned} \tag{52}$$

By considering and following the approach which resulted in (27) and substituting the results presented by (52) into the system of nonlinear FPDEs that we would like to solve, which is presented by (49), we get

$$\begin{aligned} D_t^\alpha u_1 &= e^{-x+y}\left(D_t^\alpha b_1(t) + 2D_t^\alpha b_2(t) + b_1(t) + 2b_2(t)\right) \equiv 0,\\ D_t^\beta D_t v_1 &= e^{x-y}\left(D_t^\beta c_1(t) - c_1(t)\right) \equiv 0,\\ D_t^\gamma w_1 &= e^{-x+y}\left(D_t^\gamma d_1(t) - d_1(t)\right) \equiv 0. \end{aligned} \tag{53}$$

By solving the latter system of FPDEs presented in (53), we obtain the following system of FODEs:

$$\begin{aligned} D_t^\alpha\left(b_1(t)+2b_2(t)\right) + \left(b_1(t)+2b_2(t)\right) &= 0,\\ D_t^\beta c_1(t) - c_1(t) &= 0,\\ D_t^\gamma d_1(t) - d_1(t) &= 0, \end{aligned} \tag{54}$$

together with the initial conditions

$$\begin{aligned} b_1(0) = 1, \qquad b_2(0) = 0,\\ c_1(0) = 1, \qquad c_2(t) = 0,\\ d_1(0) = 1, \qquad d_2(t) = 0. \end{aligned} \tag{55}$$

By solving (54) and (55) by considering and applying Definition 5 which is discussed in Section 2 of this paper, we obtain the following results:

$$\begin{aligned} b_1(t) - 2b_2(t) &= E_\alpha(-t^\alpha),\\ c_1(t) &= E_\beta(t^\beta),\\ d_1(0) &= E_\gamma(t^\gamma). \end{aligned} \tag{56}$$

By substituting (56) into (52), we obtain the exact solution for the system of nonlinear FPDEs in (49), as follows:

$$\begin{aligned} u(x,y,t) &= e^{x+y}E_\alpha(-t^\alpha),\\ v(x,y,t) &= e^{x-y}E_\beta(t^\beta),\\ w(x,y,t) &= e^{-x+y}E_\gamma(t^\gamma). \end{aligned} \tag{57}$$

If we now put $\alpha,\beta,\gamma \to 1$ in (57), or solve (49) and (50) for $\alpha=\beta=\gamma=1$, we obtain the following exact solution for the corresponding system of nonlinear PDEs:

$$\begin{aligned} u(x,y,t) &= e^{x+y-t},\\ v(x,y,t) &= e^{x-y+t},\\ w(x,y,t) &= e^{-x+y+t}. \end{aligned} \tag{58}$$

5. Conclusions

In this paper, a novel approach was introduced and utilized to solve linear and nonlinear systems of FPDEs. The new technique was coined by the authors of this paper as the homo-separation of variables method. In this research work, it was demonstrated through different examples how this new method can be used for solving various systems of FPDEs.

When compared with the existing published methods, it is easy to notice that the new method has many advantages. It is straightforward, easy to understand, and fast, requiring much less computations to perform a limited number of steps of the simple procedure that can be applied to find the exact solution of a wide range of types of FPDE equations' systems. Furthermore, there is no need for using linearization or restrictive assumptions when employing this new method.

The basic principle of the current method is to perform an elegant combination (which is simple and smart) of two existing techniques. These two techniques are the popular homotopy perturbation method (HPM) that was originally proposed by He [25, 26] and another popular approach called separation of variables. Finally, the resulting homo-separation of variables method, which is analytical, can be used to solve various systems of PDEs with integer and fractional order, for example, with respect to time.

Acknowledgments

The authors would like to thank Maryam Shabani and Mehran Shafaiee for their help and support as well as the inspiration the authors got through friendly scientific discussions.

References

[1] R. L. Magin, *Fractional Calculus in Bioengineering,*, Begell House, West Redding, Conn, USA, 2006.

[2] R. L. Magin and M. Ovadia, "Modeling the cardiac tissue electrode interface using fractional calculus," *JVC/Journal of Vibration and Control*, vol. 14, no. 9-10, pp. 1431–1442, 2008.

[3] J. Keener and J. Sneyd, *Mathematical Physiology*, Springer, New York, NY, USA, 2004.

[4] A. A. M. Arafa, S. Z. Rida, and M. Khalil, "Fractional modeling dynamics of HIV and CD4+ T-cells during primary infection," *Nonlinear Biomedical Physics*, vol. 6, no. 1, article 1, 2012.

[5] I. S. Jesus, J. A. Tenreiro MacHado, and J. Boaventure Cunha, "Fractional electrical impedances in botanical elements," *Journal of Vibration and Control*, vol. 14, no. 9-10, pp. 1389–1402, 2008.

[6] K. S. Cole, "Sketches of general and comparative demography," in *Cold Spring Harbor Symposia on Quantitative Biology*, vol. 1, pp. 107–116, Cold Spring Harbor Laboratory Press, 1933.

[7] V. D. Djordjevi, J. Jari, B. Fabry, J. J. Fredberg, and D. Stamenovi, "Fractional derivatives embody essential features of cell rheological behavior," *Annals of Biomedical Engineering*, vol. 31, no. 6, pp. 692–699, 2003.

[8] C. Coussot, S. Kalyanam, R. Yapp, and M. Insana, "Fractional derivative models for ultrasonic characterization of polymer and breast tissue viscoelasticity," *IEEE Transactions on Ultrasonics, Ferroelectrics, and Frequency Control*, vol. 56, no. 4, pp. 715–725, 2009.

[9] F. Mainardi, "Fractional diffusive waves in viscoelastic solids," in *Nonlinear Waves in Solids*, pp. 93–97, Fairfield, 1995.

[10] V. G. Gupta and B. Sharma, "Application of Sumudu transform in reaction-diffusion systems and nonlinear waves," *Applied Mathematical Sciences*, vol. 4, no. 9–12, pp. 435–446, 2010.

[11] I. Podlubny, *Fractional Differential Equations: An Introduction to Fractional Derivatives, Fractional Differential Equations, to Methods of Their Solution and Some of Their Applications*, vol. 198 of *Mathematics in Science and Engineering*, Academic Press, San Diego, Calif, USA, 1999.

[12] S. Momani and Z. Odibat, "Numerical approach to differential equations of fractional order," *Journal of Computational and Applied Mathematics*, vol. 207, no. 1, pp. 96–110, 2007.

[13] O. Abdulaziz, I. Hashim, and S. Momani, "Solving systems of fractional differential equations by homotopy-perturbation method," *Physics Letters A*, vol. 372, no. 4, pp. 451–459, 2008.

[14] M. Ganjiani and H. Ganjiani, "Solution of coupled system of nonlinear differential equations using homotopy analysis method," *Nonlinear Dynamics of Nonlinear Dynamics and Chaos in Engineering Systems*, vol. 56, no. 1-2, pp. 159–167, 2009.

[15] H. Jafari, M. Nazari, D. Baleanu, and C. M. Khalique, "A new approach for solving a system of fractional partial differential equations," *Computers & Mathematics with Applications*, 2012.

[16] S. Kumar, A. Yildirim, and L. Wei, "A fractional model of the diffusion equation and its analytical solution using Laplace transform," *Scientia Iranica*, vol. 19, no. 4, pp. 1117–1123, 2012.

[17] D. Kumar, J. Singh, and S. Rathore, "Sumudu decomposition method for nonlinear equations," *International Mathematical Forum*, vol. 7, no. 9–12, pp. 515–521, 2012.

[18] J. Singh, D. Kumar, and Sushila, "Homotopy perturbation sumudu transform method for nonlinear equations," *Advances in Theoretical and Applied Mechanics*, vol. 4, no. 4, pp. 165–175, 2011.

[19] A. Karbalaie, M. M. Montazeri, and H. H. Muhammed, "New approach to find the exact solution of fractional partial differential equation," *WSEAS Transactions on Mathematics*, vol. 11, no. 10, pp. 908–917, 2012.

[20] S. G. Samko, A. A. Kilbas, and O. I. Marichev, *Fractional Integrals and Derivatives: Theory and Applications*, Gordon and Breach Science Publishers, Yverdon, Switzerland, 1993.

[21] A. A. M. Arafa, S. Z. Rida, and H. Mohamed, "Homotopy analysis method for solving biological population model," *Communications in Theoretical Physics*, vol. 56, no. 5, pp. 797–800, 2011.

[22] J. Peng and K. Li, "A note on property of the Mittag-Leffler function," *Journal of Mathematical Analysis and Applications*, vol. 370, no. 2, pp. 635–638, 2010.

[23] J.-F. Cheng and Y.-M. Chu, "Solution to the linear fractional differential equation using Adomian decomposition method," *Mathematical Problems in Engineering*, vol. 2011, Article ID 587068, 14 pages, 2011.

[24] Z. M. Odibat, "Analytic study on linear systems of fractional differential equations," *Computers & Mathematics with Applications*, vol. 59, no. 3, pp. 1171–1183, 2010.

[25] J. He, "Application of homotopy perturbation method to nonlinear wave equations," *Chaos, Solitons & Fractals*, vol. 26, no. 3, pp. 695–700, 2005.

[26] J. H. He, "Homotopy perturbation method for bifurcation of nonlinear problems," *International Journal of Nonlinear Sciences and Numerical Simulation*, vol. 6, pp. 207–208, 2005.

9

A Bordism Viewpoint of Fiberwise Intersections

Gun Sunyeekhan

50 Department of Mathematics, Faculty of Science, Kasetsart University, Bangkok 10900, Thailand

Correspondence should be addressed to Gun Sunyeekhan; fscigsy@ku.ac.th

Academic Editor: Nawab Hussain

We use the geometric data to define a bordism invariant for the fiberwise intersection theory. Under some certain conditions, this invariant is an obstruction for the theory. Moreover, we prove the converse of fiberwise Lefschetz fixed point theorem.

1. Introduction

In topological fixed point theory, there is a classical questions that people may ask; given a smooth self-map $f : M \to M$ of a smooth compact manifold M, when is f homotopic to a fixed point free map?

The famous theorem, Lefschetz fixed point theorem, gave a sufficient condition to answer the above question as follows.

Theorem 1 (Lefschetz fixed point theorem). *Let $f : M \to M$ be a smooth self-map of a compact smooth manifold M.*

If f has no fixed point, then the Lefschetz number $L(f) = 0$, where $L(f) = \sum_{k\geq 0}(-1)^k \operatorname{Tr}(f_{*|}H_k(M;\mathbb{Q}))$.

In general, the converse of the above theorem does not hold. It requires a more refined invariant than the Lefschetz number to make the converse hold (see [1–3]).

For this work, we focus on the similar arguments as above for the family of smooth maps over a compact base space B. The proof of the main theorem depends heavily on the intersection problem as follows.

From now on, the notations X^a means the smooth manifold X of dimension a and I means the unit interval $[0,1]$. If Y^b is a submanifold of X^a, $f : X \to Z$, and η is a bundle over Z, then $\nu_{Y\subseteq X}$ means the normal bundle of Y in X and $f^*(\eta)$ is a pull-back bundle of η along the map f.

Later we define "framed bordism with coefficient in a bundle" as follows. Let X be a smooth manifold with a bundle ξ over it. Define $\Omega_*^{fr}(X;\xi)$ to be the bordism groups of manifolds mapping to X, together with a stable isomorphism of the normal bundle with the pullback of ξ. This framed bordism group will be a home for our invariants $L^{\text{bord}}(f)$ (described in the last section) which detects more fixed point information than the regular Lefschetz number.

(I) Suppose that E_P^{p+k}, E_Q^{q+k}, and E_M^{m+k} are smooth fiber bundles over a compact manifold B^k. Let $f : E_P^{p+k} \to E_M^{m+k}$ be a bundle map, and let E_Q be a subbundle of E_M with the inclusion bundle map $i_Q : E_Q \hookrightarrow E_M$.

We have a commutative diagram

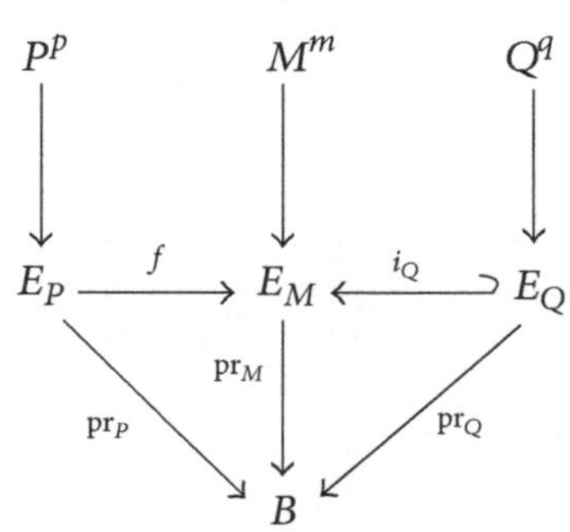

(1)

where P, Q, and M are the fibers of pr_P, pr_Q, and pr_M, respectively.

We may assume that $f \pitchfork E_Q$ in E_M (see [4]).

The homotopy pullback is

$$E(f, i_Q) := \left\{(x,\lambda,y) \in E_P \times E_M^I \times E_Q \mid \lambda(0) = f(x), \lambda(1) = y\right\}. \quad (2)$$

We have a diagram which commutes up to homotopy

$$\begin{array}{ccc} E(f,i_Q) & \xrightarrow{\pi_Q} & E_Q \\ {\scriptstyle \pi_P}\downarrow & & \downarrow{\scriptstyle i_Q} \\ E_P & \xrightarrow{f} & E_M \end{array} \qquad (3)$$

where π_P and π_Q are the trivial projections; that is, we have a homotopy $E(f,i_Q)\times I \overset{K}{\longrightarrow} E_M$ defined by $K(x,\lambda,y,t)=\lambda(t)$, $t\in I$.

We also have a map $c: f^{-1}(E_Q)\rightarrow E(f,i_Q)$ defined by $x\mapsto (x, c_{f(x)}, f(x))$ where $c_{f(x)}$ = constant path in E_M at $f(x)$.

Transversality yields a bundle map

$$\begin{array}{ccc} \nu_{f^{-1}(E_Q)\subseteq E_P} & \longrightarrow & \nu_{E_Q\subseteq E_M} \\ \downarrow & & \downarrow \\ f^{-1}(E_Q) & \xrightarrow{f_{|f^{-1}(E_Q)}} & E_Q \end{array} \qquad (4)$$

Choose an embedding $E_P^{p+k}\subseteq S^L$, for sufficiently large L, where S^L is a sphere of dimension L. Then, we have $f^{-1}(E_Q)\overset{i}{\hookrightarrow} E_P^{p+k}\subseteq S^L$. So $\nu_{f^{-1}(E_Q)\subseteq S^L}\cong \nu_{f^{-1}(E_Q)\subseteq E_P}\oplus i^*\nu_{E_P\subseteq S^L}$.

The commutative diagram

$$\begin{array}{ccc} f^{-1}(E_Q) & \xrightarrow{f_{|f^{-1}(E_Q)}} & E_Q \\ & {\scriptstyle c}\searrow \quad\quad {\scriptstyle i}\searrow & \\ & E(f,i_Q) \xrightarrow{\pi_Q} E_Q & \\ & {\scriptstyle \pi_P}\downarrow & \\ & E_P & \end{array} \qquad (5)$$

yields a bundle map

$$\begin{array}{ccc} \nu_{f^{-1}(E_Q)\subseteq S^L} & \xrightarrow{\widehat{c}} & \pi^*(\nu_{E_Q\subseteq E_M})\oplus\pi^*(\nu_{E_P\subseteq S^L}) := \zeta \\ \downarrow & & \downarrow \\ f^{-1}(E_Q) & \xrightarrow{c} & E(f,i_Q) \end{array} \qquad (6)$$

Thus, $(c,\widehat{c})$ determines an element $[c,\widehat{c}]\in\Omega^{fr}_{p+q+k-m}(E(f,i_Q);\xi)$.

(II) Suppose that $(N\overset{c_1}{\longrightarrow}E(f,i_Q), \nu_{N\subseteq S^L}\overset{\widehat{c}_1}{\longrightarrow}\xi)$ is another representative of $[c,\widehat{c}]$, where $N^{p+q+k-m}\subseteq E_P^{p+k}\subseteq S^L$. This means we have a normal bordism $(W\overset{\mathscr{C}}{\longrightarrow}E(f,i_Q), \nu_W\overset{\widehat{\mathscr{C}}}{\longrightarrow}\xi)$ between $(c,\widehat{c})$ and $(c_1,\widehat{c}_1)$; that is

(i) $W^{p+q+k-m+1}\subseteq(S^L\times I)$,

(ii) $\partial W\subseteq(S^L\times\partial I)$,

(iii) $W \pitchfork (S^L\times\partial I)$,

(iv) $W\cap(S^L\times 0)=f^{-1}(E_Q)$ and $W\cap(S^L\times 1)=N$

such that

$$\begin{aligned} \mathscr{C}_{|f^{-1}(E_Q)} &= c: f^{-1}(E_Q)\longrightarrow E(f,i_Q),\\ \mathscr{C}_{|N} &= c_1: N\longrightarrow E(f,i_Q),\\ \widehat{\mathscr{C}}_{|\nu_{f^{-1}(E_Q)\subseteq S^L}} &= \widehat{c}, \qquad \widehat{\mathscr{C}}_{|\nu_{N\subseteq S^L}}=\widehat{c}_1. \end{aligned} \qquad (7)$$

Theorem 2 (main theorem). *Assume $m>q+((p+k)/2)+1$. Then, there exists a smooth fiber-preserving map over B*

$$\Psi: E_P\times I\longrightarrow E_M \qquad (8)$$

such that $\Psi_{|E_P\times\{0\}}=f$, $\Psi\pitchfork E_Q$ and $\Psi^{-1}_{|E_P\times\{1\}}(E_Q)=N$.

Note that if we let $g=\Psi_{|E_P\times\{1\}}$, then g is fiber-preserving homotopic to f and $g^{-1}(E_Q)=N$.

In 1974, Hatcher and Quinn [5] showed an interesting result as follows.

Theorem 3 (Hatcher-Quinn). *Given smooth bundle maps $f: E_P\rightarrow E_M$ and $i_Q: E_Q\rightarrow E_M$ over a compact base space B which are immersions in each fiber, assume $m>q+((p+k)/2)+1$ and $m>p+((q+k)/2)+1$. Then there is a fiber-preserving map $g: E_P\rightarrow E_M$ over B homotopic to f such that $g^{-1}(E_Q)=N$.*

Their theorem required a map f to be an immersion on each fiber which cannot be applied to our main theorem. We decided to give an alternative technique to proof the main theorem by constructing the required homotopy step by step. This techniques could be used as a model to achieve the similar result for the equivariant setting (see [6]).

2. Proof of the Main Theorem

Theorem 4 (Whitney embedding theorem). *Let M^m and N^n be smooth manifolds, and let $f: M\rightarrow N$ be a smooth map. If $n>2m$, then f is homotopic to an embedding $g: M\rightarrow N$.*

Lemma 5. *Let $f: M^m\rightarrow N^n$ be a map between two smooth manifolds. Let A be a closed submanifold of M. Assume that $f_{|A}$ is an embedding.*

If $n>2m$, then f is homotopic to an embedding g relative to A.

Proof. Let T be a tubular neighborhood of A in M.

Step 1. Extend the embedding $f_{|A}: A\rightarrow N$ to an embedding $f_T: T\rightarrow N$.

Let $\nu_{A\subseteq M}$ be the normal bundle of A in M and $D(\nu)$ denote the disc bundle of ν. Then the tubular neighborhood theorem implies $D(\nu_{A\subseteq M})\cong T$.

Claim. For any given embedding $A \overset{g}{\hookrightarrow} N$ and vector bundle η over A, then

$$(g \text{ extends to an embedding of } D(\eta) \text{ into } N) \Longleftrightarrow (\text{there exists a bundle monomorphism } \phi : \eta \longrightarrow \nu(g)), \quad (9)$$

where $\nu(g)$ is the normal bundle of A in N via g.

Proof of Claim. ($\Leftarrow$) We have a diagram

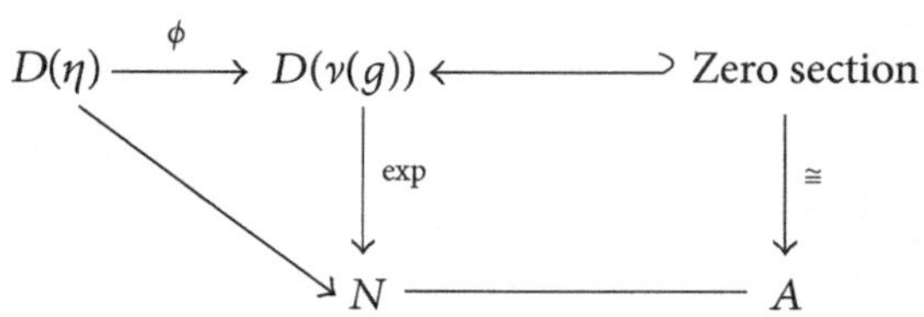

(10)

where exp is the exponential map.

Note that $\exp(D(\nu(g)) \cong$ tubular neighborhood of A in N via g. Then $\exp\circ\phi : D(\eta) \hookrightarrow N$ is a desired embedding.

($\Rightarrow$) Assume there exists an embedding g_T so that the following diagram commutes

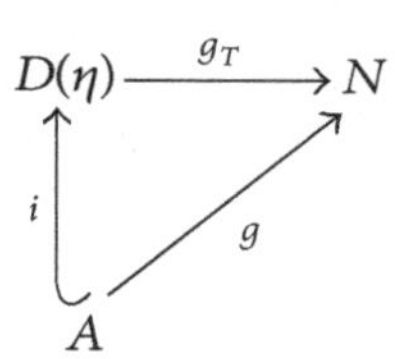

(11)

Then $\nu(g) \cong \eta \oplus i^*\nu(g_T)$.

We are in the situation where we have a commutative diagram

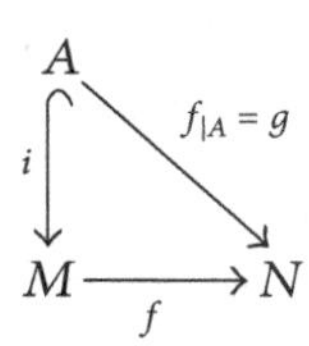

(12)

Let $\nu(f) := f^*\tau_N - \tau_M$. Then $i^*(\nu(f)) \oplus \nu_{A\subseteq M} \overset{\text{stable}}{\cong} \nu(g)$, where $\overset{\text{stable}}{\cong}$ denotes the stable isomorphism between 2 vector bundles.

If $n-a > a$, then $i^*(\nu(f)) \oplus \nu_{A\subseteq M} \cong \nu(g)$, so there exists a bundle monomorphism

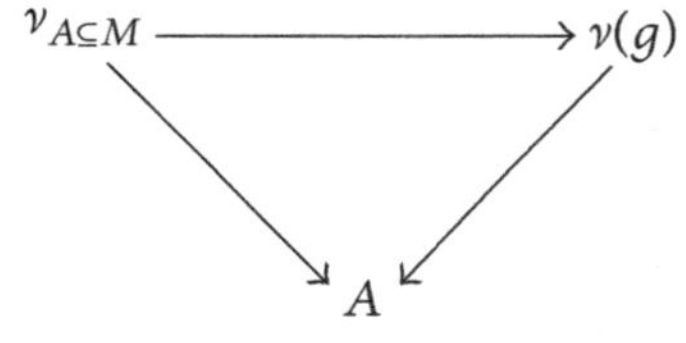

(13)

Apply the Claim when $g = f_{|A}$ and $\eta = \nu_{A\subseteq M}$. Then we have an extension embedding of g from $D(\nu_{A\subseteq M}) \cong T \overset{f_T}{\hookrightarrow} N$.

Step 2. We have a map $f_{|M-\text{int}\,T} : M \setminus \text{int}(T) \to N$ and $\partial(M \setminus \text{int}(T)) = \partial T$.

The condition $b > (c + a/2 + 1)$ and Theorem 4 imply that $f_{|M-\text{int}(T)}$ is homotopic to an embedding $g_{M-\text{int}(T)}$.

Define $g : M \to N$ by

$$g(x) = \begin{cases} g_{M-\text{int}(T)}(x) & \text{if } x \in M \setminus \text{int}(T), \\ g_T(x) & \text{if } x \in T. \end{cases} \quad (14)$$

Then f is homotopic to g relative to A.

2.1. Proof of the Main Theorem. We divide the proof into 3 steps.

Step 1. Goal: fiber-preserving homotoped the map $W \xrightarrow{a:=\pi_P\circ\mathscr{C}} E_P$ to an embedding over B.

By assumption, we have

$$W \xrightarrow{\mathscr{C}} E(f, i_Q) \subseteq E_P \times E_M^I \times E_Q, \quad (15)$$

and we also have maps

$$W \xrightarrow{a:=\pi_P\circ\mathscr{C}} E_P, \qquad W \xrightarrow{b:=\pi_Q\circ\mathscr{C}} E_Q, \qquad W \times I \xrightarrow{H} E_M, \quad (16)$$

where $H := K \circ (\mathscr{C} \times \text{id}_I)$, so $H_{|W\times 0} = f \circ a$, $H_{|W\times 1} = b$.

Recall that $\partial W = f^{-1}(E_Q) \sqcup N$, $a_{|\partial W}$ is just the inclusion of $f^{-1}(E_Q)$ and N into E_P.

Apply the condition $m > q + ((p+k)/2) + 1$ to Lemma 5, there exists an embedding $\mathscr{A} \simeq a$ (rel ∂W). That is, we have a commutative diagram

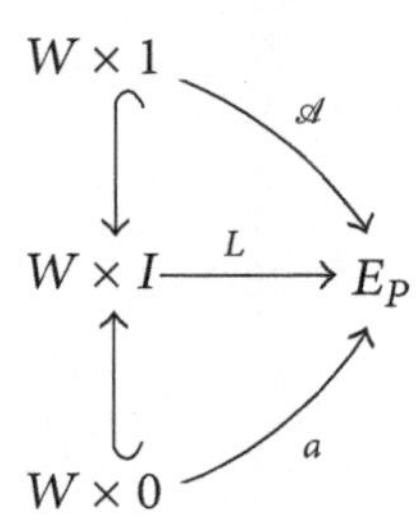

(17)

We have a map $W \xrightarrow{b:=\pi_Q\circ\mathscr{C}} E_Q$. By concatenating the homotopy H and $f \circ L$ together, we have a homotopy $V : W \times [1,2] \to E_M$ such that $V(\_, 1) = \mathscr{A}$ and $V(\_, 2) = b$. Hence, $i_Q \circ b$ is homotopic to $f \circ \mathscr{A}$.

Next, we want to modify the homotopy V such that it is fiber preserving with respect to pr_M.

Note that we have a commutative diagram

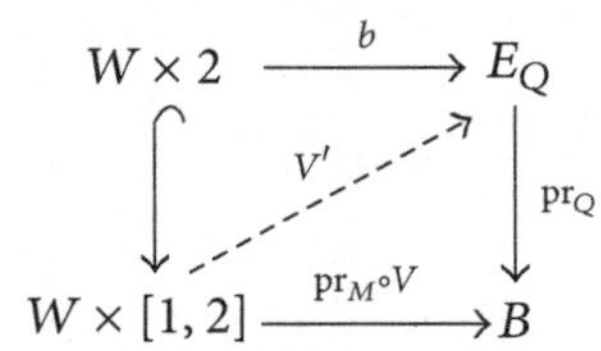

(18)

We can apply the homotopy lifting property for pr_Q to get a homotopy of b to b' through V' such that the following diagram commute:

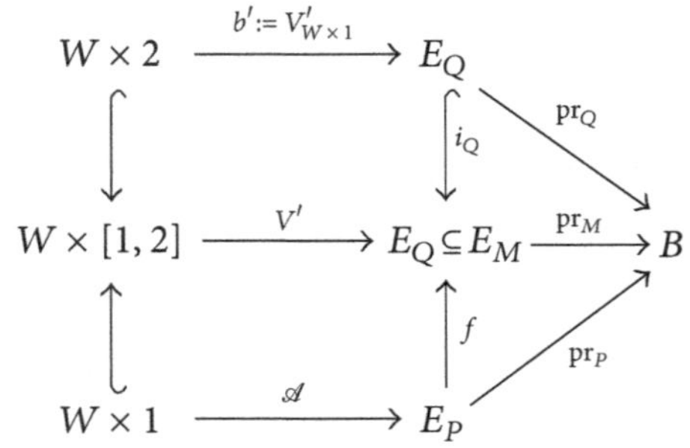

(19)

Let $\Psi_W := b' : W \to E_Q$. Then Ψ_W is a fiber-preserving map over B through the lifting V'.

Step 2. Goal: construct a bundle isomorphism

$$\nu(\mathscr{A}) \oplus \epsilon^1 \cong b'^*\left(\nu_{E_Q \subseteq E_M}\right), \quad (20)$$

where ϵ^1 is the trivial bundle.

Since $\dim W < \operatorname{rank} \nu(\mathscr{A})$, it is enough to give a stable equivalence between such bundles.

Now, we have

$$W \overset{\mathscr{A}}{\hookrightarrow} E_P \subseteq S^L \Longrightarrow \nu_{W \subseteq S^L} \cong \nu(\mathscr{A}) \oplus \mathscr{A}^*\left(\nu_{E_P \subseteq S^L}\right). \quad (21)$$

We also have a commutative diagram

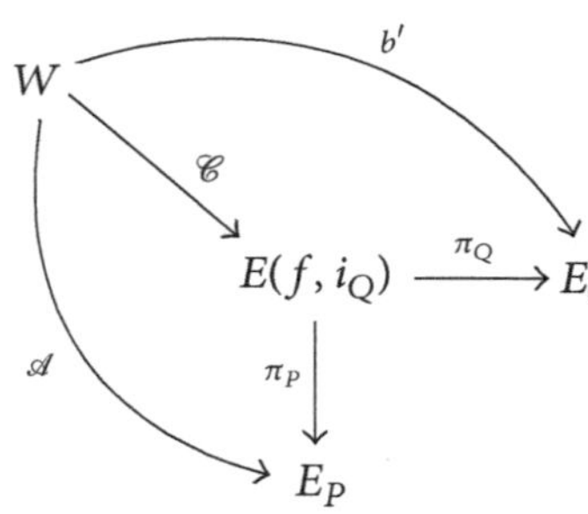

(22)

$$\mathscr{A} \simeq a = \pi_P \circ \mathscr{C} \Longrightarrow \mathscr{A}^*\left(\nu_{E_P \subseteq S^L}\right) \cong (\pi_P \circ \mathscr{C})^*\left(\nu_{E_P \subseteq S^L}\right),$$
$$b' \simeq b = \pi_Q \circ \mathscr{C} \Longrightarrow b'^*\left(\nu_{E_Q \subseteq E_M}\right) \cong (\pi_Q \circ \mathscr{C})^*\left(\nu_{E_Q \subseteq E_M}\right). \quad (23)$$

Thus, the bundle map $\widehat{\mathscr{C}} : \nu_{W \subseteq S^L \times I} \to \xi = \pi_P^*(\nu_{E_P \subseteq S^L}) \oplus \pi_Q^*(\nu_{E_Q \subseteq E_M})$ yields the following stable isomorphism

$$\nu_{W \subseteq S^L} \oplus \epsilon^1 \overset{\text{stable}}{\cong} (\pi_P \circ \mathscr{C})^*\left(\nu_{E_P \subseteq S^L}\right) \oplus (\pi_Q \circ \mathscr{C})^*\left(\nu_{E_Q \subseteq E_M}\right). \quad (24)$$

Putting (21), (23), and (24) together, we get

$$\nu(\mathscr{A}) \oplus (\pi_P \circ \mathscr{C})^*\left(\nu_{E_P \subseteq S^L}\right) \oplus \epsilon^1 \overset{\text{stable}}{\cong} (\pi_P \circ \mathscr{C})^*\left(\nu_{E_P \subseteq S^L}\right) \oplus b'^*\left(\nu_{E_Q \subseteq E_M}\right). \quad (25)$$

Consequently, we have

$$\nu(\mathscr{A}) \oplus \epsilon^1 \cong b'^*\left(\nu_{E_Q \subseteq E_M}\right). \quad (26)$$

This implies that we did construct a bundle map

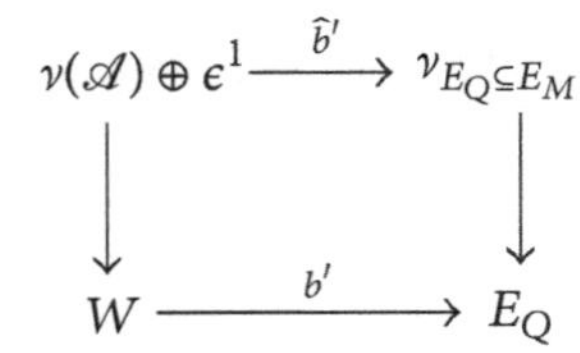

(27)

which gives us the extension of map b' to the tubular neighborhood of W in E_P. More precisely,

$$\Psi_T : D(\nu(\mathscr{A})) \hookrightarrow D\left(\nu(\mathscr{A}) \oplus \epsilon^1\right) \overset{\widehat{b'}}{\longrightarrow} D\left(\nu_{E_Q \subseteq E_M}\right), \quad (28)$$

where D denotes the disc bundle. Note that $\Psi_T \pitchfork E_Q$ and $\Psi_T(\partial D(\eta_1)) \subseteq E_M \setminus E_Q$.

Since $\nu(\mathscr{A}) \oplus \epsilon^1 \cong b'^*(\nu_{E_Q \subseteq E_M})$, we can find a subbundle η_2 of $b'^*(\nu_{E_Q \subseteq E_M})$ such that $\eta_2 \cong \nu(\mathscr{A})$. For simplicity, let $\eta_1 := \nu(\mathscr{A})$.

Step 3. Goal: construct the fiber-preserving smooth map $\Psi : E_P \times I \to E_M$ over B.

Recall that we have

$$W \hookrightarrow D(\nu(\mathscr{A})) \simeq D\left(\nu(\mathscr{A}) \oplus \epsilon^1\right) \cong D\left(b'^*\left(\nu_{E_Q \subseteq E_M}\right)\right). \quad (29)$$

Then there exists a neighborhood $\overline{D}$ of W in $D(b'^*(\nu_{E_Q \subseteq E_M}))$ such that $\overline{D} \simeq D(b'^*(\nu_{E_Q \subseteq E_M}))$ and $\overline{D} \cong D(\nu(\mathscr{A}))$.

According to (19), we have a commutative diagram

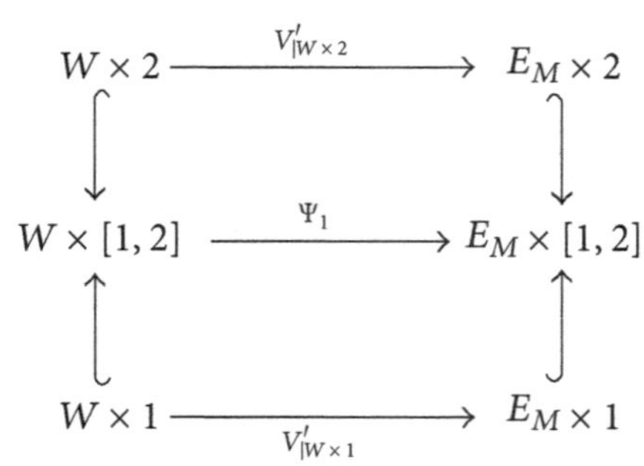

(30)

where $\Psi_1(w,t) = (V'(w,t), t)$.

Let $\eta_1 = \nu(\mathscr{A} \oplus \epsilon^1)$, $\eta_2 = b^*(\nu_{E_Q \subseteq E_M})$.

According to (27), there exists a bundle η over $W \times I$ such that $\eta_{|W\times i} = \eta_i$ for $i = 1, 2$.

Let $D_1 := D(\nu(\mathscr{A}))$, $D_2 := \overline{D}$. Then $D_i \hookrightarrow D(\eta_i)$ is a homotopy equivalence for $i = 1, 2$ and also $D_1 \cong D_2$.

Since $D_1 \cup W \times [1,2] \cup D_2 \hookrightarrow D_1 \times [1,2]$ is a cofibration and a homotopy equivalence, there exist an extension

$D_1 \times [1,2] \xrightarrow{\widehat{\Psi}_1} E_M \times [1,2]$ such that the following diagram commutes

$$\begin{array}{ccccc} D_2 & \longrightarrow & D(\nu_{E_Q \subseteq E_M}) & \xhookrightarrow{\exp} & E_M \times 2 \\ \downarrow & & & & \downarrow \\ D_1 \times [1,2] & & \xrightarrow{\widehat{\Psi}_1} & & E_M \times [1,2] \\ \uparrow & & & & \uparrow \\ D_1 & & \xrightarrow{f_{|D_1}} & & E_M \cong E_M \times 1 \end{array} \quad (31)$$

Next we want to construct an embedding $W \xhookrightarrow{\mathscr{A}'} D_2 \times [2,3]$ such that the following hold:

(i) $\mathscr{A}'(W) \cap \{D_2 \times 2\} = f^{-1}(E_Q)$,

(ii) $\mathscr{A}'(W) \cap \{D_2 \times 3\} = N$,

(iii) $\mathscr{A}' \pitchfork D_2 \times \partial[2,3]$.

We start by letting $\alpha : W \to [2,3]$ be a smooth map such that $\alpha \pitchfork \partial[2,3]$, $\alpha^{-1}(2) = f^{-1}(E_Q)$ and $\alpha^{-1}(3) = N$, and we also have an inclusion $W \xhookrightarrow{i_W} D_2$.

Let $\mathscr{A}' := i_W \times \alpha$. Then $\mathscr{A}'$ is such a required map.

By the construction, we have

$$D\left(\nu\left(\mathscr{A}'\right)\right) \cong D_2 \times [2,3]. \quad (32)$$

Let ψ_2 be the composition of the maps

$$W \xhookrightarrow{\mathscr{A}'} D_2 \times [2,3] \xrightarrow{\Psi_{T|D_2} \times \mathrm{id}_{[2,3]}} M \times [2,3] \xrightarrow{\text{proj}} M. \quad (33)$$

Define a map $\Psi_2 := \psi_2 \times \alpha : W \to E_M \times [2,3]$.

Using the fact that $D(\nu(\mathscr{A}')) \cong D_2 \times [2,3]$, then $D_2 \times \partial[2,3] \cup \mathscr{A}'(W) \hookrightarrow D_2 \times [2,3]$ is a cofibration and homotopy equivalence. Hence there exists an extension $D_2 \times [2,3] \xrightarrow{\widehat{\Psi}_2} E_M \times [2,3]$.

Note that for $(x,3) \in D_2 \times 3$ such that $\widehat{\Psi}_2(x,3) \in E_Q \times 3$, the map $\widehat{\Psi}_{2|D_2 \times 3} = \Psi_T$ forces that x has to be in W, so the definition of $\widehat{\Psi}_2$ implies $x \in N$. Thus

$$\widehat{\Psi}^{-1}_{2|D_2 \times 3}(E_Q) = N. \quad (34)$$

We define a map $\widetilde{\Psi} : \{E_P \times I\} \cup \{D_1 \times [1,2]\} \cup \{D_2 \times [2,3]\} \to E_M \times [0,3]$ by

$$\widetilde{\Psi}(p,t) = \begin{cases} (f(p), t) & \text{if } t \in [0,1], \\ (\widehat{\Psi}_1(p), t) & \text{if } t \in [1,2], \\ (\widehat{\Psi}_2(p), t) & \text{if } t \in [2,3]. \end{cases} \quad (35)$$

Then $\widetilde{\Psi}$ is well-defined map over B by the construction.

It's not hard to see that $\{E_P \times I\} \cup \{D_1 \times [1,2]\} \cup \{D_2 \times [2,3]\}$ is diffeomorphic to $E_P \times I$. Define the map Ψ to be the composition of maps

$$\begin{aligned} E_P \times I &\xrightarrow{\cong} \{E_P \times I\} \cup \{D_1 \times [1,2]\} \cup \{D_2 \times [2,3]\} \\ &\xrightarrow{\widetilde{\Psi}} E_M \times [0,3] \xrightarrow{\text{proj}} E_M, \end{aligned} \quad (36)$$

where proj is the projection to the first factor.

Thus, we get a map $\Psi : E_P \times I \to E_M$ over B so that $\Psi_{|E_P \times 0} = f$. By construction, $\Psi \pitchfork E_Q$ and $\Psi^{-1}_{|E_P \times 1}(E_Q) = N$ as required.

Corollary 6. *Assume $m > q + ((p+k)/2) + 1$. Then the map f can be fiber-preserving homotoped to a map whose image does not intersect E_Q if and only if $[c, \widetilde{c}] = 0 \in \Omega^{fr}_{p+q+k-m}(E(f, i_Q); \xi)$.*

3. Application to Fixed Point Theory

Let $p : M^{m+k} \to B^k$ be a smooth fiber bundle with compact fibers and $k > 2$. Assume that B is a closed manifold. Let $f : M \to M$ be a smooth map over B. That is, $p \circ f = p$.

The fixed point set of f is

$$\mathrm{Fix}(f) := \{x \in M \mid f(x) = x\}. \quad (37)$$

We have a homotopy pull-back diagram

$$\begin{array}{ccc} \mathscr{L}_f M & \xrightarrow{ev_0} & M \\ {\scriptstyle ev_1}\downarrow & & \downarrow{\scriptstyle \Delta} \\ M & \xrightarrow[\Delta_f]{} & M \times_B M \end{array} \quad (38)$$

where

(i) $\mathscr{L}_f M := \{\alpha \in M^I \mid f(\alpha(0)) = \alpha(1)\}$,

(ii) ev_0 and ev_1 are the evaluation map at 0 and 1 respectively,

(iii) $\Delta :=$ the diagonal map, defined by $x \mapsto (x,x)$,

(iv) $\Delta_f :=$ the twisted diagonal map, defined by $x \mapsto (x, f(x))$,

(v) $M \times_B M :=$ the fiber bundle over B with fiber over $b \in B$, given by $F_b \times F_b$ where F_b is the fiber of p over b.

Proposition 7. *There exists a homotopy from f to f_1 such that $\Delta_{f_1} \pitchfork \Delta$.*

The proof relies on the work of Koźniewski [4], relating to B-manifolds. Let B be a smooth manifold. A B-manifold is a manifold X together with a locally trivial submersion $p : X \to B$. A B-map is a smooth fiber-preserving map.

Lemma 8. *Let X and Y be B-manifolds, and let Z be a B-submanifold of Y. Let $g : X \to Y$ be a B-map. Then, there* *is a fiber-preserving smooth B-homotopy $H_t : X \to Y$ such that $H_0 = g$ and $H_1 \pitchfork Z$.*

Proof. See [7] for the proof.

We have a transversal (pull-back) square

$$\begin{array}{ccc} \mathrm{Fix}(f_1) & \xrightarrow{\ i\ } & M \\ {\scriptstyle i}\downarrow & & \downarrow{\scriptstyle \Delta} \\ M & \xrightarrow[\Delta_{f_1}]{} & M \times M \end{array} \tag{39}$$

where i is the inclusion. Transversality yields that $\nu(i) \cong i^*(\nu(\Delta)) \cong i^*(\tau M)$.

Choose an embedding $M \hookrightarrow S^{m+k}$ for sufficiently large k. Then we have

$$\begin{aligned} \nu_{\mathrm{Fix}(f_1)\subseteq S^{m+k}} &\cong \nu(i) \oplus i^*\left(\nu_{M\subseteq S^{m+k}}\right) \\ &\cong i^*(\tau M) \oplus i^*\left(\nu_{M\subseteq S^{p+k}}\right) \cong \epsilon. \end{aligned} \tag{40}$$

We denote this bundle isomorphism by $\nu_{\mathrm{Fix}(f_1)\subseteq S^{m+k}} \xrightarrow{\widehat{g}} \epsilon$. We also have a map $\mathrm{Fix}(f_1) \xrightarrow{g} \mathscr{L}_f M$ defined by $x \mapsto c_x$, where c_x is the constant map at x. Thus $L^{\mathrm{bord}}(f) := [\mathrm{Fix}(f_1), g, \widehat{g}]$ determines the element in $\Omega_k^{fr}(\mathscr{L}_f M; \epsilon)$.

Applying Theorem 2, We obtain the following corollary.

Corollary 9 (converse of fiberwise Lefschetz fixed point theorem). *Let $f : M^{m+k} \to M^{m+k}$ be a smooth bundle map over the closed manifold B^k. Assume that $m \geq k + 3$. Then, f is fiber homotopic to a fixed point free map if and only if $L^{bord}(f) = 0 \in \Omega_k^{fr}(\mathscr{L}_f M; \epsilon)$.*

Corollary 10. *Let $f : M^{m+k} \to M^{m+k}$ be a smooth bundle map over the closed manifold B^k. Assume that $m \geq k+3$. If there is $[N, N \xrightarrow{g} \mathscr{L}_f M, \nu_{N\subseteq S^{m+k}} \xrightarrow{\widehat{g}} \epsilon] \in \Omega_k^{fr}(\mathscr{L}_f M; \epsilon) f$ where N is a finite subset of M such that $L^{bord}(f) = [N, g, \widehat{g}]$, then the map f can be fiber-preserving homotoped to a map h such that* $\mathrm{Fix}(h) = N$.

Acknowledgments

The author gratefully acknowledge the Coordinating Center for Thai Government Science and Technology Scholarship Students (CSTS) and the National Science and Technology Development Agency (NSTDA) for funding this research. The author would also like to thank the referees, P.M. Akhmet'ev Akhmet'ev and Hichem Ben-El-Mechaiekh, for all the comments and suggestions.

References

[1] F. Wecken, "Fixpunktklassen. I," *Mathematische Annalen*, vol. 117, pp. 659–671, 1941 (German).

[2] J. Nielsen, "Über die Minimalzahl der Fixpunkte bei den Abbildungstypen der Ringflächen," *Mathematische Annalen*, vol. 82, no. 1-2, pp. 83–93, 1920 (German).

[3] B. J. Jiang, "Fixed point classes from a differential viewpoint," in *Fixed Point Theory (Sherbrooke, Que., 1980)*, vol. 886 of *Lecture Notes in Mathematics*, pp. 163–170, Springer, Berlin, Germany, 1981.

[4] T. Koźniewski, "The category of submersions," *L'Académie Polonaise des Sciences. Bulletin. Série des Sciences Mathématiques*, vol. 27, no. 3-4, pp. 321–326, 1979.

[5] A. Hatcher and F. Quinn, "Bordism invariants of intersections of submanifolds," *Transactions of the American Mathematical Society*, vol. 200, pp. 327–344, 1974.

[6] G. Sunyeekhan, *Equivariant intersection theory [Ph.D. thesis]*, University of Notre Dame, Notre Dame, Indiana, 2010.

[7] V. Coufal, *A family version of Lefschetz-Nielsen fixed point theory [Ph.D. thesis]*, University of Notre Dame, Notre Dame, Indiana, 2004.

10

Geometric Methods to Investigate Prolongation Structures for Differential Systems with Applications to Integrable Systems

Paul Bracken

Department of Mathematics, University of Texas, Edinburg, TX 78541-2999, USA

Correspondence should be addressed to Paul Bracken; bracken@utpa.edu

Academic Editor: Aloys Krieg

A type of prolongation structure for several general systems is discussed. They are based on a set of one forms for which the underlying structure group of the integrability condition corresponds to the Lie algebra of $SL(2,\mathbb{R})$, $O(3)$, and $SU(3)$. Each will be considered in turn and the latter two systems represent larger 3×3 cases. This geometric approach is applied to all of the three of these systems to obtain prolongation structures explicitly. In both 3×3 cases, the prolongation structure is reduced to the situation of three smaller 2×2 problems.

1. Introduction

Geometric approaches have been found useful in producing a great variety of results for nonlinear partial differential equations [1]. A specific geometric approach discussed here has been found to produce a very elegant, coherent, and unified understanding of many ideas in nonlinear physics by means of fundamental differential geometric concepts. In fact, relationships between a geometric interpretation of soliton equations, prolongation structure, Lax pairs, and conservation laws can be clearly realized and made use of. The interest in the approach, its generality, and the results it produces do not depend on a specific equation at the outset. The formalism in terms of differential forms [2] can encompass large classes of nonlinear partial differential equation, certainly the AKNS systems [3, 4], and it allows the production of generic expressions for infinite numbers of conservation laws. Moreover, it leads to the consequence that many seemingly different equations turn out to be related by a gauge transformation.

Here the discussion begins by studying prolongation structures for a 2×2 $SL(2,\mathbb{R})$ system discussed first by Sasaki [5, 6] and Crampin [7] to present and illustrate the method. This will also demonstrate the procedure and the kind of prolongation results that emerge. It also provides a basis from which to work out larger systems since they can generally be reduced to 2×2 problems. Of greater complexity are a pair of 3×3 problems which will be considered next. It is shown how to construct an $O(3)$ system based on three constituent one forms as well as an $SU(3)$ system composed of eight fundamental one forms. The former has not appeared. The 3×3 problem is less well known than the 2×2 problem; however, some work has appeared in [8, 9]. In the first of these, the 3×3 solitons degenerate to the AKNS solitons in three ways, while in the latter nondegenerate case, they do not. All of the results are presented explicitly; that is, the coefficients of all the forms and their higher exterior derivatives are calculated and given explicitly. Maple is used to do this in the harder cases. The approach is quite unified, and so once the formalism is established for the $SL(2,\mathbb{R})$ system, the overall procedure can be carried over to the other Lie algebra cases as well.

Of course, the 3×3 problems will yield more types of conservation laws. The prolongation structures of the 3×3 problems will be observed to be reduceable to the same type as the 2×2 system considered at the beginning. However, the Riccati representations become much more complicated [10]. The generalization of this formalism to an $n\times n$ problem then becomes straightforward after completing the study of these smaller cases [11]. The prolongation structure of an $n\times n$ problem is reduced to a collection of n smaller $(n-1)\times(n-1)$ cases and finally at the end to a set of 2×2 problems such as the $SL(2,\mathbb{R})$ case at the beginning. Finally, to summarize at the end, a collection of conservation laws are developed from one of the results and some speculation as to how this type relates to other procedures [12, 13].

2. $SL(2,\mathbb{R})$ Prolongations

The procedure begins by associating a system of Pfaffian equations to the nonlinear system to be studied, namely,

$$\alpha_i = 0, \quad \alpha_i = dy_i - \Omega_{ij} y_j, \quad i, j = 1, 2. \tag{1}$$

In (1), Ω is a traceless 2×2 matrix which consists of a family of one forms, ω_i. At the end, it is desired to express these one forms in terms of independent variables, which are called x and t, the dependent variables and their derivatives. However, no specific equation need be assumed at the start. The underlying structure group is $SL(2,\mathbb{R})$. Explicitly, the matrix of one forms is given by the 2×2 matrix

$$\Omega = \begin{pmatrix} \omega_1 & \omega_2 \\ \omega_3 & -\omega_1 \end{pmatrix}. \tag{2}$$

The integrability conditions are expressed as the vanishing of a traceless 2×2 matrix of two forms Θ given by

$$\Theta = d\Omega - \Omega \wedge \Omega. \tag{3}$$

In terms of components, the left-hand side of (3) has the form

$$\Theta = \left(\Theta_{ij}\right) = \begin{pmatrix} \vartheta_1 & \vartheta_2 \\ \vartheta_3 & -\vartheta_1 \end{pmatrix}. \tag{4}$$

Substituting (2) into (3), a matrix which contains the integrability equations in terms of the basis one forms is obtained. The three equations are

$$\begin{gathered} \vartheta_1 = d\omega_1 - \omega_2 \wedge \omega_3, \qquad \vartheta_2 = d\omega_2 - 2\omega_1 \wedge \omega_2, \\ \vartheta_3 = d\omega_3 + 2\omega_1 \wedge \omega_3. \end{gathered} \tag{5}$$

Of use in the theorems which follow, it is also useful to have the $d\omega_i$ in (5) expressed explicitly in terms of the ϑ_i as

$$\begin{gathered} d\omega_1 = \vartheta_1 + \omega_2 \wedge \omega_3, \qquad d\omega_2 = \vartheta_2 + 2\omega_1 \wedge \omega_2, \\ d\omega_3 = \vartheta_3 - 2\omega_1 \wedge \omega_3. \end{gathered} \tag{6}$$

Therefore, by selecting a particular choice for the set $\{\omega_i\}$, the nonlinear equation of interest can be written simply as

$$\Theta = 0, \qquad \vartheta_i = 0, \quad i = 1, 2, 3. \tag{7}$$

It is easy to see that the system is closed. Upon differentiating Θ and substituting (3), we obtain

$$d\Theta = \Omega \wedge \Theta - \Theta \wedge \Omega. \tag{8}$$

This implies that the exterior derivatives of the set of $\{\vartheta_i\}$ are contained in the ring of two forms $\{\vartheta_i\}$.

Now the differential ideal can be prolonged by including the forms $\{\alpha_1, \alpha_2\}$ given by (1).

Theorem 1. *The differential one forms*

$$\alpha_1 = dy_1 - \omega_1 y_1 - \omega_2 y_2, \qquad \alpha_2 = dy_2 - \omega_3 y_1 + \omega_1 y_2, \tag{9}$$

have the following exterior derivatives:

$$\begin{aligned} d\alpha_1 &= \omega_1 \wedge \alpha_1 + \omega_2 \wedge \alpha_2 - y_1 \vartheta_1 - y_2 \vartheta_2, \\ d\alpha_2 &= -\omega_1 \wedge \alpha_2 + \omega_3 \wedge \alpha_1 + y_2 \vartheta_1 - y_1 \vartheta_3. \end{aligned} \tag{10}$$

Hence, the exterior derivatives of the $\{\alpha_i\}$ are contained in the ring of forms $\{\vartheta_i, \alpha_j\}$.

Proof. The forms (9) follow by substituting (2) into (1). Differentiating α_1 in (9) it is found that

$$d\alpha_1 = -y_1 d\omega_1 + \omega_1 \wedge dy_1 - y_2 d\omega_2 + \omega_2 \wedge dy_2. \tag{11}$$

Obtaining dy_1 and dy_2 from (9) and $d\omega_1$, $d\omega_2$ from (6), $d\alpha_1$ becomes

$$\begin{aligned} d\alpha_1 &= -y_1 \vartheta_1 - y_1 \omega_2 \wedge \omega_3 + \omega_1 \wedge \alpha_1 + y_2 \omega_1 \wedge \omega_2 - y_2 \vartheta_2 \\ &\quad - 2y_2 \omega_1 \wedge \omega_2 + \omega_2 \wedge \alpha_2 + y_1 \omega_2 \wedge \omega_3 - y_2 \omega_2 \wedge \omega_1 \\ &= \omega_1 \wedge \alpha_1 + \omega_2 \wedge \alpha_2 - y_1 \vartheta_1 - y_2 \vartheta_2. \end{aligned} \tag{12}$$

Similarly, beginning with α_2,

$$\begin{aligned} d\alpha_2 &= -y_1 d\omega_3 + \omega_3 \wedge dy_1 + y_2 d\omega_1 - \omega_1 \wedge dy_2 \\ &= -y_1 \vartheta_3 + 2y_1 \omega_1 \wedge \omega_3 + \omega_3 \wedge \alpha_1 \\ &\quad + y_1 \omega_3 \wedge \omega_1 + y_2 \omega_3 \wedge \omega_2 \\ &\quad + y_2 \vartheta_1 + y_2 \omega_2 \wedge \omega_3 - \omega_1 \wedge \alpha_2 - y_1 \omega_1 \wedge \omega_3 \\ &= -\omega_1 \wedge \alpha_2 + \omega_3 \wedge \alpha_1 + y_2 \vartheta_1 - y_1 \vartheta_3, \end{aligned} \tag{13}$$

as required.

Corollary 2. *The exterior derivatives of the α_i in (9) can be expressed concisely in terms of the matrix elements of Θ and Ω as*

$$d\alpha_i = -\Theta_{ij} y_j + \Omega_{ij} \wedge \alpha_j, \quad i, j = 1, 2. \tag{14}$$

The corollary is Theorem 1 after using the definitions of the matrices in (2) and (4).

The forms α_1, α_2 given in (9) lead to a natural Riccati representation called α_3, α_4 for this differential system. This arises by taking the following linear combinations of α_1 and α_2:

$$\begin{aligned} y_1^2 \alpha_3 &= y_1 \alpha_2 - y_2 \alpha_1 \\ &= y_1 dy_2 - y_2 dy_1 - y_1^2 \omega_3 + 2y_1 y_2 \omega_1 + y_2^2 \omega_2, \\ y_2^2 \alpha_4 &= y_2 \alpha_1 - y_1 \alpha_2 \\ &= y_2 dy_1 - y_1 dy_2 - y_2^2 \omega_2 - 2y_1 y_2 \omega_1 + y_1^2 \omega_3. \end{aligned} \tag{15}$$

Two new functions, or pseudopotentials, y_3 and y_4 can be introduced which are defined by the transformation

$$y_3 = \frac{y_2}{y_1}, \qquad y_4 = \frac{y_1}{y_2}. \tag{16}$$

Then, from (15), the following forms which have a Riccati structure result:

$$\begin{aligned} \alpha_3 &= dy_3 - \omega_3 + 2y_3\omega_1 + y_3^2\omega_2, \\ \alpha_4 &= dy_4 - \omega_2 - 2y_4\omega_1 + y_4^2\omega_3. \end{aligned} \tag{17}$$

These equations could also be thought of as Riccati equations for y_3 and y_4.

Theorem 3. *(i) The one forms α_3, α_4 given by (17) have exterior derivatives*

$$\begin{aligned} d\alpha_3 &= 2y_3\vartheta_1 + y_3^2\vartheta_2 - \vartheta_3 + 2\alpha_3 \wedge (\omega_1 + y_3\omega_2), \\ d\alpha_4 &= -2y_4\vartheta_1 - \vartheta_2 + y_4^2\vartheta_3 + 2\alpha_4 \wedge (-\omega_1 + y_4\omega_3). \end{aligned} \tag{18}$$

Consequently, results (18) are contained in the ring spanned by $\{\vartheta_i\}$ and $\{\alpha_j\}$.

(ii) Define the forms $\sigma_1 = \omega_1 + y_3\omega_2$ and $\sigma_2 = -\omega_1 + y_4\omega_3$ appearing in (18). The exterior derivatives of σ_1 and σ_2 are found to be

$$\begin{aligned} d\sigma_1 &= \vartheta_1 + y_3\vartheta_2 + \alpha_3 \wedge \omega_2, \\ d\sigma_2 &= -\vartheta_1 + y_3\vartheta_2 + \alpha_4 \wedge \omega_2. \end{aligned} \tag{19}$$

Therefore $d\sigma_1 \equiv 0$ and $d\sigma_2 \equiv 0 \bmod \{\vartheta_i, \alpha_j\}$.

The proof of this Theorem proceeds along exactly the same lines as Theorem 1. The $SL(2,\mathbb{R})$ structure and the connection interpretation can be based on the forms α_3, α_4.

At this point, the process can be continued. Two more pseudopotentials y_5 and y_6 can be introduced and the differential system can be extended by including two more one forms α_5 and α_6. The specific structure of these forms is suggested by Theorem 3.

Theorem 4. *Define two one forms α_5 and α_6 as*

$$\alpha_5 = dy_5 - \omega_1 - y_3\omega_2, \qquad \alpha_6 = dy_6 + \omega_1 - y_4\omega_3. \tag{20}$$

The exterior derivatives of the one forms (20) are given by the expressions

$$d\alpha_5 = -\vartheta_1 - y_3\vartheta_3 - \alpha_3 \wedge \omega_2, \qquad d\alpha_6 = \vartheta_1 - y_4\vartheta_3 - \alpha_4 \wedge \omega_3. \tag{21}$$

Results (21) specify the closure properties of the forms α_5 and α_6. This theorem is proved along similar lines keeping in mind that the results for dy_3 and dy_4 are obtained from (17).

A final extension can be made by adding two additional pseudopotentials y_7 and y_8. At each stage, the ideal of forms is being enlarged and is found to be closed over the underlying subideal which does not contain the two new forms.

Theorem 5. *Define the one forms α_7 and α_8 by means of*

$$\alpha_7 = dy_7 - e^{-2y_5}\omega_2, \qquad \alpha_8 = dy_8 - e^{-2y_6}\omega_3. \tag{22}$$

The exterior derivatives of the forms α_7 and α_8 are given by

$$\begin{aligned} d\alpha_7 &= 2e^{-2y_5}\alpha_5 \wedge \omega_2 - e^{-2y_5}\vartheta_2, \\ d\alpha_8 &= 2e^{-2y_6}\alpha_6 \wedge \omega_3 - e^{-2y_6}\vartheta_3, \end{aligned} \tag{23}$$

and are closed over the prolonged ideal.

3. $O(3)$ Prolongations

The first 3×3 problem we consider is formulated in terms of a set of one forms ω_i for $i = 1, 2, 3$. The following system of Pfaffian equations is to be associated to a specific nonlinear equation

$$\alpha_i = 0, \quad \alpha_i = dy_i - \Omega_{ij}y_j, \quad i, j = 1, 2, 3. \tag{24}$$

In (24), Ω is a traceless 3×3 matrix of one forms. The nonlinear equation to emerge is expressed as the vanishing of a traceless 3×3 matrix of two forms Θ,

$$\Theta = 0, \qquad \Theta = d\Omega - \Omega \wedge \Omega. \tag{25}$$

These can be considered as integrability conditions for (24). The closure property, the gauge transformation, and the gauge theoretic interpretation can also hold for 3×3 systems. Explicitly the matrix Ω is given by

$$\Omega = \begin{pmatrix} 0 & -\omega_1 & \omega_2 \\ \omega_1 & 0 & -\omega_3 \\ -\omega_2 & \omega_3 & 0 \end{pmatrix}. \tag{26}$$

Using (26) in (24) and (25), the following prolonged differential system is obtained:

$$\begin{aligned} dy_1 &= \alpha_1 - y_2\omega_1 + y_3\omega_2, & d\omega_1 &= \vartheta_1 - \omega_2 \wedge \omega_3, \\ dy_2 &= \alpha_2 + y_1\omega_1 - y_3\omega_3, & d\omega_2 &= \vartheta_2 - \omega_3 \wedge \omega_1, \\ dy_3 &= \alpha_3 - y_1\omega_2 + y_2\omega_3, & d\omega_3 &= \vartheta_3 - \omega_1 \wedge \omega_2. \end{aligned} \tag{27}$$

By straightforward exterior differentiation of each α_i and using (27), the following Theorem results.

Theorem 6. *The α_i given in (27) have exterior derivatives which are contained in the ring spanned by $\{\vartheta_i\}_1^3$ and $\{\alpha_j\}_1^3$ and given explicitly as*

$$\begin{aligned} d\alpha_1 &= y_2\vartheta_1 - y_3\vartheta_2 - \omega_1 \wedge \alpha_2 + \omega_2 \wedge \alpha_3, \\ d\alpha_2 &= -y_1\vartheta_1 + y_3\vartheta_3 + \omega_1 \wedge \alpha_1 - \omega_3 \wedge \alpha_3, \\ d\alpha_3 &= y_1\vartheta_2 - y_2\vartheta_3 - \omega_2 \wedge \alpha_1 + \omega_3 \wedge \alpha_2. \end{aligned} \tag{28}$$

The one forms $\{\alpha_i\}_1^3$ admit a series of Riccati representations which can be realized by defining six new one forms,

$$\begin{aligned} &y_1^2\alpha_4 = y_1\alpha_2 - y_2\alpha_1, \qquad y_1^2\alpha_5 = y_1\alpha_3 - y_3\alpha_1, \\ &y_2^2\alpha_6 = y_2\alpha_1 - y_1\alpha_2, \\ &y_2^2\alpha_7 = y_2\alpha_3 - y_3\alpha_2, \qquad y_3^2\alpha_8 = y_3\alpha_1 - y_1\alpha_3, \\ &y_3^2\alpha_9 = y_3\alpha_2 - y_2\alpha_3. \end{aligned} \tag{29}$$

In effect, the larger 3×3 system is breaking up into several 2×2 systems. To write the new one forms explicitly, introduce the new functions $\{y_j\}_4^9$ which are defined in terms of the original $\{y_i\}_1^3$ as follows:

$$y_4 = \frac{y_2}{y_1}, \quad y_5 = \frac{y_3}{y_1}, \quad y_6 = \frac{y_1}{y_2}, \quad y_7 = \frac{y_3}{y_1}, \quad y_8 = \frac{y_1}{y_3}, \quad y_9 = \frac{y_2}{y_3}. \tag{30}$$

In terms of the functions defined in (30), the $\{\alpha_j\}_4^9$ are given as

$$\begin{aligned}
\alpha_4 &= dy_4 - \left(1 + y_4^2\right)\omega_1 + y_4 y_5 \omega_2 + y_5 \omega_3,\\
\alpha_5 &= dy_5 - y_4 y_5 \omega_1 + \left(1 + y_5^2\right)\omega_2 - y_4 \omega_3,\\
\alpha_6 &= dy_6 + \left(1 + y_6^2\right)\omega_1 - y_7 \omega_2 - y_6 y_7 \omega_3,\\
\alpha_7 &= dy_7 + y_6 y_7 \omega_1 + y_6 \omega_2 - \left(1 + y_7^2\right)\omega_3,\\
\alpha_8 &= dy_8 + y_9 \omega_1 - \left(1 + y_8^2\right)\omega_2 + y_8 y_9 \omega_3,\\
\alpha_9 &= dy_9 - y_8 \omega_1 - y_8 y_9 \omega_2 + \left(1 + y_9^2\right)\omega_3.
\end{aligned} \tag{31}$$

To state the next theorem, we need to define the following three matrices:

$$\begin{aligned}
\Omega_1 &= \begin{pmatrix} 2y_4\omega_1 - y_5\omega_2 & -y_4\omega_2 - \omega_3 \\ y_5\omega_1 + \omega_3 & y_4\omega_1 - 2y_5\omega_2 \end{pmatrix},\\
\Omega_2 &= \begin{pmatrix} -2y_6\omega_1 + y_7\omega_3 & \omega_2 + y_6\omega_3 \\ -y_7\omega_1 - \omega_6 & -y_6\omega_1 + 2y_7\omega_3 \end{pmatrix},\\
\Omega_3 &= \begin{pmatrix} 2y_8\omega_2 - y_9\omega_3 & -\omega_1 - y_9\omega_3 \\ \omega_1 + y_9\omega_2 & y_8\omega_2 - 2y_9\omega_3 \end{pmatrix}.
\end{aligned} \tag{32}$$

Theorem 7. *The closure properties for the forms* $\{\alpha_i\}_4^9$ *given in* (31) *can be summarized in the form*

$$\begin{aligned}
d\begin{pmatrix}\alpha_4\\ \alpha_5\end{pmatrix} &= \begin{pmatrix} -\left(1+y_4^2\right)\vartheta_1 + y_4y_5\vartheta_2 + y_5\vartheta_3 \\ -y_4y_5\vartheta_1 + \left(1+y_5^2\right)\vartheta_2 - y_4\vartheta_3 \end{pmatrix} + \Omega_1 \wedge \begin{pmatrix}\alpha_4\\ \alpha_5\end{pmatrix},\\
d\begin{pmatrix}\alpha_6\\ \alpha_7\end{pmatrix} &= \begin{pmatrix} \left(1+y_6^2\right)\vartheta_1 - y_7\vartheta_2 - y_6y_7\vartheta_3 \\ y_6y_7\vartheta_1 + y_6\vartheta_2 - \left(1+y_7^2\right)\vartheta_3 \end{pmatrix} + \Omega_2 \wedge \begin{pmatrix}\alpha_6\\ \alpha_7\end{pmatrix},\\
d\begin{pmatrix}\alpha_8\\ \alpha_9\end{pmatrix} &= \begin{pmatrix} y_1\vartheta_1 - \left(1+y_8^2\right)\vartheta_2 + y_8y_9\vartheta_3 \\ -y_8\vartheta_1 - y_8y_9\vartheta_2 + \left(1+y_9^2\right)\vartheta_3 \end{pmatrix} + \Omega_3 \wedge \begin{pmatrix}\alpha_8\\ \alpha_9\end{pmatrix}.
\end{aligned} \tag{33}$$

The exterior derivatives are therefore contained in the ring spanned by $\{\vartheta_i\}_1^3$ *and the* $\{\alpha_j\}_4^9$.

In terms of Ω_i, a corresponding matrix of two forms Θ_i can be defined in terms of the corresponding Ω_i defined in (32),

$$\Theta_i = d\Omega_i - \Omega_i \wedge \Omega_i, \quad i = 1, 2, 3. \tag{34}$$

Theorem 8. *The* 2×2 *matrix of two forms* Θ_i *defined by* (34) *is contained in the ring of forms spanned by* $\{\vartheta_i\}_1^3$ *coupled with either* $\{\alpha_4, \alpha_5\}$, $\{\alpha_6, \alpha_7\}$, *or* $\{\alpha_8, \alpha_9\}$ *when* $i = 1, 2, 3$, *respectively.*

Proof. Differentiating Ω_1 and simplifying, we have

$$\begin{aligned}
\Theta_1 &= d\Omega_1 - \Omega_1 \wedge \Omega_1\\
&= \begin{pmatrix} 2y_4\vartheta_1 - y_5\vartheta_2 + 2\alpha_4 \wedge \omega_1 - \alpha_5 \wedge \omega_2 & -y_4\vartheta_2 - \vartheta_3 - \alpha_4 \wedge \omega_2 \\ y_5\vartheta_1 + \vartheta_3 + \alpha_5 \wedge \omega_1 & y_4\vartheta_1 - 2y_5\vartheta_2 + \alpha_4 \wedge \omega_1 - 2\alpha_5 \wedge \omega_2 \end{pmatrix}.
\end{aligned} \tag{35}$$

Similar matrix expressions can be found for the cases in which Ω_1 is replaced by Ω_2 or Ω_3, respectively.

Unlike Ω given in (26), the Ω_i in (32) are not traceless. Define their traces to be

$$\kappa_i = \operatorname{tr}\Omega_i, \quad i = 1, 2, 3. \tag{36}$$

These traces provide convenient ways of generating conservation laws on account of the following Theorem.

Theorem 9. *The exterior derivatives of traces* (36) *are given by*

$$\begin{aligned}
\frac{1}{3}d\kappa_1 &= y_4\vartheta_1 - y_5\vartheta_2 - \omega_1 \wedge \alpha_4 + \omega_2 \wedge \alpha_5,\\
\frac{1}{3}d\kappa_2 &= -y_6\vartheta_1 + y_7\vartheta_3 + \omega_1 \wedge \alpha_6 - \omega_3 \wedge \alpha_7,\\
\frac{1}{3}d\kappa_3 &= y_8\vartheta_2 - y_9\vartheta_3 - \omega_2 \wedge \alpha_8 + \omega_3 \wedge \alpha_9.
\end{aligned} \tag{37}$$

Results (37) *are contained in the ring spanned by* $\{\vartheta_i\}_1^3$ *and* $\{\alpha_j\}_4^9$.

4. $SU(3)$ Prolongations

To formulate a 3×3 $SU(3)$ problem, the set of Pfaffian equations,

$$\alpha_i = 0, \quad \alpha_i = dy_i - \Omega_{ij} y_j, \quad i, j = 1, 2, 3, \tag{38}$$

are associated with the nonlinear equation. In (38), Ω is a traceless 3×3 matrix consisting of a system of one forms. The nonlinear equation to be considered is expressed as the vanishing of a traceless 3×3 matrix of two forms Θ exactly as in (25) which constitute the integrability condition for (38).

A 3×3 matrix representation of the Lie algebra for $SU(3)$ is introduced by means of generators λ_j for $j = 1, \dots, 8$, which satisfy the following set of commutation relations:

$$[\lambda_l, \lambda_m] = 2i\, f_{lmn} \lambda_n. \tag{39}$$

The structure constants f_{lmn} are totally antisymmetric in l, m, n. The one form Ω is expressed in terms of the generators λ_i as

$$\Omega = \sum_{l=1}^{8} \omega_l \lambda_l. \tag{40}$$

Then the two forms Θ can be written as

$$\Theta = \sum_{l=1}^{8} \vartheta_l \lambda_l, \quad \vartheta_l = d\omega_l - i f_{lmn} \omega_m \wedge \omega_n. \tag{41}$$

The nonlinear equation to be solved has the form $\vartheta_l = 0$ for $l = 1, \dots, 8$.

It will be useful to display (40) and (41) by using the nonzero structure constants given in (39), since these may not be readily accessible. An explicit representation for the matrix Ω used here is given by

$$\Omega = \begin{pmatrix} \omega_3 + \frac{1}{\sqrt{3}}\omega_8 & \omega_1 - i\omega_2 & \omega_4 - i\omega_5 \\ \omega_1 + i\omega_2 & -\omega_3 + \frac{1}{\sqrt{3}}\omega_8 & \omega_6 - i\omega_7 \\ \omega_4 + i\omega_5 & \omega_6 + i\omega_7 & -\frac{2}{\sqrt{3}}\omega_8 \end{pmatrix}. \tag{42}$$

Moreover, in order that the presentation can be easier to follow, the eight forms $d\omega_i$ specified by (41) will be given explicitly:

$$\begin{aligned}
d\omega_1 &= \vartheta_1 + 2i\omega_2 \wedge \omega_3 + i\omega_4 \wedge \omega_7 - i\omega_5 \wedge \omega_6, \\
d\omega_2 &= \vartheta_2 - 2i\omega_1 \wedge \omega_3 + i\omega_4 \wedge \omega_6 + i\omega_5 \wedge \omega_7, \\
d\omega_3 &= \vartheta_3 + 2i\omega_1 \wedge \omega_2 + i\omega_4 \wedge \omega_5 - i\omega_6 \wedge \omega_7, \\
d\omega_4 &= \vartheta_4 - i\omega_1 \wedge \omega_7 - i\omega_2 \wedge \omega_6 - i\omega_3 \wedge \omega_5 + \sqrt{3}i\omega_5 \wedge \omega_8, \\
d\omega_5 &= \vartheta_5 + i\omega_1 \wedge \omega_6 - i\omega_2 \wedge \omega_7 + i\omega_3 \wedge \omega_4 - \sqrt{3}i\omega_4 \wedge \omega_8, \\
d\omega_6 &= \vartheta_6 - i\omega_1 \wedge \omega_5 + i\omega_2 \wedge \omega_4 + i\omega_3 \wedge \omega_7 + \sqrt{3}i\omega_7 \wedge \omega_8, \\
d\omega_7 &= \vartheta_7 + i\omega_1 \wedge \omega_4 + i\omega_2 \wedge \omega_5 - i\omega_3 \wedge \omega_6 - \sqrt{3}i\omega_6 \wedge \omega_8, \\
d\omega_8 &= \vartheta_8 + \sqrt{3}i\omega_4 \wedge \omega_5 + \sqrt{3}i\omega_6 \wedge \omega_7.
\end{aligned} \tag{43}$$

Substituting (42) into (38), the following system of one forms $\{\alpha_i\}_1^3$ is obtained,

$$\begin{aligned}
\alpha_1 &= dy_1 - \left(\omega_3 + \frac{1}{\sqrt{3}}\omega_8\right) y_1 - (\omega_1 - i\omega_2)\, y_2 - (\omega_4 - i\omega_5)\, y_3, \\
\alpha_2 &= dy_2 - (\omega_1 + i\omega_2)\, y_1 + \left(\omega_3 - \frac{1}{\sqrt{3}}\omega_8\right) y_2 - (\omega_6 - i\omega_7)\, y_3, \\
\alpha_3 &= dy_3 - (\omega_4 + i\omega_5)\, y_1 - (\omega_6 + i\omega_7)\, y_2 + \frac{2}{\sqrt{3}}\omega_8 y_3.
\end{aligned} \tag{44}$$

The increase in complexity of this system makes it often necessary to resort to the use of symbolic manipulation to carry out longer calculations, and for the most part, only results are given.

Theorem 10. *The exterior derivatives of the α_i in (44) are given by*

$$\begin{aligned}
d\alpha_1 = &- y_2 (\vartheta_1 - i\vartheta_2) - y_1 \vartheta_3 - y_3 (\vartheta_4 - i\vartheta_5) \\
&- \frac{1}{\sqrt{3}} y_1 \vartheta_8 + \left(\omega_3 + \frac{1}{\sqrt{3}}\omega_8\right) \wedge \alpha_1 \\
&+ (\omega_1 - i\omega_2) \wedge \alpha_2 + (\omega_4 - i\omega_5) \wedge \alpha_3, \\
d\alpha_2 = &- y_1 (\vartheta_1 + i\vartheta_2) + y_2 \vartheta_3 \\
&- \frac{1}{\sqrt{3}} y_2 \vartheta_8 - y_3 (\vartheta_6 - i\vartheta_7) + (\omega_1 + i\omega_2) \wedge \alpha_1 \\
&- \left(\omega_3 - \frac{1}{\sqrt{3}}\omega_8\right) \wedge \alpha_2 + (\omega_6 - i\omega_7) \wedge \alpha_3, \\
d\alpha_3 = &- y_1 (\vartheta_4 + i\vartheta_5) - y_2 (\vartheta_6 + i\vartheta_7) \\
&+ \frac{2}{\sqrt{3}} y_3 \vartheta_8 + (\omega_4 + i\omega_5) \wedge \alpha_1 \\
&+ (\omega_6 + i\omega_7) \wedge \alpha_2 - \frac{2}{\sqrt{3}} \omega_8 \wedge \alpha_3.
\end{aligned} \tag{45}$$

In order to keep the notation concise, an abbreviation will be introduced. Suppose that $\{\gamma_i\}$ is a system of forms, and i, j are integers, then we make the abbreviation

$$\gamma_{ij\pm} = \gamma_i \pm i\gamma_j. \tag{46}$$

At this point, quadratic pseudopotentials can be introduced in terms of the homogeneous variables of the same form as

(30). In the same way that (31) was produced, the following Pfaffian equations based on the set $\{\alpha_i\}_1^3$ in (44) are obtained:

$$\begin{aligned}
\alpha_4 &= dy_4 - \omega_{12+} + 2y_4\omega_3 + y_4^2\omega_{12-} \\
&\quad - y_5\omega_{67-} + y_4y_5\omega_{45-}, \\
\alpha_5 &= dy_5 - \omega_{45+} + y_5\left(\omega_3 + \sqrt{3}\omega_8\right) \\
&\quad + y_5^2\omega_{45-} - y_4\omega_{67+} + y_4y_5\omega_{12-}, \\
\alpha_6 &= dy_6 - \omega_{12-} - 2y_6\omega_3 + y_6^2\omega_{12+} \\
&\quad - y_7\omega_{45-} + y_6y_7\omega_{67-}, \\
\alpha_7 &= dy_7 - \omega_{67+} - y_7\left(\omega_3 - \sqrt{3}\omega_8\right) \\
&\quad + y_7^2\omega_{67-} - y_6\omega_{45+} + y_6y_7\omega_{12+}, \\
\alpha_8 &= dy_8 - \omega_{45-} - y_8\left(\omega_3 + \sqrt{3}\omega_8\right) \\
&\quad + y_8^2\omega_{45+} - y_9\omega_{12-} + y_8y_9\omega_{67+}, \\
\alpha_9 &= dy_9 - \omega_{67-} + y_9\left(\omega_3 - \sqrt{3}\omega_8\right) \\
&\quad + y_9^2\omega_{67+} - y_8\omega_{12+} + y_8y_9\omega_{45+}.
\end{aligned} \tag{47}$$

These are coupled Riccati equations for the pairs of pseudopotentials (y_4, y_5), (y_6, y_7), and (y_8, y_9), respectively.

Define the following set of 2×2 matrices of one forms Ω_i for $i = 1, 2, 3$ as

$$\begin{aligned}
\Omega_1 &= \begin{pmatrix} -2\omega_3 - 2y_4\omega_{12-} - y_5\omega_{45-} & \omega_{67-} - y_4\omega_{45-} \\ \omega_{67+} - y_5\omega_{12-} & -\omega_3 - \sqrt{3}\omega_8 - y_4\omega_{12-} - 2y_5\omega_{45-} \end{pmatrix}, \\
\Omega_2 &= \begin{pmatrix} 2\omega_3 - 2y_6\omega_{12+} - y_7\omega_{67-} & \omega_{45-} - y_6\omega_{67-} \\ \omega_{45+} - y_7\omega_{12+} & \omega_3 - \sqrt{3}\omega_8 - y_6\omega_{12+} - 2y_7\omega_{67-} \end{pmatrix}, \\
\Omega_3 &= \begin{pmatrix} \omega_3 + \sqrt{3}\omega_8 - 2y_8\omega_{45+} - y_9\omega_{67+} & \omega_{12-} - y_8\omega_{67+} \\ \omega_{12+} - y_9\omega_{45+} & -\omega_3 + \sqrt{3}\omega_8 - y_8\omega_{45+} - 2y_9\omega_{67+} \end{pmatrix}.
\end{aligned} \tag{48}$$

Making use of the matrices Ω_i defined in (48), the following result can be stated.

Theorem 11. *The closure properties of the set of forms $\{\alpha_i\}_4^9$ given by (47) can be expressed in terms of the Ω_i as follows:*

$$\begin{aligned}
d\begin{pmatrix}\alpha_4 \\ \alpha_5\end{pmatrix}
&= \begin{pmatrix} \left(y_4^2 - 1\right)\vartheta_1 - i\left(y_4^2 + 1\right)\vartheta_2 + 2y_4\vartheta_3 + y_4y_5\vartheta_{45-} - y_5\vartheta_{67-} \\ y_4y_5\vartheta_{12-} + y_5\vartheta_3 + \left(y_4^2 - 1\right)\vartheta_4 - i\left(y_5^2 + 1\right)\vartheta_5 - y_4\vartheta_{67+} + \sqrt{3}y_5\vartheta_8 \end{pmatrix} \\
&\quad + \Omega_1 \wedge \begin{pmatrix}\alpha_4 \\ \alpha_5\end{pmatrix}, \\
d\begin{pmatrix}\alpha_6 \\ \alpha_7\end{pmatrix}
&= \begin{pmatrix} \left(y_6^2 - 1\right)\vartheta_1 + i\left(y_6^2 + 1\right)\vartheta_2 - 2y_6\vartheta_3 - y_7\vartheta_{45-} + y_6y_7\vartheta_{67-} \\ y_6y_7\vartheta_{12+} - y_7\vartheta_3 - y_6\vartheta_{45+} + \left(y_7^2 - 1\right)\vartheta_6 - i\left(y_7^2 + 1\right)\vartheta_7 + \sqrt{3}y_7\vartheta_8 \end{pmatrix} \\
&\quad + \Omega_2 \wedge \begin{pmatrix}\alpha_6 \\ \alpha_7\end{pmatrix}, \\
d\begin{pmatrix}\alpha_8 \\ \alpha_9\end{pmatrix}
&= \begin{pmatrix} -y_9\vartheta_{12-} - y_8\vartheta_3 + \left(y_8^2 - 1\right)\vartheta_4 + i\left(y_8^2 + 1\right)\vartheta_5 + y_8y_9\vartheta_{67+} - \sqrt{3}y_8\vartheta_8 \\ -y_8\vartheta_{12+} + y_9\vartheta_3 + y_8y_9\vartheta_{45+} + \left(y_9^2 - 1\right)\vartheta_6 + i\left(y_9^2 + 1\right)\vartheta_7 - \sqrt{3}y_9\vartheta_8 \end{pmatrix} \\
&\quad + \Omega_3 \wedge \begin{pmatrix}\alpha_8 \\ \alpha_9\end{pmatrix}.
\end{aligned} \tag{49}$$

Based on the forms Ω_i given in (48), we can differentiate to define the following matrices of two forms ξ_i given by

$$\xi_i = d\Omega_i - \Omega_i \wedge \Omega_i. \tag{50}$$

Theorem 12. *The forms ξ_i defined in (50) are given explicitly by the matrices*

$$\begin{aligned}
\xi_1 &= \begin{pmatrix} 2\omega_{12+} \wedge \alpha_4 + \omega_{45-} \wedge \alpha_5 - 2y_4\vartheta_{12-} - 2\vartheta_3 - y_5\vartheta_{45-} & \omega_{45-} \wedge \alpha_4 - y_4\vartheta_{45-} + \vartheta_{67-} \\ \omega_{12-} \wedge \alpha_5 - y_5\vartheta_{12-} + \vartheta_{67+} & \omega_{12-} \wedge \alpha_4 + 2\omega_{45-} \wedge \alpha_5 - y_4\vartheta_{12-} - \vartheta_3 - 2\vartheta_{45-} - \sqrt{3}\vartheta_8 \end{pmatrix}, \\
\xi_2 &= \begin{pmatrix} 2\omega_{12+} \wedge \alpha_6 + \omega_{67-} \wedge \alpha_7 - 2y_6\vartheta_{12+} + \vartheta_3 - y_7\vartheta_{67-} & \omega_{67-} \wedge \alpha_6 + \vartheta_{45-} - y_6\vartheta_{67-} \\ \omega_{12+} \wedge \alpha_7 - y_7\vartheta_{12+} + \vartheta_{45+} & \omega_{12+} \wedge \alpha_6 + 2\omega_{67-} \wedge \alpha_7 - y_6\vartheta_{12+} + \vartheta_3 - 2y_7\vartheta_{67-} - \sqrt{3}\vartheta_8 \end{pmatrix}, \\
\xi_3 &= \begin{pmatrix} 2\omega_{45+} \wedge \alpha_8 + \omega_{67+} \wedge \alpha_9 + \vartheta_3 - 2y_8\vartheta_{45+} - y_9\vartheta_{67+} + \sqrt{3}\vartheta_8 & \omega_{67+} \wedge \alpha_8 + \vartheta_{12-} - y_8\vartheta_{67+} \\ \omega_{45+} \wedge \alpha_9 + \vartheta_{12+} - y_9\vartheta_{45+} & \omega_{45+} \wedge \alpha_8 + 2\omega_{67+} \wedge \alpha_9 - \vartheta_3 - y_8\vartheta_{45+} - 2y_9\vartheta_{67+} + \sqrt{3}\vartheta_8 \end{pmatrix}.
\end{aligned} \tag{51}$$

From these results, it is concluded that ξ_1 is contained in the ring of $\{\alpha_4, \alpha_5\}$ and the $\{\vartheta_l\}$, ξ_2 is contained in the ring of $\{\alpha_6, \alpha_7\}$ and $\{\vartheta_l\}$, and ξ_3 is in the ring of $\{\alpha_8, \alpha_9\}$ and $\{\vartheta_l\}$.

The two forms $\{\xi_i\}_1^3$ therefore vanish for the solutions of the nonlinear equations (39) and of the coupled Riccati equations $\{\alpha_j = 0\}_4^9$. As in the case of O(3), the one forms Ω_i are not traceless. Denoting the trace of Ω_i by

$$\tau_i = \operatorname{tr}\Omega_i, \tag{52}$$

the following theorem then follows.

Theorem 13. *The closure properties of the exterior derivatives of the traces τ_i are given by*

$$\begin{aligned}
\frac{1}{3}d\tau_1 &= \omega_{12-} \wedge \alpha_4 + \omega_{45-} \wedge \alpha_5 \\
&\quad - y_4\vartheta_{12-} - \vartheta_3 - y_5\vartheta_{45-} - \frac{1}{\sqrt{3}}\vartheta_8,
\end{aligned}$$

$$\frac{1}{3}d\tau_2 = \omega_{12+} \wedge \alpha_6 + \omega_{67-} \wedge \alpha_7 - y_6\vartheta_{12+} + \vartheta_3 - y_7\vartheta_{67-} - \frac{1}{\sqrt{3}}\vartheta_8,$$
$$\frac{1}{3}d\tau_3 = \omega_{45+} \wedge \alpha_8 + \omega_{67+} \wedge \alpha_9 - y_8\vartheta_{45+} - y_9\vartheta_{67+} + \frac{2}{\sqrt{3}}\vartheta_8. \quad (53)$$

The one forms (52) then can be used to generate conservation laws. Note that as in the $O(3)$ case, a 3×3 problem Ω has been reduced to three separate 2×2 problems in terms of the matrices Ω_l. Further prolongation can be continued beginning with the one forms Ω_l given in (48). To briefly outline the further steps in the procedure, a system of Pfaffians $\{\widetilde{\alpha}_{lj}\}$ is introduced which are based on the one forms Ω_l

$$\widetilde{\alpha}_{lj} = 0, \qquad \widetilde{\alpha}_{lj} = dy_{lj} - (\Omega_l)_{jk}y_{lk}. \quad (54)$$

The subscript l in y_{lj} and $\widetilde{\alpha}_{lj}$ indicates that they belong to the subsystem defined by Ω_l, and it is not summed over. Exterior differentiation of $\widetilde{\alpha}_{lj}$ yields

$$d\widetilde{\alpha}_{lj} = (\Omega_l)_{jk} \wedge \widetilde{\alpha}_{lk} - (\widetilde{\Theta})_{jk}y_{lk}. \quad (55)$$

The $\{y_{lj}\}$ are pseudopotentials for the original nonlinear equation (39). Next, quadratic pseudopotentials are included as the homogeneous variables and the prolongation can be continued.

5. A Conservation Law and Summary

These types of results turn out to be very useful for further study of integrable equations. They can be used for generating infinite numbers of conservation laws. In addition, the results are independent of any further structure of the forms ω_i in all cases. In fact, the one form Ω need not be thought of as unique. This is due to the fact that Ω and Θ are form invariant under the gauge transformation $\Omega \to \omega' = dA\ A^{-1} + A\Omega A^{-1}$ and $\Theta \to \Theta' = A\Theta A^{-1}$ where, in the $SL(2,\mathbb{R})$ case, A is an arbitrary 2×2 space-time-dependent matrix of determinant one. The gauge transformation property holds in the 3×3 case as well. The pseudopotentials serve as potentials for conservation laws in a generalized sense. They can be defined under a choice of Pfaffian forms such as

$$\alpha_i = 0, \qquad \alpha_i = dy_i + F_i\,dx + G_i\,dt, \quad (56)$$

with the property that the exterior derivatives $d\alpha_i$ are contained in the ring spanned by $\{\alpha_i\}$ and $\{\vartheta_l\}$

$$d\alpha_i = \sum_j A_{ij} \wedge \alpha_j + \sum_l \Gamma_{il}\vartheta_l. \quad (57)$$

This can be thought of as a generalization of the Frobenius Theorem for complete integrability of Pfaffian systems.

Theorem 3 implies that $d\sigma_i = 0$ when $i = 1, 2$ for solutions of the original equation. If either of these forms is expressed as $\sigma = \mathcal{I}dx + \mathcal{J}dt$, then $d\sigma = 0$ implies that

$$\frac{\partial \mathcal{I}}{\partial t} - \frac{\partial \mathcal{J}}{\partial t} = 0. \quad (58)$$

Thus $\mathcal{I}$ is a conserved density and $\mathcal{J}$ a conserved current. Let us finally represent $\omega_1 = a_1dx + b_1dt$, $\omega_2 = a_2dx + b_2dt$, and $\omega_3 = \eta dx + b_3dt$, where η is a parameter. The x-dependent piece of α_3 in (17) implies

$$y_{3,x} + 2a_1y_3 + a_2y_3^2 - \eta = 0. \quad (59)$$

By substituting an asymptotic expansion around $\eta = \infty$ for y_3 in (59) of the form,

$$y_3 = \sum_0^{\infty} \eta^{-n}Y_n, \quad (60)$$

a recursion for the Y_n is obtained. Thus (59) becomes,

$$\sum_n \eta^{-n}Y_{n,x} + 2a_1\sum_n \eta^{-n}Y_n + a_2\left(\sum_n \eta^{-n}Y_n\right)^2 - \eta = 0. \quad (61)$$

Expanding the power, collecting coefficients of η, and equating coefficients of η to zero, we get

$$Y_{1,x} + 2a_1Y_1 = 1, \qquad Y_{n,x} + 2a_1Y_n + a_2\sum_{k=1}^{n-1} Y_{n-k}Y_k = 0. \quad (62)$$

The consistency of solving α_3 by using just the x part is guaranteed by complete integrability. The x part of the form σ_1 is expressed as

$$(\sigma_1)_x = \left(a_1 + \sum_{n=1}^{\infty} \eta^{-n}Y_na_2\right)dx. \quad (63)$$

Consequently, from (58) the nth conserved density is

$$\mathcal{I}_n = a_2Y_n. \quad (64)$$

The existence of links between the types of prolongation here and other types of prolongations which are based on closed differential systems is a subject for further work.

References

[1] F. B. Estabrook and H. D. Wahlquist, "Classical geometries defined by exterior differential systems on higher frame bundles," *Classical and Quantum Gravity*, vol. 6, no. 3, pp. 263–274, 1989.

[2] H. Cartan, *Differential Forms*, Dover, Mineola, NY, USA, 2006.

[3] M. J. Ablowitz, D. J. Kaup, A. C. Newell, and H. Segur, "Method for solving the sine-Gordon equation," *Physical Review Letters*, vol. 30, pp. 1262–1264, 1973.

[4] M. J. Ablowitz, D. K. Kaup, A. C. Newell, and H. Segur, "Nonlinear-evolution equations of physical significance," *Physical Review Letters*, vol. 31, pp. 125–127, 1973.

[5] R. Sasaki, "Pseudopotentials for the general AKNS system," *Physics Letters A*, vol. 73, no. 2, pp. 77–80, 1979.

[6] R. Sasaki, "Geometric approach to soliton equations," *Proceedings of the Royal Society. London*, vol. 373, no. 1754, pp. 373–384, 1980.

[7] M. Crampin, "Solitons and $SL(2, \mathbb{R})$," *Physics Letters A*, vol. 66, no. 3, pp. 170–172, 1978.

[8] D. J. Kaup and J. Yang, "The inverse scattering transform and squared eigenfunctions for a degenerate 3×3 operator," *Inverse Problems*, vol. 25, no. 10, Article ID 105010, 2009.

[9] D. J. Kaup and R. A. Van Gorder, "The inverse scattering transform and squared eigenfunctions for the nondegenerate 3×3 operator and its soliton structure," *Inverse Problems*, vol. 26, no. 5, Article ID 055005, 2010.

[10] P. Bracken, "Intrinsic formulation of geometric integrability and associated Riccati system generating conservation laws," *International Journal of Geometric Methods in Modern Physics*, vol. 6, no. 5, pp. 825–837, 2009.

[11] D. J. Kaup and R. A. Van Gorder, "Squared eigenfunctions and the perturbation theory for the nondegenerate $N \times N$ operator: a general outline," *Journal of Physics A*, vol. 43, Article ID 434019, 2010.

[12] H. D. Wahlquist and F. B. Estabrook, "Prolongation structures of nonlinear evolution equations," *Journal of Mathematical Physics*, vol. 16, pp. 1–7, 1975.

[13] P. Bracken, "On two-dimensional manifolds with constant Gaussian curvature and their associated equations," *International Journal of Geometric Methods in Modern Physics*, vol. 9, no. 3, Article ID 1250018, 2012.

11

Characterizations of Ideals in Intermediate C-Rings $A(X)$ via the A-Compactifications of X

Joshua Sack[1] and Saleem Watson[2]

[1] *Institute of Logic, Language, and Computation, Universiteit van Amsterdam, P.O. Box 94242, 1090 GE Amsterdam, The Netherlands*
[2] *Department of Mathematics, California State University, Long Beach, CA 90840, USA*

Correspondence should be addressed to Joshua Sack; joshua.sack@gmail.com

Academic Editor: David Dobbs

Let X be a completely regular topological space. An intermediate ring is a ring $A(X)$ of continuous functions satisfying $C^*(X) \subseteq A(X) \subseteq C(X)$. In Redlin and Watson (1987) and in Panman et al. (2012), correspondences $\mathscr{Z}_A$ and $\mathfrak{Z}_A$ are defined between ideals in $A(X)$ and z-filters on X, and it is shown that these extend the well-known correspondences studied separately for $C^*(X)$ and $C(X)$, respectively, to any intermediate ring. Moreover, the inverse map $\mathscr{Z}_A^{\leftarrow}$ sets up a one-one correspondence between the maximal ideals of $A(X)$ and the z-ultrafilters on X. In this paper, we define a function $\mathfrak{K}_A$ that, in the case that $A(X)$ is a C-ring, describes $\mathfrak{Z}_A$ in terms of extensions of functions to realcompactifications of X. For such rings, we show that $\mathfrak{Z}_A^{\leftarrow}$ maps z-filters to ideals. We also give a characterization of the maximal ideals in $A(X)$ that generalize the Gelfand-Kolmogorov theorem from $C(X)$ to $A(X)$.

1. Introduction

Let X be a completely regular space and $A(X)$ an intermediate ring of continuous real-valued functions; that is, $C^*(X) \subseteq A(X) \subseteq C(X)$. It is well known that there is a natural correspondence $\mathbf{Z}$ between ideals of $C(X)$ and z-filters on X as described in [1, pages 26-27]. Such a correspondence $\mathbf{E}$ also exists for $C^*(X)$ [1, Problem 2L]. In [2], a correspondence $\mathscr{Z}_A$ between the ideals of any $A(X)$ and the z-filters on X was introduced, and its properties were further investigated in [3–5]. In [6], another correspondence $\mathfrak{Z}_A$ between ideals of any $A(X)$ and z-filters on X is introduced. It is shown in [6] that the correspondences $\mathscr{Z}_A$ and $\mathfrak{Z}_A$ extend the correspondences $\mathbf{E}$ and $\mathbf{Z}$ from $C^*(X)$ and $C(X)$, respectively, to all intermediate rings, and an explicit formula is stated that relates the two correspondences. In this paper, we give a characterization (Definition 3 and Theorem 6) of the correspondence $\mathfrak{Z}_A$ for intermediate C-rings $A(X)$ in terms of the A-compactifications of X introduced in [7]. In this setting, we show (Theorem 14) that the inverse map $\mathfrak{Z}_A^{\leftarrow}$ of the set map $\mathfrak{Z}_A$ maps ideals in $A(X)$ to z-filters on X. We also give a characterization of the maximal ideals in $A(X)$. This characterization generalizes from $C(X)$ to $A(X)$ the Gelfand-Kolmogorov theorem (Theorem 8). We follow the notation in [1, 6].

2. Preliminaries

For convenience we state some of the definitions and results needed in this paper.

Following the notation in [1], we set

$$\begin{aligned} \mathbf{Z}(X) &= \mathbf{Z}[C(X)] \\ &= \{\mathbf{Z}(f) \mid f \in C(X)\} \end{aligned} \quad (1)$$

to be the collection of the zero sets $\mathbf{Z}(f) = f^{-1}(\{0\})$ of all functions $f \in C(X)$. In this paper, we generally work with functions f on a fixed set X, as well as some extensions g of f to larger domains. As expected, $\mathbf{Z}(g)$ then denotes the zero set of g on the larger domain.

A z-filter (z-ultrafilter, resp.) on X is the intersection of $\mathbf{Z}(X)$ with a filter (ultrafilter, resp.) on X. The *kernel* k of a set $\mathcal{S}$ of z-ultrafilters is defined by

$$k(\mathcal{S}) = \bigcap \{\mathscr{U} \mid \mathscr{U} \in \mathcal{S}\}. \quad (2)$$

One can verify that the kernel of a set of z-ultrafilters is a z-filter. The *hull* h of a z-filter $\mathscr{F}$ is defined by

$$h(\mathscr{F}) = \{\mathscr{U} \mid \mathscr{U} \text{ is a } z\text{-ultrafilter on } X,\ \mathscr{U} \supseteq \mathscr{F}\}. \quad (3)$$

Given a set $E \subseteq X$, let $\langle E \rangle$ denote the set of all zero sets $F \subseteq X$, such that $E \subseteq F$.

Given any two functions f and g on a set X and $H \subseteq X$, we write $f \equiv g$ on H if $f(x) = g(x)$ for all $x \in H$. For each intermediate ring $A(X)$ of continuous functions, each noninvertible $f \in A(X)$, and each $H \subseteq X$, we say that f is *H-regular in $A(X)$* if there exists $g \in A(X)$ such that $fg \equiv 1$ on H. We just say f is H-regular if $A(X)$ is understood by context. For any set $E \subseteq X$, let E^c be the complement of E in X.

For each $f \in A(X)$, set

$$\mathcal{Z}_A(f) = \{E \in \mathbf{Z}(X) \mid f \text{ is } E^c\text{-regular}\},$$
$$\mathfrak{Z}_A(f) = \{E \in \mathbf{Z}(X) \mid f \text{ is } H\text{-regular for each zero-set } H \subseteq E^c\}. \quad (4)$$

For an ideal $I \subset A(X)$, we set $\mathcal{Z}_A[I] = \bigcup\{\mathcal{Z}_A(f) \mid f \in I\}$ and $\mathfrak{Z}_A[I] = \bigcup\{\mathfrak{Z}_A(f) \mid f \in I\}$. Several properties of $\mathcal{Z}_A$ and $\mathfrak{Z}_A$ are proved in [3, 5, 6]. In particular, we have the following lemma, which we state here for convenience.

Lemma 1. *Let $A(X)$ be an intermediate ring. Then the following hold.*

(a) *For any noninvertible $f \in A(X)$, both $\mathcal{Z}_A(f)$ and $\mathfrak{Z}_A(f)$ are z-filters on X. If f is invertible, then $\mathcal{Z}_A(f) = \mathfrak{Z}_A(f) = \mathbf{Z}[X]$, the set of all zero sets in X.*

(b) *For an ideal $I \subset A(X)$, both $\mathcal{Z}_A[I]$ and $\mathfrak{Z}_A[I]$ are z-filters on X.*

(c) *For any $f \in A(X)$, we have $\mathfrak{Z}_A(f) = kh\mathcal{Z}_A(f)$.*

(d) *If $\mathcal{F}$ is a z-filter on X, then $\mathcal{Z}_A(f) \subseteq \mathcal{F}$ if and only if $\lim_{\mathcal{F}} fh = 0$ for all $h \in A(X)$.*

Item (a) is from [2, 6], Item (b) is from [3, 6], Item (c) is from [6], and Item (d) is from [2].

The *Stone-Čech compactification* of X, denoted βX, is any topological space homeomorphic to the space of z-ultrafilters on X topologized with the hull-kernel closure operator as follows: the closure of any set U of z-ultrafilters is $hk(U)$. Throughout this paper, we will in particular take βX to consist of a superset of X, whose points, which we denote $p, q, \ldots$, can be viewed as indices of z-ultrafilters on X. Let $\mathcal{U}$ be the map which associates every $p \in \beta X$ with a z-ultrafilter $\mathcal{U}_p$:

$$p \overset{\mathcal{U}}{\longmapsto} \mathcal{U}_p, \quad (5)$$

such that for each $p \in X$, $\mathcal{U}_p = \langle\{p\}\rangle$ is the fixed z-ultrafilter containing p, and for each $p \in \beta X \setminus X$, $\mathcal{U}_p$ is a unique free z-ultrafilter, such that $\mathcal{U}$ is a one-to-one correspondence between βX and the set of z-ultrafilters on X. The topology on βX is defined in such a way that the map $\mathcal{U}$ is a homeomorphism. Making use of the fact that the zero sets form a base for the collection of closed sets [1, page 38], one can check that $\mathcal{U}$ maps X homeomorphically onto the subspace of fixed z-ultrafilters, and hence X is a subspace of βX.

A z-filter $\mathcal{F}$ on X is called *A-stable* if for every $f \in A(X)$ there exists a set in $\mathcal{F}$ on which f is bounded (see [7, 8]). Following [7], for each $A(X)$ we define the *A-compactification* $\upsilon_A X$ of X as the subspace of βX where

$$\upsilon_A X = \left\{p \in \beta X \mid \mathcal{U}_p \text{ is } A\text{-stable}\right\}. \quad (6)$$

From [5, Theorem 4.6], it holds that $\upsilon_A X$ is a realcompactification of X. Note that if $A(X) = C^*(X)$, then $\upsilon_A X = \beta X$. If $A(X) = C(X)$, then $\upsilon_A X = \upsilon X$ is the *Hewitt realcompactification*.

To refine our understanding of the topology on $\upsilon_A X$, we define the *A-stable hull* h^A of a z-filter $\mathcal{F}$ by

$$h^A(\mathcal{F}) = \{\mathcal{U} \mid \mathcal{U} \text{ is an } A\text{-stable } z\text{-ultrafilter on } X, \mathcal{U} \supseteq \mathcal{F}\}. \quad (7)$$

It is immediate from the definition of the subspace topology that $\upsilon_A X$ is homeomorphic (via $\mathcal{U}|_{\upsilon_A X}$) to the hull-kernel topology restricted to A-stable z-ultrafilters, that is, the topology with the following closure operator: the closure of any set U is $h^A k(U)$. It follows that

$$p \in \mathrm{cl}_{\upsilon_A X} E \quad \text{iff } E \in \mathcal{U}_p. \quad (8)$$

From [7], we have that the space $\upsilon_A X$ consists of the points of βX to which every function $f \in A(X)$ can be continuously extended. We denote the extension of f to $\upsilon_A X$ by f^{υ_A}. From [7, Theorem 9], we have the value of f^{υ_A} at a point $p \in \upsilon_A X$ is given by

$$f^{\upsilon_A}(p) = \lim_{\mathcal{U}_p} f. \quad (9)$$

In [7], a ring $A(X)$ of continuous functions is called a *C-ring* if there is a completely regular space Y such that $A(X)$ is isomorphic to $C(Y)$. Clearly $C(X)$ and $C^*(X)$ are C-rings (with $C^*(X)$ isomorphic to $C(\beta X)$). We use the following result from [7, Theorem 7].

Lemma 2. *Let $A(X)$ be an intermediate ring. Then the following hold.*

(a) *$A(X)$ is a C-ring if and only if $A(X)$ is isomorphic to $C(\upsilon_A X)$.*

(b) *If $A(X)$ is a C-ring, then $f \in A(X)$ is invertible in $A(X)$ if and only if $\mathbf{Z}(f^{\upsilon_A}) = \emptyset$.*

In addition, it is shown in [5, Theorem 4.7] that there is a bijective correspondence between the realcompactifications of X and the C-rings on X.

3. Characterizations Using Realcompactifications

In this section, we utilize the realcompactifications of X to provide a new description of the function $\mathfrak{Z}_A$ and of maximal ideals of $A(X)$, when $A(X)$ is an intermediate C-ring. The new description of the maximal ideals of $A(X)$ generalizes the Gelfand-Kolmogorov theorem [1, page 102] for $C(X)$ to all intermediate C-rings.

3.1. A New Description of $\mathfrak{Z}_A$ for C-Rings. While the definition of $\mathfrak{Z}_A$ is essentially algebraic (using the property of local invertibility), we now define a function $\mathfrak{K}_A$ that provides a "topological description" (using realcompactifications of X) of a mapping from ideals of an intermediate C-ring $A(X)$ to z-filters on X, and we will show (Theorem 6) that $\mathfrak{K}_A$ coincides with $\mathfrak{Z}_A$ when $A(X)$ is an intermediate C-ring.

Definition 3. Let $A(X)$ be an intermediate ring and $f \in A(X)$. We set

$$\mathfrak{K}_A(f) = \{E \in \mathbf{Z}(X) \mid \mathbf{Z}(f^{\upsilon_A}) \subseteq \text{cl}_{\upsilon_A X} E\}. \quad (10)$$

For an ideal $I \subset A(X)$, we set $\mathfrak{K}_A[I] = \bigcup\{\mathfrak{K}_A(f) \mid f \in I\}$.

That $\mathfrak{K}_A$ is indeed a mapping from ideals of an intermediate C-ring $A(X)$ to z-filters on X will follow from our main result of the section that establishes that $\mathfrak{K}_A = \mathfrak{Z}_A$ when $A(X)$ is an intermediate C-ring. We also show that $\mathfrak{K}_A$ does not necessarily map ideals in $A(X)$ to z-filters on X when $A(X)$ is not an intermediate C-ring (Example 7).

First, to illustrate some connections that motivated the development of $\mathfrak{K}_A$, let us now observe a similarity between $\mathfrak{K}_A$ and a simple function $\widehat{\mathbf{Z}}$ that we define in terms of $\mathbf{Z}$. We define $\widehat{\mathbf{Z}}_X$ on $C(X)$ by

$$\widehat{\mathbf{Z}}_X : f \longmapsto \langle \mathbf{Z}(f) \rangle = \{E \in \mathbf{Z}(X) \mid \mathbf{Z}(f) \subseteq E\}. \quad (11)$$

We drop the subscript X when it is understood by context. It is easy to see that given an ideal I, $\widehat{\mathbf{Z}}[I] = \mathbf{Z}[I]$.

We now turn our attention to $C^*(X)$ and we note that $C^*(X)$ is an intermediate C-ring, as $C^*(X)$ is isomorphic to $C(\beta X)$. In light of this, we have for $f \in C(X)$ that

$$\widehat{\mathbf{Z}}_{\beta X} : f^{\beta} \longmapsto \langle \mathbf{Z}(f^{\beta}) \rangle = \{E_{\beta} \in \mathbf{Z}(\beta X) \mid \mathbf{Z}(f^{\beta}) \subseteq E_{\beta}\}. \quad (12)$$

Ultimately, we would like to find a map defined on $C^*(X)$ itself rather than $C(\beta X)$ and that maps to z-filters on X rather than βX (it is possible that a z-filter on βX may contain a set that does not meet X; for example, $\beta\mathbb{N} \setminus \mathbb{N}$ is the zero set of the function $f^{\beta\mathbb{N}}$, where $f(n) = 1/n$). Observe that

$$\begin{aligned}\bigcap \{E_{\beta} \in \mathbf{Z}(\beta X) \mid \mathbf{Z}(f^{\beta}) \subseteq E_{\beta}\} &= \mathbf{Z}(f^{\beta}) \\ &= \bigcap \{\text{cl}_{\beta X} E \mid E \in \mathbf{Z}(X), \mathbf{Z}(f^{\beta}) \subseteq \text{cl}_{\beta X} E\},\end{aligned} \quad (13)$$

where the first equality is immediate from the fact that $\mathbf{Z}(f^{\beta}) \in \mathbf{Z}(\beta X)$, and the second equality holds because the sets of the form $\text{cl}_{\beta X} E$ are a base for the closed sets in βX (see [1, page 94]). This motivates the following definition that relates $\mathbf{Z}(f^{\beta})$ to X:

$$\mathfrak{K}_{C^*} : f \longmapsto \{E \in \mathbf{Z}(X) \mid \mathbf{Z}(f^{\beta}) \subseteq \text{cl}_{\beta X} E\}. \quad (14)$$

We then generalize $\mathfrak{K}_{C^*}$ to all intermediate rings $A(X)$ to arrive at Definition 3.

In order to prove our main theorem that $\mathfrak{K}_A = \mathfrak{Z}_A$ whenever A is an intermediate C-ring, we need some lemmas. We first show that the zero set of f^{υ_A}, viewed as a set of z-ultrafilters, is the A-stable hull of $\mathcal{Z}_A(f)$.

Lemma 4. *If $A(X)$ is an intermediate ring of continuous functions,*

$$\mathbf{Z}(f^{\upsilon_A}) \equiv h^A(\mathcal{Z}_A(f)), \quad (15)$$

where the symbol $\equiv$ indicates that one set is the homeomorphic image of the other. In particular, if $f \in A(X)$, then $p \in \mathbf{Z}(f^{\upsilon_A})$ if and only if $\mathcal{U}_p \in h^A(\mathcal{Z}_A(f))$.

Proof. We observe that the following are equivalent:

(i) $p \in \mathbf{Z}(f^{\upsilon_A})$,

(ii) $\mathcal{U}_p$ is A-stable and $\lim_{\mathcal{U}_p} f = 0$,

(iii) $\mathcal{U}_p$ is A-stable and $\lim_{\mathcal{U}_p} fh = 0$ for all $h \in A(X)$,

(iv) $\mathcal{U}_p$ is A-stable and $\mathcal{U}_p \supseteq \mathcal{Z}_A(f)$, that is, $\mathcal{U}_p \in h^A(\mathcal{Z}_A(f))$.

The equivalence (i)⇔(ii) follows from (6) and (9). The equivalence (ii)⇔(iii) follows from the fact that $\mathcal{U}_p$ is A-stable, and hence h is bounded on some set in $\mathcal{U}_p$. The equivalence (iii)⇔(iv) follows from Lemma 1(d).

We use the notation $kh\mathcal{F}$ to denote the kernel of the hull of the z-filter $\mathcal{F}$, that is, the intersection of the set of z-ultrafilters containing $\mathcal{F}$.

Lemma 5. *Let $A(X)$ be an intermediate C-ring and $f \in A(X)$. Then*

$$kh\mathcal{Z}_A(f) = kh^A\mathcal{Z}_A(f). \quad (16)$$

Proof. Without loss of generality, we may assume that $f \geq 0$ (because $\mathcal{Z}_A(f) = \mathcal{Z}_A(f^2)$). Of course $kh\mathcal{Z}_A(f) \subseteq kh^A\mathcal{Z}_A(f)$. For the other containment, suppose $E \in kh^A\mathcal{Z}_A(f)$, so E belongs to every A-stable z-ultrafilter containing $\mathcal{Z}_A(f)$. We first note that by Lemma 4, $\mathcal{U}_p \in h^A\mathcal{Z}_A(f)$ if and only if $p \in \mathbf{Z}(f^{\upsilon_A})$. In other words, for any $E \in \mathbf{Z}[X]$, we have $E \in kh^A\mathcal{Z}_A(f)$ if and only if $E \in \mathcal{U}_p$ for every p in $\mathbf{Z}(f^{\upsilon_A})$ if and only if $\text{cl}_{\upsilon_A X} E \supseteq \mathbf{Z}(f^{\upsilon_A})$. Now suppose $\mathcal{U}_q$ is any z-ultrafilter (A-stable or not) containing $\mathcal{Z}_A(f)$. We show that E belongs to $\mathcal{U}_q$. If not, then there exists $F \in \mathcal{U}_q$ such that

$$E \cap F = \emptyset. \quad (17)$$

Since E and F are disjoint zero sets in X, they are completely separated (see [1, page 17]). So there is a continuous function h that takes the value 1 on E and 0 on F and $0 \leq h \leq 1$, so $h \in A(X)$. Consider the function $k = f + h \in A(X)$. Note that $k(x) = f(x)$ for $x \in F$. Also, since $k \geq 1$ on E and since $\text{cl}_{\upsilon_A X} E \supseteq \mathbf{Z}(f^{\upsilon_A})$, it follows that $\mathbf{Z}(k^{\upsilon_A}) = \emptyset$. By Lemma 2(b), k is invertible, and as $k \equiv f$ on F, it follows that $k^{-1}f \equiv 1$ on F. Since $F \in \mathcal{U}_q$, it also follows that $\lim_{\mathcal{U}_q} k^{-1}f \neq 0$, but this contradicts the fact that $\mathcal{U}_q \supseteq \mathcal{Z}_A(f)$ because of Lemma 1(d). This contradiction stems from the assumption that E does not belong to $\mathcal{U}_q$. So $E \in \mathcal{U}_q$ for every $\mathcal{U}_q \supseteq \mathcal{Z}_A(f)$, that is, $E \in kh\mathcal{Z}_A(f)$.

Recall that for any intermediate ring $A(X)$, Lemma 1(c) gives the relationship $\mathfrak{Z}_A(f) = kh\mathcal{Z}_A(f)$. For intermediate C-rings, we can characterize $\mathfrak{Z}_A$ topologically as $\mathfrak{K}_A$.

Theorem 6. *Let $A(X)$ be an intermediate C-ring and $f \in A(X)$. Then $\mathfrak{Z}_A(f) = \mathfrak{K}_A(f)$.*

Proof. Suppose $E \in \mathfrak{Z}_A(f)$. By Lemma 1(c), $E \in kh\mathcal{Z}_A(f)$. We show that in this case $p \in \mathbf{Z}(f^{\upsilon_A})$ implies $p \in \mathrm{cl}_{\upsilon_A X}E$. Now if $p \in \mathbf{Z}(f^{\upsilon_A})$, then by Lemma 4, $\mathcal{U}_p \in h^A(\mathcal{Z}_A(f))$. Since $E \in kh\mathcal{Z}_A(f)$, it follows that $E \in \mathcal{U}_p$, and hence by (8), we have that $p \in cl_{\upsilon_A X}E$.

For the other containment, suppose $E \in \mathfrak{K}_A(f)$, that is, $\mathbf{Z}(f^{\upsilon_A}) \subseteq \mathrm{cl}_{\upsilon_A X}E$. We show that E belongs to every A-stable z-ultrafilter containing $\mathcal{Z}_A(f)$. First, let $\mathcal{U}_p$ be an A-stable z-ultrafilter containing $\mathcal{Z}_A(f)$. Then by Lemma 4, $p \in \mathbf{Z}(f^{\upsilon_A})$, so by hypothesis, $p \in \mathrm{cl}_{\upsilon_A X}E$, and hence $E \in \mathcal{U}_p$. Thus E belongs to every A-stable z-ultrafilter containing $\mathcal{Z}_A(f)$; in other words, $E \in kh^A\mathcal{Z}_A(f)$. Since $A(X)$ is an intermediate C-ring, we can apply Lemma 5 so that $E \in kh\mathcal{Z}_A(f)$, and hence by Lemma 1(c), $E \in \mathfrak{Z}_A(f)$.

The theorem does not hold if the assumption that $A(X)$ being an intermediate C-ring is removed. In fact, $\mathfrak{K}_A$ need not map ideals in $A(X)$ to z-filters on X when $A(X)$ is not a C-ring, as the following example shows.

Example 7. Let $X = [1, \infty)$. Let $A(X)$ be the smallest ring of continuous functions containing both $C^*(X)$ and $f(x) = x$. Then each function $h \in A(X)$ has the form $h(x) = \sum_{k=0}^{n} f_k(x)x^k$ for $f_k \in C^*(X)$. Note that any function $h \in A(X)$ is therefore bounded by some function of the form Mx^n for $M \in \mathbb{R}$ and $n \in \mathbb{N}$, where M is $n + 1$ times a common bound for all of f_k. We now observe that $e^x \notin A(X)$, since e^x cannot be bounded by any function of the form Mx^n for $M \in \mathbb{R}$ and $n \in \mathbb{N}$. Let $g = e^{-x}$. Then g is in $A(X)$ but is not invertible in $A(X)$. Thus $A(X)$ and $C(X)$ are not isomorphic.

Since every set of a free z-ultrafilter on $X = [1, \infty)$ must be unbounded, the identity function f is not bounded on any such set either. Hence any free z-ultrafilter on X is not A-stable, and $\upsilon_A(X) = X$. Then by Lemma 2(a), $A(X)$ is not a C-ring, since $C(\upsilon_A(X)) = C(X)$ and $C(X) \neq A(X)$.

Finally, since $g^{\upsilon_A} = g$ and $\mathbf{Z}(g) = \emptyset$, the set $\mathfrak{K}_A(g) = \{E \in Z(X) \mid \mathbf{Z}(g^{\upsilon_A}) \subseteq \mathrm{cl}_{\upsilon_A X}E\}$ consists of all zero sets of X (hence is not a z-filter), while $\mathfrak{Z}_A(g)$ is the z-filter consisting of all zero sets in X whose complement in X has an upper bound. Thus $\mathfrak{K}_A(g) \neq \mathfrak{Z}_A(g)$, which is in contrast to the conclusion of Theorem 6.

We leave open the question as to precisely what rings $A(X)$ are such that $\mathfrak{K}_A = \mathfrak{Z}_A$. We also leave open the question as to whether there exists a ring $A(X)$ such that $\mathfrak{K}_A$ does map ideals in $A(X)$ to z-filters on X, but $\mathfrak{K}_A \neq \mathfrak{Z}_A$.

3.2. Characterizing Maximal Ideals in C-Rings. The following characterization for maximal ideals in $A(X)$ is proved in [6]. Every maximal ideal in $A(X)$ is of the form

$$M_A^p = \{f \in A(X) \mid \mathfrak{Z}_A(f) \subseteq \mathcal{U}_p\}, \tag{18}$$

for $p \in \beta X$. By Lemma 1(c), $\mathfrak{Z}_A(f) \subseteq \mathcal{U}_p$ whenever $\mathcal{Z}_A(f) \subseteq \mathcal{U}_p$, and hence by (9) and Lemma 1(d), we have

$$M_{C^*}^p = \{f \in C^*(X) \mid f^\beta(p) = 0\}. \tag{19}$$

Since $\mathfrak{Z}_C(f) = \langle \mathbf{Z}(f) \rangle$, we have by (8) that

$$\begin{aligned} M_C^p &= \{f \in C(X) \mid \langle \mathbf{Z}(f) \rangle \subseteq \mathcal{U}_p\} \\ &= \{f \in C(X) \mid p \in \mathrm{cl}_{\beta X}\mathbf{Z}(f)\}. \end{aligned} \tag{20}$$

The characterizations in (19) and (20) that we obtained from (18) agree with those given in [1, pages 101-102]; the latter characterization of M_C^p is called the Gelfand-Kolmogorov theorem [1, page 102]. The following theorem provides a characterization of M_A^p that we see extends both (20) and (19) to all intermediate C-rings $A(X)$.

Theorem 8. *Let $A(X)$ be an intermediate C-ring. Then each maximal ideal in $A(X)$ is of the form*

$$M_A^p = \{f \in A(X) : p \in \mathrm{cl}_{\beta X}\mathbf{Z}(f^{\upsilon_A})\}, \tag{21}$$

where $p \in \beta X$.

Proof. $f \in M_A^p$ if and only if $\mathcal{Z}_A(f) \subseteq \mathcal{U}_p$ if and only if $\mathcal{U}_p \in h(\mathcal{Z}_A(f))$. By identifying a set in βX with its image under $\mathcal{U}$, it follows from the definition of closure in βX, Lemmas 4 and 5, that

$$\begin{aligned} \mathrm{cl}_{\beta X}\mathbf{Z}(f^{\upsilon_A}) &= hk\mathbf{Z}(f^{\upsilon_A}) = hkh^A\mathcal{Z}_A(f) \\ &= hkh\mathcal{Z}_A(f) = h\mathcal{Z}_A(f). \end{aligned} \tag{22}$$

The result now follows from the fact that $h(\mathcal{Z}_A(f)) = \mathrm{cl}_{\beta X}\mathbf{Z}(f^{\upsilon_A})$.

We now verify that this theorem generalizes the Gelfand-Kolmogorov Theorem [1, page 102] to intermediate C-rings. If $A(X) = C(X)$, then $\upsilon_A X = \upsilon X$, the Hewitt realcompactification of X. Now using the fact that $\mathrm{cl}_{\upsilon X}\mathbf{Z}(f) = \mathbf{Z}(f^\upsilon)$ [1, page 118] and Theorem 8, we have

$$\begin{aligned} M_C^p &= \{f \in C(X) \mid p \in \mathrm{cl}_{\beta X}\mathbf{Z}(f^\upsilon)\} \\ &= \{f \in C(X) \mid p \in \mathrm{cl}_{\beta X}cl_{\upsilon X}\mathbf{Z}(f)\} \\ &= \{f \in C(X) \mid p \in \mathrm{cl}_{\beta X}\mathbf{Z}(f)\}. \end{aligned} \tag{23}$$

For the case where $A(X) = C^*(X)$, we have

$$\begin{aligned} M_{C^*}^p &= \{f \in C^*(X) \mid p \in \mathrm{cl}_{\beta X}\mathbf{Z}(f^\beta)\} \\ &= \{f \in C^*(X) \mid f^\beta(p) = 0\}. \end{aligned} \tag{24}$$

So Theorem 8 simultaneously generalizes the results of (20) for $C(X)$ and (19) for $C^*(X)$ to all intermediate C-rings.

4. The Map $\mathfrak{Z}_A^{\leftarrow}$ and Ideals in $A(X)$

Recall that $\mathfrak{Z}_A$ maps ideals to z-filters (Lemma 1(a)). Here we show that for an intermediate C-ring $A(X)$, the inverse map $\mathfrak{Z}_A^{\leftarrow}$, defined by

$$\mathfrak{Z}_A^{\leftarrow}[\mathcal{F}] = \{f \in A(X) : \mathfrak{Z}_A(f) \subseteq \mathcal{F}\} \quad (25)$$

maps z-filters on X to ideals in $A(X)$ (Theorem 14). The corresponding result for the maps $\mathbf{Z}$ (for $C(X)$) and $\mathcal{Z}_A$ (for any intermediate ring $A(X)$) is proved, respectively, in [1] and [3]. Our proof for $\mathfrak{Z}_A$ makes use of Theorem 6.

We need some lemmas concerning meets and joins on the lattice of z-filters. Recall that $\mathcal{F} \vee \mathcal{G}$ is the smallest z-filter containing both the z-filters $\mathcal{F}$ and $\mathcal{G}$. Similarly, $\mathcal{F} \wedge \mathcal{G}$ is the largest z-filter contained in both the z-filters $\mathcal{F}$ and $\mathcal{G}$.

The following lemma is from [6].

Lemma 9. *Let $A(X)$ be an intermediate ring and $f, g \in A(X)$.*

(a) $\mathcal{Z}_A(fg) = \mathcal{Z}_A(f) \wedge \mathcal{Z}_A(g)$.

(b) $\mathcal{Z}_A(f+g) \subseteq \mathcal{Z}_A(f) \vee \mathcal{Z}_A(g)$.

(c) *If $f, g \geq 0$, then* $\mathcal{Z}_A(f+g) = \mathcal{Z}_A(f) \vee \mathcal{Z}_A(g)$.

The following is a special case of [6, Lemma 4.2(a)].

Lemma 10. *For all $f, g \in A(X)$, $kh(\mathcal{Z}_A(f) \wedge \mathcal{Z}_A(g)) = kh(\mathcal{Z}_A(f)) \wedge kh(\mathcal{Z}_A(g))$.*

To obtain the analog of Lemma 10 for joins, we need the following lemma.

Lemma 11. *If E is a zero set in X and if $\mathbf{Z}(f^{\upsilon_A}) \cap \mathbf{Z}(g^{\upsilon_A}) \subseteq \mathrm{cl}_{\upsilon_A X} E$ then there exist zero sets Z_1 and Z_2 in X such that $\mathbf{Z}(f^{\upsilon_A}) \subseteq \mathrm{cl}_{\upsilon_A X} Z_1$, $\mathbf{Z}(g^{\upsilon_A}) \subseteq \mathrm{cl}_{\upsilon_A X} Z_2$, and $Z_1 \cap Z_2 = E$.*

Proof. Let $H = \mathrm{cl}_{\upsilon_A X} E$ and let $Y = \upsilon_A X \setminus H$. Now, the sets $\mathbf{Z}(f^{\upsilon_A}) \cap Y$ and $\mathbf{Z}(g^{\upsilon_A}) \cap Y$ are disjoint zero sets in Y, so are contained in disjoint zero set neighborhoods W_1 and W_2 in Y. Moreover, since W_1 is a neighborhood of $\mathbf{Z}(f^{\upsilon_A}) \cap Y$, it follows that

$$\mathbf{Z}(f^{\upsilon_A}) \cap Y \subseteq \mathrm{cl}_{\upsilon_A X}(W_1) = \mathrm{cl}_{\upsilon_A X}(W_1 \cap X). \quad (26)$$

Similarly $\mathbf{Z}(g^{\upsilon_A}) \cap Y \subseteq \mathrm{cl}_{\upsilon_A X}(W_2 \cap X)$. Let

$$Z_1 = (W_1 \cap X) \cup E, \qquad Z_2 = (W_2 \cap X) \cup E. \quad (27)$$

Since $W_1 \cap X$ is a zero set in $X \setminus E$, it follows that Z_1 is a zero set in X by the fact that the cozero set of a cozero set is a cozero set [9, Proposition 1.1]. Similarly, Z_2 is a zero set in X. Also, $\mathbf{Z}(f^{\upsilon_A}) \subseteq \mathrm{cl}_{\upsilon_A X} Z_1$, because

$$\begin{aligned} \mathbf{Z}(f^{\upsilon_A}) &= (\mathbf{Z}(f^{\upsilon_A}) \cap Y) \cup (\mathbf{Z}(f^{\upsilon_A}) \cap H) \\ &\subseteq \mathrm{cl}_{\upsilon_A X}(W_1 \cap X) \cup \mathrm{cl}_{\upsilon_A X} E \\ &= \mathrm{cl}_{\upsilon_A X}((W_1 \cap X) \cup E) \\ &= \mathrm{cl}_{\upsilon_A X} Z_1. \end{aligned} \quad (28)$$

Similarly, $\mathbf{Z}(g^{\upsilon_A}) \subseteq \mathrm{cl}_{\upsilon_A X} Z_2$. Finally, $Z_1 \cap Z_2 = E$ because $W_1 \cap X$ and $W_2 \cap X$ are disjoint.

The next lemma shows that the kernel-hull operation distributes over the join operation on the lattice of z-filters.

Theorem 12. *If $A(X)$ is an intermediate C-ring, then*

$$kh(\mathcal{Z}_A(f) \vee \mathcal{Z}_A(g)) = kh(\mathcal{Z}_A(f)) \vee kh(\mathcal{Z}_A(g)). \quad (29)$$

Proof. The containment $kh(\mathcal{Z}_A(f) \vee \mathcal{Z}_A(g)) \supseteq kh(\mathcal{Z}_A(f)) \vee kh(\mathcal{Z}_A(g))$ is a special case of [6, Lemma 4.2(b)].

We show that the other containment is equivalent to Lemma 11 and hence must hold. First, we show the equivalence of the premises by showing that the following are equivalent. Recall that $\mathcal{Z}_A(f) = \mathcal{Z}_A(f^2)$, and hence we can assume without loss of generality that $f, g \geq 0$.

(i) $E \in kh(\mathcal{Z}_A(f) \vee \mathcal{Z}_A(g))$.

(ii) $E \in kh(\mathcal{Z}_A(f+g))$.

(iii) $\mathbf{Z}((f+g)^{\upsilon_A}) \subseteq \mathrm{cl}_{\upsilon_A X} E$.

(iv) $\mathbf{Z}(f^{\upsilon_A}) \cap \mathbf{Z}(g^{\upsilon_A}) \subseteq \mathrm{cl}_{\upsilon_A X} E$.

The equivalence (i)⇔(ii) follows from Lemma 9(c). The equivalence (ii)⇔(iii) follows from Theorem 6. The equivalence (iii)⇔(iv) follows from the assumption that $f, g \geq 0$. This establishes the equivalence between the premises of left-to-right containment of this lemma and Lemma 11.

For the equivalence of the conclusions, note that $E \in kh(\mathcal{Z}_A(f)) \vee kh(\mathcal{Z}_A(g))$ if and only if E is an intersection of a set in $kh(\mathcal{Z}_A(f))$ with a set in $kh(\mathcal{Z}_A(g))$. In other words, there must exist $Z_1 \in kh(\mathcal{Z}_A(f))$ and $Z_2 \in kh(\mathcal{Z}_A(g))$ such that $Z_1 \cap Z_2 = E$. By Lemmas 4 and 5, this is equivalent to the statement that there exist zero sets Z_1, Z_2 such that $\mathbf{Z}(f^{\upsilon_A}) \subseteq \mathrm{cl}_{\upsilon_A X} Z_1$, $\mathbf{Z}(g^{\upsilon_A}) \subseteq \mathrm{cl}_{\upsilon_A X} Z_2$, and $Z_1 \cap Z_2 = E$, which is the conclusion of Lemma 11.

Corollary 13. *Let $A(X)$ be an intermediate ring of continuous functions and $f, g \in A(X)$. Then*

(a) $\mathfrak{Z}_A(fg) = \mathfrak{Z}_A(f) \wedge \mathfrak{Z}_A(g)$,

(b) $\mathfrak{Z}_A(f+g) \subseteq \mathfrak{Z}_A(f) \vee \mathfrak{Z}_A(g)$.

Proof. Item (a) immediately follows from Lemmas 10 and 9(a).

Item (b) immediately follows from Theorem 12 and Lemma 9(b).

We are now ready to prove the main result of this section.

Theorem 14. *Let $A(X)$ be an intermediate C-ring. If $\mathcal{F}$ is a z-filter on X, then $\mathfrak{Z}_A^{\leftarrow}[\mathcal{F}]$ is an ideal in $A(X)$.*

Proof. Let $\mathcal{F}$ be a z-filter and $I = \mathfrak{Z}_A^{\leftarrow}[\mathcal{F}]$. If $f \in I$ and $g \in A(X)$, then $\mathfrak{Z}_A(f) \subseteq \mathcal{F}$. By Corollary 13(a), $\mathfrak{Z}_A(fg) = \mathfrak{Z}_A(f) \wedge \mathfrak{Z}_A(g) \subseteq \mathcal{F}$, so $fg \in I$. If $f, g \in I$, then both $\mathfrak{Z}_A(f) \subseteq \mathcal{F}$ and $\mathfrak{Z}_A(g) \subseteq \mathcal{F}$. By Corollary 13(b), $\mathfrak{Z}_A(f+g) \subseteq \mathfrak{Z}_A(f) \vee \mathfrak{Z}_A(g) \subseteq \mathcal{F}$, so $f+g \in I$. Thus I is an ideal.

Acknowledgment

Joshua Sack was partially supported by The Netherlands Organisation for Scientific Research VIDI Project 639.072.904.

References

[1] L. Gillman and M. Jerison, *Rings of Continuous Functions*, Springer, New York, NY, USA, 1976.

[2] L. Redlin and S. Watson, "Maximal ideals in subalgebras of $C(X)$," *Proceedings of the American Mathematical Society*, vol. 100, no. 4, pp. 763–766, 1987.

[3] H. L. Byun and S. Watson, "Prime and maximal ideals in subrings of $C(X)$," *Topology and its Applications*, vol. 40, no. 1, pp. 45–62, 1991.

[4] L. H. Byun, L. Redlin, and S. Watson, "Local bounded inversion in rings of continuous functions," *Commentationes Mathematicae*, vol. 37, pp. 55–63, 1997.

[5] L. Redlin and S. Watson, "Structure spaces for rings of continuous functions with applications to realcompactifications," *Fundamenta Mathematicae*, vol. 152, no. 2, pp. 151–163, 1997.

[6] P. Panman, J. Sack, and S. Watson, "Correspondences between ideals and z-filters for rings of continuous functions between C^* and C," *Commentationes Mathematicae*, vol. 52, no. 1, pp. 11–20, 2012.

[7] H. L. Byun, L. Redlin, S. Watson, and "A-compactications and rings of continuous functions between C^* and C," *Topological Vector Spaces, Algebras, and Related Areas*, vol. 316, pp. 130–139, 1995.

[8] M. Mandelker, "Supports of continuous functions," *Transactions of the American Mathematical Society*, vol. 156, pp. 73–83, 1971.

[9] R. L. Blair and A. W. Hager, "Extensions of zero-sets and of real-valued functions," *Mathematische Zeitschrift*, vol. 136, pp. 41–52, 1974.

12

Oscillation in Pest Population and Its Management: A Mathematical Study

Samit Bhattacharyya and Suma Ghosh

Departments of Mathematics and Biology, University of Utah, Salt Lake City, UT 84112, USA

Correspondence should be addressed to Samit Bhattacharyya; samit@math.utah.edu

Academic Editor: H. Srivástava

We study the role of predation dynamics in oscillation of pest population in insect ecology. A two-dimensional pest control model (under the use of insecticides) with time delay in predation is considered in this paper. By the Hopf bifurcation theory, we prove the existence of the stable oscillation of the system. We also consider the economic viability of the control process. First we improve the Pontryagin maximum principle (PMP) where the delay in the system is sufficiently small and control function is linear, and then we apply the improved version of PMP to perform the optimal analysis of the pest control model as a special case.

1. Introduction

Long-term forecasts of pest pressure are central to the effective management of many agricultural and forest insect pests, because prediction of abundance and distribution (pest pressure) of a pest species is crucial for both strategic planning and tactical decision-making [1]. For example, timely forecasts would be useful for determining insecticide budgets. However, forecasting pest pressure is more problematic, because many factors influence the abundance of pests. For instance, abundance of predators interacting in the food web associated with the pest species or the effects of weather, rainfall, and temperature (e.g., [2]) may lead to a sustained oscillation in the pest population. There are a number of insects (mostly insects attacking forest trees) which exhibit fluctuating population density [3–5], although usually regular and occurring repeatedly between well-defined lower and upper boundaries. As these insects cause widespread economic damage, the reasons of their outbreaks have been a focus of intensive research.

There are various factors which cause the population changes in insect ecology such as density-dependent birth-death mechanism [5] or external factors like weather. For example, Stenseth et al. [6] found that the dynamics of Canadian lynx (*Lynx canadensis*) populations is mostly driven by weather condition. Since 1950s, ecologists have been proposing that "populations are limited either by extrinsic factors—such as weather, especially extremes of cold, drought, or rainfall—or by intrinsic factors—such as birth and death rates, or interactions with other species" [3, 5, 7, 8]. But the biological mechanisms that drive the oscillations are not yet well investigated.

1.1. A Case Study. Turchin and his group [9] designed a long-term experiment to test the hypothesis in favor of the existence of cycles in one forest insect, the southern pine beetle (SPB) *Dendroctonus frontalis*. They suspected that the mechanism behind this oscillation was beetle's population interaction with its predators. Time-series analysis indicated that SPB fluctuations were driven primarily by endogenous (density-dependent) factors: approximately 80% of the variance in the rate of population change was explained by a joint action of current and lagged population densities. The evidence for second-order dynamics, that is, delayed density dependence [5], was strong, because regression of the rate of population change on lagged density was highly significant and it alone explained 55% of the variance [9].

However, they have demonstrated [10] and drawn three broad outcomes corresponding to the hypotheses that predation was (i) an exogenous, (ii) a first-order endogenous, or (iii) a second-order endogenous factor (see Figure 1 in Supplementary Material: SI_figures available online at http://dx.doi.org/10.1155/2013/653080). In the first case, there is no dynamical feedback between prey density and the predation impact. The average predator-induced mortality may be very high, and still predators would have no dynamical impact, simply reducing the intrinsic rate of population increase to a lower value. Fluctuations in predator-imposed mortality affect prey density in a stochastic manner, but they cannot drive a regular oscillation. In the second case, predators respond to changes in prey population without a significant lag time. Therefore the dynamical role of predators, in this case, was stabilizing rather than causing oscillations. Generalist predators may act in this manner, reducing the amplitude of oscillations or preventing diverging oscillations. Only in the third case, when acting in a delayed density-dependent manner, the predators are actually causing the oscillation. Thus their experimental study suggests that predation with time delay may be one of the fundamental causes of population oscillation in insect ecology.

1.2. Our Motivation. The experimental study by Turchin and his colleagues motivated us to formulate a mathematical framework to describe this oscillatory phenomenon in insect ecosystem. The goal of the paper is twofold. The first goal is to characterize the delay involved in predation that leads to oscillation in population and also what makes the oscillation stable! To mention, our study on describing oscillation for such kind of biological mechanism is not only to support Turchin's experiments on SPB, but also important in other aspects of biological questions such as whether any control measure can stabilize the oscillation and if any, how it happens.

The second goal relates to the economic viability of the control process using optimization theory. We develop Pontryagin's maximum principle (PMP) for delay differential equation where the delay is sufficiently small and underlined control function in the system is linear, and then we apply this theorem in our model. This is however the simplest condition necessary to develop PMP under delay. Similar things may be generalized with measurable functions under suitable restrictions, but this is outside the scope of current study. However, the second part on optimal control theory has its own importance independent of the first part.

2. Mathematical Model and Preliminaries

The mathematical formulation of pest control problem and its management (i.e., bionomic aspects of the pest control model under various control measure, like spraying of insecticide, release of additional predator or sterile males) have been taken up very recently by Bhattacharya and Karan [11], Bhattacharyya and Bhattacharya [12, 13], Ghosh et al. [14, 15]. But ecological aspect of the pest control problem has not been emphasized yet very much [16]. We consider a simple pest control model of a density dependent pest population under

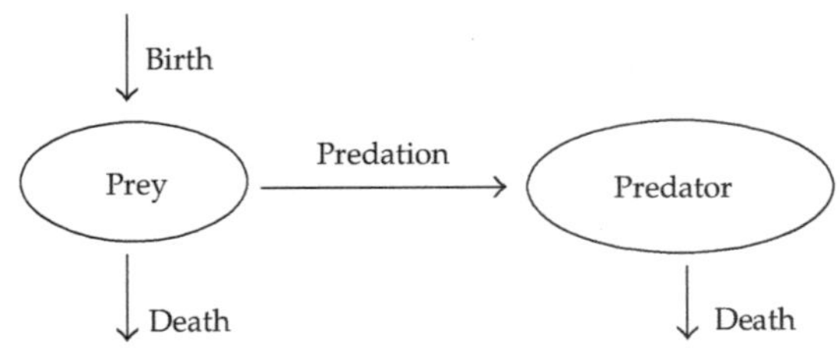

FIGURE 1: Transition diagram of prey-predator system. Arrows and their directions indicate the underlined biological process and its positive or negative effect on the species.

application of insecticide at the constant rate "u" given by Figure 1

$$\frac{dx}{dt} = rx\left(1-\frac{x}{K}\right) - \frac{\alpha xy}{\gamma + x} - q_1 ux,$$
$$\frac{dy}{dt} = y\left(-d - \epsilon y + \frac{\beta x}{\gamma + x}\right) - q_2 uy, \tag{1}$$

where x and y represent the density of the population of pest and its natural enemy, respectively. r denotes the positive intrinsic birth rate of pest, K is its carrying capacity, d is the natural mortality of predators, and ϵ is the density dependence factor. The terms in the model "$q_1 ux$" and "$q_2 uy$" define the pest and predator mortality due to application of insecticide, where q_1 and q_2 denote proportional damage of species due to per unit application of pesticide. We consider the Holling-type II predational form for describing the grazing phenomenon [17, 18].

The initial condition for model (1) may be taken as any point in the region R_+^2, where by R_+^2 we mean

$$R_+^2 = \{(x, y) \in R^2 : x \geq 0,\ y \geq 0\} \tag{2}$$

and R denotes the set of real numbers.
Equilibria. The system has three equilibria, namely, the trivial equilibrium $E_0 = (0,0)$, axial equilibrium $E_K = (K,0)$, and interior equilibrium $E_* = (x^*, y^*)$, where

$$y^* = \frac{(\beta - d - q_2 u)\, x^* - (d + q_2 u)\, \gamma}{\epsilon (\gamma + x^*)}, \tag{3}$$

and x^* is given by the following equation:

$$a_1 x^3 + a_2 x^2 + a_3 x - a_4 = 0, \tag{4}$$

where $a_1 = \epsilon r$, $a_2 = 2\gamma\epsilon r - K(r - q_1 u)\epsilon$, $a_3 = \epsilon r\gamma^2 + \alpha K(\beta - d - q_2 u) - 2\epsilon\gamma K(r - q_1 u)$, and $a_4 = \alpha K\gamma(d + q_2 u) + (r - q_1 u)K\epsilon\gamma^2$.

When carrying capacity "K" is high, the coefficient a_2 is negative and a_4 is positive, when $u < r/q_1$. Hence, by Decarte's rule of sign, (5) has either three or one positive root depending on values of other parameters. Moreover, if we notice the expression of the coefficients, it reveals that application of insecticide decreases the pest population in the system.

Now, the constant term $a_4 = 0 \Rightarrow u = u_0 := (\alpha d + r\epsilon\gamma)/(q_1\epsilon\gamma - q_2\alpha)$, which shows that if we apply the insecticide of amount u_0, pest will disappear from the system and the system reaches the trivial equilibrium $(0,0)$. We will also show later that under this condition the trivial equilibrium $(0,0)$ becomes a stable one. Again, for the persistence of the predator in the system, it is necessary that $u < (\beta - d)/q_2$ and moreover

$$x^* > \frac{(d + q_2 u)\gamma}{\beta - d - q_2 u}. \quad (5)$$

Equation (6) indicates that the minimum number of pest $(d + q_2 u)\gamma/(\beta - d - q_2 u)$ is required for persistence of predator in the system. However, $u_0 > r/q_1$, and so it would be enough to keep the balance of the ecosystem, if we maintain

$$u < u_{\max} := \min\left(\frac{r}{q_1}, \frac{\beta - d}{q_2}\right). \quad (6)$$

3. Model with Distributed Delay

We assume that the predation activity of predators involves some time lag. We also assume that this predation activity follows a gamma distribution. This form of distributed delay kernels are widely studied in many biological modelings ([19, 20] and references therein). Under the use of such kernels, our system (1) can be expressed as

$$\begin{aligned}\frac{dx}{dt} &= rx\left(1 - \frac{x}{K}\right) - \frac{\alpha xy}{\gamma + x} - q_1 ux,\\ \frac{dy}{dt} &= y(-d - \epsilon y)\\ &\quad + \beta\left[\int_{\infty}^{t} a^{k+1}\frac{(t-s)^k}{k!}e^{-a(t-s)}\frac{x(s)}{\gamma + x(s)}ds\right]y - q_2 uy,\end{aligned} \quad (7)$$

where k, a nonnegative integer, is the order of the delay kernel and a is a nonnegative real number. Both k and a are related with the mean time lag by $T = (k+1)/a$. However we shall study the system for $k = 0$. With the transformation $x_1(t) = a\int_{\infty}^{t} e^{-a(t-s)}(x(s)/(\gamma + x(s)))ds$, the above system reduces to the following equivalent coupled differential equations:

$$\begin{aligned}\frac{dx}{dt} &= rx\left(1 - \frac{x}{K}\right) - \frac{\alpha xy}{\gamma + x} - q_1 ux,\\ \frac{dy}{dt} &= y(-d - \epsilon y + \beta x_1) - q_2 uy,\\ \frac{dx_1}{dt} &= a\left(\frac{x}{\gamma + x} - x_1\right).\end{aligned} \quad (8)$$

The system (7) possesses the same equilibria as obtained for the system (1). It is easy to see that trivial equilibrium E_0 becomes asymptotically stable if $u > r/q_1$, otherwise it becomes unstable for $u < r/q_1$. Again, the axial equilibrium E_K is locally asymptotically stable if $u > (\beta - d)/q_2$. But this violates the positivity of y^* in (4). Thus, existence of positive interior equilibrium E^* implies that the axial equilibrium E_K is an unstable saddle in character. The characteristic equation around $E^* = (x^*, y^*)$ is given by

$$\lambda^3 + p_0\lambda^2 + p_1\lambda + p_2 = 0, \quad (9)$$

where

$$\begin{aligned}p_0 &= \left(\frac{rx^*}{K} - \frac{\alpha x^* y^*}{(\gamma + x^*)^2}\right) + \epsilon y^* + a,\\ p_1 &= \left(\frac{rx^*}{K} - \frac{\alpha x^* y^*}{(\gamma + x^*)^2}\right)(\epsilon y^* + a) + a\epsilon y^*,\\ p_2 &= \left(\frac{rx^*}{K} - \frac{\alpha x^* y^*}{(\gamma + x^*)^2}\right)a\epsilon y^* + \frac{a\alpha\beta\gamma x^* y^*}{(\gamma + x^*)^3}.\end{aligned} \quad (10)$$

Using Routh-Hurewitz criterion, we see that real parts of all roots are negative if

$$\frac{rx^*}{K} - \frac{\alpha x^* y^*}{(\gamma + x^*)^2} > 0 \quad (11)$$

holds, and thus E^* is locally asymptotically stable. Hence, in this case, there is no possibility of exchange of stability.

Next we show by applying Benedixson-Dulac criteria [21] that the system (8) around E^* has no periodic solutions.

Let $h(x, y, x_1) = (\gamma + x)/xy^{m+1}$, where m is to be chosen suitably. Obviously, $h(x, y, x_1) > 0$, whenever $x, y, x_1 > 0$. We define

$$\begin{aligned}g_1(x, y, x_1) &= rx\left(1 - \frac{x}{K}\right) - \frac{\alpha xy}{\gamma + x} - q_1 ux,\\ g_2(x, y, x_1) &= y(-d - \epsilon y + \beta x_1) - q_2 uy\\ g_3(x, y, x_1) &= a\left(\frac{x}{\gamma + x} - x_1\right),\end{aligned} \quad (12)$$

$$\Omega(x, y, x_1) = \frac{\partial}{\partial x}(g_1 h) + \frac{\partial}{\partial y}(g_2 h) + \frac{\partial}{\partial x_1}(g_3 h).$$

Then

$$\begin{aligned}&\Omega(x, y, x_1)\\ &\le \frac{1}{xy^{m+1}}\left[r\left(1 - \frac{\gamma}{K}\right)x + (\gamma + x)\{md + (m-1)\epsilon y - a\}\right].\end{aligned} \quad (13)$$

So, if we take $m < 1$, $md - a < 0$, and $r(1 - \gamma/K) - (md - a)\gamma$, then $\Omega(x, y, x_1) < 0$ for all $x, y, x_1 > 0$. However, these imply together $m < \min((a - r/\gamma + r/K)/d, 1)$. Therefore, there will be no periodic solution of the system (8).

Hence we find that the cyclic nature of the pest population in such forestry or agriculture ecosystem cannot be explained by this type of distribution of predation activity of natural enemies. It is also worth noticing that if the order k of the delay kernel goes to infinity, while keeping the mean delay $T = (k+1)/a$ fixed, then the distributed delay can be viewed as a discrete delay [22]. Hence to explain the periodic nature of pest population, we shall assume that the process of predation activity follows a discrete time variation.

4. Model with Discrete Delay

Under this assumption that predation activity follows a discrete time variation, the system (1) takes the following form:

$$\begin{aligned} \frac{dx}{dt} &= rx\left(1 - \frac{x}{K}\right) - \frac{\alpha xy}{\gamma + x} - q_1 ux, \\ \frac{dy}{dt} &= y\left(-d - \epsilon y + \frac{\beta x(t-\tau)}{\gamma + x(t-\tau)}\right) - q_2 uy, \end{aligned} \tag{14}$$

where τ denotes the discrete time delay.

There are series of papers published by Chaplain and his colleagues on delayed predator-prey system. For example, Xu and Chaplain [23] studied the following delayed Gause-type predator-prey system without dominating instantaneous negative feedbacks mechanism:

$$\begin{aligned} \dot{x}_1 &= x_1\left(a_1 - a_{11}x_1(t-\tau_1) - a_{12}\frac{x_2(t)}{m + x_1(t)}\right), \\ \dot{x}_2 &= x_2\left(-a_2 + a_{21}\frac{x_1(t-\tau_2)}{m + x_1(t-\tau_2)} - a_{22}x_2(t-\tau_3)\right). \end{aligned} \tag{15}$$

They investigated the uniform persistence of the system under some suitable parametric conditions. In another paper, Xu et al. [24] considered a ratio-dependent predator-prey model with distributed time delay. There is another recent paper by Lin and Yuan [25] on prey-predator system with distributed time delay, where the delay kernel was general gamma distribution:

$$\begin{aligned} \dot{u} &= \gamma u\left(1 - \frac{u}{K}\right) - \frac{m}{y}\left(\frac{vu}{a+u}\right), \\ \dot{v} &= v\left[m\left(g * \left(\frac{u}{a+u}\right)\right) - d\right], \end{aligned} \tag{16}$$

where the convolution is defined by $g * (u/(a+u))(t) = \int_{-\infty}^{t} g(t-s)(u(s)/(a+u(s)))ds$. However, both of the studies do not consider the intraspecific competition of the predator species.

One special case of system (14) is the following model studied by Freedman and Ruan [26]:

$$\begin{aligned} \dot{u} &= \gamma u\left(1 - \frac{u}{K}\right) - \frac{m}{y}\left(\frac{vu}{a+u}\right), \\ \dot{v} &= v\left[\frac{mu(t-\tau)}{a + u(t-\tau)} - d\right]. \end{aligned} \tag{17}$$

Zhao et al. [27] establish the existence of periodic solution of system by constructing an appropriate map and showing that the map has a nontrivial fixed point. Faria [28] considered the prey-predator system with delay in predation function:

$$\begin{aligned} \dot{x} &= x(t)\left[a_1 - a_{11}x(t) - a_{12}y(t-\tau_1)\right], \\ \dot{y} &= y(t)\left[-a_2 + a_{21}x(t-\tau_2) - a_{22}y(t)\right], \end{aligned} \tag{18}$$

where τ_1 and $\tau_2 \geq 0$ are constants. There is another most recent work on delayed system by Yan and Li [29] as a modification of the above system. This paper also discusses the global existence of periodic solution in the system. However, results in these works may be found as a special case of our system.

Another work that studied the effect of delay in prey-predator system is done by Chattopadhyay et al. [30]:

$$\begin{aligned} \frac{dP}{dt} &= rP\left(1 - \frac{P}{K}\right) - \alpha PZ, \\ \frac{dZ}{dt} &= \beta PZ - \mu Z - \frac{\theta P(t-\tau)}{\gamma + P(t-\tau)}Z. \end{aligned} \tag{19}$$

They considered the phytoplankton-zooplankton system to understand the mechanism of harmful algal blooms in presence of toxic substance and incorporated the delay in the zooplankton mortality term due to liberation of toxic substance by phytoplankton.

The novelty of our paper lies in the fact that the results build on the current literature offering insight into population control strategies; and to best of our knowledge, all the studies mentioned above do not reflect this scenario. Our paper also gives important insight into the effects of control measures showing that a control agent can stabilize a population provided the delay in predator response function is sufficiently small. Thus, our paper discusses the delay problem in mathematical ecology in the light of control theory. This aspect indeed makes our paper important as well as different from earlier works on delayed problem.

However, for our system (14), as in the previous case, it has three equilibria, namely, a trivial equilibrium E_0, an axial equilibrium E_K, and a positive interior equilibrium E^*. The trivial equilibrium is an unstable saddle point, and existence of E^* makes the the axial equilibrium E_K also an unstable saddle.

To investigate the nature of the system around E^*, we perturb system (14) around $E^* = (x^*, y^*)$ and apply Nyquist theorem. A detailed discussion is given in Section 1 of Supplementary Material.

Bifurcation of the Solution. We consider the parameter time delay τ as bifurcation parameter of the system (14). Assuming $x' = e^{\lambda}t$ as a solution of the characteristic equation of the perturbed system, we derive the following theorem.

Theorem 1. *Suppose that (11) holds and $A_1B_1 + A_2B_2 < 0$. Then the real parts of the solutions of characteristic equation (15) (see Supplementary Material) are negative for $\tau < \tau_0$, where τ_0 is the smallest value for which there is a solution of equation (15) with real part zero. For $\tau > \tau_0$, E^* is unstable. Further τ increases through τ_0, and E^* bifurcates into a small amplitude of periodic oscillation.*

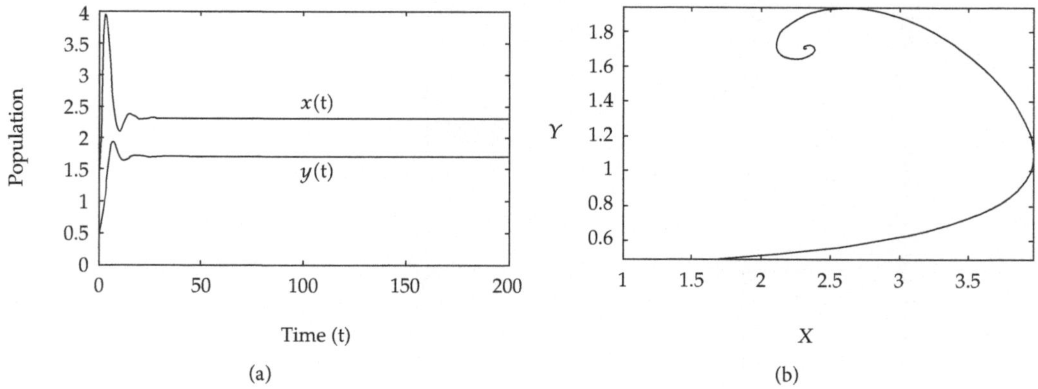

FIGURE 2: The dynamics of the pest and predator population without any delay in predation activity of predators. (a) The top curve depicts $x(t)$, and the bottom one depicts $y(t)$. Clearly the solution approaches a positive asymptotically stable equilibrium (2.32034, 1.70924), and there is no oscillation into the system. (b) A solution curve with initial condition (1, 0.5) that approaches above equilibrium.

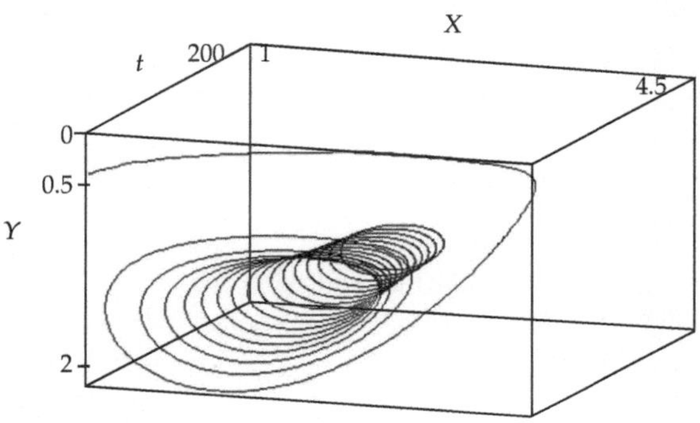

FIGURE 3: Natural dynamics of the system when $\tau = 1.5$ gives the delay in the predation activity. Solution is clearly creating stable limit cycle. So we can observe that an oscillatory nature comes into the system due to the presence of delay in the predation activity of natural enemies.

A detailed discussion on proof of the above theorem is given in Section 2 of Supplementary Material.

Hence we can observe that introduction of a delay into the predation activity of the natural enemy brings an oscillatory nature in the system (1). Thus our analysis supports the experimental findings of Turchin et al. [10].

5. Stability of the Bifurcation

In the previous section we obtained the condition that guarantees that the system (1) undergoes the Hopf bifurcation at the critical values given by Theorem 1. In this section we determine the formula that establishes the stability of bifurcating periodic orbits using the approach adapted in Hassard et al. [31]. The detailed calculation of the normal form and the specific condition on delay τ are derived in Supplementary Material. We see that stability of the bifurcation of the periodic solution depends on the system parameters. For example, the bifurcation is supercritical if $\mu_2 > 0$ and subcritical if $\mu_2 < 0$. Also, if $\tau_2 > 0$, the period of the solution increases with increase of τ, where

$$\begin{aligned} \mu_2 &= -\frac{\operatorname{Re} C_1(0)}{\alpha'(0)}, \\ \tau_2 &= -\frac{\operatorname{Im} C_1(0) + \mu_2 \omega'(0)}{\omega_+}, \\ \beta_2 &= 2 \operatorname{Re} C_1(0), \end{aligned} \tag{20}$$

where

$$\alpha'(0) = \left.\frac{dv}{d\tau}\right|_{\tau=\tau_0}, \qquad \omega'(0) = \left.\frac{d\omega}{d\tau}\right|_{\tau=\tau_0}. \tag{21}$$

The terms and signs are defined in the Supplementary Material.

6. Optimization of Net Profit: The Second Goal

So far we have seen that under suitable threshold limits of model parameters and control parameters "u," the system has unique asymptotically stable equilibrium when delay "τ" is small, and for further increase of delay, the system enters into a Hopf bifurcation and gives a stable oscillation. But during the process considering spraying of insecticides as a control measure, there is an obvious question of incurring some cost and allied profit in the whole process. Thus the objective is to quantify the units to express the net profit during the given time of experiment. In other words, this precisely means that we are to construct an economic model out of the given dynamic model of pest population control. In this case the problem reduces to an optimal control problem. Our task is then to formulate an optimal policy when forms of control measure in the system are defined in a mathematical form and finally to find out the restrictions on the economic parameters of the model for making the policy viable. In fact, we perform the optimal analysis of the model (14) when delay is small enough. In this section we first improve the Pontryagin maximum principle (PMP) for linear control functions and

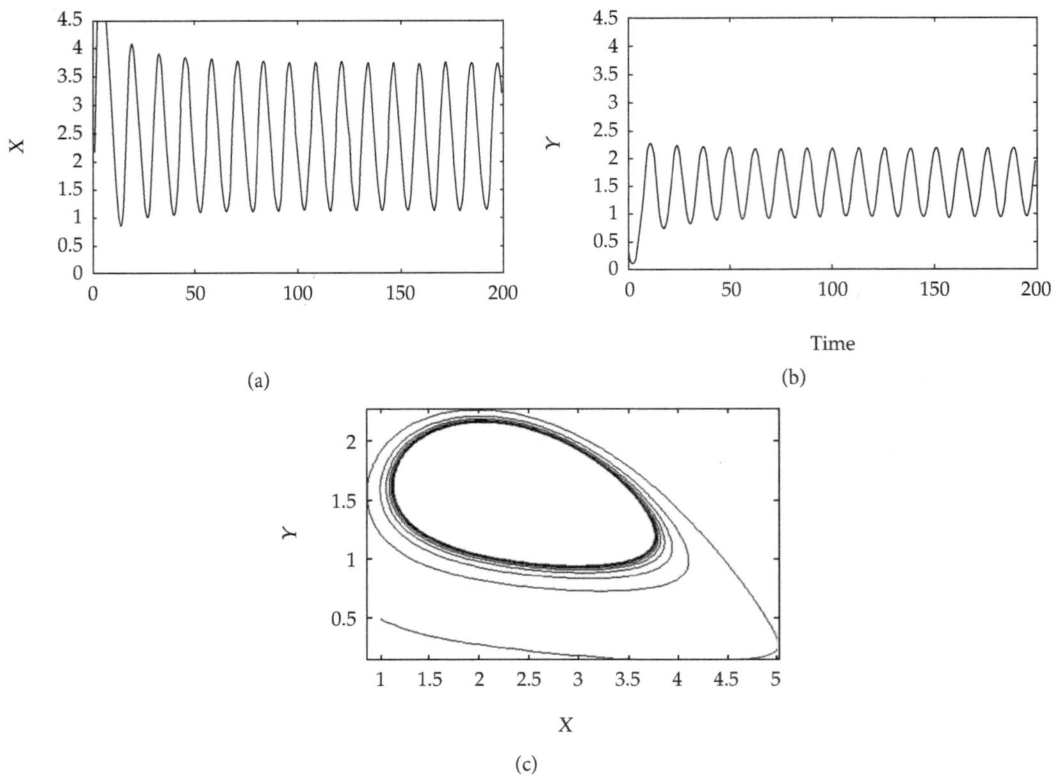

FIGURE 4: The behaviors of the population dynamics of pest and natural enemies have been shown in the above figures for delay $\tau = 1.7$. First two figures give the time series graphs of the populations, and the bottom one shows the solution curve of the system for initial condition (1, 0.5). These show that solution is clearly a stable limit cycle.

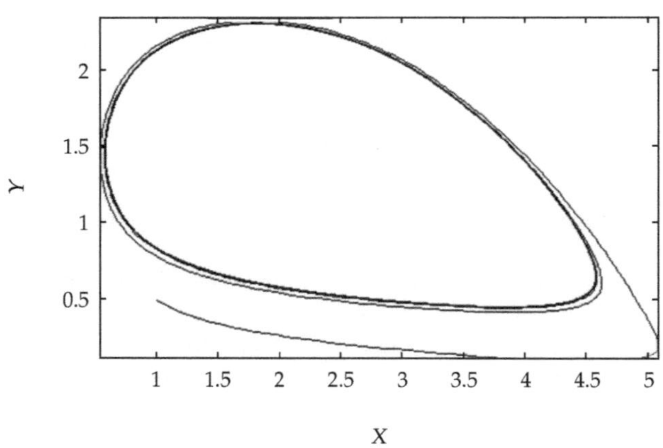

FIGURE 5: Behavior of the solution of the system when delay increases, that is, $\tau = 2$. The solution curve for the initial condition (1, 0.5) has been shown in the figure. It is observed that period of the closed orbit increases with increase of delay in the predation activity.

delay is sufficiently small. However, we apply PMP to our model (14) of the system with constant control only to find out the optimal biomass of the species conveniently.

The most general form of the control problem involving small delay of discrete type is given by

$$\dot{x} = f(x, x_\tau, u, u_\tau), \tag{22}$$

where $x(t)$ is an n-dimensional state vector and u is an m-dimensional control vector. Moreover, $x_\tau = x(t - \tau)$ and $u_\tau = u(t - \tau)$. We assume that the vector valued function $f : R^n \times R^n \times R^m \times R^m \to R^n$ has continuous partial derivatives of all orders with respect to all its arguments. The set of all permissible control is denoted by U. Moreover, it is assumed for definiteness that the control in questions is continuous from the left at their discontinuities, that is,

$$u(t) = u(t - 0). \tag{23}$$

We assume that the system (22) has a unique solution $x_t(0, \phi)$ with initial condition $\phi \in C([-\tau, 0], R^n)$ and under the permissible control $u(t)$, for $t \geq 0$.

Hence under the above conditions, we now formulate the standard optimal control problem with delay of discrete type as follows:

$$\begin{aligned} &\text{System}: \dot{x} = f(x, x_\tau, u, u_\tau), \\ &\text{Initially}: x(0) = \phi, \quad \phi \in C([-\tau, 0], R^n), \\ &\text{Constraint}: u \in U, \\ &\text{Objective}: \text{Optimize } J(u) = g[x(T), T] \\ &\qquad + \int_0^T \pi(x, x_\tau, u, u_\tau). \end{aligned} \tag{24}$$

We may mention here that the above optimal control problem is a relaxed optimal control problem, in the sense

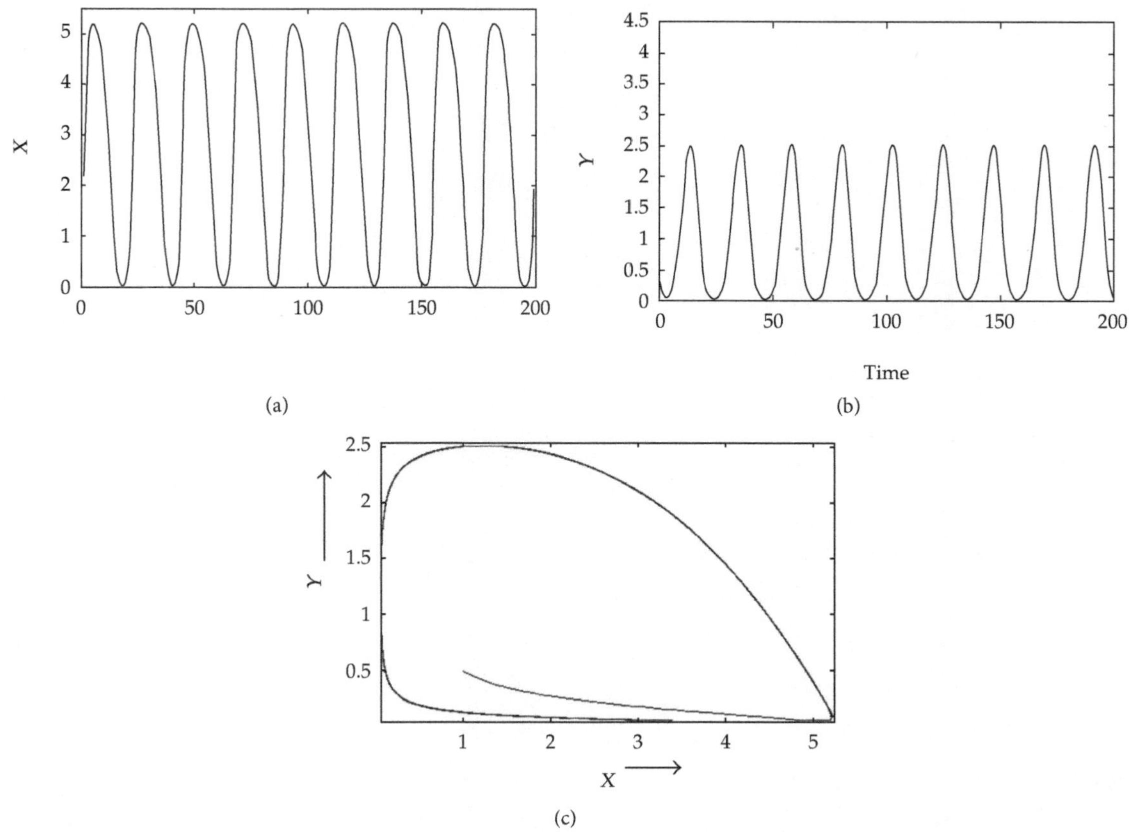

FIGURE 6: Behaviors of the system have been shown for more increasing delay $\tau = 3$. We may note that period of the orbit still increases for increase of delay. This precisely means that there is a less "peak" of pest population than earlier in a certain amount of time (compare Figure 5).

that we have not included the target in the decision problem. In ecological problem much work remains to be done on how we should specify the target in an optimal control problem which arises in ecology. In our case, we assume that the control measure, that is, insecticide is applied for certain period of time T and the rate of control measure is linear function of time. The objective function consists of some desirable criterion at the end of the planning period plus a criterion involving the profit function $\pi(x, x_\tau, u, u_\tau)$ which describes some desirable performance index during the planning period. Necessary conditions can be stated compactly in terms of Hamiltonian function. By definition Hamiltonian function is

$$H(x, x_\tau, u, u_\tau) = \pi(x, x_\tau, u, u_\tau) + \sum_{i=1}^{n} \lambda_i f_i(x, x_\tau, u, u_\tau). \tag{25}$$

The functions $\lambda_i(t)$ are called costate variables. The set of necessary conditions that we have deduced for the above optimal control problem is as follows.

Theorem 2. *Whenever delay in the system is sufficiently small and control function is linear in time t, a necessary condition for an admissible set (x^*, u^*) to be optimal is that there exists $\lambda(t) = (\lambda_1(t), \lambda_2(t), \ldots, \lambda_n(t))$ such that*

$$\begin{aligned} \lambda^{\mathrm{tr}} &= \frac{\partial g}{\partial x}, \quad \text{at } t = T, \\ \dot{\lambda}^{\mathrm{tr}} &= -\frac{\partial H}{\partial x} - \frac{\partial H}{\partial x_\tau}, \\ \frac{\partial H}{\partial u} + \frac{\partial H}{\partial u_\tau} &= \tau \frac{\partial H}{\partial x_\tau}\frac{\partial f}{\partial u}, \quad \text{at } u = u^*, \end{aligned} \tag{26}$$

where λ^{tr} denotes the transpose of the vector $\lambda(t) = (\lambda_1(t), \lambda_2(t) \ldots \lambda_n(t))$.

The details of the proof are given in Appendix C of Supplementary Material.

Applying the necessary conditions (26) in Theorem 2, we will now continue the optimal analysis for our system. It may be mentioned once again that the control in questions of our particular system is constant. The profit function π for our system is given by

$$\pi(x, y, u) = (p_1 q_1 x - p_2 q_2 y - c)\, u(t), \tag{27}$$

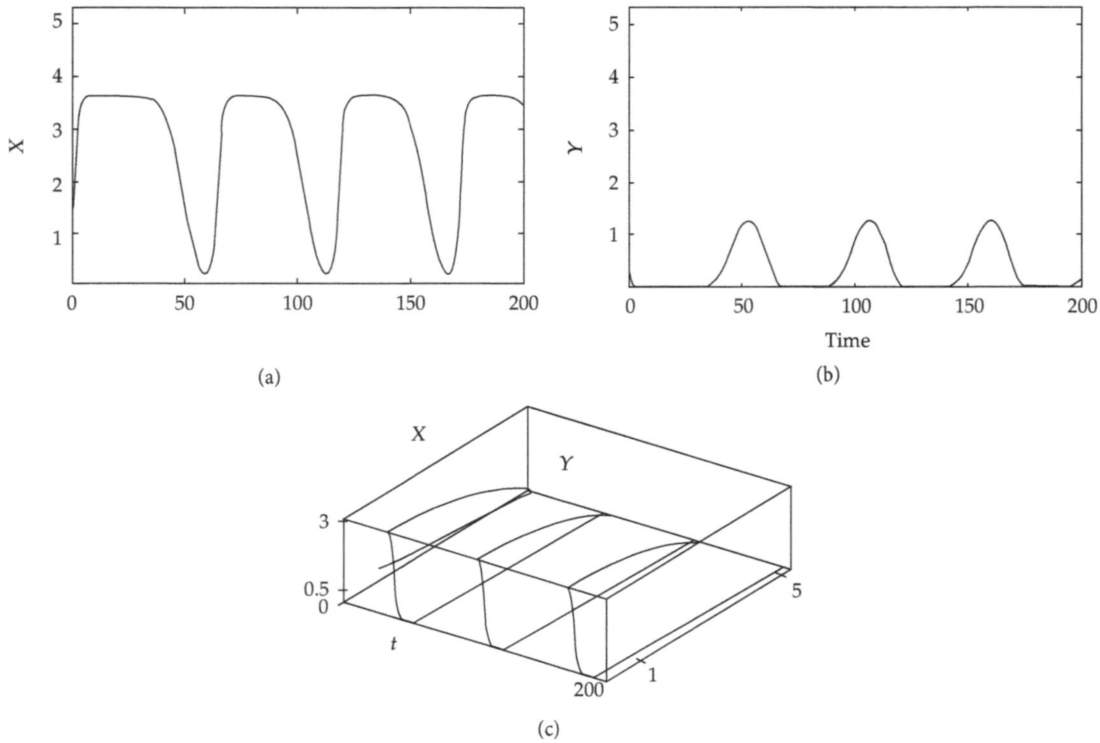

FIGURE 7: Behavior of the system for larger delay $\tau = 10$. First two figures indicate the time series graphs of the pest and predator population, whereas the bottom one shows the solution curve for initial condition (1, 0.5). These figures show that period of oscillation increases when delay increases. It may be clearly observed from the bottom one that oscillation of pest (X) is higher than its natural enemy (Y).

where p_1 is the projected profit per unit killing of pest and p_2 is the projected loss per unit killing of predator. The problem is then to optimize the integral

$$J = \int_0^T \pi(x, y, u)\, dt, \tag{28}$$

over the control parameter u, where $u \in (0, u_{\max})$. The detailed calculation of the optimal biomass of prey and predator and also optimal value of the control measure are derived in Supplementary Material.

7. Discussion and Conclusion

This paper considers the following two major issues:

(a) the cyclic nature of pest population in an agroecosystem or a forestry ecosystem,

(b) the improvement of Pontryagin's maximum principle (PMP) for system of differential equations with small delay of discrete type and its application to the pest management problem.

The experimental study of Turchin and his colleague [10] suggests that predation by natural enemy is one fundamental cause driving oscillation of pest population in the insect ecology. As the process behind the oscillation is still not clear, we have investigated the phenomenon through a simple model consisting of insect pest and its predator under two types of distribution of predation activity of the predator. We have observed that the cyclic nature of pest population, which is very common in agro- or forestry ecosystem, cannot be explained if the predation activity follows Holling-type II rule with gamma distribution. However, if it follows a delay of discrete type, we have observed that the system around the positive equilibrium enters a Hopf bifurcation and exhibits the cyclic nature for certain amount of time delay. To ascertain the local behavior, we have performed the stability analysis of bifurcating periodic solutions and have obtained the conditions for supercritical or subcritical bifurcations.

We have employed center manifold theory to show the stability of bifurcation of the oscillation. The approach of center manifold theory developed by Hassard and colleagues [31] shows that characteristic exponent determining the stability of the bifurcated periodic orbit is the same whether computed for original system or the system restricted on the associated center manifold. However, there is another existing approach showing that stability of bifurcating cycle is application of Poore's condition [32]. The stability information of the bifurcating periodic orbits is contained in the following theorem of Poore, coupled with an algebraic expression, which completely reduces the determination of stability to an algebraic problem. In Poore's theorem, the restricted original differential equation on center manifold reduced to a perturbed differential equation of a small parameter. Thus stability of the original periodic solution depends on characteristic multiplier of the variational matrix

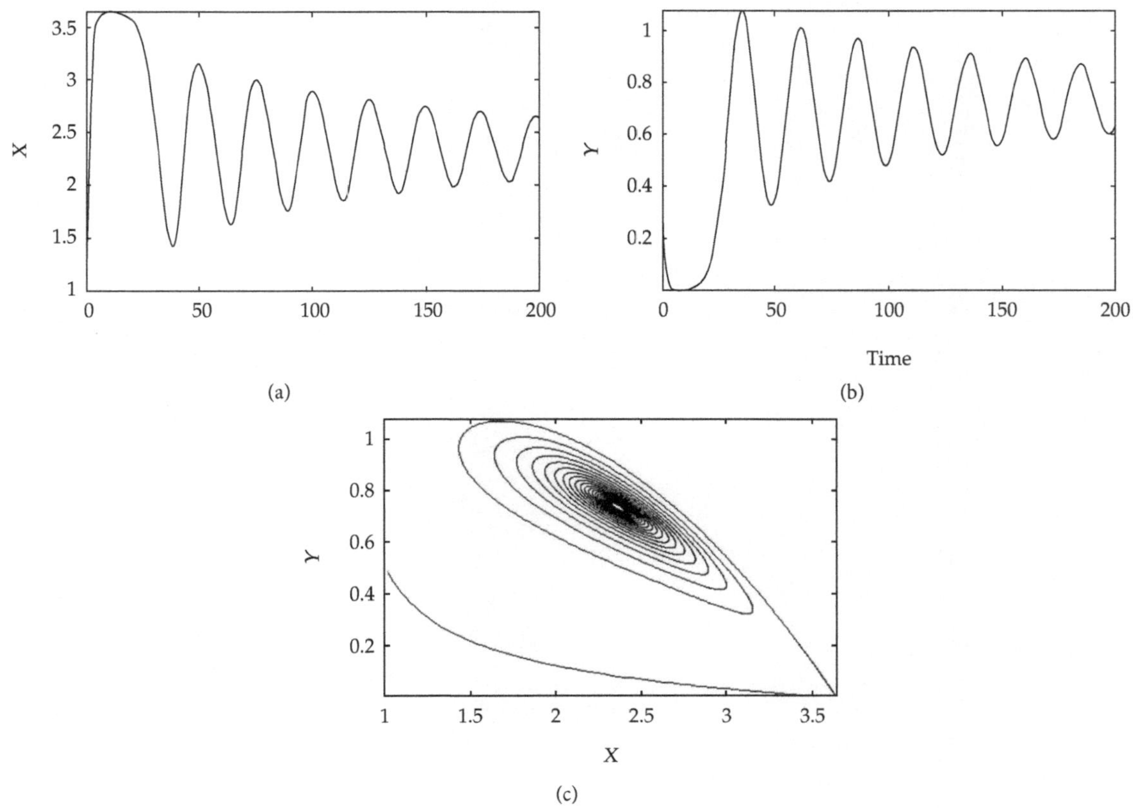

FIGURE 8: Behavior of the controlled system, that is, when pesticide is applied into the system ($u = 1$). The first two figures show the time series graphs of the populations of pest and natural enemy, whereas the bottom one depicts the solution curve starting from initial point (1, 0.5). Though the system is with a delay of amount $\tau = 6$ in the predation function, it shows that after a bit oscillation solution approaches an asymptotically stable equilibrium (2.3607, 0.743821). This implies that application of certain amount of control measure could normalize the oscillatory nature of the pest population and brings it into stable position.

TABLE 1: The parameter values of the model.

Parameters	Values
r	40 week^{-1}
α	100 week^{-1}
γ	1.48
ϵ	1.585
K	2×10^5 lit^{-1}
β	50 week^{-1}
d	5 week^{-1}

of perturbed system. This is a more sophisticated way of showing periodic stability compared to technique developed by Hassard and colleagues.

However, we have also performed a numerical experiment to substantiate the analytical findings with the following set of values (Table 1). Numerical solutions of equations were carried out using the modified fourth-order Runga-Kutta method. The parameter values are taken in such a way that it reflects the most realistic scenario in context of pest control problem.

For convenience of the simulation we have used the logarithmic scale. For those sets of values, we see that, under the condition in the Theorem 1, we have only one positive root $\omega_+ = 0.542$ and this implies $\tau_0 = 1.495$. In fact, it is observed from Figure S2 that for $\tau = 1.5$ the system creates an stable limit cycle bifurcating from the positive asymptotically stable equilibrium. In fact, it is seen that, for $\tau = 0$, the system remains stable and approaches an asymptotically stable equilibrium (2.32034, 1.70924) (Figure 2). Further we obtained the value of $D = 0.580 - 2.088i$ in the eigenvector $q(\theta)$ in (36) (see Supplementary Material) and $\overline{v}_1 = -0.014 + 2.156i$ and $\overline{v}_2 = 0.502 - 0.104i$ in eigenvector $q^*(\theta)$. Moreover, it was found that $g_{20} = 0.721 + 2.891i$, $g_{11} = 2.345 - 0.912i$, $g_{02} = 13.234 + 4.213i$, and $g_{21} = -7.952 - 0.801i$. With these values of g_{ij}'s, we obtained the real part $\operatorname{Re} C_1(0) = -9.444$. This implies that $\mu_2 = 376.281$ and further $\tau_2 = 131.711$. Thus we see that for this set of parameters, $\mu_2 > 0$, that is, the bifurcation is supercritical, and the system exhibits stable limit cycle (Figures 3 and 4).

Further since $\tau_2 > 0$, the period of the oscillations increases with increase of τ (Figures 5 and 6). The results indicate that the system has asymptotic equilibrium for $0 \leq \tau < 1.495$ and becomes unstable (by growing oscillations) for $\tau > 1.495$. The system exhibits a stable periodic solution for

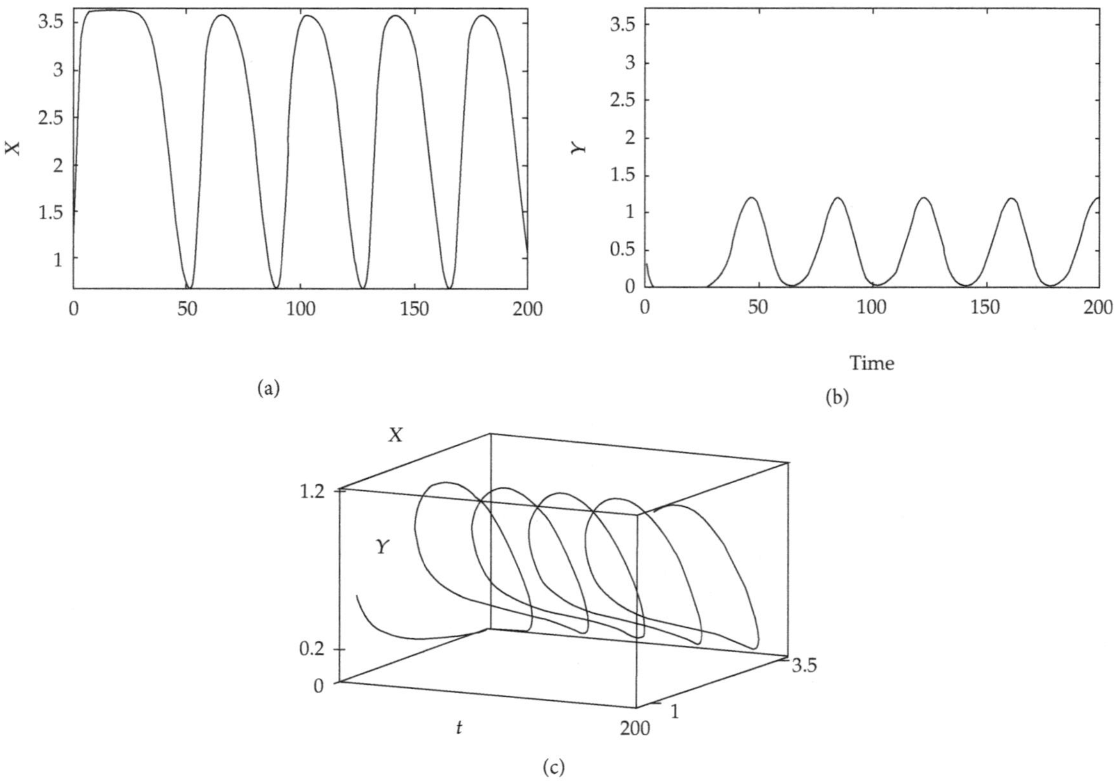

FIGURE 9: Behavior of the controlled system when delay is taken as $\tau = 15.5$. System again becomes oscillatory, even if the control measure is applied in a certain amount ($u = 1$). First two figures denote the time series graphs of the population, and bottom one depicts the solution curve of the system with initial condition (1, 0.5).

$\tau > 1.495$ (Figures 7, 8, and 9). These observations indicate that there is threshold limit of the delay τ in the predation activity of the natural enemy, below which the system shows no excitability and above which the system enters into excitable range. These findings demonstrate the delay effect of predation activity and cyclic nature of pest population, which was suspected through experimental observations by Turchin and his colleagues [10]. We may mention here that delayed density-dependent predation is not only the single reason for oscillation of pest population in the insect ecology, but there are other factors like weather, especially extremes of cold, drought, or rainfall [33]. In spite of that our mathematical findings cannot be ignored, as it supports the experimental conclusion of Turchin and his colleagues. In fact, this will motivate the biologist to perform more explicit study in the laboratory in this direction.

Another important fact that we investigate in our study is control of this oscillation. A careful observation of the bifurcation analysis indicates that the coefficients which are responsible for rise and fall of pest population, its upper and lower boundaries, time period, even the value of bifurcation parameter τ(23) (see Supplementary Material) are highly dependent on the control parameter u. In fact, we perform some numerical experiments whenever $u = 1$, $q_1 = 1$, and $q_2 = 0.2$. It is seen that with these parameters estimation, $\omega_+ = 0.0999$ and $\tau_+ = 15.276$. This means that system will remain stable for $\tau < 15.276$ (Figure S7). Thus one can control the oscillation of pest population up to certain extent by applying control measures such as use of insecticide and release of additional predators. However, for larger delay, the system again enters into Hopf bifurcation and exhibits stable periodic solution (Figure S8).

An allied problem with this control policy in pest control program is its economic viability. Some of the recent works on control theory in the respect of pest management problem could be found in the references like Bhattacharya and Karan [11, 34], and Bhattacharyya and Bhattacharya [35]. But all the earlier works involve application of Pontryagin's maximum principle (PMP) on system of ordinary differential equations (under single control parameter or multicontrol parameters). In fact, control theory for ordinary differential equation is well known. But it is less developed for delay differential equations. The well-known PMP, which is an essential tool for solving the corresponding optimal control problem in ODE, has no exactly convenient form for delay differential equations. This part of the paper studies the corresponding problem with delay differential equations with small delay. It is better to mention here that system remains stable and solutions approach an asymptotically stable equilibrium for small delay. However, we have been able to obtain a set of necessary conditions for optimality whenever the delay in system is sufficiently small and control is linear function in time t and then apply it suitably in our ecosystem model to get the optimal biomass under optimal control.

Finally, we would like to mention that dynamics of insect ecology is very complex one. There are indeed many factors, as we stated earlier, extrinsic factors such as weather, especially extremes of cold, drought, or rainfall; or by intrinsic factors such as birth and death rates, or interactions with other species may cause oscillation. Recent theoretical and empirical studies have started to combine the extrinsic and intrinsic hypotheses to explain the fundamental problem in insect ecology [36]. However, the present work on one hand gives some insight on this important ecological phenomenon and on the other hand throws some light on the control of the situation.

Acknowledgment

The authors are thankful to the anonymous reviewers for their comments and suggestions that greatly improved the paper.

References

[1] D. A. Maelzer and M. P. Zalucki, "Long range forecasts of the numbers of Helicoverpa punctigera and H. armigera (Lepidoptera: Noctuidae) in Australia using the Southern Oscillation Index and the Sea Surface Temperature," *Bulletin of Entomological Research*, vol. 90, no. 2, pp. 133–146, 2000.

[2] T. Yonow et al., "Modelling the population dynamics of the Queensland fruit fly, Bactrocera (Dacus) tryoni: a cohort-based approach incorporating the effects of weather," *Ecological Modelling*, vol. 173, pp. 9–30, 2004.

[3] C. Elton, "Periodic fluctuations in the numbers of animals," *British Journal of Experimental Biology*, vol. 2, pp. 119–163, 1924.

[4] A. J. Nicholson, "Supplement: the balance of animal populations," *Journal of Animal Ecology*, vol. 2, no. 1, pp. 131–178, 1933.

[5] T. Royama, *Analytical Population Dynamics*, Chapman & Hall, London, UK, 1992.

[6] N. C. Stenseth, K.-S. Chan, and H. Tong, "Common dynamic structure of Canada lynx populations within three climatic regions," *Science*, vol. 285, no. 5430, pp. 1071–1073, 1999.

[7] P. Turchin, "Population regulation: a synthetic view," *Oikos*, vol. 84, no. 1, pp. 153–159, 1999.

[8] A. R. E. Sinclair and R. P. Pech, "Density dependence, stochasticity, compensation and predator regulation," *Oikos*, vol. 75, pp. 164–173, 1996.

[9] P. Turchin, P. L. Lorio, A. D. Taylor, and R. F. Billings, "Why do populations of southern pine beetles (Coleoptera: Scolytidae) fluctuate?" *Environmental Entomology*, vol. 20, pp. 401–409, 1991.

[10] P. Turchin, A. D. Taylor, and J. D. Reeve, "Dynamical role of predators in population cycles of a forest insect: An experimental test," *Science*, vol. 285, no. 5430, pp. 1068–1071, 1999.

[11] D. K. Bhattacharya and S. Karan, "Pest management of two non-interacting pests in the presence of a common predator," *Journal of Applied Mathematics and Computing*, vol. 13, pp. 301–322, 2003.

[12] S. Bhattacharyya and D. K. Bhattacharya, "A more realistic approach to pest-management problem," *Bulletin of Mathematical Biology*, vol. 69, no. 4, pp. 1277–1310, 2007.

[13] S. Bhattacharyya and D. K. Bhattacharya, "An improved integrated pest management model under 2-control parameters (sterile male and pesticide)," *Mathematical Biosciences*, vol. 209, no. 1, pp. 256–281, 2007.

[14] S. Ghosh and D. K. Bhattacharya, "Optimization in microbial pest control: an integrated approach," *Applied Mathematical Modelling*, vol. 34, no. 5, pp. 1382–1395, 2010.

[15] S. Ghosh, S. Bhattacharyya, and D. K. Bhattacharya, "The role of viral infection in pest control: a mathematical study," *Bulletin of Mathematical Biology*, vol. 69, pp. 2649–2691, 2007.

[16] M. B. Thomas, "Ecological approaches and the development of truly integrated pest management," *Proceedings of the National Academy of Sciences of the United States of America*, vol. 96, pp. 5944–5951, 1999.

[17] D. Ludwig, D. D. Jones, and C. S. Holling, "Qualitative analysis of an insect outbreak system: the spruce budwarm and forest," *Journal of Animal Ecology*, vol. 47, pp. 315–332, 1978.

[18] C. B. Huffacker, "Experimental study on predation: dispersion factors and prey-predator oscillation," *Hilgardia*, vol. 27, pp. 343–383, 1958.

[19] J. M. Cushing, *Integrodifferential equations and delay models in population dynamics*, Springer, Berlin, Germany, 1977.

[20] N. MacDonald, *Time Lags in Biological Models*, vol. 27, Springer, Berlin, Germany, 1978.

[21] L. Perko, *Differential equations and dynamical systems*, vol. 7, Springer, New York, NY, USA, 2nd edition, 1996.

[22] G. S. K. Wolkowicz, H. Xia, and S. Ruan, "Competition in the chemostat: a distributed delay model and its global asymptotic behavior," *SIAM Journal on Applied Mathematics*, vol. 57, no. 5, pp. 1281–1310, 1997.

[23] R. Xu and M. A. J. Chaplain, "Persistence and global stability in a delayed Gause-type predator-prey system without dominating instantaneous negative feedbacks," *Journal of Mathematical Analysis and Applications*, vol. 265, no. 1, pp. 148–162, 2002.

[24] R. Xu, F. A. Davidson, and M. A. J. Chaplain, "Persistence and stability for a two-species ratio-dependent predator-prey system with distributed time delay," *Journal of Mathematical Analysis and Applications*, vol. 269, no. 1, pp. 256–277, 2002.

[25] G. Lin and R. Yuan, "Periodic solution for a predator-prey system with distributed delay," *Mathematical and Computer Modelling*, vol. 42, no. 9-10, pp. 959–966, 2005.

[26] H. I. Freedman and S. G. Ruan, "Uniform persistence in functional-differential equations," *Journal of Differential Equations*, vol. 115, no. 1, pp. 173–192, 1995.

[27] T. Zhao, Y. Kuang, and H. L. Smith, "Global existence of periodic solutions in a class of delayed Gause-type predator-prey systems," *Nonlinear Analysis: Theory, Methods & Applications A*, vol. 28, no. 8, pp. 1373–1394, 1997.

[28] T. Faria, "Stability and bifurcation for a delayed predator-prey model and the effect of diffusion," *Journal of Mathematical Analysis and Applications*, vol. 254, no. 2, pp. 433–463, 2001.

[29] X.-P. Yan and W.-T. Li, "Hopf bifurcation and global periodic solutions in a delayed predator-prey system," *Applied Mathematics and Computation*, vol. 177, no. 1, pp. 427–445, 2006.

[30] J. Chattopadhyay, R. R. Sarkar, and A. El Abdllaoui, "A delay differential equation model on harmful algal blooms in the presence of toxic substances," *Journal of Mathematics Applied in Medicine and Biology*, vol. 19, no. 2, pp. 137–161, 2002.

[31] B. D. Hassard, N. D. Kazarinoff, and Y. H. Wan, *Theory and Applications of Hopf Bifurcation*, vol. 41, Cambridge University Press, Cambridge, UK, 1981.

[32] A. B. Poore, "On the theory and application of the Hopf-Friedrichs bifurcation theory," *Archive for Rational Mechanics and Analysis*, vol. 60, pp. 371–393, 1976.

[33] E. Ranta, V. Kaitala, and P. Lundberg, "A tale of big game and small bugs," *Science*, vol. 285, no. 5430, pp. 1022–1023, 1999.

[34] D. K. Bhattacharya and S. Karan, "On bionomic model of integrated pest management of a single pest population," *Differential Equations and Dynamical Systems*, vol. 12, no. 3-4, pp. 301–330, 2004.

[35] S. Bhattacharyya and D. K. Bhattacharya, "A new approach to pest management problem," *Journal of Biological Systems*, vol. 13, no. 2, pp. 117–130, 2005.

[36] P. A. P. Moran, "The statistical analysis of the Canadian Lynx cycle," *Australian Journal of Zoology*, vol. 1, pp. 291–298, 1953.

On Pseudohyperbolical Smarandache Curves in Minkowski 3-Space

Esra Betul Koc Ozturk,[1,2] Ufuk Ozturk,[1,2] Kazim İlarslan,[3] and Emilija Nešović[4]

[1] *Department of Mathematics, Faculty of Sciences, University of Cankiri Karatekin, Cankiri 18100, Turkey*
[2] *School of Mathematics & Statistical Sciences, Arizona State University, Room PSA442, Tempe, AZ 85287-1804, USA*
[3] *Department of Mathematics, Faculty of Sciences and Art, University of Kırıkkale, Kırıkkale 71450, Turkey*
[4] *University of Kragujevac, Faculty of Science, Department of Mathematics and Informatics, Kragujevac 34000, Serbia*

Correspondence should be addressed to Ufuk Ozturk; uuzturk@asu.edu

Academic Editor: B. N. Mandal

We define pseudohyperbolical Smarandache curves according to the Sabban frame in Minkowski 3-space. We obtain the geodesic curvatures and the expression for the Sabban frame vectors of special pseudohyperbolic Smarandache curves. Finally, we give some examples of such curves.

1. Introduction

In the theory of curves in the Euclidean and Minkowski spaces, one of the interesting problems is the problem of characterization of a regular curve. In the solution of the problem, the curvature functions κ and τ of a regular curve have an effective role. It is known that we can determine the shape and size of a regular curve by using its curvatures κ and τ. Another approach to the solution of the problem is to consider the relationship between the corresponding Frenet vectors of two curves. For instance, Bertrand curves and Mannheim curves arise from this relationship. Another example is the Smarandache curves. They are the objects of Smarandache geometry, that is, a geometry which has at least one Smarandachely denied axiom [1]. An axiom is said to be Smarandachely denied if it behaves in at least two different ways within the same space. Smarandache geometries are connected with the theory of relativity and the parallel universes.

If the position vector of a regular curve α is composed by the Frenet frame vectors of another regular curve β, then the curve α is called a Smarandache curve [2]. Special Smarandache curves in Euclidean and Minkowski spaces are studied by some authors [3–8]. The curves lying on a pseudohyperbolic space H_0^2 in Minkowski 3-space $\mathbb{E}_1^3$ are characterized in [9].

In this paper, we define pseudohyperbolical Smarandache curves according to the Sabban frame $\{\alpha, T, \xi\}$ in Minkowski 3-space. We obtain the geodesic curvatures and the expressions for the Sabban frame's vectors of special pseudohyperbolical Smarandache curves. In particular, we prove that special $T\xi$-pseudohyperbolical Smarandache curves do not exist. Besides, we give some examples of special pseudohyperbolical Smarandache curves in Minkowski 3-space.

2. Basic Concepts

The Minkowski 3-space $\mathbb{R}_1^3$ is the Euclidean 3-space $\mathbb{R}^3$ provided with the standard flat metric given by

$$\langle \cdot, \cdot \rangle = -dx_1^2 + dx_2^2 + dx_3^2, \quad (1)$$

where (x_1, x_2, x_3) is a rectangular Cartesian coordinate system of $\mathbb{R}_1^3$. Since g is an indefinite metric, recall that a nonzero vector $\vec{x} \in \mathbb{R}_1^3$ can have one of the three Lorentzian causal characters: it can be spacelike if $\langle \vec{x}, \vec{x} \rangle > 0$, timelike if $\langle \vec{x}, \vec{x} \rangle < 0$, and null (lightlike) if $\langle \vec{x}, \vec{x} \rangle = 0$. In particular, the norm (length) of a vector $\vec{x} \in \mathbb{R}_1^3$ is given by $\|\vec{x}\| = \sqrt{|\langle \vec{x}, \vec{x} \rangle|}$ and two vectors $\vec{x}$ and $\vec{y}$ are said to be orthogonal if $\langle \vec{x}, \vec{y} \rangle = 0$. Next, recall that an arbitrary curve $\alpha = \alpha(s)$ in $\mathbb{E}_1^3$ can locally be *spacelike, timelike,* or *null (lightlike)* if all of its velocity vectors $\alpha'(s)$ are, respectively, *spacelike, timelike,* or *null (lightlike)* for every $s \in I$ [10]. If $\|\alpha'(s)\| \neq 0$ for every $s \in I$, then α is a *regular curve* in $\mathbb{R}_1^3$. A spacelike (timelike) regular curve α is parameterized by pseudo-arclength parameter s which is given by $\alpha : I \subset \mathbb{R} \to \mathbb{R}_1^3$, and then the tangent vector $\alpha'(s)$ along α has unit length, that is, $\langle \alpha'(s), \alpha'(s) \rangle = 1(\langle \alpha'(s), \alpha'(s) \rangle = -1)$ for all $s \in I$, respectively.

For any $\vec{x} = (x_1, x_2, x_3)$ and $\vec{y} = (y_1, y_2, y_3)$ in the space $\mathbb{R}_1^3$, the *pseudovector product* of $\vec{x}$ and $\vec{y}$ is defined by

$$\vec{x} \times \vec{y} = (-x_2 y_3 + x_3 y_2, x_3 y_1 - x_1 y_3, x_1 y_2 - x_2 y_1). \quad (2)$$

Remark 1. Let $\vec{x} = (x_1, x_2, x_3)$, $\vec{y} = (y_1, y_2, y_3)$, and $\vec{z} = (z_1, z_2, z_3)$ be vectors in $\mathbb{R}_1^3$. Then,

(i) $\langle \vec{x} \times \vec{y}, \vec{z} \rangle = \begin{vmatrix} x_1 & x_2 & x_3 \\ y_1 & y_2 & y_3 \\ z_1 & z_2 & z_3 \end{vmatrix}$,

(ii) $\vec{x} \times (\vec{y} \times \vec{z}) = -\langle \vec{x}, \vec{z} \rangle \vec{y} + \langle \vec{x}, \vec{y} \rangle \vec{z}$,

(iii) $\langle \vec{x} \times \vec{y}, \vec{x} \times \vec{y} \rangle = -\langle \vec{x}, \vec{x} \rangle \langle \vec{y}, \vec{y} \rangle + \langle \vec{x}, \vec{y} \rangle^2$,

where $\times$ is the pseudovector product in the space $\mathbb{R}_1^3$.

Lemma 2. *In the Minkowski 3-space $\mathbb{R}_1^3$, the following properties are satisfied [10]:*

(i) *two timelike vectors are never orthogonal;*

(ii) *two null vectors are orthogonal if and only if they are linearly dependent;*

(iii) *timelike vector is never orthogonal to a null vector.*

Pseudohyperbolic space in the Minkowski 3-space $\mathbb{R}_1^3$ is a quadric defined by

$$H_0^2 = \left\{ \vec{x} \in \mathbb{R}_1^3 \mid -x_1^2 + x_2^2 + x_3^2 = -1 \right\}. \quad (3)$$

Let $\alpha : I \subset \mathbb{R} \to H_0^2$ be a regular unit speed curve lying fully in H_0^2 in $\mathbb{R}_1^3$. Then its position vector α is timelike vector, which implies that tangent vector $\vec{T} = \alpha'$ is the unit spacelike vector for all $s \in I$. Hence we have orthonormal Sabban frame $\{\alpha(s), T(s), \xi(s)\}$ along the curve α, where $\xi(s) = \alpha(s) \times T(s)$ is the unit spacelike vector. The corresponding Frenet formulae of α, according to the Sabban frame, read

$$\begin{bmatrix} \alpha' \\ T' \\ \xi' \end{bmatrix} = \begin{bmatrix} 0 & 1 & 0 \\ 1 & 0 & k_g(s) \\ 0 & -k_g(s) & 0 \end{bmatrix} \begin{bmatrix} \alpha \\ T \\ \xi \end{bmatrix}, \quad (4)$$

where $k_g(s) = \det(\alpha(s), T(s), T'(s))$ is the geodesic curvature of α on H_0^2 in $\mathbb{R}_1^3$ and s is arc length parameter of *alpha*. In particular, the following relations hold:

$$\alpha \times T = \xi, \qquad T \times \xi = -\alpha, \qquad \xi \times \alpha = T. \quad (5)$$

3. Pseudohyperbolical Smarandache Curves in Minkowski 3-Space

In this section, we define and investigate pseudohyperbolical Smarandache curves in Minkowski 3-space according to the Sabban frame.

Let $\alpha = \alpha(s)$ and $\beta = \beta(s^*)$ be two regular unit speed curves lying fully in pseudohyperbolic space H_0^2 in $\mathbb{R}_1^3$ and let $\{\alpha, T, \xi\}$ and $\{\beta, T_\beta, \xi_\beta\}$ be the moving Sabban frames of these curves, respectively. Then we have the following definitions of pseudohyperbolical Smarandache curves.

Definition 3. Let $\alpha : I \subset \mathbb{R} \mapsto H_0^2$ be a regular unit speed curve lying fully in H_0^2. Then $\alpha\xi$-pseudohyperbolical Smarandache curve $\beta : I \subset \mathbb{R} \mapsto H_0^2$ of α is defined by

$$\beta(s^*(s)) = \frac{1}{\sqrt{2}} (a\alpha(s) + b\xi(s)), \quad (6)$$

where $a, b \in \mathbb{R}_0$ and $a^2 - b^2 = 2$.

Definition 4. Let $\alpha : I \subset \mathbb{R} \mapsto H_0^2$ be a regular unit speed curve lying fully in H_0^2.Then αT-pseudohyperbolical Smarandache curve $\beta : I \subset \mathbb{R} \mapsto H_0^2$ of α is defined by

$$\beta(s^*(s)) = \frac{1}{\sqrt{2}} (a\alpha(s) + bT(s)), \quad (7)$$

where $a, b \in R_0$ and $a^2 - b^2 = 2$.

Definition 5. Let $\alpha : I \subset R \mapsto H_0^2$ be a regular unit speed curve lying fully in H_0^2. Then $\alpha T\xi$ -pseudohyperbolical Smarandache curve $\beta : I \subset R \mapsto H_0^2$ of α is defined by

$$\beta(s^*(s)) = \frac{1}{\sqrt{3}} (a\alpha(s) + bT(s) + c\xi(s)), \quad (8)$$

where $a, b, c \in \mathbb{R}_0$ and $a^2 - b^2 - c^2 = 3$.

Theorem 6. *Let $\alpha : I \subset \mathbb{R} \mapsto H_0^2$ be a regular unit speed curve lying fully in H_0^2. Then $T\xi$-pseudohyperbolical Smarandache curve $\beta : I \subset \mathbb{R} \mapsto H_0^2$ of α does not exist.*

Proof. Assume that there exists $T\xi$-pseudohyperbolical Smarandache curve of α. Then it can be written as

$$\beta(s^*(s)) = \frac{1}{\sqrt{2}} (aT(s) + b\xi(s)), \quad (9)$$

where $a^2 + b^2 = -2$, which is a contradiction.

In the theorems which follow, we obtain Sabban frame $\{\beta, T_\beta, \xi_\beta\}$ and geodesic curvature κ_g^β of pseudohyperbolical Smarandache curve β.

Theorem 7. *Let $\alpha : I \subset \mathbb{R} \mapsto H_0^2$ be a regular unit speed curve lying fully in H_0^2 with the Sabban frame $\{\alpha, T, \xi\}$ and* *the geodesic curvature* k_g. *If* $\beta : I \subset \mathbb{R} \mapsto H_0^2$ *is* $\alpha\xi$-*pseudohyperbolical Smarandache curve of* α, *then its frame* $\{\beta, T_\beta, \xi_\beta\}$ *is given by*

$$\begin{bmatrix} \beta \\ T_\beta \\ \xi_\beta \end{bmatrix} = \begin{bmatrix} \frac{a}{\sqrt{2}} & 0 & \frac{b}{\sqrt{2}} \\ 0 & \epsilon & 0 \\ \epsilon\frac{b}{\sqrt{2}} & 0 & \epsilon\frac{a}{\sqrt{2}} \end{bmatrix} \begin{bmatrix} \alpha \\ T \\ \xi \end{bmatrix}, \tag{10}$$

and the corresponding geodesic curvature k_g^β *reads*

$$k_g^\beta = \frac{ak_g - b}{\left|a - bk_g\right|}, \quad a, b \in \mathbb{R}_0, \tag{11}$$

where $a^2 - b^2 = 2$ *and* $\epsilon = \pm 1$.

Proof. Differentiating (6) with respect to s and using (4), we obtain

$$\beta'(s) = \frac{d\beta}{ds^*}\frac{ds^*}{ds} = \frac{a - bk_g}{\sqrt{2}}T,$$
$$T_\beta\frac{ds^*}{ds} = \frac{a - bk_g}{\sqrt{2}}T, \tag{12}$$

where

$$\frac{ds^*}{ds} = \frac{\left|a - bk_g\right|}{\sqrt{2}}. \tag{13}$$

Therefore, the unit spacelike tangent vector of the curve β is given by

$$T_\beta = \epsilon T, \tag{14}$$

where $\epsilon = +1$ if $a - bk_g > 0$ for all s and $\epsilon = -1$ if $a - bk_g < 0$ for all s.

Differentiating (14) with respect to s, we find

$$\frac{dT_\beta}{ds^*}\frac{ds^*}{ds} = \epsilon\left(\alpha + k_g\xi\right), \tag{15}$$

and from (13) and (15) we get

$$T'_\beta = \frac{\sqrt{2}\epsilon}{\left|a - bk_g\right|}\left(\alpha + k_g\xi\right). \tag{16}$$

On the other hand, from (6) and (14) it can be easily seen that

$$\xi_\beta = \beta \times T_\beta$$
$$= \epsilon\frac{b}{\sqrt{2}}\alpha + \epsilon\frac{a}{\sqrt{2}}\xi \tag{17}$$

is a unit spacelike vector.

Consequently, the geodesic curvature k_g^β of the curve $\beta = \beta(s^*)$ is given by

$$k_g^\beta = \det\left(\beta, T_\beta, T'_\beta\right)$$
$$= \frac{ak_g - b}{\left|a - bk_g\right|}. \tag{18}$$

Theorem 8. *Let* $\alpha : I \subset \mathbb{R} \mapsto H_0^2$ *be a regular unit speed curve lying fully in* H_0^2 *with the Sabban frame* $\{\alpha, T, \xi\}$ *and the geodesic curvature* k_g. *If* $\beta : I \subset \mathbb{R} \mapsto H_0^2$ *is* αT-*pseudohyperbolical Smarandache curve of* α, *then its frame* $\{\beta, T_\beta, \xi_\beta\}$ *is given by*

$$\begin{bmatrix} \beta \\ T_\beta \\ \xi_\beta \end{bmatrix} = \begin{bmatrix} \frac{a}{\sqrt{2}} & \frac{b}{\sqrt{2}} & 0 \\ \frac{b}{\sqrt{2 + (bk_g)^2}} & \frac{a}{\sqrt{2 + (bk_g)^2}} & \frac{bk_g}{\sqrt{2 + (bk_g)^2}} \\ -\frac{b^2k_g}{\sqrt{4 + 2(bk_g)^2}} & -\frac{abk_g}{\sqrt{4 + 2(bk_g)^2}} & \frac{\sqrt{2}}{\sqrt{2 + (bk_g)^2}} \end{bmatrix} \times \begin{bmatrix} \alpha \\ T \\ \xi \end{bmatrix}, \tag{19}$$

and the corresponding geodesic curvature k_g^β *reads*

$$k_g^\beta = \frac{b^2k_g\varepsilon_1 - abk_g\varepsilon_2 + 2\varepsilon_3}{\left(2 + b^2k_g^2\right)^{5/2}}, \quad a, b \in \mathbb{R}_0, \tag{20}$$

where

$$\varepsilon_1 = -b^3k_gk'_g + a\left(2 + \left(bk_g\right)^2\right),$$
$$\varepsilon_2 = -ab^2k_gk'_g + \left(b - bk_g^2\right)\left(2 + \left(bk_g\right)^2\right), \tag{21}$$
$$\varepsilon_3 = -b^3k_g^2k'_g + \left(ak_g + bk'_g\right)\left(2 + \left(bk_g\right)^2\right),$$

and $a^2 - b^2 = 2$.

Proof. Differentiating (7) with respect to s and using (4), we obtain

$$\beta'(s) = \frac{d\beta}{ds^*}\frac{ds^*}{ds} = \frac{1}{\sqrt{2}}\left(b\alpha + aT + bk_g\xi\right),$$
$$T_\beta\frac{ds^*}{ds} = \frac{1}{\sqrt{2}}\left(b\alpha + aT + bk_g\xi\right), \tag{22}$$

where

$$\frac{ds^*}{ds} = \sqrt{\frac{2 + \left(bk_g\right)^2}{2}}. \tag{23}$$

Therefore, the unit spacelike tangent vector of the curve β is given by

$$T_\beta = \frac{1}{\sqrt{2 + \left(bk_g\right)^2}}\left(b\alpha + aT + bk_g\xi\right). \tag{24}$$

Differentiating (24) with respect to s, it follows that

$$\frac{dT_\beta}{ds^*}\frac{ds^*}{ds} = \frac{1}{\left(2+\left(bk_g\right)^2\right)^{3/2}}\left(\varepsilon_1\alpha+\varepsilon_2 T+\varepsilon_3\xi\right). \tag{25}$$

From (4) and (23), we get

$$T'_\beta = \frac{\sqrt{2}}{\left(2+\left(bk_g\right)^2\right)^2}\left(\varepsilon_1\alpha+\varepsilon_2 T+\varepsilon_3\xi\right), \tag{26}$$

where

$$\begin{aligned}\varepsilon_1 &= -b^3k_gk'_g + a\left(2+\left(bk_g\right)^2\right),\\ \varepsilon_2 &= -ab^2k_gk'_g + \left(b-bk_g^2\right)\left(2+\left(bk_g\right)^2\right),\\ \varepsilon_3 &= -b^3k_g^2k'_g + \left(ak_g+bk'_g\right)\left(2+\left(bk_g\right)^2\right).\end{aligned} \tag{27}$$

On the other hand, from (7) and (24), it can be easily seen that

$$\begin{aligned}\xi_\beta &= \beta\times T_\beta\\ &= -\frac{b^2k_g}{\sqrt{4+2\left(bk_g\right)^2}}\alpha - \frac{abk_g}{\sqrt{4+2\left(bk_g\right)^2}}T + \frac{\sqrt{2}}{\sqrt{2+\left(bk_g\right)^2}}\xi.\end{aligned} \tag{28}$$

Hence ξ_β is a unit spacelike vector.

Therefore, the geodesic curvature k_g^β of the curve $\beta = \beta(s^*)$ is given by

$$\begin{aligned}k_g^\beta &= \det\left(\beta, T_\beta, T'_\beta\right)\\ &= \frac{b^2k_g\varepsilon_1 - abk_g\varepsilon_2 + 2\varepsilon_3}{\left(2+b^2k_g^2\right)^{5/2}}.\end{aligned} \tag{29}$$

Theorem 9. *Let $\alpha : I \subset \mathbb{R} \mapsto H_0^2$ be a regular unit speed curve lying fully in H_0^2 with the Sabban frame $\{\alpha, T, \xi\}$ and the geodesic curvature k_g. If $\beta : I \subset \mathbb{R} \mapsto H_0^2$ is $\alpha T\xi$-pseudohyperbolical Smarandache curve of α, then its frame $\{\beta, T_\beta, \xi_\beta\}$ is given by*

$$\begin{bmatrix}\beta\\ T_\beta\\ \xi_\beta\end{bmatrix} = \begin{bmatrix}\frac{a}{\sqrt{3}} & \frac{b}{\sqrt{3}} & \frac{c}{\sqrt{3}}\\ \frac{b}{\sqrt{\left|\left(ak_g-c\right)^2-3k_g^2+3\right|}} & \frac{a-ck_g}{\sqrt{\left|\left(ak_g-c\right)^2-3k_g^2+3\right|}} & \frac{bk_g}{\sqrt{\left|\left(ak_g-c\right)^2-3k_g^2+3\right|}}\\ \frac{-\left(a^2-3\right)k_g+ac}{\sqrt{\left|3\left(ak_g-c\right)^2-9k_g^2+9\right|}} & \frac{bc-abk_g}{\sqrt{\left|3\left(ak_g-c\right)^2-9k_g^2+9\right|}} & \frac{3+c^2-ack_g}{\sqrt{\left|3\left(ak_g-c\right)^2-9k_g^2+9\right|}}\end{bmatrix}\begin{bmatrix}\alpha\\ T\\ \xi\end{bmatrix} \tag{30}$$

and the corresponding geodesic curvature k_g^β reads

$$\begin{aligned}k_g^\beta &= \left(-\left(ac-\left(a^2-3\right)k_g\right)\varepsilon_1\right.\\ &\quad \left.+\left(bc-abk_g\right)\varepsilon_2 + \left(3+c^2-ack_g\right)\varepsilon_3\right)\\ &\quad \times\left(\left(\left(ak_g-c\right)^2-3k_g^2+3\right)^{5/2}\right)^{-1},\\ &\qquad a,b,c\in\mathbb{R}_0,\end{aligned} \tag{31}$$

where

$$\begin{aligned}\varepsilon_1 &= -\left(\left(ak_g-c\right)ak'_g - 3k'_gk_g\right)b\\ &\quad + \left(\left(ak_g-c\right)^2-3k_g^2+3\right)\left(a-ck_g\right),\\ \varepsilon_2 &= -\left(\left(ak_g-c\right)ak'_g - 3k'_gk_g\right)\left(a-ck_g\right)\\ &\quad + \left(\left(ak_g-c\right)^2-3k_g^2+3\right)\left(b-ck'_g-bk_g^2\right),\\ \varepsilon_3 &= -\left(\left(ak_g-c\right)ak'_g - 3k'_gk_g\right)bk_g\\ &\quad + \left(\left(ak_g-c\right)^2-3k_g^2+3\right)\left(\left(a-ck_g\right)k_g+bk'_g\right)\end{aligned} \tag{32}$$

and $a^2-b^2-c^2=3$.

Proof. Differentiating (8) with respect to s and by using (4), we find

$$\beta'(s) = \frac{d\beta}{ds^*}\frac{ds^*}{ds} = \frac{1}{\sqrt{3}}\left(b\alpha+\left(a-ck_g\right)T+bk_g\xi\right), \tag{33}$$

and thus

$$T_\beta \frac{ds^*}{ds} = \frac{1}{\sqrt{3}}\left(b\alpha + \left(a - ck_g\right)T + bk_g\xi\right), \quad (34)$$

where

$$\frac{ds^*}{ds} = \sqrt{\frac{\left|\left(ak_g - c\right)^2 - 3k_g^2 + 3\right|}{3}}. \quad (35)$$

The geodesic curvature $k_g \neq (ac \pm \sqrt{a^2c^2 - (a^2 - 3)(c^2 + 3)})/(a^2 - 3)$ for all s, since $\beta = \beta(s^*)$ is a unit speed regular curve in $\mathbb{R}^3_1$.

Therefore, the unit spacelike tangent vector of the curve β is given by

$$T_\beta = \frac{1}{\sqrt{\left|\left(ak_g - c\right)^2 - 3k_g^2 + 3\right|}}\left(b\alpha + \left(a - ck_g\right)T + bk_g\xi\right). \quad (36)$$

Differentiating (36) with respect to s and from (4) and (35), it follows that

$$T'_\beta = \frac{\sqrt{3}}{\left(\left(ak_g - c\right)^2 - 3k_g^2 + 3\right)^2}\left(\varepsilon_1\alpha + \varepsilon_2 T + \varepsilon_3\xi\right), \quad (37)$$

where

$$\begin{aligned}
\varepsilon_1 &= -\left(\left(ak_g - c\right)ak'_g - 3k'_g k_g\right)b \\
&\quad + \left(\left(ak_g - c\right)^2 - 3k_g^2 + 3\right)\left(a - ck_g\right), \\
\varepsilon_2 &= -\left(\left(ak_g - c\right)ak'_g - 3k'_g k_g\right)\left(a - ck_g\right) \\
&\quad + \left(\left(ak_g - c\right)^2 - 3k_g^2 + 3\right)\left(b - ck'_g - bk_g^2\right), \\
\varepsilon_3 &= -\left(\left(ak_g - c\right)ak'_g - 3k'_g k_g\right)bk_g \\
&\quad + \left(\left(ak_g - c\right)^2 - 3k_g^2 + 3\right)\left(\left(a - ck_g\right)k_g + bk'_g\right).
\end{aligned} \quad (38)$$

On the other hand, from (8) and (36) it can be easily seen that

$$\begin{aligned}
\xi_\beta &= \beta \times T_\beta \\
&= \frac{-\left(a^2 - 3\right)k_g + ac}{\sqrt{\left|3\left(ak_g - c\right)^2 - 9k_g^2 + 9\right|}}\alpha \\
&\quad + \frac{bc - abk_g}{\sqrt{\left|3\left(ak_g - c\right)^2 - 9k_g^2 + 9\right|}}T \\
&\quad + \frac{3 + c^2 - ack_g}{\sqrt{\left|3\left(ak_g - c\right)^2 - 9k_g^2 + 9\right|}}\xi.
\end{aligned} \quad (39)$$

Hence ξ_β is a unit spacelike vector.

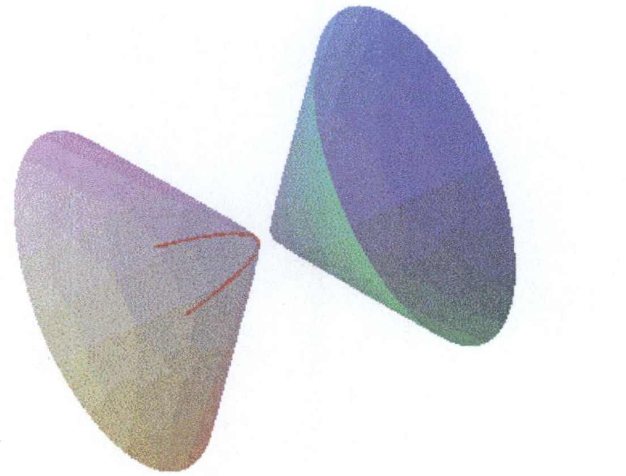

FIGURE 1: The curve α on $H_0^2(1)$.

Therefore, the geodesic curvature k_g^β of the curve $\beta = \beta(s^*)$ is given by

$$\begin{aligned}
k_g^\beta &= \det\left(\beta, T_\beta, T'_\beta\right) \\
&= \left(-\left(ac - \left(a^2 - 3\right)k_g\right)\varepsilon_1 + \left(bc - abk_g\right)\varepsilon_2 \right. \\
&\quad \left. + \left(3 + c^2 - ack_g\right)\varepsilon_3\right) \\
&\quad \times \left(\left(\left(ak_g - c\right)^2 - 3k_g^2 + 3\right)^{5/2}\right)^{-1}.
\end{aligned} \quad (40)$$

Corollary 10. *If $\alpha : I \subset \mathbb{R} \mapsto H_0^2$ is a geodesic curve on H_0^2 in Minkowski 3-space $\mathbb{E}^3_1$, then*

(1) *αT-pseudohyperbolic Smarandache curve is also geodesic on H_0^2;*

(2) *$\alpha\xi$-pseudohyperbolic and $\alpha T\xi$-pseudohyperbolic Smarandache curves have constant geodesic curvatures on H_0^2.*

4. Examples

Example 1. Let α be a unit speed curve lying in pseudohyperbolic space $H_0^2(1)$ in the Minkowski 3-space $\mathbb{E}^3_1$ with parameter equation (see Figure 1)

$$\alpha(s) = \left(\frac{s^2}{2} + 1, \frac{s^2}{2}, s\right). \quad (41)$$

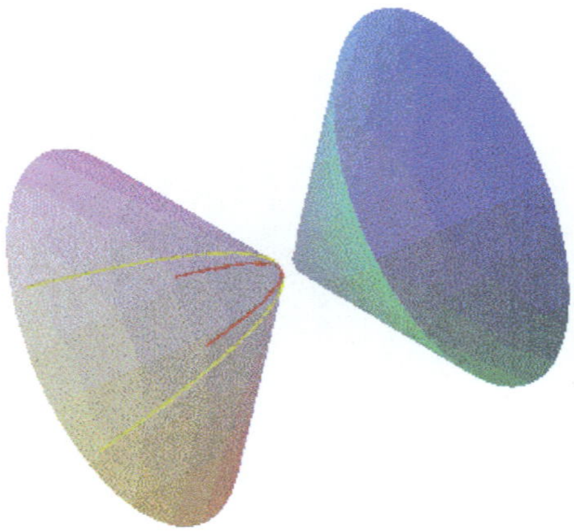

FIGURE 2: The $\alpha\xi$-pseudohyperbolical Smarandache curve β and the curve α on $H_0^2(1)$.

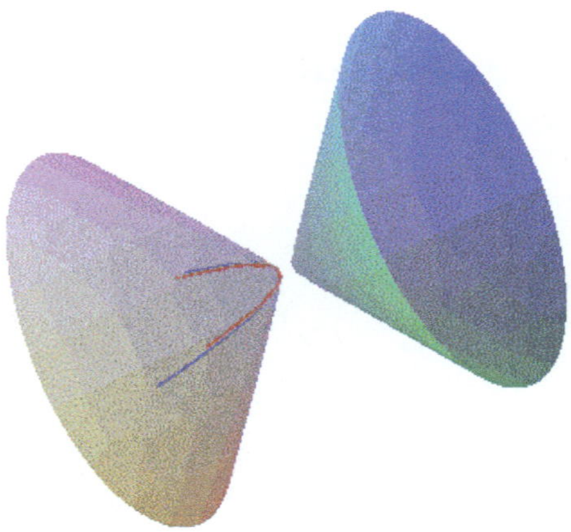

FIGURE 3: The αT-pseudohyperbolical Smarandache curve β and the curve α on $H_0^2(1)$.

The orthonormal Sabban frame $\{\alpha(s), T(s), \xi(s)\}$ along the curve α is given by

$$\alpha(s) = \left(\frac{s^2}{2}+1, \frac{s^2}{2}, s\right),$$
$$T(s) = \alpha'(s) = (s, s, 1), \quad (42)$$
$$\xi(s) = \alpha(s) \times T(s) = \left(\frac{s^2}{2}, \frac{s^2}{2}-1, s\right).$$

In particular, the geodesic curvature k_g of the curve α has the form

$$k_g(s) = -1. \quad (43)$$

Case 1. If we take $a = 2$, $b = \sqrt{2}$, then from (6) the $\alpha\xi$-pseudohyperbolical Smarandache curve β is given by (see Figure 2)

$$\beta(s^*(s)) = \frac{1}{2}\left(\left(\sqrt{2}+1\right)s^2 + 2\sqrt{2}, \left(\sqrt{2}+1\right)s^2 - 2, \left(2\sqrt{2}+2\right)s\right). \quad (44)$$

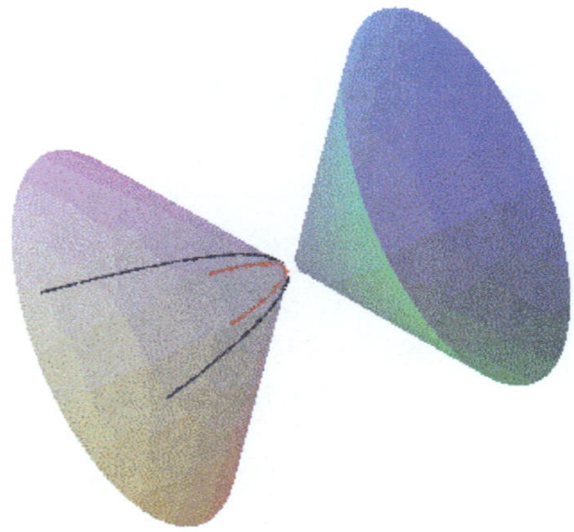

FIGURE 4: The $\alpha T\xi$-pseudohyperbolical Smarandache curve β and the curve α on $H_0^2(1)$.

From Theorem 7, its frame $\{\beta, T_\beta, \xi_\beta\}$ is given by

$$\begin{bmatrix}\beta \\ T_\beta \\ \xi_\beta\end{bmatrix} = \begin{bmatrix}\sqrt{2} & 0 & 1 \\ 0 & 1 & 0 \\ -1 & 0 & \sqrt{2}\end{bmatrix}\begin{bmatrix}\alpha \\ T \\ \xi\end{bmatrix}, \quad (45)$$

and the corresponding geodesic curvature k_g^β reads

$$k_g^\beta = -1. \quad (46)$$

Case 2. If we take $a = \sqrt{3}$, $b = -1$, then from (7) the αT-pseudohyperbolical Smarandache curve is given by (see Figure 3)

$$\beta(s^*(s)) = \frac{\sqrt{2}}{4}\left(\sqrt{3}s^2 - 2s + 2\sqrt{3}, \sqrt{3}s^2 - 2s, 2\sqrt{3}s - 2\right). \quad (47)$$

According to Theorem 8, its frame $\{\beta, T_\beta, \xi_\beta\}$ is given by

$$\begin{bmatrix}\beta \\ T_\beta \\ \xi_\beta\end{bmatrix} = \begin{bmatrix}\frac{\sqrt{6}}{2} & -\frac{\sqrt{2}}{2} & 0 \\ -\frac{\sqrt{3}}{3} & 1 & \frac{\sqrt{3}}{3} \\ \frac{\sqrt{6}}{6} & -\frac{\sqrt{2}}{2} & \frac{\sqrt{6}}{3}\end{bmatrix}\begin{bmatrix}\alpha \\ T \\ \xi\end{bmatrix}, \quad (48)$$

and the corresponding geodesic curvature k_g^β reads

$$k_g^\beta = -1. \quad (49)$$

Case 3. If $a = 3$, $b = \sqrt{3}$, and $c = \sqrt{3}$, then from (7) the $\alpha T\xi$-pseudohyperbolical Smarandache curve β is given by (see Figure 4)

$$\beta(s^*(s)) = \frac{1}{2}\left(\left(\sqrt{3}+1\right)s^2 + 2s + 2\sqrt{3}, \left(\sqrt{3}+1\right)s^2 + 2s - 2, \left(2\sqrt{3}+2\right)s + 2\right). \quad (50)$$

FIGURE 5: Pseudohyperbolical Smarandache curves of α and the curve α on $H_0^2(1)$.

By using Theorem 9, it follows that the frame $\{\beta, T_\beta, \xi_\beta\}$ is given by (see Figure 4)

$$\begin{bmatrix} \beta \\ T_\beta \\ \xi_\beta \end{bmatrix} = \begin{bmatrix} \sqrt{3} & 1 & 1 \\ \dfrac{\sqrt{3}-1}{2} & 1 & \dfrac{1-\sqrt{3}}{2} \\ \dfrac{\sqrt{3}+1}{2} & 1 & \dfrac{\sqrt{3}+1}{2} \end{bmatrix} \begin{bmatrix} \alpha \\ T \\ \xi \end{bmatrix}, \tag{51}$$

and the corresponding geodesic curvature k_g^β reads

$$k_g^\beta = -6\left(2+\sqrt{3}\right)\left(3+\sqrt{3}\right)^{1/2}. \tag{52}$$

Also, Pseudohyperbolical Smarandache curves of α and the curveα on H_0^2 (1) with Figure 5 are shown.

Acknowledgment

The authors would like to thank the referees for their helpful suggestions and comments which significantly improved the first version of the paper.

References

[1] C. Ashbacher, "Smarandache geometries," *Smarandache Notions Journal*, vol. 8, no. 1–3, pp. 212–215, 1997.

[2] M. Turgut and S. Yılmaz, "Smarandache curves in Minkowski space-time," *International Journal of Mathematical Combinatorics*, vol. 3, pp. 51–55, 2008.

[3] A. T. Ali, "Special smarandache curves in the euclidean space," *International Journal of Mathematical Combinatorics*, vol. 2, pp. 30–36, 2010.

[4] K. Taşköprü and M. Tosun, "Smarandache curves on S^2," *Boletim da Sociedade Paranaense de Matemática*, vol. 32, no. 1, pp. 51–59, 2014.

[5] N. Bayrak, O. Bektas, and S. Yuce, "Special Smarandache a Curves in $\mathbb{E}_1^3$," In press, http://arxiv.org/abs/1204.5656.

[6] T. Körpinar and E. Turhan, "A new approach on timelike biharmonic slant helices according to Bishop frame in Lorentzian Heisenberg group $Heis^3$," *Kochi Journal of Mathematics*, vol. 6, pp. 8–15, 2011.

[7] T. Körpinar and E. Turhan, "Characterization of Smarandache M_1M_2-curves of spacelike biharmonic B-slant helices according to Bishop frame in $\mathbb{E}(1,1)$," *Advanced Modeling and Optimization*, vol. 14, no. 2, pp. 327–333, 2012.

[8] Y. Yayli and E. Ziplar, "Frenet-serret motion and ruled surfaces with constant slope," *International Journal of Physical Sciences*, vol. 6, no. 29, pp. 6727–6734, 2011.

[9] M. Petrović-Torgašev and E. Šućurović, "Some characterizations of the spacelike, the timelike and the null curves on the pseudohyperbolic space $\mathbb{H}_0^2$ in $\mathbb{E}_1^3$," *Kragujevac Journal of Mathematics*, vol. 22, pp. 71–82, 2000.

[10] B. O'Neill, *Semi-Riemannian Geometry*, Academic Press, New York, NY, USA, 1983.

Optimal Consumption in a Stochastic Ramsey Model with Cobb-Douglas Production Function

Md. Azizul Baten[1] and Anton Abdulbasah Kamil[2]

[1] *Department of Decision Science, School of Quantitative Sciences, Universiti Utara Malaysia (UUM), Sintok, 06010 Kedah, Malaysia*
[2] *Mathematics Program, School of Distance Education, Universiti Sains Malaysia (USM), 11800 Penang, Malaysia*

Correspondence should be addressed to Md. Azizul Baten; baten_math@yahoo.com

Academic Editor: Aloys Krieg

A stochastic Ramsey model is studied with the Cobb-Douglas production function maximizing the expected discounted utility of consumption. We transformed the Hamilton-Jacobi-Bellman (HJB) equation associated with the stochastic Ramsey model so as to transform the dimension of the state space by changing the variables. By the viscosity solution method, we established the existence of viscosity solution of the transformed Hamilton-Jacobi-Bellman equation associated with this model. Finally, the optimal consumption policy is derived from the optimality conditions in the HJB equation.

1. Introduction

In financial decision-making problems, Merton's [1, 2] papers seemed to be pioneering works. In his seminal work, Merton [2] showed how a stochastic differential for the labor supply determined the stochastic processes for the short-term interest rate and analyzed the effects of different uncertainties on the capital-to-labor ratio. The existence and uniqueness of solutions to the state equation of the Ramsay problem [2] is not yet available. In this study, we turned to Merton's [2] original problem that is revisited considering the growth model for the Cobb-Douglas production function in the finite horizon. Let us define the following quantities:

τ_k = $\inf\{t \geq 0 : K_t = 0\}$,

K_t = capital stock at time $t \geq 0$,

L_t = labor supply at time $t \geq 0$,

λ = constant rate of depreciation, $\lambda \geq 0$,

$c_t K_t$ = consumption rate at time $t \geq 0, 0 \leq c_t \leq 1$,

$c_t K_t / L_t$ = totality of consumption rate per labor;

$F(K, L) = AK^{\alpha} L^{1-\alpha}$ with $0 < \alpha < 1$ and A is a constant, production function producing the commodity for the capital stock $K > 0$ and the labor supply $L > 0$,

n = rate of labor growth (nonzero constant),

σ = non-zero constant coefficients,

ρ = discount rate $\rho > 0$,

$U(c)$ = utility function for the consumption rate $c \geq 0$,

W_t = one-dimensional standard Brownian motion on a complete probability space $(\Omega, \mathcal{F}, P)$ endowed with the natural filtration $\mathcal{F}_t$ generated by $\sigma(W_s, s \leq t)$.

Let us assume that $c = \{c_t\}$ is a consumption policy per capita such that c_t is nonnegative $\mathcal{F}_t = \sigma(W_s, s \leq t)$, a progressively measurable process,

$$\int_0^t c_s ds < \infty \quad \text{a.s. } \forall t \geq 0, \tag{1}$$

and we denote by $\mathcal{A}$ the set of all consumption policies $\{c_t\}$ per capita.

The utility function $U(c)$ is assumed to have the following properties:

$$\begin{aligned} &U \in C[0,\infty) \bigcap C^2(0,\infty), \\ &U(c): \text{strictly concave on } [0,\infty), \\ &U'(c): \text{strictly decreasing}, \quad U'' < 0 \text{ for } c > 0, \\ &U'(\infty) = U'(0+) = 0, \qquad U'(0+) = U(\infty) = \infty. \end{aligned} \tag{2}$$

Following Merton [2], we make the following assumption on the Cobb-Douglas production function $F(K, L)$:

$$\begin{aligned} &F(\gamma K, \gamma L) = \gamma F(K, L), \quad \text{for } \gamma > 0, \\ &F_K(K, L) > 0, \qquad F_L(K, L) > 0, \qquad F_{KK}(K, L) < 0 \\ &F_{LL}(K, L) < 0, \\ &F(0, L) = F(K, 0) = 0, \qquad F_K(0+, L) < \infty, \\ &F_K(\infty, L) = 0, \quad L > 0. \end{aligned} \tag{3}$$

We are concerned with the economic growth model to maximize the expected discount utility of consumption

$$J(c) = E\left[\int_0^T e^{-\rho t} U\left(\frac{c_t K_t}{L_t}\right) dt\right] \tag{4}$$

per labor with a horizon T over the class $c \in \mathscr{A}$ subject to the capital stock K_t, and the labor supply L_t is governed by the stochastic differential equation

$$\begin{aligned} dK_t &= [F(K_t, L_t) - \lambda K_t - c_t K_t]\, dt \quad K_0 = K,\ K > 0, \\ dL_t &= nL_t dt + \sigma L_t dW_t, \quad L_0 = L,\ L > 0. \end{aligned} \tag{5}$$

This optimal consumption problem has been studied by Merton [2], Kamien and Schwartz [3], Koo [4], Morimoto and Kawaguchi [5], Morimoto [6], and Zeldes [7]. Recently, this kind of problem is treated by Baten and Sobhan [8] for one-sector neoclassical growth model with the constant elasticity of substitution (CES) production function in the infinite time horizon case. The studies of Ramsey-type stochastic growth models are also available in Amilon and Bermin [9], Bucci et al. [10], Posch [11], and Roche [12]; comprehensive coverage of this subject can be found, for example, in the books of Chang [13], Malliaris et al., [14], Turnovsky [15, 16], and Walde [17–19]. Continuous-time steady-state studies under lower-dimensional uncertainty carried out, for example, by Merton [2] and Smith [20] within a Ramsey-type setup, and, for example, by Bourguignon [21], Jensen and Richter [22], and Merton [2] within a Solow-Swan-type setup. But these papers did not deal with establishing the existence of viscosity solution of the transformed Hamilton-Jacobi-Bellman equation, and they did not derive the optimal consumption policy from the optimality conditions in the HJB equation associated with the stochastic Ramsey problem, which we have dealt with in this paper.

On the other hand, Oksendal [23] considered a cash flow modeled with geometric Brownian motion to maximize the expected discounted utility of consumption rate for a finite horizon with the assumption that the consumer has a logarithmic utility for his/her consumption rate. He added a jump term (represented by a Poissonian random measure) in a cash flow model. The problem discussed in Oksendal [23] is related to the optimal consumption and portfolio problems associated with a random time horizon studied in Blanchet-Scalliet et al., [24], Bouchard and Pham [25], and Blanchet-Scalliet et al., [26]. However, our paper's approach is different.

By the principle of optimality, it is natural that u solves the general (two-dimensional) Hamilton-Jacobi-Bellman (in short, HJB) equation

$$\begin{aligned} &-\rho u(K, L) + \frac{1}{2}\sigma^2 L^2 u_{LL}(K, L) + nLu_L(K, L) \\ &\quad + \{F(K, L) - \lambda K\} u_K(K, L) \\ &\quad + U^*(u_K(K, L), L) = 0, \\ &u(0, L) = 0, \quad K > 0,\ L > 0, \end{aligned} \tag{6}$$

where $U^*(u_K, L) = \max_{c>0}\{U(cK/L) - cKu_K\}$ and u_K, u_L, and u_{LL} are partial derivatives of $u(K, L)$ with respect to K and L.

The technical difficulty in solving the problem lies in the fact that the HJB equation (6) is a parabolic PDE with two spatial variables K and L. We apply the viscosity method of Fleming and Soner [27] and Soner [28] to this problem to show that the transformed one-dimensional HJB equation admits a viscosity solution v and the optimal consumption policy can be represented in a feedback from the optimality conditions in the HJB equation.

This paper is organized as follows. In Section 2, we transform the two-dimensional HJB equation (6) associated with the stochastic Ramsey model. In Section 3, we show the existence of viscosity solution of the transformed HJB equation. In Section 4, a synthesis of the optimal consumption policy is presented in the feedback from the optimality conditions. Finally, Section 5 concludes with some remarks.

2. Transformed Hamilton-Jacobi-Bellman Equation

In order to transform the HJB equation (6) to one-dimensional second-order differential equation, that is, from the two-dimensional state space form (one state K for capital stock and the other state L for labor force), it has been transformed to a one-dimensional form, for ($x = K/L$) the ratio of capital to labor. Let us consider the solution $u(K, L)$ of (6) of the form

$$u(K, L) = v\left(\frac{K}{L}\right), \quad L > 0. \tag{7}$$

Clearly

$$\begin{aligned} Lu_K &= v_K, \qquad Lu_L = -Kv_K, \\ L^2 u_{LL} &= K^2 v_{KK} + 2Kv_K. \end{aligned} \tag{8}$$

Setting $x = K/L$ and substituting these above in (6), we have the HJB equation of v of the following form:

$$-\rho v(x) + \frac{1}{2}\sigma^2 x^2 v''(x) + \left(f(x) - \tilde{\lambda}x\right) v'(x) + U^*\left(x, v'(x)\right) = 0, \qquad v(0) = 0, \quad x > 0, \tag{9}$$

where $\tilde{\lambda} = n + \lambda - \sigma^2$, $f(x) = F(x,1)$, and $U^*(x,q) = \max_{0 \le c \le 1}\{U(cx) - cxq\}, q \in \mathbf{R}$.

We found that (9) is the transformed HJB equation associated with the stochastic utility consumption problem so as to maximize

$$\tilde{J}(c) = E\left[\int_0^T e^{-\rho t} U(c_t x_t)\, dt\right] \tag{10}$$

over the class $c \in \tilde{\mathscr{A}}$, subject to

$$dx_t = \left[f(x_t) - \tilde{\lambda}x_t - c_t\right] dt - \sigma x_t dW_t, \quad x_0 = x, \ x \ge 0, \tag{11}$$

where $c \in \tilde{\mathscr{A}}$ denotes the class $\mathscr{A}$ with $\{x_t\}$ replacing $\{K_t\}$. We choose $\delta > 0$ and rewrite (9) as

$$-\left(\rho + \frac{1}{\delta}\right) v(x) + \frac{1}{2}\sigma^2 x^2 v''(x) + \left(f(x) - \tilde{\lambda}x\right) v'(x) + U^*\left(x, v'(x)\right) + \frac{1}{\delta} v(x) = 0, \qquad v(0) = 0, \quad x > 0. \tag{12}$$

The value function can be defined as a function whose value is the maximum value of the objective function of the consumption problem, that is,

$$V(x_t) = \sup_{c \in \tilde{\mathscr{A}}} E\left[\int_0^\tau e^{-(\rho + 1/\delta)t} \left\{U(c_t x_t) + \frac{1}{\delta} v(x_t)\right\} dt\right], \tag{13}$$

where $\tau = \tau(x) = \inf\{t \ge 0 : x_t = 0\}$ and $c = \{c_t\}$ is the element of the class $\tilde{\mathscr{A}}$ consisting of $\mathscr{F}_t$ progressively measurable processes such that

$$\int_0^t c_s ds < \infty \quad \text{a.s. } \forall t \ge 0. \tag{14}$$

3. Viscosity Solutions

In this section, we will show the existence results on the viscosity solution v of the HJB equation (9).

3.1. Definition. Let $v \in C[0,\infty)$ and $v(0) = 0$. Then v is called a viscosity solution of the reduced (one-dimensional) HJB equation (9) if the following relations hold:

$$\begin{aligned} &-\rho v(x) + \frac{1}{2}\sigma^2 x^2 q + p\left(f(x) - \tilde{\lambda}x\right) + U^*(x,p) \ge 0, \\ &\qquad \forall (p,q) \in J^{2,+} v(x), \ \forall x > 0, \\ &-\rho v(x) + \frac{1}{2}\sigma^2 x^2 q + p\left(f(x) - \tilde{\lambda}x\right) + U^*(x,p) \le 0, \\ &\qquad \forall (p,q) \in J^{2,-} v(x), \ \forall x > 0, \end{aligned} \tag{15}$$

where $J^{2,+}$ and $J^{2,-}$ are defined by

$$\begin{aligned} &J^{2,+} v(x) \\ &= \left\{(p,q) \in \mathbf{R}^2 : \limsup_{\bar{x} \to x} \frac{v(\bar{x}) - v(x) - p(\bar{x} - x) - (1/2) q |\bar{x} - x|^2}{|\bar{x} - x|^2} \le 0\right\}, \\ &J^{2,-} v(x) \\ &= \left\{(p,q) \in \mathbf{R}^2 : \liminf_{\bar{x} \to x} \frac{v(\bar{x}) - v(x) - p(\bar{x} - x) - (1/2) q |\bar{x} - x|^2}{|\bar{x} - x|^2} \ge 0\right\}. \end{aligned} \tag{16}$$

We assume that

$$\rho + \tilde{\lambda} > 0, \tag{17}$$

$$-\rho + \left(f'(0) + |\tilde{\lambda}|\right) + \frac{1}{2}\sigma^2 < 0. \tag{18}$$

Take $0 < \alpha < \rho$, and we choose $P_1 > 0$ by concavity such that

$$f(x) - \tilde{\lambda}x \le P_1. \tag{19}$$

Taking sufficiently large $P_0 > P_1$, we observe by (2) and (19) that $\phi(t,x) = e^{\tau - t}(x + P_0)$ fulfills

$$\begin{aligned} &-\alpha\phi(x) + \frac{1}{2}\sigma^2 x^2 \phi''(x) + \left(f(x) - \tilde{\lambda}x\right)\phi'(x) + U^*\left(x, \phi'(x)\right) \\ &\quad \le e^{\tau - t}\left(-x - P_0 + P_1\right) + Uo\left(U'\right)^{-1} e^{\tau - t} \\ &\quad < -x - P_0 + P_1 + Uo\left(U'\right)^{-1}(1) < 0, \end{aligned} \tag{20}$$

for some constant $P_0 > 0$.

Lemma 1. *One assumes (2), (17), (11), and (20), then the value function $V(x)$ fulfills*

$$0 \le V(x_t) \le \phi(x_t), \tag{21}$$

$$\sup_{c \in \tilde{\mathscr{A}}} E\left[e^{-(\rho + 1/\delta)\tau} |x(\tau) - \bar{x}(\tau)|\right] \le |x - \bar{x}| \tag{22}$$

for any stopping time τ, *where* $\{\tilde{x}(\tau)\}$ *is the solution of* (11) *to* $c \in \mathscr{A}$ *with* $\tilde{x}(0) = \tilde{x}$.

Proof. Itô's formula gives

$$\begin{aligned} 0 &\leq e^{-(\rho+1/\delta)(t\wedge\tau)}\phi\left(x_{(t\wedge\tau)}\right) \\ &= \phi(x) + \int_0^{t\wedge\tau} e^{-(\rho+1/\delta)s} \\ &\quad \times \left\{-\left(\rho+\frac{1}{\delta}\right)\phi(x_s) \right. \\ &\quad + \left(f(x) - \tilde{\lambda}x_s - c_s x_s\right)\phi'(x_s) \\ &\quad \left. + \frac{1}{2}\sigma^2 x_s^2 \phi''(x_s)\right\} ds \\ &\quad - \int_0^{t\wedge\tau} e^{-(\rho+1/\delta)s}\sigma x_s \phi'(x_s)\, dW_s. \end{aligned} \tag{23}$$

Since

$$\begin{aligned} E\left[\int_0^t \left|x_s\phi'(x_s)\right|^2 ds\right] &\leq e^{2(\tau-t)} E\left[\int_0^t |x_s|^2 ds\right] \\ &\leq e^{2(\tau-t)} E\left[\int_0^t \left(1+|x_s|^2\right) ds\right] \\ &= e^{2(\tau-t)} t + e^{2(\tau-t)} E\left[\int_0^t |x_s|^2 ds\right], \end{aligned} \tag{24}$$

and by considering $c_t = 0$, for all $t \geq 0$, then (11) becomes

$$d\breve{x}_t = \left[f(\breve{x}_t) - \tilde{\lambda}\breve{x}_t\right] dt - \sigma\breve{x}_t dw_t, \quad \breve{x}_0 = x,\ x > 0, \tag{25}$$

by the comparison theorem of Ikeda and Watanabe [29]; we see that $0 < x_t \leq \breve{x}_t$, for all $t \geq 0$. Hence, by applying the existence and uniqueness theorem for (11), we have

$$E\left[\int_0^t |x_s|^2 ds\right] < \infty. \tag{26}$$

Therefore, from (24), we have

$$E\left[\int_0^t \left|x_s\phi'(x_s)\right|^2 ds\right] < \infty, \tag{27}$$

which yields that $\int_0^t e^{-(\rho+1/\delta)s}\sigma x_s\phi'(x_s)dW_s$ is a martingale, and again by (11), we can take sufficiently small $\delta > 0$ such that

$$\begin{aligned} &E\left[\sup_t \int_0^t e^{-(\rho+1/\delta)s}\sigma x_s\phi'(x_s)\, dW_s\right] \\ &\quad \leq 2|\sigma|\, E\left[\sup_t \int_0^\infty e^{-(\rho+1/\delta)s} x_s^2 ds\right]^{1/2} < \infty. \end{aligned} \tag{28}$$

Hence,

$$E\left[\int_0^{t\wedge\tau} e^{-(\rho+1/\delta)s}\sigma x_s\phi'(x_s)\, dw_s\right] = 0. \tag{29}$$

Therefore, by (20), (11) and taking expectation on both sides of (23), we obtain

$$E\left[\int_0^{t\wedge\tau} e^{-(\rho+1/\delta)s}\left\{U(c_s x_s) + \frac{1}{\delta}v(x_s)\right\} ds\right] \leq \phi(x), \tag{30}$$

from which we deduce (21).

We set $z_t = x_t - \tilde{x}_t$ and by (11), it is clear that

$$\begin{aligned} dz_t = d(x_t - \tilde{x}_t) &= \left[f(x_t) - f(\tilde{x}_t) - \tilde{\lambda}(x_t - \tilde{x}_t)\right] dt \\ &\quad - \sigma(x_t - \tilde{x}_t)\, dW_t. \end{aligned} \tag{31}$$

Since by (3), $f(z)$ is Lipschitz continuous and concave and $f(0) = 0$, then we have

$$dz_t \leq \left(f'(0) + |\tilde{\lambda}|\right) z_t dt - \sigma z(t)\, dW_t, \quad z(0) = x - \tilde{x}. \tag{32}$$

Take $\mu > 0$ such that $g_\mu(z) = (z^2 + \mu)^{1/2}$, and we can find from (18) and (20) that

$$\begin{aligned} &-\left(\rho+\frac{1}{\delta}\right)g_\mu(z) + \frac{1}{2}\sigma^2 z^2 g_\mu''(z) + \left(f'(0) + |\tilde{\lambda}|\right)|z g_\mu'(z)| \\ &\quad \leq \left(z^2+\mu\right)^{1/2}\left\{-\left(\rho+\frac{1}{\delta}\right) + \frac{1}{2}\sigma^2 \right. \\ &\quad\quad \left. + \left(f'(0) + |\tilde{\lambda}|\right)\right\} < 0, \quad \forall z \in \mathbf{R}. \end{aligned} \tag{33}$$

Using Itô's formula and by (20), we have

$$\begin{aligned} &E\left[e^{-(\rho+1/\delta)\tau} g_\mu(z_\tau)\right] \\ &= g_\mu(x - \tilde{x}) \\ &\quad + E\left[\int_0^\tau e^{-(\rho+1/\delta)s} \right. \\ &\quad\quad \times \left\{-\left(\rho+\frac{1}{\delta}\right)g_\mu(z_s) \right. \\ &\quad\quad + \left(f(x_s) - f(\tilde{x}_s) - \tilde{\lambda}z_s\right)g_\mu'(z_s) \\ &\quad\quad \left. + \frac{1}{2}\sigma^2 z_s^2 g_\mu''(z_s)\right\} ds \\ &\quad\quad \left. - \int_0^\tau e^{-(\rho+1/\delta)s}\sigma z_s g_\mu'(z_s)\, dW_s\right] \\ &\leq g_\mu(x - \tilde{x}). \end{aligned} \tag{34}$$

Letting $\delta \to 0$, and by Fatou's lemma, we obtain

$$E\left[e^{-(\rho+1/\delta)\tau}|z_\tau|\right] \leq |x - \tilde{x}|, \tag{35}$$

which implies (22).

Theorem 2. *One assumes* (2), (3), (17), *and* (18), *then the value function is a viscosity solution of the reduced (one-dimension) HJB equation* (9) *such that* $0 \leq v(x) \leq \phi(x)$.

Proof. Following (13) and (21) we have

$$v(x) = \sup_{c\in\widetilde{\mathscr{A}}} E\left[\int_0^{\tau} e^{-(\rho+1/\delta)t}\left\{U(c_t x_t) + \frac{1}{\delta}v(x_t)\right\}dt\right] \le \phi(x), \quad \forall x \ge 0, \tag{36}$$

for any stopping time τ. By (13) and for any $\varepsilon > 0$, there exists $c \in \widetilde{\mathscr{A}}$ such that

$$\begin{aligned} v(x) - \varepsilon &< E\left[\int_0^{\tau} e^{-(\rho+1/\delta)t}\left\{U(c_t x_t) + \frac{1}{\delta}v(x_t)\right\}dt\right] \\ &= E\left[\int_0^{\tau} e^{-(\rho+1/\delta)t}U(c_t x_t)\,dt\right] \\ &+ E\left[\int_0^{\tau} e^{-(\rho+1/\delta)t}\frac{1}{\delta}v(x_t)\,dt\right]. \end{aligned} \tag{37}$$

Since $f(z)$ is Lipschitz continuous, it follows that

$$\begin{aligned} dz_t &= \left[f(x_t) - f(\widetilde{x}_t) - \widetilde{\lambda}(x_t - \widetilde{x}_t)\right]dt - \sigma(x_t - \widetilde{x}_t)\,dW_t \\ &\le \left(f'(0) + |\widetilde{\lambda}|\right)z_t dt - \sigma z_t dW_t, \quad z(0) = x - \widetilde{x}. \end{aligned} \tag{38}$$

By (11), we can consider that $\widetilde{z}_t$ is the solution of

$$d\widetilde{z}_t = \left(f'(0) + |\widetilde{\lambda}|\right)\widetilde{z}_t dt - \sigma\widetilde{z}_t dW_t, \quad \widetilde{z}(0) = x > 0. \tag{39}$$

So by the comparison theorem Ikeda and Watanabe [29], we have

$$\tau_z \downarrow \widetilde{\tau}, \quad z_t \le \widetilde{z}_t \downarrow 0, \quad \text{a.s. as } z \downarrow 0. \tag{40}$$

Since $E[\sup_{0\le t\le L}\widetilde{z}_t^2] < \infty$ for all $L > 0$, now by (11) we have

$$E\left[z_{\tau\wedge L}\right] = z + E\left[\int_0^{\tau\wedge L}\left\{f(z_t) - \widetilde{\lambda}z_t - c_t\right\}dt\right]. \tag{41}$$

Letting $z \downarrow 0$ and then $L \to 0$, we obtain

$$E\left[\int_0^{\widetilde{\tau}} c_t dt\right] = E\left[\int_0^{\widetilde{\tau}}\left\{f(0) - c_t\right\}dt\right] \ge 0, \tag{42}$$

so that

$$E\left[\int_0^{\widetilde{\tau}} e^{-(\rho+1/\delta)t}U(c_t x_t)\,dt\right] = 0. \tag{43}$$

Passing to the limit to (37) and applying (43), we obtain

$$\begin{aligned} v(0+) - \varepsilon &< E\left[\int_0^{\infty} e^{-(\rho+1/\delta)t}\frac{1}{\delta}v(0+)\,dt\right] \\ &= \frac{v(0+)}{\rho\delta + 1}, \end{aligned} \tag{44}$$

which implies $v(0+) = 0$. Thus, $v \in C[0,\infty)$. So by the standard stability results of Fleming and Soner [27], we deduce that v is a viscosity solution of (9).

4. Optimal Consumption Policy

Under the assumption (1) and (2), Lemma 3 has revealed that the value function of the representative household assets must approach zero as time approaches infinity.

Lemma 3. *One assumes (2), (3), and (17). Then for any $(c_t) \in \mathscr{A}$. One has*

$$\lim_{t\to\infty}\inf E\left[e^{-\rho t}u(K_t, L_t)\right] = 0. \tag{45}$$

Proof. By (17) and (18), we take $\mu \in (0,\rho)$ such that

$$-\mu + \frac{1}{2}\sigma^2 + \left(f'(0) + |\widetilde{\lambda}|\right) - \widetilde{\lambda} < 0. \tag{46}$$

Take $\mu > 0$ and $x = K/L > 0$ such that $g_\mu(x) = (x^2 + \mu)^{1/2}$, and by (33) and (46)

$$\begin{aligned} &-\mu g_\mu(x) + \frac{1}{2}\sigma^2 x^2 g_\mu''(x) + \left(f'(0) + |\widetilde{\lambda}|\right)\left|x g_\mu'(x)\right| \\ &\le \left(x^2 + \mu\right)^{1/2}\left\{-\mu + \frac{1}{2}\sigma^2 + \left(f'(0) + |\widetilde{\lambda}|\right)\right\} < 0. \end{aligned} \tag{47}$$

Setting $\Phi(K,L) = A(K/L)^\gamma$, where $A, \gamma > 0$, we have by (6) and (47)

$$\begin{aligned} &-\mu\Phi(K,L) + \frac{1}{2}\sigma^2 L^2\Phi_{LL} + nL\Phi_L + \Phi_K(F(K,L) - \lambda K) \\ &+ U^*(L\Phi_K) < 0 \quad K, L > 0. \end{aligned} \tag{48}$$

By Itô's formula and (48), we obtain

$$\begin{aligned} &e^{-\rho t}\Phi(K_t, L_t) \\ &= \Phi(K,L) + \int_0^t e^{-\rho s} \\ &\quad\times\left\{(-\rho + \mu)\Phi\right. \\ &\quad + \Phi_K(F(K,L) - \lambda K - c_s L) + nL\Phi_L \\ &\quad \left.+ \frac{1}{2}\sigma^2 L^2\Phi_{LL} - \mu\Phi\right\}\Big|_{(K=K_s, L=L_s)} ds \end{aligned}$$

$$- \int_0^t e^{-\rho s}\sigma L_s \Phi_L dw_s$$
$$\le \Phi(K,L) + \int_0^t e^{-\rho s}\{(-\rho+\mu)\Phi - c_s L\Phi_K - U^*(L\Phi_K)\}|_{(K=K_s,L=L_s)}ds$$
$$- \int_0^t e^{-\rho s}\sigma L_s \Phi_L dw_s$$
$$\le \Phi(K,L) + \int_0^t e^{-\rho s}\{(-\rho+\mu)\Phi - U(c_s)\}|_{(K=K_s,L=L_s)}ds$$
$$- \int_0^t e^{-\rho s}\sigma L_s \Phi_L dw_s,$$
$$0 \le E\left[e^{-\rho t}\Phi(K_t,L_t)\right] \le \Phi(K,L) + E\left[\int_0^t e^{-\rho s}(-\rho+\mu)\Phi(K_s,L_s)\,ds\right]. \tag{49}$$

Letting $t \to \infty$, we have

$$E\left[\int_0^\infty e^{-\rho t}\Phi(K_t,L_t)\,dt\right] \le \frac{1}{\rho-\mu}\Phi(K,L) < \infty, \tag{50}$$

which implies $\liminf_{t\to\infty} E[e^{-\rho t}\Phi(K_t,L_t)] = 0$. By (21), we have

$$u(K,L) \le \Phi(K,L), \tag{51}$$

which completes the proof.

We give a synthesis of the optimal policy $c^* = \{c_t^*\}$ for the optimization problem (4) subject to (5).

Theorem 4. *Under (2) and (3), there exists a unique solution $K_t^* \ge 0$ of*

$$dK_t^* = [F(K_t^*,L_t) - \lambda K_t^* - c_t^* K_t^*]\,dt \quad K_0^* = K,\ K > 0, \tag{52}$$

and the optimal consumption policy $c_t^* \in \mathcal{A}$ is given by

$$c_t^* = g\left(\frac{K_t}{L_t}, L_t u_K(K_t^*,L_t)\right). \tag{53}$$

Proof. Let us consider $G(K,L) = F(K_t,L_t) - \lambda K_t - g(K_t/L_t, L_t u_K(K_t,L_t))K_t$ and since $F(K,L)$. $g(K_t/L_t, L_t u_K(K_t^*,L_t))$ are continuous and $G(0,L) = 0$, there exists an $\mathcal{F}_t$ progressively measurable solution κ_t of

$$d\kappa_t = G(K_t,L_t)\,dt, \quad \kappa_0 = K > 0. \tag{54}$$

Now we shall show $K_t^* \ge 0$ for all $t \ge 0$ a.s. Suppose $0 < v'(0+) < \infty$. By (9), we have $v'(x) \ge 0$ for all $x > 0$, since $U^*(v'(x)) = \infty$ if $v'(x) < 0$. Moreover, by L'Hospital's rule this gives

$$\lim_{x\to 0+} x^2 v''(x) = \lim_{x\to 0+} x^2 v''(x) + 2xv'(x) = \lim_{x\to 0+} \frac{x^2 v'(x)}{x} = 0. \tag{55}$$

Letting $x \to 0+$ in (9), we have $U^*(v'(0+)) = 0$, and this is contrary with (2). Therefore, we get $v'(0+) = \infty$, which implies $g(0, Lu_K(0+,L)) = 0$. We note by the concavity of F that $G(K,L) \le C_1 K + C_2 L$ for $C_1, C_2 > 0$. Then applying the comparison theorem to (54) and

$$d\widetilde{K}_t = \left(C_1\widetilde{K}_t + C_2 L_t\right)dt, \quad \widetilde{K}_0 = K > 0, \tag{56}$$

we obtain $0 \le K_t^* \le \widetilde{K}_t$ for all $t \ge 0$. Further, in case $\tau_K^* = \inf\{t \ge 0 : K_t^* = 0\} = \tau_\kappa < \infty$, we have at $t = \tau_K^*$

$$\frac{dK_t^*}{dt} = F(K_t^*,L_t) - \lambda K_t^* - g\left(\frac{K_t}{L_t}, L_t u_K(K_t^*,L_t)\right)L_t = 0. \tag{57}$$

Therefore, K_t^* solves (52) and $K_t^* \ge 0$. To prove uniqueness, let $K_i^*(t)$, $i = 1,2$, be two solutions of (52). Then $K_1^*(t) - K_2^*(t)$ satisfies

$$d(K_1^*(t) - K_2^*(t)) = \Big[(F(K_1^*(t),L_t) - F(K_2^*(t),L_t)) - \lambda(K_1^*(t) - K_2^*(t)) - \left(g\left(\frac{K_t}{L_t}, L_t u_K(K_t^*,L_t)\right)L_t - g\left(\frac{K_t}{L_t}, L_t u_K(K_t^*,L_t)\right)L_t\right)\Big]dt, \tag{58}$$

$K_1^*(0) - K_2^*(0) = 0$. We have

$$\begin{aligned} d(K_1^*(t) - K_2^*(t))^2 &= 2(K_1^*(t) - K_2^*(t))\,d(K_1^*(t) - K_2^*(t)) \\ &= 2(K_1^*(t) - K_2^*(t)) \\ &\quad\times\Big[(F(K_1^*(t),L_t) - F(K_2^*(t),L_t)) - \lambda(K_1^*(t) - K_2^*(t)) \\ &\quad - \left(g\left(\frac{K_t}{L_t}, L_t u_K(K_t^*,L_t)\right)L_t - g\left(\frac{K_t}{L_t}, L_t u_K(K_t^*,L_t)\right)L_t\right)\Big]dt. \end{aligned} \tag{59}$$

Note that the function $x \rightarrow -g(K_t/L_t, L_t u_K(K_t^*, L_t))L_t$ is decreasing. Hence,

$$\begin{aligned}(K_1^*(t) - K_2^*(t))^2 &\le 2\int_0^t (K_1^*(s) - K_2^*(s)) \\ &\quad \times [(F(K_1^*(s), L_s) - F(K_2^*(s), L_s)) \\ &\quad - \lambda(K_1^*(s) - K_2^*(s))]\, ds \\ &\le 2\left(F'(0) + |\tilde{\lambda}|\right)\int_0^t (K_1^*(s) - K_2^*(s))^2 ds.\end{aligned} \tag{60}$$

By Gronwall's lemma, we have

$$K_1^*(t) = K_2^*(t), \quad \forall t > 0. \tag{61}$$

So, the uniqueness of (52) holds.

Now by (6), (52), and Itô's formula, we have

$$\begin{aligned}e^{-\rho t}u(K_t^*, L_t) &= u(K, L) + \int_0^t e^{-\rho s} \\ &\quad \times \Big\{-\rho u(K, L) + u_K(K, L) \\ &\quad \times (F(K, L) - \lambda K - c_s^* K) \\ &\quad + nLu_L(K, L) \\ &\quad + \frac{1}{2}\sigma^2 L^2 u_{LL}(K, L)\Big\}\Big|_{(K=K_s^*, L=L_s)} ds \\ &\quad + \int_0^t e^{-\rho s}\sigma L_s u_L(K, L)\, dW_s.\end{aligned} \tag{62}$$

By the HJB equation (6), we have

$$\begin{aligned}e^{-\rho t}u(K_t^*, L_t) &= u(K, L) - \int_0^t e^{-\rho s}U\left(\frac{c_s^* K_s^*}{L_s}\right) ds \\ &\quad + \int_0^t e^{-\rho s}\sigma L_s u_L(K, L)\, dW_s,\end{aligned} \tag{63}$$

from which

$$\begin{aligned}&E\left[e^{-\rho(t\wedge\tau_n)}u\left(K_{t\wedge\tau_n}^*, L_{t\wedge\tau_n}\right)\right] \\ &\quad + E\left[\int_0^{t\wedge\tau_n} e^{-\rho s}U\left(\frac{c_s^* K_s^*}{L_s}\right) ds\right] = u(K, L),\end{aligned} \tag{64}$$

where $\{\tau_n\}$ is a sequence of localizing stopping times for the local martingale. From (11), (51), and Doob's inequalities for martingales, it follows that

$$\begin{aligned}&E\left[\sup_n e^{-\rho(t\wedge\tau_n)}u\left(K_{t\wedge\tau_n}^*, L_{t\wedge\tau_n}\right)\right] \\ &\le E\left[\sup_{0\le r\le t} e^{-\alpha r}\phi\left(K_{(r)}^*, L_{(r)}\right)\right] \\ &\le \phi(K, L) + E\left[\sup_{0\le r\le t}\left|\int_0^r e^{-\alpha t}\sigma K_t^* dW_t\right|\right] \\ &\le \phi(K, L) + 2|\sigma| E\left[\int_0^t \left(e^{-\alpha t}K_t^*\right)^2 dt\right]^{1/2} \\ &\le \phi(K, L) + 2|\sigma| E\left[\int_0^t \widetilde{K}_t^2 dt\right]^{1/2}.\end{aligned} \tag{65}$$

Letting $n \rightarrow \infty$ and $t \rightarrow \infty$, hence, we obtain by the dominated convergence theorem

$$\begin{aligned}&E\left[e^{-\rho\tau^*}u(K_{\tau^*}^*, L_{\tau^*})\right] \\ &\quad + E\left[\int_0^{\tau^*} e^{-\rho s}U\left(\frac{c_s^* K_s^*}{L_s}\right) ds\right] = u(K, L).\end{aligned} \tag{66}$$

We deduce by Lemma 3

$$J(c^*) = E\left[\int_0^{\tau^*} e^{-\rho s}U\left(\frac{c_s^* K_s^*}{L_s}\right) ds\right] = u(K, L). \tag{67}$$

Following the same calculation as above, we have

$$\begin{aligned}e^{-\rho t}u(K_t, L_t) &= u(K, L) + \int_0^t e^{-\rho s} \\ &\quad \times \Big\{-\rho u(K, L) \\ &\quad + u_K(K, L)(F(K, L) - \lambda K - c_s K) \\ &\quad + nLu_L(K, L)\frac{1}{2}\sigma^2 \\ &\quad \times L^2 u_{LL}(K, L)\Big\}\Big|_{(K=K_s, L=L_s)} ds \\ &\quad + \int_0^t e^{-\rho s}\sigma L_s u_L(K, L)\, dW_s.\end{aligned} \tag{68}$$

Again by the HJB equation (6), we can obtain

$$0 \le E\left[e^{-\rho\tau}u(K_\tau, L_\tau)\right] \le u(K, L) - E\left[\int_0^\tau e^{-\rho s}U\left(\frac{c_s K_s}{L_s}\right) ds\right] \tag{69}$$

from which

$$J(c) = E\left[\int_0^\tau e^{-\rho s}U\left(\frac{c_s K_s}{L_s}\right) ds\right] \le u(K, L) \tag{70}$$

for any $(c_t) \in \mathscr{A}$. The proof is complete.

Remark 5. From the proof of Theorem 4, it follows that

$$\inf_{c\in\mathcal{A}} E\left[\int_s^T e^{-\rho t} U\left(\frac{c_t K_t}{L_t}\right) dt\right] = V(s,K,L). \qquad (71)$$

Thus, under (2), we observe that the smooth solution of u of the HJB equation (6). Furthermore, let v be the solution of (9) on the entire domain $[0,T) \times (0,\infty)$ with $v(T,x) = 0$, $x > 0$. Setting $x = K/L$ and $u(t,K,L) = v(t,K/L)$, $K, L > 0$, by (8), we have that u satisfies (6). Therefore, we obtain the uniqueness of v.

5. Concluding Remarks

In this paper we have studied the optimal consumption problem of maximizing the expected discounted value of consumption utility in the context of one-sector neoclassical economic growth with Cobb-Douglas production function. We have derived a transformed (one-dimensional) Hamilton-Jacobi-Bellman equation associated with the optimization problem. By the technique of viscosity method we established the viscosity solution to the transformed (one-dimensional) Hamilton-Jacobi-Bellman equation. Finally we have derived the optimal consumption feedback form from the optimality conditions in the two-dimensional HJB equation.

Acknowledgment

The authors wish to thank the anonymous referees for their comments that have led to an improved version of this paper.

References

[1] R. C. Merton, "Lifetime portfolio selection under uncertainity: the continuous time case," *Review of Economics and Statistics*, vol. 51, no. 3, pp. 247–257, 1969.

[2] R. C. Merton, "An asymptotic theory of growth under uncertainty," *Review of Economic Studies*, vol. 42, no. 3, pp. 375–393, 1975.

[3] M. I. Kamien and N. L. Schwartz, *Dynamic Optimization, The Calculus of Variations and Optimal Control in Economics and Management*, North-Holland Publishing Co., Amsterdam, The Netherlands, 1981.

[4] H. K. Koo, "Consumption and portfolio selection with labor income: a continuous time approach," *Mathematical Finance*, vol. 8, no. 1, pp. 49–65, 1988.

[5] H. Morimoto and K. Kawaguchi, "Optimal exploitation of renewable resources by the viscosity solution method," *Stochastic Analysis and Applications*, vol. 20, no. 5, pp. 927–946, 2002.

[6] H. Morimoto, "Optimal consumption models in economic growth," *Journal of Mathematical Analysis and Applications*, vol. 337, no. 1, pp. 480–492, 2008.

[7] S. P. Zeldes, "Optimal consumption with stochastic income: deviations from certainty equivalence," *Quarterly Journal of Economics*, vol. 104, no. 2, pp. 275–298, 1989.

[8] Md. A. Baten and A. Sobhan, "Optimal consumption in a growth model with the CES production function," *Stochastic Analysis and Applications*, vol. 25, no. 5, pp. 1025–1042, 2007.

[9] H. Amilon and H.-P. Bermin, "Welfare effects of controlling labor supply: an application of the stochastic Ramsey model," *Journal of Economic Dynamics & Control*, vol. 28, no. 2, pp. 331–348, 2003.

[10] A. Bucci, C. Colapinto, M. Forster, and D. L. Torre, "On human capital and economic growth with random technology shocks," Departemental Working Paper 2008-36, University of Milan, Department of Economics, Milan, Italy, 2008.

[11] O. Posch, "Structural estimation of jump-diffusion processes in macroeconomics," *Journal of Econometrics*, vol. 153, no. 2, pp. 196–210, 2009.

[12] H. Roche, "Stochastic growth: a duality approach," *Journal of Economic Theory*, vol. 113, no. 1, pp. 131–143, 2003.

[13] F.-R. Chang, *Stochastic Optimization in Continuous Time*, Cambridge University Press, Cambridge, UK, 2004.

[14] A. G. Malliaris, Malliare, and Haltiwanger, *Stochastic Methods in Economics and Finance*, North-Holland Publishing Co., Amsterdam, The Netherlands, 1982.

[15] S. J. Turnovsky, "Government policy in a stochastic growth model with elastic labor supply," *Journal of Public Economic Theory*, vol. 2, no. 4, pp. 389–433, 2000.

[16] S. J. Turnovsky, *Methods of Macroeconomic Dynamics*, MIT Press, Cambridge, Mass, USA, 2nd edition, 2000.

[17] K. Walde, "Endogenous growth cycles," *International Economic Review*, vol. 46, no. 3, pp. 867–894, 2005.

[18] K. Walde, *Applied Intertemporal Optimization. Lecture Notes*, University of Mainz, Mainz, Germany, 2009.

[19] K. Walde, "Production technologies in stochastic continuous time models," CESifo Working Paper 2831, CESifo Group Munich, Munchen, Germany, 2009.

[20] W. Smith, "Inspecting the mechanism exactly: a closed-form solution to a stochastic growth model," vol. 7, no. 1, p. 30, 2007.

[21] F. Bourguignon, "A particular class of continuous-time stochastic growth models," *Journal of Economic Theory*, vol. 9, no. 2, pp. 141–158, 1974.

[22] B. S. Jensen and M. Richter, "Stochastic one-sector and two-sector growth models in continuous time," in *Stochastic Economic Dynamics, Part 5*, B. S. Jensen and T. Palokangas, Eds., pp. 167–216, Business School Press, Copenhagen, Denmark, 2007.

[23] B. Oksendal, *A Universal Optimal Consumption Rate for an Insider*, Pure Mathematics, Department of Mathematics, University of Oslo, Oslo, Norway, 2004.

[24] C. Blanchet-Scalliet, N. El Karoui, M. Jeanblanc, and L. Martellini, "Optimal investment decisions when time-horizon is uncertain," *Journal of Mathematical Economics*, vol. 44, no. 11, pp. 1100–1113, 2008.

[25] B. Bouchard and H. Pham, "Wealth-path dependent utility maximization in incomplete markets," *Finance and Stochastics*, vol. 8, no. 4, pp. 579–603, 2004.

[26] C. Blanchet-Scalliet, N. Karoui, M. Jeanblanc, and L. Martinelli, "Optimal investment and consumption decisions when time-horizon is uncertain," *Journal of Economics Dynamics and Control*, vol. 29, no. 10, pp. 1737–1764, 2005.

[27] W. H. Fleming and H. M. Soner, *Controlled Markov Processes and Viscosity Solutions*, Springer-Verlag, New York, NY, USA, 1993.

[28] H. M. Soner, "Controlled Markov processes, viscosity solutions and applications to mathematical finance," in *Viscosity Solutions and Applications*, vol. 1660 of *Lecture Notes in Mathematics*, pp. 134–185, Springer, Berlin, Germany, 1997.

[29] N. Ikeda and S. Watanabe, *Stochastic Differential Equations and Diffusion Processes*, North-Holland, Amsterdam, The Netherlands, 1981.

15

Some Elementary Aspects of Means

Mowaffaq Hajja

Department of Mathematics, Yarmouk University, Irbid, Jordan

Correspondence should be addressed to Mowaffaq Hajja; mowhajja@yahoo.com

Academic Editor: Peter Bullen

We raise several elementary questions pertaining to various aspects of means. These questions refer to both known and newly introduced families of means, and include questions of characterizations of certain families, relations among certain families, comparability among the members of certain families, and concordance of certain sequences of means. They also include questions about internality tests for certain mean-looking functions and about certain triangle centers viewed as means of the vertices. The questions are accessible to people with no background in means, and it is also expected that these people can seriously investigate, and contribute to the solutions of, these problems. The solutions are expected to require no more than simple tools from analysis, algebra, functional equations, and geometry.

1. Definitions and Terminology

In all that follows, $\mathbb{R}$ denotes the set of real numbers and $\mathbb{J}$ denotes an interval in $\mathbb{R}$.

By a data set (or a list) in a set S, we mean a finite subset of S in which repetition is allowed. Although the order in which the elements of a data set are written is not significant, we sometimes find it convenient to represent a data set in S of size n by a point in S^n, the cartesian product of n copies of S.

We will call a data set $A = (a_1, \dots, a_n)$ in $\mathbb{R}$ *ordered* if $a_1 \leq \cdots \leq a_n$. Clearly, every data set in $\mathbb{R}$ may be assumed ordered.

A *mean* of k variables (or a *k-dimensional mean*) on $\mathbb{J}$ is defined to be any function $\mathscr{M} : \mathbb{J}^k \to \mathbb{J}$ that has the *internality* property

$$\min\{a_1, \dots, a_k\} \leq \mathscr{M}(a_1, \dots, a_k) \leq \max\{a_1, \dots, a_k\} \quad (1)$$

for all a_j in $\mathbb{J}$. It follows that a mean $\mathscr{M}$ must have the property $\mathscr{M}(a, \dots, a) = a$ for all a in $\mathbb{J}$.

Most means that we encounter in the literature, and all means considered below, are also symmetric in the sense that

$$\mathscr{M}(a_1, \dots, a_k) = \mathscr{M}(a_{\sigma(1)}, \dots, a_{\sigma(n)}) \quad (2)$$

for all permutations σ on $\{1, \dots, n\}$, and *1-homogeneous* in the sense that

$$\mathscr{M}(\lambda a_1, \dots, \lambda a_k) = \lambda \mathscr{M}(a_{\sigma(1)}, \dots, a_{\sigma(n)}) \quad (3)$$

for all permissible $\lambda \in \mathbb{R}$.

If $\mathscr{M}$ and $\mathscr{N}$ are two k-dimensional means on $\mathbb{J}$, then we say that $\mathscr{M} \leq \mathscr{N}$ if $\mathscr{M}(a_1, \dots, a_k) \leq \mathscr{N}(a_1, \dots, a_k)$ for all $a_j \in \mathbb{J}$. We say that $\mathscr{M} < \mathscr{N}$ if $\mathscr{M}(a_1, \dots, a_k) < \mathscr{N}(a_1, \dots, a_k)$ for all $a_j \in \mathbb{J}$ for which $a_1, \dots, a_k$ are not all equal. This exception is natural since $\mathscr{M}(a, \dots, a)$ and $\mathscr{N}(a, \dots, a)$ must be equal, with each being equal to a. We say that $\mathscr{M}$ and $\mathscr{N}$ are *comparable* if $\mathscr{M} \leq \mathscr{N}$ or $\mathscr{N} \leq \mathscr{M}$.

A *distance* (or *a distance function*) on a set S is defined to be any function $d : S \times S \to [0, \infty)$ that is *symmetric* and *positive definite*, that is,

$$d(a, b) = d(b, a), \quad \forall a, b \in S,$$
$$d(a, b) = 0 \Longleftrightarrow a = b. \quad (4)$$

Thus a *metric* is a distance that satisfies the triangle inequality

$$d(a, b) + d(b, c) \geq d(a, c), \quad \forall a, b, c \in S, \quad (5)$$

a condition that we find too restrictive for our purposes.

2. Examples of Means

The *arithmetic*, *geometric*, and *harmonic* means of two positive numbers were known to the ancient Greeks; see [1, pp. 84–90]. They are usually denoted by $\mathscr{A}$, $\mathscr{G}$, and $\mathscr{H}$, respectively, and are defined, for $a, b > 0$, by

$$\mathscr{A}(a,b) = \frac{a+b}{2},$$
$$\mathscr{G}(a,b) = \sqrt{ab},$$
$$\mathscr{H}(a,b) = \frac{2}{1/a+1/b} = \frac{2ab}{a+b}. \tag{6}$$

The celebrated inequalities

$$\mathscr{H}(a,b) < \mathscr{G}(a,b) < \mathscr{A}(a,b) \quad \forall a,b > 0 \tag{7}$$

were also known to the Greeks and can be depicted in the well-known figure that is usually attributed to Pappus and that appears in [2, p. 364]. Several other less well known means were also known to the ancient Greeks; see [1, pp. 84–90].

The three means above, and their natural extensions to any number n of variables, are members of a large two-parameter family of means, known now as the *Gini* means and defined by

$$G_{r,s}(x_1,\dots,x_n) = \left(\frac{N_r(x_1,\dots,x_n)}{N_s(x_1,\dots,x_n)}\right)^{1/(r-s)}, \tag{8}$$

where N_j are the Newton polynomials defined by

$$N_j(x_1,\dots,x_n) = \sum_{k=1}^{n} x_k^j. \tag{9}$$

Means of the type $G_{r,r-1}$ are known as *Lehmer*'s means, and those of the type $G_{r,0}$ are known as *Hölder* or *power* means. Other means that have been studied extensively are the *elementary symmetric polynomial* and *elementary symmetric polynomial ratio* means defined by

$$\left(\frac{\sigma_r}{C_r^n}\right)^{1/r}, \qquad \frac{\sigma_r/C_r^n}{\sigma_{r-1}/C_{r-1^n}}, \tag{10}$$

where σ_r is the rth elementary symmetric polynomial in n variables, and where

$$C_r^n = \binom{n}{r}. \tag{11}$$

These are discussed in full detail in the encyclopedic work [3, Chapters III and V].

It is obvious that the power means $\mathscr{P}_r$ defined by

$$\mathscr{P}_r(a_1,\dots,a_n) = G_{r,0}(a_1,\dots,a_n) = \left(\frac{a_1^r+\cdots+a_n^r}{n}\right)^{1/r} \tag{12}$$

that correspond to the values $r = -1$ and $r = 1$ are nothing but the harmonic and arithmetic means $\mathscr{H}$ and $\mathscr{A}$, respectively. It is also natural to set

$$\mathscr{P}_0(a_1,\dots,a_n) = \mathscr{G}(a_1,\dots,a_n) = (a_1\dots a_n)^{1/n}, \tag{13}$$

since

$$\lim_{r\to 0}\left(\frac{a_1^r+\cdots+a_n^r}{n}\right)^{1/r} = (a_1\dots a_n)^{1/n} \tag{14}$$

for all $a_1,\dots,a_n > 0$.

The inequalities (7) can be written as $\mathscr{P}_{-1} < \mathscr{P}_0 < \mathscr{P}_1$. These inequalities hold for any number of variables and they follow from the more general fact that $\mathscr{P}_r(a_1,\dots,a_n)$, for fixed $a_1,\dots,a_n > 0$, is strictly increasing with r. Power means are studied thoroughly in [3, Chapter III].

3. Mean-Producing Distances and Distance Means

It is natural to think of the mean of any list of points in any set to be the point that is *closest* to that list. It is also natural to think of a point as closest to a list of points if the sum of its distances from these points is minimal. This mode of thinking associates means to distances.

If d is a distance on S, and if $A = (a_1,\dots,a_n)$ is a data set in S, then a *d-mean* of A is defined to be any element of S at which the function

$$f(x) = \sum_{i=1}^{n} d(x,a_i) \tag{15}$$

attains its minimum. It is conceivable that (15) attains its minimum at many points, or nowhere at all. However, we shall be mainly interested in distances d on $\mathbb{J}$ for which (15) attains its minimum at a unique point x_A that, furthermore, has the property

$$\min\{a : a \in A\} \le x_A \le \max\{a : a \in A\} \tag{16}$$

for every data set A. Such a distance is called a *mean-producing* or a *mean-defining* distance, and the point x_A is called the *d-mean* of A or the *mean of A arising from the distance d* and will be denoted by $\mu_d(A)$. A mean $\mathscr{M}$ is called a *distance mean* if it is of the form μ_d for some distance d.

Problem Set 1. (1-a) Characterize those distances on $\mathbb{J}$ that are mean-producing.

(1-b) Characterize those pairs of mean producing distances on $\mathbb{J}$ that produce the same mean.

(1-c) Characterize distance means.

4. Examples of Mean-Producing Distances

If d_0 is the discrete metric defined on $\mathbb{R}$ by

$$d_0(a,b) = \begin{cases} 1 & \text{if } a \ne b, \\ 0 & \text{if } a = b, \end{cases} \tag{17}$$

then the function $f(x)$ in (15) is nothing but the number of elements in the given data set A that are different from x, and therefore every element having maximum frequency in A minimizes (15) and is hence a d_0-mean of A. Thus the discrete metric gives rise to what is referred to in statistics as "the" *mode* of A. Due to the nonuniqueness of the mode, the discrete metric is not a mean-producing distance.

Similarly, the usual metric $d = d_1$ defined on $\mathbb{R}$ by

$$d_1(a, b) = |a - b| \tag{18}$$

is not a mean-producing distance. In fact, it is not very difficult to see that if $A = (a_1, \ldots, a_n)$ is an ordered data set of even size $n = 2m$, then any number in the closed interval $[a_m, a_{m+1}]$ minimizes

$$\sum_{j=1}^{n} |x - a_j| \tag{19}$$

and is therefore a d_1-mean of A. Similarly, one can show that if A is of an odd size $n = 2m - 1$, then a_m is the unique d_1-mean of A. Thus the usual metric on $\mathbb{R}$ gives rise to what is referred to in statistics as "the" *median* of A.

On the other hand, the distance d_2 defined on $\mathbb{R}$ by

$$d_2(a, b) = (a - b)^2 \tag{20}$$

is a mean-producing distance, although it is not a metric. In fact, it follows from simple derivative considerations that the function

$$\sum_{j=1}^{n} (x - a_j)^2 \tag{21}$$

attains its minimum at the unique point

$$x = \frac{1}{n}\left(\sum_{j=1}^{n} a_j\right). \tag{22}$$

Thus d_2 is a mean-producing distance, and the corresponding mean is nothing but the arithmetic mean.

It is noteworthy that the three distances that come to mind most naturally give rise to the three most commonly used "means" in statistics. In this respect, it is also worth mentioning that a fourth mean of statistics, the so-called *midrange*, will be encountered below as a very natural *limiting* distance mean.

The distances d_1 and d_2 (and in a sense, d_0 also) are members of the family d_p of distances defined by

$$d_p(a, b) = |a - b|^p. \tag{23}$$

It is not difficult to see that if $p > 1$, then d_p is a mean-producing distance. In fact, if $A = (a_1, \ldots, a_n)$ is a given data set, and if

$$f(x) = \sum_{j=1}^{n} |x - a_j|^p, \tag{24}$$

then

$$f''(x) = p(p-1) \sum_{j=1}^{n} |x - a_j|^{p-2} \geq 0, \tag{25}$$

with equality if and only if $a_1 = \cdots = a_n = x$. Thus f is convex and cannot attain its minimum at more than one point. That it attains its minimum follows from the continuity of $f(x)$, the compactness of $[a_1, a_n]$, and the obvious fact that $f(x)$ is increasing on $[a_n, \infty)$ and is decreasing on $(-\infty, a_1]$. If we denote the mean that d_p defines by μ_p, then $\mu_p(A)$ is the unique zero of

$$\sum_{j=1}^{n} \operatorname{sign}(x - a_j) |x - a_j|^{p-1}, \tag{26}$$

where $\operatorname{sign}(t)$ is defined to be 1 if t is nonnegative and -1 otherwise.

Note that no matter what $p > 1$ is, the two-dimensional mean μ_p arising from d_p is the arithmetic mean. Thus when studying μ_p, we confine our attention to the case when the number k of variables is greater than two. For such k, it is impossible in general to compute $\mu_p(A)$ in closed form.

Problem 2. It would be interesting to investigate comparability among $\{\mu_p : p > 1\}$.

It is highly likely that no two means μ_p are comparable.

5. Deviation and Sparseness

If d is a mean-producing distance on S, and if μ_d is the associated mean, then it is natural to define the d-deviation $\mathcal{D}_d(A)$ of a data set $A = (a_1, \ldots, a_n)$ by an expression like

$$\mathcal{D}_d(A) = \mu_d\{d(\mu_d(A), a_i) : 1 \leq i \leq n\}. \tag{27}$$

Thus if d is defined by

$$d(x, y) = (x - y)^2, \tag{28}$$

then μ_d is nothing but the arithmetic mean or ordinary *average* μ defined by

$$\mu = \mu(a_1, \ldots, a_n) = \frac{a_1 + \cdots + a_n}{n}, \tag{29}$$

and $\mathcal{D}_d$ is the (squared) standard deviation $\sigma^{(2)}$ given by

$$\sigma^{(2)}(a_1, \ldots, a_n) = \frac{|a_1 - \mu|^2 + \cdots + |a_n - \mu|^2}{n}. \tag{30}$$

In a sense, this provides an answer to those who are puzzled and mystified by the choice of the exponent 2 (and not any other exponent) in the standard definition of the standard deviation given in the right-hand side of (30). In fact, distance means were devised by the author in an attempt to remove that mystery. Somehow, we are saying that the ordinary average μ and the standard deviation $\sigma^{(2)}$ must be taken or discarded together, being both associated with the same distance d given in (28). Since few people question the sensibility of the definition of μ given in (29), accepting the standard definition of the standard deviation given in (30) as is becomes a *must*.

It is worth mentioning that choosing an exponent other than 2 in (30) would result in an *essentially* different notion of deviations. More precisely, if one defines $\sigma^{(k)}$ by

$$\sigma^{(k)}(a_1,\ldots,a_n) = \frac{|a_1-\mu|^k+\cdots+|a_n-\mu|^k}{n}, \quad (31)$$

then $\sigma^{(k)}$ and $\sigma^{(2)}$ would of course be unequal, but more importantly, they would not be *monotone* with respect to each other, in the sense that there would exist data sets A and B with $\sigma^{(2)}(A) > \sigma^{(k)}(B)$ and $\sigma^{(2)}(A) < \sigma^{(k)}(B)$. Thus the choice of the exponent k in defining deviations is not as arbitrary as some may feel. On the other hand, it is (27) and not (31) that is the natural generalization of (30). This raises the following, expectedly hard, problem.

Problem 3. Let d be the distance defined by $d(x,y) = |x-y|^k$, and let the associated deviation $\mathscr{D}_d$ defined in (27) be denoted by $\mathscr{D}_k$. Is $\mathscr{D}_k$ monotone with respect to $\mathscr{D}_2$ for any $k \neq 2$, in the sense that

$$\mathscr{D}_k(A) > \mathscr{D}_k(B) \Longrightarrow \mathscr{D}_2(A) > \mathscr{D}_2(B)? \quad (32)$$

We end this section by introducing the notion of *sparseness* and by observing its relation with deviation. If d is a mean-producing distance on $\mathbb{J}$, and if μ_d is the associated mean, then the *d-sparseness* $\mathscr{S}_d(A)$ of a data set $A = (a_1,\ldots,a_n)$ in $\mathbb{J}$ can be defined by

$$\mathscr{S}_d(A) = \mu_d\left\{d\left(a_i,a_j\right) : 1 \le i < j \le n\right\}. \quad (33)$$

It is interesting that when d is defined by (28), the standard deviation coincides, up to a constant multiple, with the sparsenss. One wonders whether this pleasant property characterizes this distance d.

Problem Set 4. (4-a) Characterize those mean-producing distances whose associated mean is the arithmetic mean.

(4-b) If d is as defined in (28), and if d' is another mean-producing distance whose associated mean is the arithmetic mean, does it follow that $\mathscr{D}_{d'}$ and $\mathscr{D}_d$ are monotone with respect to each other?

(4-c) Characterize those mean-producing distances δ for which the deviation $\mathscr{D}_\delta(A)$ is determined by the sparseness $\mathscr{S}_\delta(A)$ for every data set A, and vice versa.

6. Best Approximation Means

It is quite transparent that the discussion in the previous section regarding the distance mean μ_p, $p > 1$, can be written in terms of best approximation in ℓ_p^n, the vector space $\mathbb{R}^n$ endowed with the p-norm $\|\cdots\|_p$ defined by

$$\|(a_1,\ldots,a_n)\|_p = \left(\sum_{j=1}^{n}|a_j|^p\right)^{1/p}. \quad (34)$$

If we denote by $\Delta = \Delta_n$ the line in $\mathbb{R}^n$ consisting of the points $(x_1,\ldots,x_n)$ with $x_1 = \cdots = x_n$, then to say that $a = \mu_p(a_1,\ldots,a_n)$ is just another way of saying that the point $(a,\ldots,a)$ is a best approximant in Δ_n of the point $(a_1,\ldots,a_n)$ with respect to the p-norm given in (34). Here, a point s_t in a subset S of a metric (or distance) space (T,D) is said to be a *best approximant* in S of $t \in T$ if $D(t,s_t) = \min\{D(t,s) : s \in S\}$. Also, a subset S of (T,D) is said to be *Chebyshev* if every t in T has exactly one best approximant in S; see [4, p. 21].

The discussion above motivates the following definition.

Definition 1. Let $\mathbb{J}$ be an interval in $\mathbb{R}$ and let D be a distance on $\mathbb{J}^n$. If the diagonal $\Delta(\mathbb{J}^n)$ of $\mathbb{J}^n$ defined by

$$\Delta(\mathbb{J}^n) = \{(a_1,\ldots,a_n) \in \mathbb{J}^n : a_1 = \cdots = a_n\} \quad (35)$$

is Chebyshev (with respect to D), then the n-dimensional mean M_D on $\mathbb{J}$ defined by declaring $M_D(a_1,\ldots,a_n) = a$ if and only if $(a,\ldots,a)$ is the best approximant of $(a_1,\ldots,a_n)$ in $\Delta(\mathbb{J}^n)$ is called the *Chebyshev* or *best approximation D-mean* or the *best approximation mean arising from D*.

In particular, if one denotes by M_p the best approximation n-dimensional mean on $\mathbb{R}$ arising from (the distance on $\mathbb{R}^n$ induced by) the norm $\|\cdots\|_p$, then the discussion above says that M_p exists for all $p > 1$ and that it is equal to μ_p defined in Section 4.

In view of this, one may also define M_∞ to be the best approximation mean arising from the ∞-norm of ℓ_∞^n, that is, the norm $\|\cdots\|_\infty$ defined on $\mathbb{R}^n$ by

$$\|(a_1,\ldots,a_n)\|_\infty = \max\left\{|a_j| : 1 \le j \le n\right\}. \quad (36)$$

It is not very difficult to see that $\mu_\infty(A)$ is nothing but what is referred to in statistics as the *mid-range* of A. Thus if $A = (a_1,\ldots,a_n)$ is an ordered data set, then

$$M_\infty(A) = \frac{a_1+a_n}{2}. \quad (37)$$

In view of the fact that d_∞ cannot be defined by anything like (23) and μ_∞ is thus meaningless, natural question arises as to whether

$$M_\infty(A) = \lim_{p\to\infty}\mu_p(A) \quad \left(\text{or equivalently} = \lim_{p\to\infty}M_p(A)\right) \quad (38)$$

for every A. An affirmative answer is established in [5, Theorem 1]. In that theorem, it is also established that

$$\lim_{p\to q}\mu_p(A) \quad \left(\text{or equivalently} \lim_{p\to q}M_p(A)\right) = M_q(A) \quad (39)$$

for all q and all A. All of this can be expressed by saying that μ_p is continuous in p for $p \in (1,\infty]$ for all A.

We remark that there is no *obvious* reason why (38) should immediately follow from the well known fact that

$$\lim_{p\to\infty}\|A\|_p = \|A\|_\infty \quad (40)$$

for all points A in $\mathbb{R}^n$.

Problem Set 5. Suppose that δ_p is a sequence of distances on a set S that converges to a distance δ_∞ (in the sense that $\lim_{p\to\infty}\delta_p(a,b) = \delta_\infty(a,b)$ for all a, b in S). Let $T \subseteq S$.

(5-a) If T is Chebyshev with respect to each δ_p, is it necessarily true that T is Chebyshev with respect to δ_∞?

(5-b) If T is Chebyshev with respect to each δ_p and with respect to δ_∞ and if x_p is the best approximant in T of x with respect to δ_p and x_∞ is the best approximant in T of x with respect to δ_∞, does it follow that x_p converges to x_∞?

We end this section by remarking that if $M = M_d$ is the n-dimensional best approximation mean arising from a distance d on $\mathbb{J}^n$, then d is significant only up to its values of the type $d(u,v)$, where $u \in \Delta(\mathbb{J}^n)$ and $v \notin \Delta(\mathbb{J}^n)$. Other values of d are not significant. This, together with the fact that

$$\text{every mean is a best approximation mean arising from a metric,} \tag{41}$$

makes the study of best approximation means less interesting. Fact (41) was proved in an unduly complicated manner in [6], and in a trivial way based on a few-line set-theoretic argument in [7].

Problem 6. Given a mean $\mathscr{M}$ on $\mathbb{J}$, a metric D on $\mathbb{J}$ is constructed in [6] so that $\mathscr{M}$ is the best approximation mean arising from D. Since the construction is extremely complicated in comparison with the construction in [7], it is desirable to examine the construction of D in [6] and see what other nice properties (such as continuity with respect to the usual metric) D has. This would restore merit to the construction in [6] and to the proofs therein and provide raison d'être for the so-called *generalized* means introduced there.

7. Towards a Unique Median

As mentioned earlier, the distance d_1 on $\mathbb{R}$ defined by (23) does not give rise to a (distance) mean. Equivalently, the 1-norm $\|\cdots\|_1$ on $\mathbb{R}^n$ defined by (34) does not give rise to a (best approximation) mean. These give rise, instead, to the many-valued function known as *the median*. Thus, following the statistician's mode of thinking, one may set

$$\begin{aligned}\mu_1(A) = M_1(A) &= \text{the median interval of } A\\ &= \text{the set of all medians of } A.\end{aligned} \tag{42}$$

From a mathematician's point of view, however, this leaves a lot to be desired, to say the least. The feasibility and naturality of defining μ_∞ as the limit of μ_p as p approaches ∞ gives us a clue on how the median μ_1 may be defined. It is a pleasant fact, proved in [5, Theorem 4], that the limit of $\mu_p(A)$ (equivalently of $M_p(A)$) as p decreases to 1 exists for every $A \in \mathbb{R}^n$ and equals one of the medians described in (42). This limit can certainly be used as *the* definition of *the* median.

Problem Set 7. Let μ_p be as defined in Section 4, and let μ^* be the limit of μ_p as p decreases to 1.

(7-a) Explore how the value of $\mu^*(A)$ compares with the common practice of taking the median of A to be the midpoint of the median interval (defined in (42) for various values of A.

(7-b) Is μ^* continuous on $\mathbb{R}^n$? If not, what are its points of discontinuity?

(7-c) Given $A \in \mathbb{R}^n$, is the convergence of $\mu_p(A)$ (as p decreases to 1) to $\mu^*(A)$ monotone?

The convergence of $\mu_p(A)$ (as p decreases to 1) to $\mu^*(A)$ is described in [5, Theorem 4], where it is proved that the convergence is ultimately monotone. It is also proved in [5, Theorem 5] that when $n = 3$, then the convergence is monotone.

It is of course legitimate to question the usefulness of defining the median to be μ^*, but that can be left to statisticians and workers in relevant disciplines to decide. It is also legitimate to question the path that we have taken the limit along. In other words, it is conceivable that there exists, in addition to d_p, a sequence d'_p of distances on $\mathbb{R}$ that converges to d_1 such that the limit μ^{**}, as p decreases to 1, of their associated distance means μ'_p is not the same as the limit μ^* of μ_p. In this case, μ^{**} would have as valid a claim as μ^* to being *the* median. However, the naturality of d_p may help accepting μ^* as a most legitimate median.

Problem Set 8. Suppose that δ_p and δ'_p, $p \in \mathbb{N}$, are sequences of distances on a set S that converge to the distances δ_∞ and δ'_∞, respectively (in the sense that $\lim_{p\to\infty}\delta_p(a,b) = \delta_\infty(a,b)$ for all a, b in S, etc.).

(8-a) If each δ_p, $p \in \mathbb{N}$, is mean producing with corresponding mean m_p, does it follow that δ_∞ is mean producing? If so, and if the mean produced by δ_∞ is m_∞, is it necessarily true that m_p converges to m_∞?

(8-b) If δ_p and δ'_p, $p \in \mathbb{N} \cup \{\infty\}$, are mean producing distances with corresponding means m_p and m'_p, and if $m_p = m'_p$ for all $p \in \mathbb{N}$, does it follow that $m_\infty = m'_\infty$?

8. Examples of Distance Means

It is clear that the arithmetic mean is the distance mean arising from the the distance d_2 given by $d_2(a,b) = (a-b)^2$. Similarly, the geometric mean on the set of positive numbers is the distance mean arising from the distance $d_{\mathscr{G}}$ given by

$$d_{\mathscr{G}}(a,b) = (\ln a - \ln b)^2. \tag{43}$$

In fact, this should not be amazing since the arithmetic mean $\mathscr{A}$ on $\mathbb{R}$ and the geometric mean $\mathscr{G}$ on $(0,\infty)$ are *equivalent* in the sense that there is a bijection $g : (0,\infty) \to \mathbb{R}$, namely $g(x) = \ln x$, for which $\mathscr{G}(a,b) = g^{-1}\mathscr{A}(g(a),g(b))$ for all a, b. Similarly, the harmonic and arithmetic means on $(0,\infty)$ are equivalent via the bijection $h(x) = 1/x$, and therefore the harmonic mean is the distance mean arising from the distance $d_{\mathscr{H}}$ given by

$$d_{\mathscr{H}}(a,b) = \left(\frac{1}{a} - \frac{1}{b}\right)^2. \tag{44}$$

The analogous question pertaining to the logarithmic mean $\mathscr{L}$ defined by

$$\mathscr{L}(a,b) = \frac{a-b}{\ln a - \ln b}, \quad a,b > 0, \tag{45}$$

remains open.

Problem 9. Decide whether the mean $\mathscr{L}$ (defined in (45)) is a distance mean.

9. Quasi-Arithmetic Means

A k-dimensional mean $\mathscr{M}$ on $\mathbb{J}$ is called a *quasi-arithmetic* mean if there is a continuous strictly monotone function g from $\mathbb{J}$ to an interval $\mathbb{I}$ in $\mathbb{R}$ such that

$$\mathscr{M}(a_1,\ldots,a_k) = g^{-1}(\mathscr{A}(g(a_1),\ldots,g(a_k))) \tag{46}$$

for all a_j in $\mathbb{J}$. We have seen that the geometric and harmonic means are quasi-arithmetic and concluded that they are distance means. To see that $\mathscr{L}$ is not quasi-arithmetic, we observe that the (two-dimensional) arithmetic mean, and hence any quasi-arithmetic mean $\mathscr{M}$, satisfies the elegant functional equation

$$\mathscr{M}(\mathscr{M}(\mathscr{M}(a,b),b),\mathscr{M}(\mathscr{M}(a,b),a)) = \mathscr{M}(a,b) \tag{47}$$

for all $a,b > 0$. However, a quick experimentation with a random pair (a,b) shows that (47) is not satisfied by $\mathscr{L}$.

This shows that $\mathscr{L}$ is not quasi-arithmetic, but does not tell us whether $\mathscr{L}$ is a distance mean, and hence does not answer Problem 9.

The functional equation (47) is a weaker form of the functional equation

$$\mathscr{M}(\mathscr{M}(a,b),\mathscr{M}(c,d)) = \mathscr{M}(\mathscr{M}(a,c),\mathscr{M}(b,d)) \tag{48}$$

for all $a,b,c,d > 0$. This condition, together with the assumption that $\mathscr{M}$ is strictly increasing in each variable, characterizes two-dimensional quasi-arithmetic means; see [8, Theorem 1, pp. 287–291]. A thorough discussion of quasi-arithmetic means can be found in [3, 8].

Problem 10. Decide whether a mean $\mathscr{M}$ that satisfies the functional equation (47) (together with any necessary smoothness conditions) is necessarily a quasi-arithmetic mean.

10. Deviation Means

Deviation means were introduced in [9] and were further investigated in [10]. They are defined as follows.

A real-valued function $E = E(x,t)$ on $\mathbb{R}^2$ is called a *deviation* if $E(x,x) = 0$ for all x and if $E(x,t)$ is a strictly decreasing continuous function of t for every x. If E is a deviation, and if $x_1,\ldots,x_n$ are given, then the *E-deviation mean* of $x_1,\ldots,x_n$ is defined to be the unique zero of

$$E(x_1,t) + \cdots + E(x_n,t). \tag{49}$$

It is direct to see that (49) has a unique zero and that this zero does indeed define a mean.

Problem 11. Characterize deviation means and explore their exact relationship with distance means.

If E is a deviation, then (following [11]), one may define d_E by

$$d_E(x,t) = \int_x^t E(x,s)\,ds. \tag{50}$$

Then $d_E(x,t) \geq 0$ and $d_E(x,t)$ is a strictly convex function in t for every x. The E-deviation mean of $x_1,\ldots,x_n$ is nothing but the unique value of t at which $d_E(x_1,t)+\cdots+d_E(x_n,t)$ attains its minimum. Thus if d_E happens to be symmetric, then d_E would be a distance and the E-deviation mean would be the distance mean arising from the distance d_E.

11. Other Ways of Generating New Means

If f and g are differentiable on an open interval $\mathbb{J}$, and if $a < b$ are points in $\mathbb{J}$ such that $f(b) \neq f(a)$, then there exists, by Cauchy's mean value theorem, a point c in (a,b), such that

$$\frac{f'(c)}{g'(c)} = \frac{g(b)-g(a)}{f(b)-f(a)}. \tag{51}$$

If f and g are such that c is unique for every a, b, then we call c the *Cauchy* mean of a and b corresponding to the functions f and g, and we denote it by $\mathscr{C}_{f,g}(a,b)$.

Another natural way of defining means is to take a continuous function F that is strictly monotone on $\mathbb{J}$, and to define the mean of $a,b \in \mathbb{J}$, $a \neq b$, to be the unique point c in (a,b) such that

$$F(c) = \frac{1}{b-a}\int_a^b F(x)\,dx. \tag{52}$$

We call c the *mean value* (mean) of a and b corresponding to F, and we denote it by $\mathscr{V}(a,b)$.

Clearly, if H is an antiderivative of F, then (53) can be written as

$$H'(c) = \frac{H(b)-H(a)}{b-a}. \tag{53}$$

Thus $\mathscr{V}_F(a,b) = \mathscr{C}_{H,E}(a,b)$, where E is the identity function.

For more on the these two families of means, the reader is referred to [12] and [13], and to the references therein.

In contrast to the attitude of thinking of the mean as the number that minimizes a certain function, there is what one may call the Chisini attitude that we now describe. A function f on $\mathbb{J}^n$ may be called a *Chisini function* if and only if the equation

$$f(a_1,\ldots,a_n) = f(x,\ldots,x) \tag{54}$$

has a unique solution $x = a \in [a_1, a_n]$ for every ordered data set $(a_1, \ldots, a_n)$ in $\mathbb{J}$. This unique solution is called the *Chisini mean* associated to f. In Chisini's own words, *x is said to be the mean of n numbers $x_1, \ldots, x_n$ with respect to a problem, in which a function of them $f(x_1, \ldots, x_n)$ is of interest, if the function assumes the same value when all the x_h are replaced by the mean value x: $f(x_1, \ldots, x_n) = f(x, \ldots, x)$*; see [14, page 256] and [1]. Examples of such Chisini means that arise in geometric configurations can be found in [15].

Problem 12. Investigate how the families of distance, deviation, Cauchy, mean value, and Chisini means are related.

12. Internality Tests

According to the definition of a mean, all that is required of a function $\mathcal{M} : \mathbb{J}^n \to \mathbb{J}$ to be a mean is to satisfy the internality property

$$\min\{a_1, \ldots, a_k\} \le \mathcal{M}(a_1, \ldots, a_k) \le \max\{a_1, \ldots, a_k\} \quad (55)$$

for all $a_j \in \mathbb{J}$. However, one may ask whether it is sufficient, for certain types of functions $\mathcal{M}$, to verify (55) for a finite, preferably small, number of well-chosen n-tuples. This question is inspired by certain elegant theorems in the theory of copositive forms that we summarize below.

12.1. Copositivity Tests for Quadratic and Cubic Forms. By a (real) *form* in n variables, we shall always mean a homogeneous polynomial $F = F(x_1, \ldots, x_n)$ in the indeterminates $x_1, \ldots, x_n$ having coefficients in $\mathbb{R}$. When the degree t of a form F is to be emphasized, we call F a t-form. Forms of degrees 1, 2, 3, 4, and 5 are referred to as *linear*, *quadratic*, *cubic*, *quartic*, and *quintic* forms, respectively.

The set of all t-forms in n variables is a vector space (over $\mathbb{R}$) that we shall denote by $\mathbb{F}_t^{(n)}$. It may turn out to be an interesting exercise to prove that the set

$$\left\{ \prod_{j=1}^{d} N_j^{e_j} : \sum_{j=1}^{d} j e_j = d \right\} \quad (56)$$

is a basis, where N_j is the Newton polynomial defined by

$$N_j = \sum_{k=1}^{n} x_k^j. \quad (57)$$

The statement above is quite easy to prove in the special case $d \le 3$, and this is the case we are interested in in this paper. We also discard the trivial case $n = 1$ and assume always that $n \ge 2$.

Linear forms can be written as aN_1, and they are not worth much investigation. Quadratic forms can be written as

$$Q = aN_1^2 + bN_2 = a\left(\sum_{k=1}^{n} x_k\right)^2 + b\left(\sum_{k=1}^{n} x_k^2\right). \quad (58)$$

Cubic and quartic forms can be written, respectively, as

$$aN_1^3 + bN_1N_2 + cN_3,$$
$$aN_1^4 + bN_1^2N_2 + cN_1N_3 + dN_2^2. \quad (59)$$

A form $F = F(x_1, \ldots, x_n)$ is said to be *copositive* if $f(a_1, \ldots, a_n) \ge 0$ for all $x_i \ge 0$. Copositive forms arise in the theory of inequalities and are studied in [14] (and in references therein). One of the interesting questions that one may ask about forms pertains to algorithms for deciding whether a given form is copositive. This problem, in full generality, is still open. However, for quadratic and cubic forms, we have the following satisfactory answers.

Theorem 2. *Let $F = F(x_1, \ldots, x_n)$ be a real symmetric form in any number $n \ge 2$ of variables. Let $\mathbf{v}_m^{(n)}$, $1 \le m \le n$, be the n-tuple whose first m coordinates are 1's and whose remaining coordinates are 0's.*

(i) *If F is quadratic, then F is copositive if and only if $F \ge 0$ at the two test n-tuples*

$$\mathbf{v}_1^{(n)} = (1, 0, \ldots, 0), \quad \mathbf{v}_n^{(n)} = (1, 1, \ldots, 1). \quad (60)$$

(ii) *If F is cubic, then F is copositive if and only if $F \ge 0$ at the n test n-tuples*

$$\mathbf{v}_m^{(n)} = \left(\overbrace{1, \ldots, 1}^{m}, \overbrace{0, \ldots, 0}^{n-m} \right), \quad 1 \le m \le n. \quad (61)$$

Part (i) is a restatement of Theorem 1(a) in [16]. Theorem 1(b) there is related and can be restated as

$$F(a_1, \ldots, a_n) \ge 0, \qquad \forall a_i \in \mathbb{R},$$
$$\Longleftrightarrow F \ge 0 \text{ at the 3 } n\text{-tuples} \quad (62)$$
$$(1, 0, \ldots, 0), (1, 1, \ldots, 1), (1, -1, 0, \ldots, 0).$$

Part (ii) was proved in [17] for $n \le 3$ and in [18] for all n. Two very short and elementary inductive proofs are given in [19].

It is worth mentioning that the n test n-tuples in (61) do not suffice for establishing the copositivity of a quartic form even when $n = 3$. An example illustrating this that uses methods from [20] can be found in [19]. However, an algorithm for deciding whether a symmetric quartic form f in n variables is copositive that consists in testing f at n-tuples of the type

$$\left(\overbrace{a, \ldots, a}^{m}, \overbrace{1, \ldots, 1}^{r}, \overbrace{0, \ldots, 0}^{n-m-r} \right),$$
$$0 \le m, \ r \le n, \ m + r \le n \quad (63)$$

is established in [21]. It is also proved there that if $n = 3$, then the same algorithm works for quintics but does not work for forms of higher degrees.

12.2. Internality Tests for Means Arising from Symmetric Forms. Let $\mathbb{F}_t^{(n)}$ be the vector space of all real t-forms in n variables, and let N_j, $1 \le j \le d$, be the Newton polynomials defined in (57). Means of the type

$$\mathcal{M} = \left(\frac{F_r}{F_s} \right)^{1/(r-s)}, \quad (64)$$

where F_j is a symmetric form of degree j, are clearly symmetric and 1-homogeneous, and they abound in the literature. These include the family of Gini means $G_{r,s}$ defined in (8) (and hence the Lehmer and Hölder means). They also include the elementary symmetric polynomial and elementary symmetric polynomial ratio means defined earlier in (10).

In view of Theorem 2 of the previous section, it is tempting to ask whether the internality of a function $\mathcal{M}$ of the type described in (64) can be established by testing it at a finite set of test n-tuples. Positive answers for some special cases of (64), and for other related types, are given in the following theorem.

Theorem 3. *Let L, Q, and C be real symmetric forms of degrees 1, 2, and 3, respectively, in any number $n \geq 2$ of nonnegative variables. Let $\mathbf{v}_k^{(n)}$, $1 \leq k \leq n$, be as defined in Theorem 2.*

(i) *$\sqrt{Q}$ is internal if and only if it is internal at the two test n-tuples: $\mathbf{v}_n^{(n)} = (1, 1, \ldots, 1)$ and $v_{n-1}^{(n)} = (1, 1, \ldots, 1, 0)$.*

(ii) *Q/L is internal if and only if it is internal at the two test n-tuples: $\mathbf{v}_n^{(n)} = (1, 1, \ldots, 1)$ and $v_1^{(n)} = (1, 0, \ldots, 0)$.*

(iii) *If $n \leq 4$, then $\sqrt[3]{C}$ is internal if and only if it is internal at the n test n-tuples*

$$\mathbf{v}_m^{(n)} = \left(\overbrace{1, \ldots, 1}^{m}, \overbrace{0, \ldots, 0}^{n-m}\right), \quad 1 \leq m \leq n. \tag{65}$$

Parts (i) and (ii) are restatements of Theorems 3 and 5 in [16]. Part (iii) is proved in [22] in a manner that leaves a lot to be desired. Besides being rather clumsy, the proof works for $n \leq 4$ only. The problem for $n \geq 5$, together with other open problems, is listed in the next problem set.

Problem Set 13. Let L, Q, and C be real symmetric cubic forms of degrees 1, 2, and 3, respectively, in n non-negative variables.

(13-a) Prove or disprove that $\sqrt[3]{C}$ is internal if and only if it is internal at the n test n-tuples

$$\mathbf{v}_m^{(n)} = \left(\overbrace{1, \ldots, 1}^{m}, \overbrace{0, \ldots, 0}^{n-m}\right), \quad 1 \leq m \leq n. \tag{66}$$

(13-b) Find, or prove the nonexistence of, a finite set T of test n-tuples such that the internality of C/Q at the n-tuples in T gurantees its internality at all nonnegative n-tuples.

(13-c) Find, or prove the nonexistence of, a finite set T of test n-tuples such that the internality of $L \pm \sqrt{Q}$ at the n-tuples in T guarantees its internality at all non-negative n-tuples.

Problem (13-b) is open even for $n = 2$. In Section 6 of [15], it is shown that the two pairs $(1, 0)$ and $(1, 1)$ do not suffice as test pairs.

As for Problem (13-c), we refer the reader to [23], where means of the type $L \pm \sqrt{Q}$ were considered. It is proved in Theorem 2 there that when Q has the special form $a\prod_{1\leq i<j\leq n}(x_i - x_j)^2$, then $L \pm \sqrt{Q}$ is internal if and only if it is internal at the two test n-tuples $\mathbf{v}_n^{(n)} = (1, 1, \ldots, 1)$ and $\mathbf{v}_{n-1}^{(n)} = (1, 1, \ldots, 1, 0)$. In the general case, sufficient and necessary conditions for internality of $L \pm \sqrt{Q}$, in terms of the coefficients of L and Q, are found in [23, Theorem 3]. However, it is not obvious whether these conditions can be rewritten in terms of test n-tuples in the manner done in Theorem 3.

13. Extension of Means, Concordance of Means

The two-dimensional arithmetic mean $\mathscr{A}^{(2)}$ defined by

$$\mathscr{A}^{(2)}(a_1, a_2) = \frac{a_1 + a_2}{2} \tag{67}$$

can be extended to any dimension k by setting

$$\mathscr{A}^{(k)}(a_1, \ldots, a_k) = \frac{a_1 + \cdots + a_k}{k}. \tag{68}$$

Although very few people would disagree on this, nobody can possibly give a mathematically sound justification of the feeling that the definition in (68) is the only (or even the best) definition that makes the sequence $A^{(k)}$ of means *harmonious* or *concordant*. This does not seem to be an acceptable definition of the notion of concordance.

In a private communication several years ago, Professor Zsolt Páles told me that Kolmogorov suggested calling a sequence $\mathcal{M}^{(k)}$ of means on $\mathbb{J}$, where $\mathcal{M}^{(k)}$ is k-dimensional, concordant if for every m and n and every a_i, b_i in $\mathbb{J}$, we have

$$\begin{aligned}\mathcal{M}^{(n+m)}&(a_1, \ldots, a_n, b_1, \ldots, b_m)\\ &= \mathcal{M}^{(2)}\left(\mathcal{M}^{(n)}(a_1, \ldots, a_n), \mathcal{M}_m(b_1, \ldots, b_m)\right).\end{aligned} \tag{69}$$

He also told me that such a definition is too restrictive and seems to confirm concordance in the case of the quasi-arithmetic means only.

Problem 14. Suggest a definition of concordance, and test it on sequences of means that you feel concordant. In particular, test it on the existing generalizations, to higher dimensions, of the logarithmic mean $\mathscr{L}$ defined in (45).

14. Distance Functions in Topology

Distance functions, which are not necessarily metrics, have appeared early in the literature on topology. Given a distance function d on any set X, one may define the *open ball* $B(a, r)$ in the usual manner, and then one may declare a subset $A \subseteq X$ *open* if it contains, for every $a \in A$, an open ball $B(a, r)$ with $r > 0$. If d has the triangle inequality, then one can proceed in the usual manner to create a topology. However, for a general distance d, this need not be the case, and distances that give rise to a coherent topology in the usual manner are called *semimetrics* and they are investigated and characterized in [24–29]. Clearly, these are the distances d for which the family $\{B(a, r) : r > 0\}$ of open balls centered at $a \in S$ forms a local base at a for every a in X.

15. Centers and Center-Producing Distances

A distance d may be defined on any set S whatsoever. In particular, if d is a distance on $\mathbb{R}^2$ and if the function $f(X)$ defined by

$$f(X) = \sum_{i=1}^{n} d(X, A_i) \tag{70}$$

attains its minimum at a unique point X_0 that lies in the convex hull of $\{A_1, \ldots, A_n\}$ for every choice of $A_1, \ldots, A_n$ in $\mathbb{R}^2$, then d will be called *a center-producing distance.*

The Euclidean metric d_1 on $\mathbb{R}^2$ produces the *Fermat-Torricelli* center. This is defined to be the point whose distances from the given points have a minimal sum. Its square, d_2, which is just a distance but not a metric, produces the *centroid*. This is the center of mass of equal masses placed at the given points. It would be interesting to explore the centers defined by d_p for other values of p.

Problem 15. Let d_p, $p > 1$, be the distance defined on $\mathbb{R}^2$ by $d_p(A, B) = \|A - B\|^p$, and let ABC be a triangle. Let $Z_p = Z_p(A, B, C)$ be the point that minimizes

$$\begin{aligned} d_p(Z, A) + d_p(Z, B) + d_p(Z, C) \\ = \|Z - A\|^p + \|Z - B\|^p + \|Z - C\|^p. \end{aligned} \tag{71}$$

Investigate how Z_p, $p \geq 1$, are related to the known triangle centers, and study the curve traced by them.

The papers [30, 31] may turn out to be relevant to this problem.

References

[1] T. Heath, *A History of Greek Mathematics*, vol. 1, Dover, New York, NY, USA, 1981.

[2] T. Heath, *A History of Greek Mathematics*, vol. 2, Dover, New York, NY, USA, 1981.

[3] P. S. Bullen, *Handbook of Means and Their Inequalities*, vol. 560, Kluwer Academic Publishers, Dordrecht, The Netherlands, 2003.

[4] F. Deutsch, *Best Approximation in Inner Product Spaces*, Springer, New York, NY, USA, 2001.

[5] A. Al-Salman and M. Hajja, "Towards a well-defined median," *Journal of Mathematical Inequalities*, vol. 1, no. 1, pp. 23–30, 2007.

[6] F. B. Saidi, "Generalized deviations and best approximation theory," *Numerical Functional Analysis and Optimization*, vol. 26, no. 2, pp. 271–283, 2005.

[7] M. Hajja, "Distance and best approximation means," preprint.

[8] J. Aczél and J. Dhombres, *Functional Equations in Several Variables*, vol. 31, Cambridge University Press, Cambridge, UK, 1989.

[9] Z. Daróczy, "Über eine Klasse von Mittelwerten," *Publicationes Mathematicae Debrecen*, vol. 19, pp. 211–217, 1972.

[10] Z. Páles, "On the convergence of means," *Journal of Mathematical Analysis and Applications*, vol. 156, no. 1, pp. 52–60, 1991.

[11] Z. Páles, Private Communications.

[12] M. E. Mays, "Functions which parametrize means," *The American Mathematical Monthly*, vol. 90, no. 10, pp. 677–683, 1983.

[13] B. Ebanks, "Looking for a few good means," *American Mathematical Monthly*, vol. 119, no. 8, pp. 658–669, 2012.

[14] M. Hall, and M. Newman, "Copositive and completely positive quadratic forms," *Proceedings of the Cambridge Philosophical Society*, vol. 59, pp. 329–339, 1963.

[15] R. Abu-Saris and M. Hajja, "Geometric means of two positive numbers," *Mathematical Inequalities & Applications*, vol. 9, no. 3, pp. 391–406, 2006.

[16] M. Hajja, "Radical and rational means of degree two," *Mathematical Inequalities & Applications*, vol. 6, no. 4, pp. 581–593, 2003.

[17] J. F. Rigby, "A method of obtaining related triangle inequalities, with applications," *Univerzitet u Beogradu. Publikacije Elektrotehničkog Fakulteta. Serija Matematika i Fizika*, no. 412–460, pp. 217–226, 1973.

[18] M. D. Choi, T. Y. Lam, and B. Reznick, "Even symmetric sextics," *Mathematische Zeitschrift*, vol. 195, no. 4, pp. 559–580, 1987.

[19] M. Hajja, "Copositive symmetric cubic forms," *The American Mathematical Monthly*, vol. 112, no. 5, pp. 462–466, 2005.

[20] K. B. Stolarsky, "The power and generalized logarithmic means," *American Mathematical Monthly*, vol. 87, pp. 545–548, 1980.

[21] W. R. Harris, "Real even symmetric ternary forms," *Journal of Algebra*, vol. 222, no. 1, pp. 204–245, 1999.

[22] R. Abu-Saris and M. Hajja, "Internal cubic symmetric forms in a small number of variables," *Mathematical Inequalities & Applications*, vol. 10, no. 4, pp. 863–868, 2007.

[23] R. Abu-Saris and M. Hajja, "Quadratic means," *Journal of Mathematical Analysis and Applications*, vol. 288, no. 1, pp. 299–313, 2003.

[24] A. V. Arhangel'skii, "Mappings and spaces," *Russian Mathematical Surveys*, vol. 21, no. 4, pp. 115–162, 1966.

[25] F. Galvin and S. D. Shore, "Completeness in semimetric spaces," *Pacific Journal of Mathematics*, vol. 113, no. 1, pp. 67–75, 1984.

[26] F. Galvin and S. D. Shore, "Distance functions and topologies," *The American Mathematical Monthly*, vol. 98, no. 7, pp. 620–623, 1991.

[27] R. Kopperman, "All topologies come from generalized metrics," *American Mathematical Monthly*, vol. 95, no. 2, pp. 89–97, 1988.

[28] D. E. Sanderson and B. T. Sims, "A characterization and generalization of semi-metrizability," *The American Mathematical Monthly*, vol. 73, pp. 361–365, 1966.

[29] W. A. Wilson, "On semi-metric spaces," *American Journal of Mathematics*, vol. 53, no. 2, pp. 361–373, 1931.

[30] D. S. Mitrinović, J. E. Pečarić, and V. Volenec, "The generalized Fermat-Torricelli point and the generalized Lhuilier-Lemoine point," *Comptes Rendus Mathématiques*, vol. 9, no. 2, pp. 95–100, 1987.

[31] P. Penning, "Expoints," *Nieuw Archief voor Wiskunde*, vol. 4, no. 1, pp. 19–31, 1986.

16

Boundary Value Problems for a Super-Sublinear Asymmetric Oscillator: The Exact Number of Solutions

Armands Gritsans and Felix Sadyrbaev

Daugavpils University, Department of Mathematics, Parades Street 1, 5400 Daugavpils, Latvia

Correspondence should be addressed to Armands Gritsans; armands.gricans@du.lv

Academic Editor: Paolo Ricci

Properties of asymmetric oscillator described by the equation $x'' = -\lambda(x^+)^p + \mu(x^-)^q$ (i), where $p \geq 1$ and $0 < q \leq 1$, are studied. A set of (λ, μ) such that the problem (i), $x(0) = 0 = x(1)$ (ii), and $|x'(0)| = \alpha$ (iii) have a nontrivial solution, is called α-spectrum. We give full description of α-spectra in terms of solution sets and solution surfaces. The exact number of nontrivial solutions of the two-parameter Dirichlet boundary value problem (i), and (ii) is given.

1. Introduction

Asymmetric oscillators were studied intensively starting from the works by Kufner and Fučík; see [1] and references therein. Simple equations like (2) given with the boundary conditions allow for complete investigation of spectra. It is known that the spectrum of the problem (2), (4) is a set of hyperbola looking curves in the (λ, μ)-plane. On the other hand, there is a plenty of works devoted to one-parameter case of equations $x'' + \lambda f(x) = 0$ given together with the two-point boundary conditions. Due to nonlinearity of f one should consider solutions $x(t;\alpha)$ with different $\alpha = x'(0)$. Bifurcation diagrams in terms of α and λ, or $\|x\|$ and λ, can serve then to evaluate the number of solutions [2–4].

In this paper we consider differential equations of the form

$$x'' = -\lambda f(x^+) + \mu g(x^-), \tag{1}$$

where $f = x^p$, $p \geq 1$, and $g = x^q$, $0 < q \leq 1$. Here λ and μ are nonnegative parameters, $x^+ = \max\{x, 0\}$, $x^- = \max\{-x, 0\}$. This equation describes asymmetric oscillator with different nonlinear restoring forces on both sides of $x = 0$. If $f = g = x$, then equation becomes famous Fučík equation

$$x'' = -\lambda x^+ + \mu x^-. \tag{2}$$

Properties of the Fučík spectrum are well known (the Fučík spectrum is a set of all pairs (λ, μ) where $\lambda, \mu \geq 0$, such that the Dirichlet problem—(2) with boundary conditions $x(0) = 0 = x(1)$—has a non-trivial solution).

The aim of our study in this paper is to describe properties of the spectrum of the problem

$$x'' = -\lambda(x^+)^p + \mu(x^-)^q, \quad 0 < q \leq 1,\ 1 \leq p, \tag{3}$$

$$x(0) = 0, \qquad x(1) = 0. \tag{4}$$

For this we study first the time maps for the related functions (Section 2), then we give the analytical description of the spectrum (Section 3), and formulate the properties of the spectrum (Section 4), including the asymptotics. In Section 5 we consider the solution sets and solution surfaces which bear information on multiplicity of solutions to the problem. Analysis of properties of solution surfaces (Section 6) can give us estimations of the number of solutions to the problem (Section 7). These estimations contain also information of properties of solutions such as the number of zeros and evaluations of $x'(0)$.

This paper continues series of publications by the authors devoted to nonlinear asymmetric oscillations [5–8].

2. Time Maps

Consider the Cauchy problem

$$x'' = -\lambda x^r, \qquad x(0) = 0, \qquad x'(0) = \alpha, \tag{5}$$

where $r > 0$. Solutions of this problem for α and λ positive have a zero. This zero will be denoted $T_r(\alpha, \lambda)$ and called time map function; more on time maps can be found in [9, 10].

Proposition 1. *Suppose $r, \alpha, \lambda > 0$, $p > 1$ and $0 < q < 1$.*

(1) For any $\alpha, \lambda > 0$ the formula is valid:

$$T_r(\alpha, \lambda) = \frac{1}{\sqrt{\lambda}} t_r\left(\frac{\alpha}{\sqrt{\lambda}}\right), \tag{6}$$

where $t_r(\gamma) = T_r(\gamma, 1)$.

(2) The function $T_r(\alpha, \lambda)$ for the problem (5) is

$$T_r(\alpha, \lambda) = 2^{r/(r+1)}(r+1)^{1/(r+1)} A(r)\, \alpha^{-(r-1)/(r+1)} \lambda^{-1/(r+1)}, \tag{7}$$

where $A(r) = \int_0^1 (ds/\sqrt{1 - s^{r+1}})$.

(3) The function $T_r(\alpha, \lambda)$ for fixed r and α is strictly decreasing function of λ and possesses the properties

$$\lim_{\lambda \to 0+} T_r(\alpha, \lambda) = +\infty, \qquad \lim_{\lambda \to +\infty} T_r(\alpha, \lambda) = 0. \tag{8}$$

(4) The function $T_p(\alpha, \lambda)$ for fixed p and λ is decreasing function of α.

(5) The function $T_q(\alpha, \lambda)$ for fixed q and λ is increasing function of α.

Proof. By standard computations.

Remark 2. $T_r(\alpha, \lambda) = \pi/\sqrt{\lambda}$ for $r = 1$, irrespective of α.

3. Spectrum

Suppose that the problems (3) and (4) are considered with the additional condition

$$|x'(0)| = \alpha > 0. \tag{9}$$

Definition 3. For given $\alpha > 0$ a set of all nonnegative (λ, μ) for which the problems (3), (4), and (9) have a nontrivial solution is called α-spectrum.

Remark 4. A solution of (3) is a C^2-function. Therefore $x'(t)$ is continuous. If z_1 and z_2 are two consecutive zeros of $x(t)$, then $|x'(z_1)| = |x'(z_2)|$ since a solution in the interval (z_1, z_2) is symmetric with respect to the middle point. Therefore $|x'(z)| = \alpha$ for any zero point z and signs of $x'(t)$ alternate.

Theorem 5. *The α-spectrum for the problem*

$$\begin{aligned} x'' &= -\lambda(x^+)^p + \mu(x^-)^q, \qquad x(0) = 0, \\ x(1) &= 0, \qquad |x'(0)| = \alpha > 0 \end{aligned} \tag{10}$$

consists of the following α-branches:

$$\begin{aligned}
F_0^+(\alpha) &= \{(\lambda, \mu) : T_p(\alpha, \lambda) = 1,\ \mu \geq 0\}, \\
F_0^-(\alpha) &= \{(\lambda, \mu) : \lambda \geq 0,\ T_q(\alpha, \mu) = 1\}, \\
F_{2i-1}^+(\alpha) &= \{(\lambda; \mu) : iT_p(\alpha, \lambda) + iT_q(\alpha, \mu) = 1\}, \\
F_{2i-1}^-(\alpha) &= \{(\lambda; \mu) : iT_q(\alpha, \lambda) + iT_p(\alpha, \mu) = 1\}, \\
F_{2i}^+(\alpha) &= \{(\lambda; \mu) : (i+1) T_p(\alpha, \lambda) + iT_q(\alpha, \mu) = 1\}, \\
F_{2i}^-(\alpha) &= \{(\lambda; \mu) : (i+1) T_q(\alpha, \mu) + iT_p(\alpha, \lambda) = 1\}.
\end{aligned} \tag{11}$$

The notation $F_i^+(\alpha)$, respectively: $F_i^-(\alpha)$, refers to solutions x which satisfy the initial conditions $x(0) = 0$, $x'(0) = \alpha > 0$ (resp.: $x'(0) = -\alpha < 0$) and have exactly i zeros in the interval $(0, 1)$.

Proof. We prove the theorem only for solutions which have exactly one zero in $(0, 1)$ and satisfy the initial condition $|x'(0)| = \alpha > 0$.

Consider the case $x'(0) = \alpha > 0$. The first zero of $x(t)$ appears at $t = T_p$ and $x'(T_p) = -\alpha$. The second zero is at $t = T_p + T_q$. One has that for solutions with exactly one zero in $(0, 1)$ the relation $T_p + T_q = 1$ holds, which defines the branch $F_1^+(\alpha)$.

Suppose $x'(0) = -\alpha$. The first zero now is at $t = T_q$. The second one is at $t = T_q + T_p$ and therefore $T_q + T_p = 1$.

The branches $F_1^+(\alpha)$ and $F_1^-(\alpha)$ are given by the equivalent relations $T_p + T_q = 1$ and $T_q + T_p = 1$, respectively, and therefore coincide.

Proof for solutions with different nodal structure is similar.

Remark 6. If $p = q = 1$, then for any $\alpha > 0$ the α-spectrum is

$$\begin{aligned}
F_0^+(\alpha) &= \left\{(\lambda, \mu) : \frac{\pi}{\sqrt{\lambda}} = 1,\ \mu \geq 0\right\}, \\
F_0^-(\alpha) &= \left\{(\lambda, \mu) : \lambda \geq 0,\ \frac{\pi}{\sqrt{\mu}} = 1\right\}, \\
F_{2i-1}^+(\alpha) &= \left\{(\lambda; \mu) : i\frac{\pi}{\sqrt{\lambda}} + i\frac{\pi}{\sqrt{\mu}} = 1\right\}, \\
F_{2i-1}^-(\alpha) &= \left\{(\lambda; \mu) : i\frac{\pi}{\sqrt{\mu}} + i\frac{\pi}{\sqrt{\lambda}} = 1\right\}, \\
F_{2i}^+(\alpha) &= \left\{(\lambda; \mu) : (i+1)\frac{\pi}{\sqrt{\lambda}} + i\frac{\pi}{\sqrt{\mu}} = 1\right\}, \\
F_{2i}^-(\alpha) &= \left\{(\lambda; \mu) : (i+1)\frac{\pi}{\sqrt{\mu}} + i\frac{\pi}{\sqrt{\lambda}} = 1\right\}.
\end{aligned} \tag{12}$$

It is classical Fučík spectrum for the problems (2) and (4), see [1].

4. Properties of the α-Spectrum

Proposition 7.

(1) *The branches* $F_{2i-1}^{+}(\alpha)$ *and* $F_{2i-1}^{-}(\alpha)$ *coincide.*

(2) *The branches* $F_{2i-1}^{\pm}(\alpha)$ *and* $F_{2i-1}^{\pm}(\alpha)$ *do not intersect unless* $i \neq j$.

(3) *The branches* $F_{2i}^{+}(\alpha)$ *and* $F_{2j}^{+}(\alpha)$ *do not intersect unless* $i \neq j$.

(4) *The branches* $F_{2i}^{-}(\alpha)$ *and* $F_{2j}^{-}(\alpha)$ *do not intersect unless* $i \neq j$.

(5) *Any branch* $F_i^s(\alpha)$ *where* $i \geq 1$ *and s is either "+" or "−" is a graph of monotonically decreasing function* $\mu = \mu(\lambda)$.

(6) *The branches* $F_{2i}^{+}(\alpha)$ *and* $F_{2i}^{-}(\alpha)$ *intersect once.*

Proof. (1) It follows from (11) that $F_{2i-1}^{+}(\alpha)$ coincides with $F_{2i-1}^{-}(\alpha)$, since both sets (branches) are defined by symmetric relations:

$$iT_p(\alpha,\lambda) + iT_q(\alpha,\mu) = 1, \qquad iT_q(\alpha,\mu) + iT_p(\alpha,\lambda) = 1. \tag{13}$$

(2), (3) and (4) follows from (11), but (5) follows from the relations (11) and (7). (6) Indeed, any point of intersection satisfies the system

$$\begin{aligned}(i+1)\,T_p(\alpha,\lambda) + iT_q(\alpha,\mu) &= 1,\\ (i+1)\,T_q(\alpha,\mu) + iT_p(\alpha,\lambda) &= 1.\end{aligned} \tag{14}$$

It follows that

$$T_p(\alpha,\lambda) = T_q(\alpha,\mu) \tag{15}$$

or

$$\begin{aligned}&2^{p/(p+1)}(p+1)^{1/(p+1)}A(p)\,\alpha^{-(p-1)/(p+1)}\lambda^{-1/(p+1)}\\ &\quad= 2^{q/(q+1)}(q+1)^{1/(q+1)}A(q)\,\alpha^{-(q-1)/(q+1)}\mu^{-1/(q+1)}.\end{aligned} \tag{16}$$

The above relation defines a curve which is a graph of monotonically increasing function $\mu = C(p,q,\alpha)\lambda^{(q+1)/(p+1)}$, where $C(p,q,\alpha)$ is a constant computable from (16). This curve emanates from the origin and intersects the graph of monotonically decreasing from $+\infty$ to 0 function $\mu = \mu(\lambda)$ from (5) only once.

Let $\lambda_i(\alpha,p)$ be a unique solution of the equation $T_p(\alpha,\lambda) = 1/i$; similarly, let $\mu_i(\alpha,q)$ be a unique solution of the equation $T_q(\alpha,\mu) = 1/i$.

Proposition 8. *Suppose* $\alpha,\lambda,\mu > 0$.

(1) *The branch* $F_{2i-1}^{\pm}(\alpha)(i \in \mathbb{N})$ *is located in the sector*

$$\{(\lambda,\mu) : \lambda > \lambda_i(\alpha,p),\ \mu > \mu_i(\alpha,q)\} \tag{17}$$

and is a hyperbola looking curve in (λ,μ) *plane with vertical asymptote* $\lambda = \lambda_i(\alpha,p)$ *and horizontal asymptote* $\mu = \mu_i(\alpha,q)$.

(2) *The branch* $F_{2i}^{+}(\alpha)$ $(i \in \mathbb{N})$ *is located in the sector*

$$\{(\lambda,\mu) : \lambda > \lambda_{i-1}(\alpha,p),\ \mu > \mu_i(\alpha,q)\} \tag{18}$$

and is a hyperbola looking curve in (λ,μ) *plane with vertical asymptote* $\lambda = \lambda_{i-1}(\alpha,p)$ *and horizontal asymptote* $\mu = \mu_i(\alpha,q)$.

(3) *The branch* $F_{2i}^{-}(\alpha)$ $(i \in \mathbb{N})$ *is located in the sector*

$$\{(\lambda,\mu) : \lambda > \lambda_i(\alpha,p),\ \mu > \mu_{i-1}(\alpha,q)\} \tag{19}$$

and is a hyperbola looking curve in (λ,μ) *plane with vertical asymptote* $\lambda = \lambda_i(\alpha,p)$ *and horizontal asymptote* $\mu = \mu_{i-1}(\alpha,q)$.

Proof. Follows from the relations (11) and Proposition 7.

Remark 9. So the positive part of the α-spectrum

$$F^{+}(\alpha) = \bigcup_{i=0}^{+\infty} F_i^{+}(\alpha) \tag{20}$$

in the extended (λ,μ)-plane may be schematically described by the chain

$$\begin{aligned}&(\lambda_1,0) \xrightarrow{F_0^+(\alpha)} (\lambda_1,+\infty) \xrightarrow{F_1^+(\alpha)} (+\infty,\mu_1) \xrightarrow{F_2^+(\alpha)} (\lambda_2,+\infty)\\ &\quad\xrightarrow{F_3^+(\alpha)} (+\infty,\mu_3) \xrightarrow{F_4^+(\alpha)} (\lambda_3,+\infty)\cdots.\end{aligned} \tag{21}$$

Similarly, the negative part of the α-spectrum

$$F^{-}(\alpha) = \bigcup_{i=0}^{+\infty} F_i^{-}(\alpha) \tag{22}$$

in the extended (λ,μ)-plane may be described as follows:

$$\begin{aligned}&(0,\mu_1) \xrightarrow{F_0^-(\alpha)} (+\infty,\mu_1) \xrightarrow{F_1^-(\alpha)} (\lambda_1,+\infty) \xrightarrow{F_2^-(\alpha)} (+\infty,\mu_2)\\ &\quad\xrightarrow{F_3^-(\alpha)} (\lambda_2,+\infty) \xrightarrow{F_4^-(\alpha)} (+\infty,\mu_3)\cdots.\end{aligned} \tag{23}$$

Proposition 10. *Let* $p > 1$ *and* $0 < q < 1$ *be fixed. It is true that*

$$\begin{aligned}&\lim_{\alpha\to+0}\lambda_1(\alpha,p) = +\infty, \qquad \lim_{\alpha\to+\infty}\lambda_1(\alpha,p) = +0,\\ &\lim_{\alpha\to+0}\mu_1(\alpha,p) = +0, \qquad \lim_{\alpha\to+\infty}\mu_1(\alpha,p) = +\infty.\end{aligned} \tag{24}$$

Proof. Follows from Propositions 1 and 10.

5. Solution Sets and Solution Surfaces

Definition 11. A solution set of the problems (3) and (4) is a set F of all triples (λ,μ,α) $(\lambda \geq 0,\ \mu \geq 0,\ \alpha > 0)$ such that there exists a nontrivial solution of the problem.

Let us distinguish between solutions of the problems (3) and (4) with different number of zeros in the interval $(0,1)$.

Let F_i^+ be a set of all triples (λ, μ, α) such that there exists a nontrivial solution of the respective problems (3) and (4), $x'(0) = \alpha > 0$ which has exactly i zeros in $(0, 1)$, $i = 0, 1, \ldots$, but F_i^- be a set of all triples (λ, μ, α) such that there exists a nontrivial solution of the respective problems (3) and (4), $x'(0) = -\alpha < 0$ which has exactly i zeros in $(0, 1)$, $i = 0, 1, \ldots$

Definition 12. F_i^+ will be called a positive i-solution surface, but F_i^--a negative i-solution surface.

5.1. Description and Properties of a Solution Set. We will identify the cross section of a solution surface F_i^s with the plane $\alpha = \alpha_0 > 0$ (α_0 is fixed) with its projection to the (λ, μ)-plane. This projection is in fact the respective branch $F_i^s(\alpha_0)$ of the spectrum of the problem (3), (4), and (9).

Theorem 13.

(1) Solution surfaces are unions of respective α-branches:

$$F_i^+ = \bigcup_{\alpha>0} F_i^+(\alpha), \qquad F_i^- = \bigcup_{\alpha>0} F_i^-(\alpha). \tag{25}$$

(2) A solution set F of the problems (3) and (4) is a union of all i-solution surfaces and is a union of all α-branches:

$$F = \bigcup_{i=1}^{\infty} F_i^{\pm} = \bigcup_{\alpha>0} F_i^{\pm}(\alpha). \tag{26}$$

(3) Solution surfaces F_{2i-1}^+ and F_{2i-1}^- coincide.

(4) Solution surfaces $F_{2i-1}^{\pm}$ and $F_{2j-1}^{\pm}$ do not intersect unless $i = j$.

(5) For given $i \neq j$ the solution surfaces $F_{2i-1}^{\pm}$ and $F_{2j-1}^{\pm}$ are centroaffine equivalent under the mapping $\Phi_{i,j} : \mathbb{R}^3 \to \mathbb{R}^3$:

$$(\lambda, \mu, \alpha) \overset{\Phi_{i,j}}{\longmapsto} (\overline{\lambda}, \overline{\mu}, \overline{\alpha}), \tag{27}$$

where $\overline{\lambda} = (j/i)^2\lambda$, $\overline{\mu} = (j/i)^2\mu$, $\overline{\alpha} = (j/i)\alpha$.

Proof. (1), (2), (3), and (4) follow from definitions of F, F_i^+, F_i^-, $F_i^+(\alpha)$, $F_i(\alpha)^-$ and Proposition 7.

(5) First observe that for $\alpha, \beta, \lambda > 0$ we have by making use of the formula (6) that

$$\begin{aligned} T(\beta, \lambda) &= \frac{1}{\sqrt{\lambda}} T\left(\frac{\beta}{\sqrt{\lambda}}, 1\right) \\ &= \frac{1}{\sqrt{\lambda}} T\left(\frac{\alpha}{\sqrt{\lambda}} \frac{\beta}{\alpha}, 1\right) = \frac{1}{\sqrt{\lambda}} T\left(\frac{\alpha}{\sqrt{\lambda\alpha^2/\beta^2}}, 1\right) \\ &= \frac{1}{\sqrt{\lambda\alpha^2/\beta^2}} \frac{\alpha}{\beta} T\left(\frac{\alpha}{\sqrt{\lambda\alpha^2/\beta^2}}, 1\right) = \frac{\alpha}{\beta} T\left(\alpha, \lambda\frac{\alpha^2}{\beta^2}\right). \end{aligned} \tag{28}$$

The above formula is applicable to T_p and to T_q.

For given $i \neq j$ and $\alpha > 0$ set $\overline{\alpha} = (j/i)\alpha$. Suppose $(\lambda, \mu, \alpha) \in F_{2i-1}^{\pm}$:

$$iT_p(\alpha, \lambda) + iT_q(\alpha, \mu) = 1. \tag{29}$$

Applying the rescaling formula (28), where $\overline{\alpha}$ replaces β, to the previous equation one has

$$\begin{aligned} i\frac{\overline{\alpha}}{\alpha} T_p\left(\overline{\alpha}, \lambda\frac{\overline{\alpha}^2}{\alpha^2}\right) + i\frac{\overline{\alpha}}{\alpha} T_q\left(\overline{\alpha}, \mu\frac{\overline{\alpha}^2}{\alpha^2}\right) &= 1, \\ jT_p\left(\overline{\alpha}, \left(\frac{j}{i}\right)^2 \lambda\right) + jT_q\left(\overline{\alpha}, \left(\frac{j}{i}\right)^2 \mu\right) &= 1, \\ jT_p(\overline{\alpha}, \overline{\lambda}) + jT_q(\overline{\alpha}, \overline{\mu}) &= 1, \end{aligned} \tag{30}$$

therefore $(\overline{\lambda}, \overline{\mu}, \overline{\alpha}) \in F_{2j-1}^+$. Since $\Phi_{i,j}^{-1} = \Phi_{j,i}$, one has $\Phi_{i,j}(F_{2i-1}^{\pm}) = F_{2j-1}^{\pm}$, and the surfaces $F_{2i-1}^{\pm}$ and $F_{2j-1}^{\pm}$ are centroaffine equivalent under the mapping $\Phi_{i,j}$.

Remark 14. Since solution surfaces $F_{2i-1}^{\pm}$ and $F_{2j-1}^{\pm}$ $(i \neq j)$ are centro-affine equivalent, they have similar shape. Therefore it is enough to study properties of the solution surface $F_1^{\pm}$, in order to know properties of other odd-numbered solution surfaces. The same is true for F_{2i-1}^-.

5.2. Cross-Sections of Solution Surfaces with the Planes $\alpha = Const$. A cross-section of any solution surface F_i^+ or F_i^- by the plane $\alpha = \text{const} > 0$ locates in the sector $Q_1(\alpha) = \{(\lambda, \mu) : \lambda > \lambda_1(\alpha), \mu > \mu_1(\alpha)\}$.

Irrespective of the choice of α no oscillatory (with at least one zero in $(0, 1)$) solution of the problems (3), (4), and (9) exists for (λ, μ) in the "dead zone" below the envelope (see Figure 1).

The analytical description of envelopes, corresponding to branches of spectra, follows.

6. Envelopes of Solution Surfaces

Any solution surface for (3) is defined by one of the following relations:

$$\begin{aligned} F_{2i-1}^{\pm} &= \{(\lambda, \mu, \alpha) : \ iT_f(\alpha, \lambda) + iT_g(\alpha, \mu) = 1\} \\ &\qquad (i = 1, 2, \ldots), \\ F_{2i}^{+} &= \{(\lambda, \mu, \alpha) : (i+1)T_f(\alpha, \lambda) + iT_g(\alpha, \mu) = 1\} \\ &\qquad (i = 0, 1, \ldots), \\ F_{2i}^{-} &= \{(\lambda, \mu, \alpha) : iT_f(\alpha, \lambda) + (i+1)T_g(\alpha, \mu) = 1\} \\ &\qquad (i = 0, 1, \ldots). \end{aligned} \tag{31}$$

We use the unifying formula

$$\mathcal{F} : kT_f(\alpha, \lambda) + mT_g(\alpha, \mu) = 1 \tag{32}$$

and consider

$$H(\lambda, \mu, \alpha) := kT_f(\alpha, \lambda) + mT_g(\alpha, \mu) - 1 = 0. \tag{33}$$

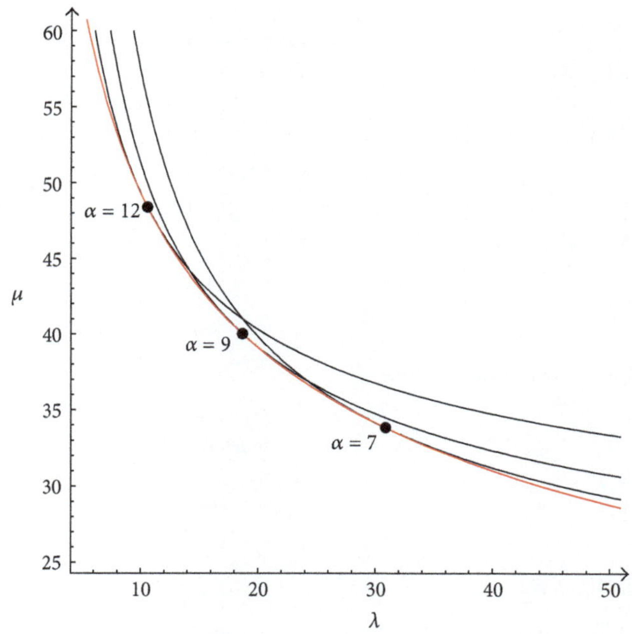

FIGURE 1: "Dead zone" beneath the envelope (in red) of branches $F_1^{\pm}(\alpha)$ of the spectrum of the problems (3) and (4) for $p = 3$, $q = 1/3$ and $\alpha = 7, 9, 12$.

Treat α as a parameter. The family of envelopes for $H(\lambda, \mu, \alpha) = 0$ can be determined from the system

$$H = 0, \qquad \frac{\partial H}{\partial \alpha} = 0. \tag{34}$$

For the case of (3) one gets

$$\begin{aligned} H(\lambda, \mu, \alpha) = {} & kh(p)\, \alpha^{-(p-1)/(p+1)} \lambda^{-1/(p+1)} \\ & + mh(q)\, \alpha^{-(q-1)/(q+1)} \mu^{-1/(q+1)} - 1, \end{aligned} \tag{35}$$

where

$$\begin{aligned} h(p) &= 2^{p/(p+1)} (p+1)^{1/(p+1)} A(p), \\ A(p) &= \int_0^1 \frac{ds}{\sqrt{1 - s^{p+1}}}, \\ \frac{\partial H}{\partial \alpha} &= -k\frac{p-1}{p+1} h(p)\, \alpha^{-2p/(p+1)} \lambda^{-1/(p+1)} \\ &\quad - m\frac{q-1}{q+1} h(q)\, \alpha^{-2q/(q+1)} \mu^{-1/(q+1)}. \end{aligned} \tag{36}$$

Consider the system

$$\begin{aligned} & kh(p)\, \alpha^{-(p-1)/(p+1)} \lambda^{-1/(p+1)} \\ & \quad + mh(q)\, \alpha^{-(q-1)/(q+1)} \mu^{-1/(q+1)} = 1, \\ & k\frac{p-1}{p+1} h(p)\, \alpha^{-2p/(p+1)} \lambda^{-1/(p+1)} \\ & \quad + m\frac{q-1}{q+1} h(q)\, \alpha^{-2q/(q+1)} \mu^{-1/(q+1)} = 0 \end{aligned} \tag{37}$$

and exclude the parameter α. One obtains that

$$\mu = \left(\frac{\psi(p,q) m^{(q+1)(p-1)}}{\psi(q,p) k^{(p+1)(q-1)}} \right)^{1/(p-1)} \lambda^{(q-1)/(p-1)}, \tag{38}$$

where

$$\psi(p,q) = \left(\frac{(q-1)(p+1)}{2(q-p)h(p)} \right)^{(p+1)(q-1)}. \tag{39}$$

(1) For

$$F_{2i-1}^{\pm} : iT_f(\alpha, \lambda) + iT_g(\alpha, \mu) = 1, \tag{40}$$

$k = i$, $m = i$, and the equation of the envelope is

$$\begin{aligned} \mathscr{E}_{2i-1}^{\pm} : \mu &= \omega_{2i-1}^{\pm}(\lambda) \\ &:= \left(\frac{\psi(p,q)\, i^{(q+1)(p-1)}}{\psi(q,p)\, i^{(p+1)(q-1)}} \right)^{1/(p-1)} \lambda^{(q-1)/(p-1)}. \end{aligned} \tag{41}$$

(2) For

$$F_{2i}^{+} : (i+1) T_f(\alpha, \lambda) + iT_g(\alpha, \mu) = 1, \tag{42}$$

$k = i + 1$, $m = i$, the equation of the envelope is

$$\begin{aligned} \mathscr{E}_{2i}^{+} : \mu &= \omega_{2i}^{+}(\lambda) \\ &:= \left(\frac{\psi(p,q)\, i^{(q+1)(p-1)}}{\psi(q,p)\, (i+1)^{(p+1)(q-1)}} \right)^{1/(p-1)} \lambda^{(q-1)/(p-1)}. \end{aligned} \tag{43}$$

(3) For

$$F_{2i}^{-} : iT_f(\alpha, \lambda) + (i+1) T_g(\alpha, \mu) = 1, \quad (44)$$

$k = i, m = i + 1$, the equation of the envelope is

$$\mathscr{E}_{2i}^{-} : \mu = \omega_{2i}^{-}(\lambda)$$

$$:= \left(\frac{\psi(p,q)(i+1)^{(q+1)(p-1)}}{\psi(q,p)\, i^{(p+1)(q-1)}} \right)^{1/(p-1)} \lambda^{(q-1)/(p-1)}. \quad (45)$$

Proposition 15. *For given $p > 1$ and $0 < q < 1$ the location of the envelopes is as follows:*

(1) $\mathscr{E}_{2i-1}^{\pm} \prec \mathscr{E}_{2i}^{+}$,

(2) $\mathscr{E}_{2i-1}^{\pm} \prec \mathscr{E}_{2i}^{-}$,

(3)

(a) *if $pq - 1 < 0$, then $\mathscr{E}_{2i}^{-} \prec \mathscr{E}_{2i}^{+}$,*

(b) *if $pq - 1 = 0$, then $\mathscr{E}_{2i}^{-} = \mathscr{E}_{2i}^{+}$,*

(c) *if $pq - 1 > 0$, then $\mathscr{E}_{2i}^{+} \prec \mathscr{E}_{2i}^{-}$,*

(4) $\mathscr{E}_{2i}^{+} \prec \mathscr{E}_{2(i+1)-1}^{\pm}$,

(5) $\mathscr{E}_{2i}^{-} \prec \mathscr{E}_{2(i+1)-1}^{\pm}$,

where $\mathscr{E}_n^s \prec \mathscr{E}_m^t$ means that $\omega_n^s(\lambda) < \omega_m^t(\lambda)$ for any $\lambda > 0$.

Proof. (1) First consider

$$\mathscr{E}_{2i-1}^{\pm} : \mu = \omega_{2i-1}^{\pm}(\lambda)$$

$$:= \left(\frac{\psi(p,q)\, i^{(q+1)(p-1)}}{\psi(q,p)\, i^{(p+1)(q-1)}} \right)^{1/(p-1)} \lambda^{(q-1)/(p-1)},$$

$$\mathscr{E}_{2i}^{+} : \mu = \omega_{2i}^{+}(\lambda)$$

$$:= \left(\frac{\psi(p,q)\, i^{(q+1)(p-1)}}{\psi(q,p)(i+1)^{(p+1)(q-1)}} \right)^{1/(p-1)} \lambda^{(q-1)/(p-1)}. \quad (46)$$

For any $\lambda > 0$,

$$\frac{\omega_{2i}^{+}(\lambda)}{\omega_{2i-1}^{\pm}(\lambda)} = \frac{i^{(q+1)(p-1)} i^{(p+1)(q-1)}}{(i+1)^{(p+1)(q-1)} i^{(q+1)(p-1)}} = \left(\frac{i}{i+1} \right)^{(p+1)(q-1)} > 1, \quad (47)$$

then $\mathscr{E}_{2i-1}^{\pm} \prec \mathscr{E}_{2i}^{+}$.

Remark 16. Layout of envelopes depends on sign$(pq-1)$ (see Figure 2).

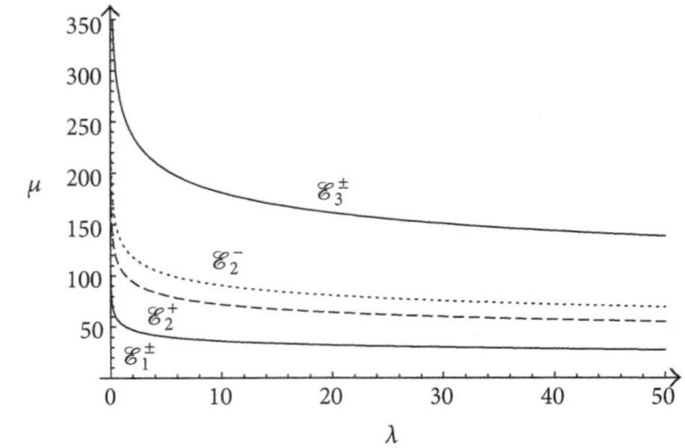

(a) the case $p = 5$ and $q = 1/3$. Since $pq - 1 > 0$, the envelopes are ordered as $\mathscr{E}_1^{\pm} \prec \mathscr{E}_2^{+} \prec \mathscr{E}_2^{-} \prec \mathscr{E}_3^{\pm} \prec \cdots$

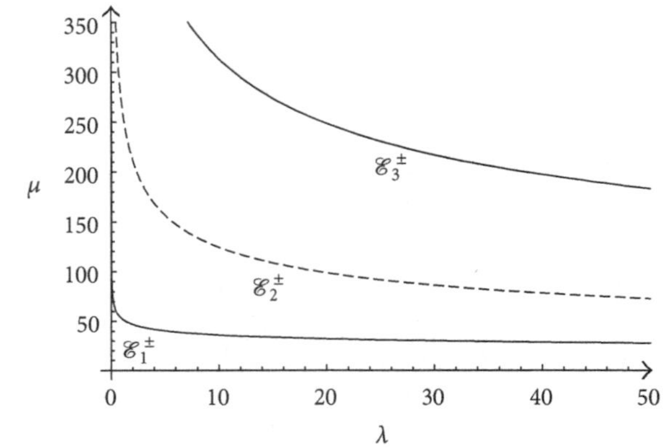

(b) the case $p = 3$ and $q = 1/3$. Since $pq - 1 = 0$, the envelopes are ordered as $\mathscr{E}_1^{\pm} \prec \mathscr{E}_2^{\pm} \prec \mathscr{E}_3^{\pm} \prec \cdots$

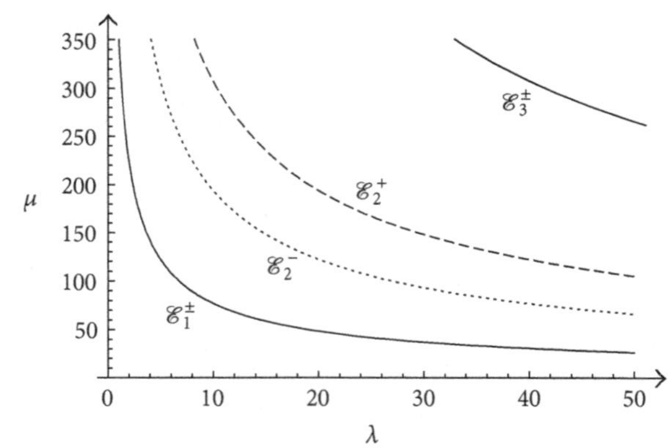

(c) the case $p = 2$ and $q = 1/3$. Since $pq - 1 < 0$, the envelopes are ordered as $\mathscr{E}_1^{\pm} \prec \mathscr{E}_2^{-} \prec \mathscr{E}_2^{+} \prec \mathscr{E}_3^{\pm} \prec \cdots$

Figure 2: Layout of envelopes depends on sign$(pq - 1)$.

7. The Number of Solutions by Geometrical Analysis of Solution Surfaces

We can detect the precise number of solutions to the problem for given positive (λ, μ). We can evaluate the initial values $x'(0)$ for solutions on a basis of geometrical analysis of solution surfaces and the respective envelopes. The nodal structure of solutions can be described also.

Let $\mu = c^2\lambda$ be the family of rays, covering the first quadrant of the (λ, μ)-plane. Consider the cross-section of a solution surface (32) by the plane $\mu = c^2\lambda$ in the (λ, μ, α)-space, for example, the curve $\mathcal{F}|_{\mu=c^2\lambda}$, which is defined by the relations $\mu = c^2\lambda$ and

$$kT_p(\alpha, \lambda) + mT_q(\alpha, c^2\lambda) = 1. \tag{48}$$

This 3D curve $\mathcal{F}|_{\mu=c^2\lambda}$ can be regularly parameterized as

$$\begin{aligned} \lambda = \lambda(r) &:= \left[\frac{ckt_f(cr) + mt_g(r)}{c}\right]^2, \\ \mu = \mu(r) &:= \left[ckt_f(cr) + mt_g(r)\right]^2, \\ \alpha = \alpha(r) &:= r\left[ckt_f(cr) + mt_g(r)\right], \end{aligned} \tag{49}$$

where $r > 0$. These formulas define homeomorphism $\mathbb{R}_+ \to \mathcal{F}|_{\mu=c^2\lambda}$, where $\mathbb{R}_+ = (0, +\infty)$.

Let $\mathcal{G}_c$ be projection of $\mathcal{F}|_{\mu=c^2\lambda}$ to the (λ, α) plane, that is:

$$\mathcal{G}_c = \left\{(\lambda, \alpha) \in \mathbb{R}_+^2 : kT_f(\alpha, \lambda) + mT_g(\alpha, c^2\lambda) = 1\right\}. \tag{50}$$

The curve $\mathcal{G}_c$ can be regularly [11] parameterized as

$$\begin{aligned} \lambda = \lambda(r) &:= \left[\frac{ckt_f(cr) + mt_g(r)}{c}\right]^2, \\ \alpha = \alpha(r) &:= r\left[ckt_f(cr) + mt_g(r)\right], \end{aligned} \tag{51}$$

where $r > 0$. These formulas define homeomorphism $\mathbb{R}_+ \to \mathcal{G}_c$.

Proposition 17. *There exists a unique parameter $r_0 > 0$ such that the line $\lambda = \lambda_* > 0$*

(a) *does not intersect the curve $\mathcal{G}_c$ if $0 < \lambda_* < \lambda(r_0)$,*

(b) *intersects the curve $\mathcal{G}_c$ only once if $\lambda_* = \lambda(r_0)$,*

(c) *intersects the curve $\mathcal{G}_c$ exactly at two points if $\lambda_* > \lambda(r_0)$.*

Proof. (a) Consider equation $\lambda'(r) = kc^2t_p'(cr) + mt_q'(r) = 0$, which turns to

$$\begin{aligned} &kc^2\frac{p-1}{p+1}h(p)c^{-2p/(p+1)}r^{-2p/(p+1)} \\ &\quad + m\frac{q-1}{q+1}h(q)r^{-2q/(q+1)} = 0. \end{aligned} \tag{52}$$

It can be found from the above that

$$r_0 = \left[-\frac{m(q-1)(p+1)h(q)}{k(p-1)(q+1)h(p)c^{2/(p+1)}}\right]^{-(p+1)(q+1)/2(p-q)} \tag{53}$$

is the only root of the equation $\lambda'(r) = 0$. Hence at the point $(\lambda_0, \alpha_0) = (\lambda(r_0), \alpha(r_0))$ the curve $\mathcal{G}_c$ has a unique tangent line parallel to the $O\alpha$ axis.

(b) Since the given parametrization of the curve $\mathcal{G}_c$ is regular, then $\alpha'(r_0) \neq 0$ and in some neighborhood of the point (λ_0, α_0) the curve $\mathcal{G}_c$ can be represented as the graph of the function $\lambda = \varphi(\alpha)$. Now find

$$\begin{aligned} \varphi''(r_0) &= \frac{\lambda''(r_0)}{(\alpha'(r_0))^2} \\ &= \frac{2\left[ckt_f(cr) + mt_g(r)\right]\left[kc^3t_f''(cr_0) + mt_g''(r_0)\right]}{c^2[\alpha'(r_0)]^2}. \end{aligned} \tag{54}$$

The next step is to detect the sign of the expression

$$\begin{aligned} &kc^3t_p''(cr) + mt_q''(r) \\ &= kc^3\frac{2p(p-1)}{(p+1)^2}h(p)c^{-(3p+1)/(p+1)}r^{-(3p+1)/(p+1)} \\ &\quad + m\frac{2q(q-1)}{(q+1)^2}h(q)r^{-(3q+1)/(q+1)} \\ &= k\frac{2p(p-1)}{(p+1)^2}h(p)c^{3-(3p+1)/(p+1)}r^{-(3p+1)/(p+1)} \\ &\quad + m\frac{2q(q-1)}{(q+1)^2}h(q)r^{-(3q+1)/(q+1)} \\ &= r^{-(3q+1)/(q+1)} \\ &\quad \times \Bigg(k\frac{2p(p-1)}{(p+1)^2}h(p)c^{3-(3p+1)/(p+1)} \\ &\quad \times r^{-(3(p+1)/(p+1))+((3q+1)/(q+1))} \\ &\quad + m\frac{2q(q-1)}{(q+1)^2}h(q)\Bigg). \end{aligned} \tag{55}$$

The routine calculations show that the expression in parentheses is equal to

$$\begin{aligned} &-m\frac{q-1}{q+1}h(q)\left(\frac{2p}{p+1} - \frac{2q}{q+1}\right) \\ &\quad = -m\frac{q-1}{q+1}h(q)\frac{2(p-q)}{(p+1)(q+1)} > 0. \end{aligned} \tag{56}$$

Hence the function $\lambda = \varphi(\alpha)$ has the strict local minimum at the point $\alpha = \alpha_0$.

(c) It follows from the relations

$$\begin{aligned} &\lim_{r\to 0+} t_p(r) = +\infty, &&\lim_{r\to +\infty} t_p(r) = 0+, \\ &\lim_{r\to 0+} rt_p(r) = 0+, &&\lim_{r\to +\infty} rt_p(r) = +\infty, \\ &\lim_{r\to 0+} t_q(r) = 0+, &&\lim_{r\to +\infty} t_q(r) = +\infty, \\ &\lim_{r\to 0+} rt_q(r) = 0+, &&\lim_{r\to +\infty} rt_q(r) = +\infty \end{aligned} \tag{57}$$

that if a parameter $r > 0$ goes from 0 to $+\infty$, then a point $(\lambda, \alpha) \in \mathcal{G}_c$ goes from the point $(+\infty, 0+)$ to the point $(+\infty, +\infty)$.

(d) It follows from the above argument that if the parameter $r > 0$ goes from 0 to $+\infty$ then a point $(\lambda, \alpha) \in \mathcal{G}_c$ goes from $(+\infty, 0+)$ to (λ_0, α_0), then turns to the right, and goes to $(+\infty, +\infty)$. Since the parametrization of the curve $\mathcal{G}_c$ is without self-intersection points, one can deduce that the curve $\mathcal{G}_c$ is a union of two branches (i.e., graphs of the functions $\alpha = \psi_1(\lambda)$ and $\alpha = \psi_2(\lambda)$, where $(\lambda_0 < \lambda < +\infty)$ which do not intersect and are continuously "glued" at the point (λ_0, α_0).

Before presenting "the exact number of solutions" result we make the following conventions:

(1) solution means a nontrivial solution of the problems (3) and (4),

(2) $\mathcal{E}_n^s$ means the envelope, where $n \in \mathbb{N}$ and s is either $+$ or $-$ or $\pm$,

(3) we mean by s-solution a solution of the problem (3), (4): (a) a solution with $x'(0) > 0$ if $s = +$; (b) a solution with $x'(0) < 0$ if $s = -$; (c) two solutions with $x'(0) = \alpha > 0$ and $x'(0) = -\alpha < 0$ if $s = \pm$.

Proposition 18. *Consider an envelope*

$$\mathcal{E}_n^s = \left\{(\lambda, \mu) \in \mathbb{R}_+^2 : \mu = \omega_n^s(\lambda)\right\}, \quad n > 0. \tag{58}$$

The problems (3) *and* (4) *have*

(a) *no s-solutions with n zeroes in* $(0, 1)$ *if* $\mu < \omega_n^s(\lambda)$,

(b) *exactly one s-solution with even number n zeroes in* $(0, 1)$ *if* $s = +$ *or* $s = -$ *and* $\mu = \omega_n^s(\lambda)$,

(c) *exactly two s-solutions with n zeroes in* $(0, 1)$ *if*

n is odd, $\mu = \omega_n^s(\lambda)$ *and* $s = \pm$

or

n is even, $\mu > \omega_n^s(\lambda)$ *and* $s = +$ *or* $s = -$,

(d) *exactly four s-solutions with odd number n of zeroes in* $(0, 1)$ *if* $s = \pm$ *and* $\mu > \omega_n^s(\lambda)$.

Proof. It can be verified analytically that $(\lambda_0, \mu_0) \in \mathcal{E}_n^s$, where

$$\lambda_0 = \lambda(r_0) = \frac{1}{c^2}\left(ckh(p)(cr_0)^{-(p-1)/(p+1)} + mh(q)r_0^{-(q-1)/(q+1)}\right)^2. \tag{59}$$

and $\mu_0 = c^2\lambda(r_0)$.

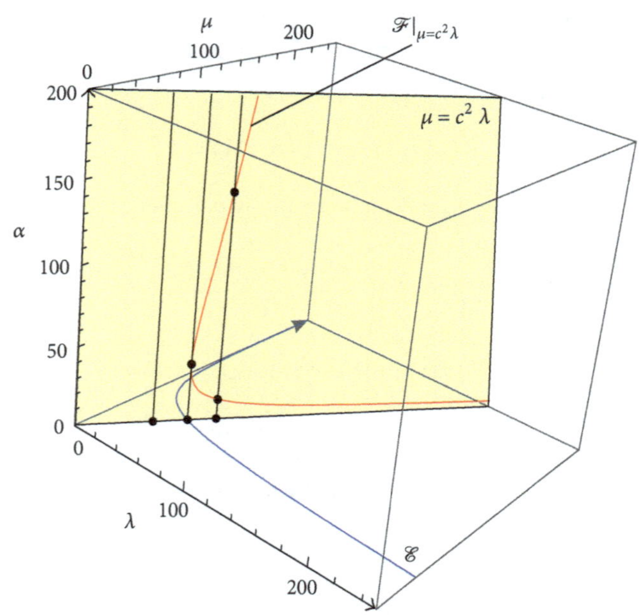

FIGURE 3: The case $p = 3$, $q = 1/2$, $k = 2$, $m = 1$ and $c = 1.2$.

Recall that the curve $\mathcal{G}_c$ is the projection of the 3D curve $\mathcal{F}|_{\mu=c^2\lambda}$ to the (λ, α)-plane. Taking in mind Proposition 17 we can assert that there exists a unique parameter $r_0 > 0$, see (53), such that the line parallel to the $O\alpha$ axis in the (λ, μ, α)-space going through the point (λ, μ), where $\lambda > 0$ and $\mu = c^2\lambda$, and the curve $\mathcal{F}|_{\mu=c^2\lambda}$ (hence the solution surface $\mathcal{F}$ also)

(i) does not intersect if $\mu < \omega_n^s(\lambda)$,

(ii) intersects only once if $\mu = \omega_n^s(\lambda)$,

(iii) intersects exactly at two points if $\mu > \omega_n^s(\lambda)$,

(see Figure 3).

Depending on the value of s one can detect the number of s-solutions as the theorem states. For example, in the case (d): if n is odd, $\mu > \omega_n^s(\lambda)$ and $s = \pm$, then the mentioned above line intersects the solution surface F_n^+ exactly twice. Since $F_n^- = F_n^+$ the mentioned above line intersects the solution surface F_n^- exactly twice also. Hence the problem (3), (4) has exactly four solutions with n zeros in $(0; 1)$.

Now we are able to prove the main theorem. In this theorem we suppose that $\omega_0^-(\lambda) = 0$ and $\omega_0^+(\lambda) = 0$ for all $\lambda > 0$ and will use auxiliary envelopes $\mathcal{E}_0^- : \mu = \omega_0^-(\lambda)$ and $\mathcal{E}_0^+ : \mu = \omega_0^+(\lambda)$.

Theorem 19. *Suppose* $0 < q < 1 < p$.

(1) *If* $pq - 1 > 0$ *and* $i \in \mathbb{N}$, *then the exact number of nontrivial solutions to the problems* (3) *and* (4) *is*

(a) $8i - 6$ *if* $\omega_{2i-2}^-(\lambda) < \mu < \omega_{2i-1}^{\pm}(\lambda)$,
(b) $8i - 4$ *if* $\mu = \omega_{2i-1}^{\pm}(\lambda)$,
(c) $8i - 2$ *if* $\omega_{2i-1}^{\pm}(\lambda) < \mu < \omega_{2i}^+(\lambda)$,
(d) $8i - 1$ *if* $\mu = \omega_{2i}^+(\lambda)$,
(e) $8i$ *if* $\omega_{2i}^+(\lambda) < \mu < \omega_{2i}^-(\lambda)$,
(f) $8i + 1$ *if* $\mu = \omega_{2i}^-(\lambda)$.

(2) *If* $pq - 1 = 0$ *and* $i \in \mathbb{N}$, *then the exact number of nontrivial solutions to the problems* (3) *and* (4) *is*

(a) $8i - 6$ *if* $\omega_{2i-2}^{+}(\lambda) < \mu < \omega_{2i-1}^{\pm}(\lambda)$,
(b) $8i - 4$ *if* $\mu = \omega_{2i-1}^{\pm}(\lambda)$,
(c) $8i - 2$ *if* $\omega_{2i-1}^{\pm}(\lambda) < \mu < \omega_{2i}^{+}(\lambda)$,
(d) $8i$ *if* $\mu = \omega_{2i}^{+}(\lambda)$.

(3) *If* $pq - 1 < 0$ *and* $i \in \mathbb{N}$ *then the exact number of nontrivial solutions to the problems* (3) *and* (4) *is*

(a) $8i - 6$ *if* $\omega_{2i-2}^{+}(\lambda) < \mu < \omega_{2i-1}^{\pm}(\lambda)$,
(b) $8i - 4$ *if* $\mu = \omega_{2i-1}^{\pm}(\lambda)$,
(c) $8i - 2$ *if* $\omega_{2i-1}^{\pm}(\lambda) < \mu < \omega_{2i}^{-}(\lambda)$,
(d) $8i - 1$ *if* $\mu = \omega_{2i}^{-}(\lambda)$,
(e) $8i$ *if* $\omega_{2i}^{-}(\lambda) < \mu < \omega_{2i}^{+}(\lambda)$,
(f) $8i + 1$ *if* $\mu = \omega_{2i}^{+}(\lambda)$.

Proof. First notice that if $\lambda > 0$ and $\mu > 0$ then there exists exactly one $\pm$-solution without zeros in $(0, 1)$, in fact, accordingly to conventions, two solutions without zeros in $(0, 1)$, and of opposite sign values of $x'(0)$.

We consider only the case $pq-1 > 0$. Similarly two other possible cases $pq - 1 < 0$ and $pq - 1 = 0$ can be treated.

It follows from Proposition 15 that the envelopes are ordered as

$$\begin{aligned} &\mathscr{E}_0^- \prec \mathscr{E}_1^\pm \prec \mathscr{E}_2^+ \prec \mathscr{E}_2^- \prec \mathscr{E}_3^\pm \prec \cdots \prec \mathscr{E}_{2i-1}^\pm \\ &\prec \mathscr{E}_{2i}^+ \prec \mathscr{E}_{2i}^- \prec \mathscr{E}_{2i+1}^\pm \prec \cdots. \end{aligned} \tag{60}$$

Let us consider the case (1), when $pq - 1 > 0$, and the subcase (a): $\omega_{2i-2}^{-}(\lambda) < \mu < \omega_{2i-1}^{\pm}(\lambda)$ for $i = 1, 2, \ldots$.

(i) if $i = 1$ then there are $2 = 8 \cdot 1 - 6$ solutions without zeros;

(ii) if $i = 2$, then there are *two* solutions without zeros, two $\pm$-solutions with 1 zero (that is, *four* solutions with 1 zero), *two* "+"-solution with 2 zeros and *two* "−"-solutions with 2 zeros; totally there are $10 = 8 \cdot 2 - 6$ (nontrivial) solutions;

(iii) if $i = 3$, then in addition to 10 solutions mentioned in the previous step, there are also two $\pm$-solutions with 3 zeros (that is, *four* solutions with 3 zeros), *two* "+"-solutions with 4 zeros and *two* "−"-solution with 4 zeros; totally there are $18 = 8 \cdot 3 - 6$ (nontrivial) solutions and so on;

(iv) if $i \in \mathbb{N}$, then totally there are $8i - 6$ (nontrivial) solutions.

The other cases can be considered analogously.

References

[1] A. Kufner and S. Fučík, *Nonlinear Differential Equations*, Elsevier, Amsterdam, The Netherlands, 1980.

[2] P. Korman, "Global solution branches and exact multiplicity of solutions for two point boundary value problems," in *Handbook of Differential Equations, Ordinary Differential Equations*, A. Canada, P. Drabek, and A. Fonda, Eds., vol. 3, pp. 547–606, Elsevier, North-Holland, Amsterdam, The Netherlands, 2006.

[3] P. Korman and Y. Li, "Generalized averages for solutions of two-point Dirichlet problems," *Journal of Mathematical Analysis and Applications*, vol. 239, no. 2, pp. 478–484, 1999.

[4] P. Korman, Y. Li, and T. Ouyang, "Exact multiplicity results for boundary value problems with nonlinearities generalising cubic," *Proceedings of the Royal Society of Edinburgh, Series A*, vol. 126, no. 3, pp. 599–616, 1996.

[5] A. Gritsans and F. Sadyrbaev, "On nonlinear Fučík type spectra," *Mathematical Modelling and Analysis*, vol. 13, no. 2, pp. 203–210, 2008.

[6] A. Gritsans and F. Sadyrbaev, "Nonlinear spectra: the Neumann problem," *Mathematical Modelling and Analysis*, vol. 14, no. 1, pp. 33–42, 2009.

[7] A. Gritsans and F. Sadyrbaev, "Nonlinear spectra for parameter dependent ordinary differential equations," *Nonlinear Analysis: Modelling and Control*, vol. 12, no. 2, pp. 253–267, 2007.

[8] F. Sadyrbaev, "Multiplicity in parameter-dependent problems for ordinary differential equations," *Mathematical Modelling and Analysis*, vol. 14, no. 4, pp. 503–514, 2009.

[9] R. Schaaf, *Global Solution Branches of Two Point Boundary Value Problems*, vol. 1458 of *Lecture Notes in Mathematics*, Springer, Berlin, Germany, 1990.

[10] A. Gritsans and F. Sadyrbaev, "Time map formulae and their applications," in *Proceedings LU MII "Mathematics, Differential Equations"*, vol. 8, pp. 72–93, Riga, Latvia, 2008.

[11] C. G. Gibson, *Elementary Geometry of Differentiable Curves*, Cambridge University Press, Cambridge, UK, 2001.

17

A Framework for Coxeter Spectral Classification of Finite Posets and Their Mesh Geometries of Roots

Daniel Simson and Katarzyna Zając

Faculty of Mathematics and Computer Science, Nicolaus Copernicus University, Ulica Chopina 12/18, 87-100 Toruń, Poland

Correspondence should be addressed to Daniel Simson; simson@mat.umk.pl

Academic Editor: Marco Squassina

Following our paper [Linear Algebra Appl. 433(2010), 699–717], we present a framework and computational tools for the Coxeter spectral classification of finite posets $J \equiv (J, \preceq)$. One of the main motivations for the study is an application of matrix representations of posets in representation theory explained by Drozd [Funct. Anal. Appl. 8(1974), 219–225]. We are mainly interested in a Coxeter spectral classification of posets J such that the symmetric Gram matrix $G_J := (1/2)[C_J + C_J^{tr}] \in \mathbb{M}_J(\mathbb{Q})$ is positive semidefinite, where $C_J \in \mathbb{M}_J(\mathbb{Z})$ is the incidence matrix of J. Following the idea of Drozd mentioned earlier, we associate to J its Coxeter matrix $\mathrm{Cox}_J := -C_J \cdot C_J^{-tr}$, its Coxeter spectrum $\mathbf{specc}_J$, a Coxeter polynomial $\mathrm{cox}_J(t) \in \mathbb{Z}[t]$, and a Coxeter number $\mathbf{c}_J$. In case G_J is positive semi-definite, we also associate to J a reduced Coxeter number $\check{\mathbf{c}}_J$, and the defect homomorphism $\partial_J : \mathbb{Z}^J \to \mathbb{Z}$. In this case, the Coxeter spectrum $\mathbf{specc}_J$ is a subset of the unit circle and consists of roots of unity. In case G_J is positive semi-definite of corank one, we relate the Coxeter spectral properties of the posets J with the Coxeter spectral properties of a simply laced Euclidean diagram $DJ \in \{\widetilde{\mathbb{D}}_n, \widetilde{\mathbb{E}}_6, \widetilde{\mathbb{E}}_7, \widetilde{\mathbb{E}}_8\}$ associated with J. Our aim of the Coxeter spectral analysis of such posets J is to answer the question when the Coxeter type $\mathbf{Ctype}_J := (\mathbf{specc}_J, \mathbf{c}_J, \check{\mathbf{c}}_J)$ of J determines its incidence matrix C_J (and, hence, the poset J) uniquely, up to a $\mathbb{Z}$-congruency. In connection with this question, we also discuss the problem studied by Horn and Sergeichuk [Linear Algebra Appl. 389(2004), 347–353], if for any $\mathbb{Z}$-invertible matrix $A \in \mathbb{M}_n(\mathbb{Z})$, there is $B \in \mathbb{M}_n(\mathbb{Z})$ such that $A^{tr} = B^{tr} \cdot A \cdot B$ and $B^2 = E$ is the identity matrix.

1. Introduction

In the present paper, we continue our Coxeter spectral study of finite posets, started in [1], in a close connection with the Coxeter spectral technique introduced in [2–4] for acyclic edge-bipartite graphs or signed graphs in the sense of [5]. We also follow some of the techniques of representation theory, graph combinatorics, and the spectral graph theory; see [6–31].

Here, we use the terminology and notation introduced in [1, 4, 26–28]. We denote by $\mathbb{N} \subseteq \mathbb{Z} \subseteq \mathbb{Q}$ the set of nonnegative integers, the ring of integers, and the rational number field. Given $m \geq 1$, we view $\mathbb{Z}^m$ as a free abelian group and denote by $e_1, \ldots, e_m$ the standard $\mathbb{Z}$-basis of $\mathbb{Z}^m$. Given an index set J, we denote by $\mathbb{Z}^J$ the abelian group of all vectors $v = (v_j)_{j\in J}$, with integer coordinates $v_j \in \mathbb{Z}$, by $\mathbb{M}_J(\mathbb{Z})$ the $\mathbb{Z}$-algebra of all square J by J integral matrices, and by $E \in \mathbb{M}_J(\mathbb{Z})$ the identity matrix. In particular, $\mathbb{M}_m(\mathbb{Z})$, with $m \geq 1$, is the $\mathbb{Z}$-algebra of all square m by m matrices. The group

$$\mathrm{Gl}(m, \mathbb{Z}) := \{A \in \mathbb{M}_m(\mathbb{Z}), \det A \in \{-1, 1\}\} \subseteq \mathbb{M}_m(\mathbb{Z}) \quad (1)$$

is called the (integral) *general linear group*. We say that two square rational matrices $A, A' \in \mathbb{M}_m(\mathbb{Q})$ are *$\mathbb{Z}$-equivalent*, or *$\mathbb{Z}$-congruent*, (and denote $A \sim_{\mathbb{Z}} A'$) if there is a matrix $B \in \mathrm{Gl}(m, \mathbb{Z})$ such that $A' = B^{tr} \cdot A \cdot B$. By a poset $J \equiv (J, \preceq)$ we mean a finite partially ordered set J with respect to a partial order relation $\preceq$. Following [26], a poset J is called a *one-peak poset* if J has a unique maximal element $*$. A finite poset J is uniquely determined by its *incidence matrix* $C_J \in \mathbb{M}_J(\mathbb{Z})$, that is, the square $J \times J$ matrix, as follows:

$$C_J = [c_{ij}]_{i,j\in J} \in \mathbb{M}_J(\mathbb{Z}), \quad \text{with } c_{ij} = \begin{cases} 1, & \text{for } i \preceq j, \\ 0, & \text{for } i \npreceq j. \end{cases} \quad (2)$$

Following an idea of Drozd [32] (developed in [27]), we have introduced in [1, 28] the *Tits matrix* $\widehat{C}_J \in \mathbb{M}_J(\mathbb{Z})$ of J to be the integral matrix

$$\widehat{C}_J = \left[\widehat{c}_{ij}\right]_{i,j\in J} \in \mathbb{M}_J(\mathbb{Z}),$$

$$\text{with } \widehat{c}_{ij} \begin{cases} 1, & i = j \text{ or } j \preceq i, i, j \notin \max J, \\ 0, & i, j \text{ incomparable, or } i \preceq j \text{ and } i, j \notin \max J, \\ -1, & \text{if } i \prec j \text{ and } j \in \max J, \end{cases} \tag{3}$$

where $\max J$ is the set of all maximal elements of J. Usually, we equip the elements of J with a numbering; that is, J is viewed as $J = \{a_1, \ldots, a_m\}$, $m = |J| \geq 1$. Throughout, we fix such a numbering and make the identifications $\mathbb{M}_m(\mathbb{Z}) \equiv \mathbb{M}_J(\mathbb{Z})$ and $\mathbb{Z}^m \equiv \mathbb{Z}^J$. The incidence matrix $C_J \in \mathbb{M}_m(\mathbb{Z}) \equiv \mathbb{M}_J(\mathbb{Z})$ and the Tits matrix $\widehat{C}_J \in \mathbb{M}_m(\mathbb{Z}) \equiv \mathbb{M}_J(\mathbb{Z})$ depend on the numbering of $a_1, \ldots, a_m$. Namely, if $J' = \{a'_1, \ldots, a'_m\}$ is obtained from $J = \{a_1, \ldots, a_m\}$ by a permutation $\sigma \in \mathbf{S}_m$ and $\widehat{\sigma} \in \mathrm{Gl}(m, \mathbb{Z})$ is the permutation matrix of σ, then

$$C_{J'} = \widehat{\sigma}^{-1} \cdot C_J \cdot \widehat{\sigma}, \qquad \widehat{C}_{J'} = \widehat{\sigma}^{-1} \cdot \widehat{C}_J \cdot \widehat{\sigma}. \tag{4}$$

Note that any poset J admits an *upper-triangular numbering* $J = \{a_1, \ldots, a_m\}$; that is, $a_i \preceq a_j$ implies $i \leq j$. In this case, $C_J \in \mathbb{M}_m(\mathbb{Z})$ is an upper-triangular matrix with 1 on the main diagonal, and, hence, $\det C_J = 1$, and $\det C_{J'} = 1$, for any numbering $J' = \{a'_1, \ldots, a'_m\}$.

Fix a numbering $a_1, \ldots, a_m$ of elements of J. Following [1, 28], by the *Euler matrix* of the poset J we mean the inverse

$$\overline{C}_J := C_J^{-1} \in \mathbb{M}_m(\mathbb{Z}) \equiv \mathbb{M}_J(\mathbb{Z}) \tag{5}$$

of C_J. Following [3, 4], we call

$$\begin{gathered} Ad_J := C_J + C_J^{\mathrm{tr}} - 2 \cdot E, \\ P_J(t) = \det(t \cdot E - Ad_J) \in \mathbb{Z}[t] \end{gathered} \tag{6}$$

the *symmetric adjacency matrix* and the *characteristic polynomial* of the poset J. The set $\mathbf{spec}_J$ of all $m = |J|$ real roots of $P_J(t)$ is defined to be the (real) *spectrum* of the poset J.

We denote by $q_J, \widehat{q}_J, \overline{q}_J : \mathbb{Z}^J \equiv \mathbb{Z}^m \to \mathbb{Z}$ the *incidence* quadratic form, the *Tits* quadratic form, and the *Euler* quadratic form of J defined by the formulae

$$\begin{gathered} q_J(x) = x \cdot C_J \cdot x^{\mathrm{tr}} = \sum_{j\in J} x_j^2 + \sum_{i\prec j} x_i x_j, \\ \widehat{q}_J(x) = x \cdot \widehat{C}_J \cdot x^{\mathrm{tr}} = \sum_{j\in J} x_j^2 + \sum_{i\prec j\in \check{J}} x_i x_j - \sum_{p\in\max J} \sum_{i\prec p} x_i x_p, \\ \overline{q}_J(x) = x \cdot \overline{C}_J \cdot x^{\mathrm{tr}} = x \cdot C_J^{-1} \cdot x^{\mathrm{tr}}, \end{gathered} \tag{7}$$

respectively, where $\check{J} = J \setminus \max J$, $\max J$ is the set of all maximal elements in J, and $\widehat{C}_J \in \mathbb{M}_J(\mathbb{Z})$ is the Tits matrix of J; see (27) and [1, 28] for a definition. The matrices

$$\begin{gathered} G_J := \frac{1}{2}\left[C_J + C_J^{\mathrm{tr}}\right], \qquad \widehat{G}_J := \frac{1}{2}\left[\widehat{C}_J + \widehat{C}_J^{\mathrm{tr}}\right], \\ \overline{G}_J := \frac{1}{2}\left[\overline{C}_J + \overline{C}_J^{\mathrm{tr}}\right] \in \mathbb{M}_J(\mathbb{Q}), \end{gathered} \tag{8}$$

with rational coefficients, are called the *symmetric incidence Gram matrix*, the *symmetric Tits-Gram matrix*, and the *symmetric Euler-Gram matrix* of J. The matrices

$$\begin{gathered} \widehat{Ad}_J := \widehat{C}_J + \widehat{C}_J^{\mathrm{tr}} - 2 \cdot E, \\ \overline{Ad}_J := \overline{C}_J + \overline{C}_J^{\mathrm{tr}} - 2 \cdot E = C_J^{-1} + C_J^{-\mathrm{tr}} - 2 \cdot E, \end{gathered} \tag{9}$$

with integer coefficients, are called the *Tits adjacency matrix*, and the *Euler adjacency matrix* of J. The polynomials

$$\begin{gathered} P_J(t) := \det(t \cdot E - Ad_J) = \det\left(t \cdot E - \widehat{Ad}_J\right), \\ \overline{P}_J(t) := \det\left(t \cdot E - \overline{Ad}_J\right) \end{gathered} \tag{10}$$

are called the *characteristic polynomial* of J and the *Euler-characteristic polynomial* of J, respectively.

Example 1. (a) If I is the poset

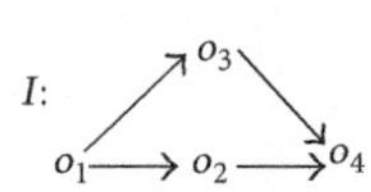

(11)

then $P_I(t) = \overline{P}_I(t) = t^4 - 5t^2 - 4t$; that is, the characteristic polynomial $P_I(t)$ of I coincides with the Euler-characteristic polynomial $\overline{P}_I(t)$ of I.

(b) If J is the poset

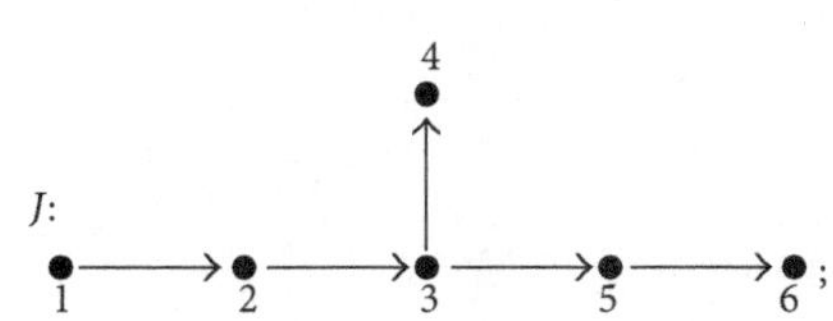

(12)

of the Dynkin type $\mathbb{E}_6$, then the characteristic polynomial $P_J(t)$ of J does not coincide with the Euler-characteristic polynomial $\overline{P}_J(t)$ of J, because

$$\begin{gathered} P_J(t) = t^6 - 13t^4 - 26t^3 - 15t^2 + 2t + 3, \\ \overline{P}_J(t) = t^6 - 5t^4 + 5t^2 - 1. \end{gathered} \tag{13}$$

Following [17, 33], we introduce the following definition.

Definition 2. (a) We define a poset J to be *positive* (resp., *nonnegative*) if the incidence form $q_J : \mathbb{Z}^J \to \mathbb{Z}$ of J is positive (resp., nonnegative); that is, $q_J(v) > 0$, for any nonzero $v \in \mathbb{Z}^J$ (resp., $q_J(v) \geq 0$, for any $v \in \mathbb{Z}^J$).

(b) We define a poset J to be *principal* if its incidence form $q_J : \mathbb{Z}^J \to \mathbb{Z}$ is principal in the sense of [34, Definition 2.1]; that is, q_J is nonnegative, not positive, and the kernel

$$\operatorname{Ker} q_J := \left\{v \in \mathbb{Z}^J;\ q_J(v) = 0\right\} \tag{14}$$

is an infinite cyclic subgroup of $\mathbb{Z}^J$.

Following the main idea of the Coxeter spectral analysis of acyclic edge-bipartite graphs (signed graphs) presented in [3, 4], we study finite posets J (with a fixed numbering $J = \{a_1, \ldots, a_m\}$) by means of the *Coxeter spectrum*

$$\mathbf{specc}_J \subseteq \mathbb{C} \tag{15}$$

of J, that is, the set $\mathbf{specc}_J$ of all $m = |J|$ eigenvalues of the *Coxeter matrix*

$$\operatorname{Cox}_J := -C_J \cdot C_J^{-\mathrm{tr}} \in \mathbb{M}_m(\mathbb{Z}) \equiv \mathbb{M}_J(\mathbb{Z}) \tag{16}$$

of J, or equivalently, the set $\mathbf{specc}_J$ of all $m = |J|$ roots of the *Coxeter polynomial*

$$\begin{aligned} \operatorname{cox}_J(t) &:= \det\left(t \cdot E - \operatorname{Cox}_J\right) = \det\left(t \cdot E - \widehat{\operatorname{Cox}}_J\right) \\ &= \det\left(t \cdot E - \overline{\operatorname{Cox}}_J\right) \in \mathbb{Z}[t]; \end{aligned} \tag{17}$$

see (31) and [1]. It follows from (4) that the Coxeter spectrum $\mathbf{specc}_J$ of J and the spectrum $\mathbf{spec}_J$ of J do not depend on the numbering of the elements of the poset J.

A motivation. We recall from [26, 27] that the problems we study in the paper have a bimodule matrix problem interpretation and have essential applications in reducing some classes of partitioned matrices with coefficients in a field K to their canonical forms. For simplicity of its presentation, we illustrate it in case when $\widehat{q}_J(x)$ is the Tits quadratic form (7) of the poset $J = \{a_1, \ldots, a_n, *, +\}$, with an upper-triangular partial order $\preceq$ such that J has precisely two maximal elements $* := *_{n+1}$ and $+ := +_{n+2}$. In this case, we have

$$\widehat{q}_J(x) = \sum_{a_i \in J} x_i^2 + \sum_{a_i \prec a_j, i, j \le n} x_i x_j - \sum_{a_i \prec *} x_i x_* - \sum_{a_j \prec +} x_j x_+. \tag{18}$$

Fix a vector $v = (v_1, \ldots, v_n, v_*, v_+) \in \mathbb{N}^{n+2} \equiv \mathbb{N}^J$, and consider the K-vector space $\mathbf{Mat}_v^J$ of all partitioned matrices of the form (compare with [27])

$$A = \begin{array}{|c|c|c|c|l} \hline A_{1*} & A_{2*} & \cdots & A_{n*} & \} v_* \\ \hline A_{1+} & A_{2+} & \cdots & A_{n+} & \} v_+ \\ \hline \underbrace{\qquad}_{v_1} & \underbrace{\qquad}_{v_2} & \cdots & \underbrace{\qquad}_{v_n} & \end{array} \tag{19}$$

with coefficients in K, where $A_{i*} = 0$ if $a_i \not\preceq *$ and $A_{j+} = 0$ if $a_j \not\preceq +$. Consider the group $\mathbf{G}_v^J$ generated by the elementary transformations of the following three types:

(a) all simultaneous transformations on rows inside each horizontal block;

(b) all simultaneous transformations on columns inside each vertical block;

(c) all simultaneous transformations on columns from the ith column block to jth column block, if the relation $a_i \preceq a_j$ holds in the poset $J \setminus \{*, +\}$ (with natural zero-adjustments).

It follows from [27, Section 2] (see also [16, 26, 32]) that the problem of finding canonical forms of matrices in $\mathbf{Mat}_v^J$, with respect to the elementary transformations from the set $\mathbf{G}_v^J$, is controlled by the Tits quadratic form $\widehat{q}_J$ in the following sense. For any $v \in \mathbb{N}^J$, there is only a finite number $\mathbf{G}_v^J$-canonical forms of matrices in $\mathbf{Mat}_v^J$ if and only if the form $\widehat{q}_J$ is weakly positive; that is, $\widehat{q}_J(v)$ is positive, for all nonzero vectors $v \in \mathbb{N}^J$. Moreover, there is one-to-one correspondence between the irreducible $\mathbf{G}_v^J$-canonical forms in $\mathbf{Mat}_v^J$ and the vectors $v \in \mathbb{N}^J$ satisfying the equation $\widehat{q}_J(v) = 1$. A solution of the problem is given in [27] and [1, Theorem 1.6]. A useful homological interpretation (in terms of the Euler characteristic) of the $\mathbb{Z}$-bilinear Tits form $\widehat{b}_J(x, y) = x \cdot \widehat{C}_J \cdot y^{\mathrm{tr}}$ (26) and $\mathbb{Z}$-bilinear Euler form $\overline{b}_J(x, y) = x \cdot \overline{C}_J \cdot y^{\mathrm{tr}}$ is given in [1, (1.3)]. The reader is referred to [6–8, 25] for a detailed study and a solution of other important matrix problems of high computational complexity that have many useful applications in representation theory; see [16, 26].

We show in Section 3 that the Coxeter spectral analysis of principal posets J essentially uses the Coxeter spectra of the simply laced Euclidean diagrams presented in Figure 1.

The nonsymmetric *Gram matrix* $\check{G}_\Delta$ of any graph $\Delta = (\Delta_0, \Delta_1) \in \{\widetilde{\mathbb{D}}_n, n \ge 4, \widetilde{\mathbb{E}}_6, \widetilde{\mathbb{E}}_7, \widetilde{\mathbb{E}}_8\}$ of Figure 1, with the set of vertices $\Delta_0 = \{1, \ldots, n, n+1\}$ and the set of edges Δ_1, is defined to be the matrix

$$\check{G}_\Delta = \begin{bmatrix} 1 & d_{12}^\Delta & \cdots & d_{1n}^\Delta & d_{1n+1}^\Delta \\ 0 & 1 & \cdots & d_{2n}^\Delta & d_{2n+1}^\Delta \\ \vdots & \vdots & \ddots & \vdots & \vdots \\ 0 & 0 & \cdots & 1 & d_{nn+1}^\Delta \\ 0 & 0 & \cdots & 0 & 1 \end{bmatrix} \in \mathbb{M}_{n+1}(\mathbb{Z}), \tag{20}$$

where $d_{ij}^\Delta = -|\Delta_1(i, j)|$, if there is an edge $\bullet_i - \bullet_j$ and $i \le j$. We set $d_{ij}^\Delta = 0$, if $\Delta_1(i, j)$ is empty or $j < i$.

The Coxeter polynomial $\operatorname{cox}_\Delta(t) := \det(t \cdot E + \check{G}_\Delta \cdot \check{G}_\Delta^{-\mathrm{tr}})$ of any diagram $\Delta = (\Delta_0, \Delta_1) \in \{\widetilde{\mathbb{D}}_n, n \ge 4, \widetilde{\mathbb{E}}_6, \widetilde{\mathbb{E}}_7, \widetilde{\mathbb{E}}_8\}$ does not depend on the numbering of the vertices in Δ_0 and is presented in (48). If $n \ge 1$ and $\Delta = \widetilde{\mathbb{A}}_n$, the Coxeter polynomial $\operatorname{cox}_\Delta(t) := \det(t \cdot E + \check{G}_\Delta \cdot \check{G}_\Delta^{-\mathrm{tr}})$ of Δ depends on the numbering of the vertices in Δ_0 and is one of

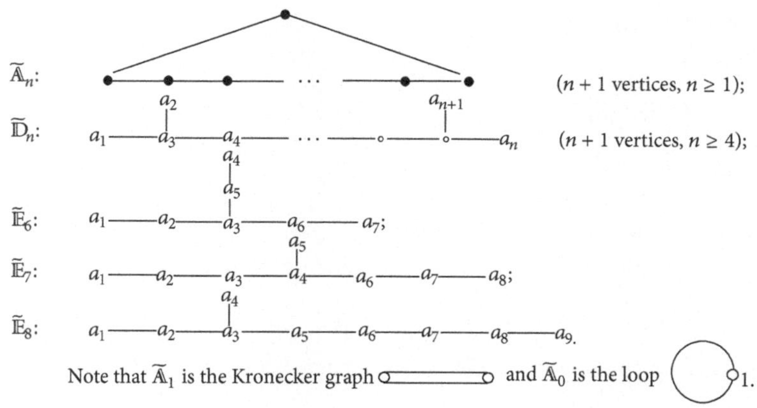

FIGURE 1: Simply laced Euclidean (extended Dynkin) diagrams.

the polynomials $F_{\Delta}^{(1)}(t), F_{\Delta}^{(2)}(t), \ldots, F_{\Delta}^{(m_n)}(t)$ presented in [4], where

$$F_{\Delta}^{(j)}(t) = t^{n+1} - t^{n-j+1} - t^{j} + 1$$
$$= (t-1)^2 \cdot \mathfrak{v}_j(t) \cdot \mathfrak{v}_{n-j+1}(t),$$
$$m_n = \begin{cases} \dfrac{n}{2}, & \text{if } n \text{ is even,} \\ \dfrac{n+1}{2}, & \text{if } n+1 \text{ is even,} \end{cases} \tag{21}$$

for $j = 1, \ldots, m_n$, and $\mathfrak{v}_m(t) = t^{m-1} + t^{m-2} + \cdots + t^2 + t + 1$. In particular, if $n + 1$ is even and $j = m_n = (n+1)/2$, then $t^{n-j+1} = t^j$ and

$$F_{\Delta}^{(m_n)}(t) = F_{\Delta}^{((n+1)/2)}(t) = t^{n+1} - 2t^{(n+1)/2} + 1. \tag{22}$$

Following [4, 21], we associate (in Section 2) to any principal poset J a simply laced Euclidean diagram $DJ \in \{\widetilde{\mathbb{A}}_n, n \geq 1, \widetilde{\mathbb{D}}_n, n \geq 4, \widetilde{\mathbb{E}}_6, \widetilde{\mathbb{E}}_7, \widetilde{\mathbb{E}}_8\}$ such that the incidence symmetric Gram matrix $G_J := (1/2)[C_J + C_J^{tr}] \in \mathbb{M}_J(\mathbb{Q})$ is $\mathbb{Z}$-congruent to the *symmetric Gram matrix*

$$G_{DJ} := \frac{1}{2}\left[\check{G}_{DJ} + \check{G}_{DJ}^{tr}\right] \in \mathbb{M}_{DJ}(\mathbb{Q}) = \mathbb{M}_J(\mathbb{Q}) \tag{23}$$

of DJ; that is, there is a $\mathbb{Z}$-invertible matrix B such that $G_{DJ} = B^{tr} \cdot G_J \cdot B$.

One of the aims of the Coxeter spectral analysis of nonnegative finite posets is to study the question when the Coxeter type

$$\mathbf{Ctype}_J := \left(\mathbf{specc}_J, \mathbf{c}_J, \check{\mathbf{c}}_J\right) \tag{24}$$

of a poset J determines the matrix C_J (and, hence, the poset J) uniquely, up to a $\mathbb{Z}$-congruency. Here, we set $\check{\mathbf{c}}_J = \mathbf{c}_J$, if J is positive. In other words, we claim that, for any pair J, I of nonnegative one-peak posets, $\mathbf{Ctype}_J = \mathbf{Ctype}_I$ if and only if the incidence matrices C_J and C_I are $\mathbb{Z}$-congruent. We also study the problem related with the results proved by Horn and Sergeichuk [35], if for any $\mathbb{Z}$-invertible matrix $A \in \mathbb{M}_n(\mathbb{Z})$, there exists $B \in \mathbb{M}_n(\mathbb{Z})$ such that $A^{tr} = B^{tr} \cdot A \cdot B$ and $B^2 = E$ is the identity matrix; see [17, 18].

The main results of the present paper on nonnegative posets J can be summarised as follows:

(1) canonical equivalences between the incidences, Tits, and Euler quadratic form (and corresponding Coxeter transformations and Coxeter spectra) of any poset J, established in Proposition 5;

(2) a characterization of principal posets given in Section 3. We show that a connected poset J is principal if and only if there exists a simply laced Euclidean diagram

$$DJ \in \left\{\widetilde{\mathbb{D}}_n, n \geq 4, \widetilde{\mathbb{E}}_6, \widetilde{\mathbb{E}}_7, \widetilde{\mathbb{E}}_8\right\} \tag{25}$$

such that the symmetric Gram matrix $G_J := (1/2)[C_J + C_J^{tr}] \in \mathbb{M}_J(\mathbb{Q})$ of J is $\mathbb{Z}$-congruent to the symmetric Gram matrix $G_{DJ} := (1/2)[\check{G}_{DJ} + \check{G}_{DJ}^{tr}] \in \mathbb{M}_{DJ}(\mathbb{Q})$ of DJ. Moreover, we show in Section 3 that, given a connected principal poset J, the Coxeter spectrum $\mathbf{specc}_J$ is a subset of a unit circle $\mathcal{S}^1 = \{z \in \mathbb{C}; |z| = 1\}$, $1 \in \mathbf{specc}_J$, and any $z \in \mathbf{specc}_J$ is a root of unity;

(3) a Coxeter spectral classification result (Corollary 11) asserting that, given a pair I, J of one-peak principal posets with at most 13 elements, the following conditions are equivalent:

(3a) $DI = DJ$,

(3b) $\mathbf{specc}_I = \mathbf{specc}_J$,

(3c) $\check{\mathbf{c}}_I = \check{\mathbf{c}}_J$ and $|I| = |J|$,

(3d) the incidence matrix $C_J \in \mathbb{M}_J(\mathbb{Z})$ is $\mathbb{Z}$-congruent to the incidence matrix $C_I \in \mathbb{M}_I(\mathbb{Z})$; that is, there is a $\mathbb{Z}$-invertible matrix B such that $C_I = B^{tr} \cdot C_J \cdot B$.

In Section 3, we study principal posets by means of the defect and the reduced Coxeter number, and in Section 4, we present a framework for the study of nonnegative posets of corank $r \geq 2$ by means of their defect and the reduced Coxeter number. Examples are given in Sections 3–5.

The reader is referred to [1, 16, 17, 26] for a background of poset representation theory and elementary introduction to the poset matrix problems.

2. A Framework for the Coxeter Spectral Analysis of Finite Posets

The quadratic wanderings on finite posets J studied in [1] are playing a key role in the representation theory of posets, algebras, and coalgebras, as well as in the algebraic combinatorics of posets; see [6, 9–14, 16, 24–26, 28, 31, 32, 36–39]. Except for the incidence wandering and the Euler wanderings defined by the incidence matrix $C_J \in \mathbb{M}_m(\mathbb{Z}) \equiv \mathbb{M}_J(\mathbb{Z})$ (2), with $\det C_J = 1$ and a fixed numbering $J = \{a_1, \ldots, a_m\}$, as well as the Euler matrix $\overline{C}_J := C_J^{-1}$, we study in [1, 26–28] the Tits wandering defined by the *Tits matrix* $\widehat{C}_J \in \mathbb{M}_m(\mathbb{Z}) \equiv \mathbb{M}_J(\mathbb{Z})$ of J (see [28, (3.6)]), that is, the Gram matrix of the *Tits* $\mathbb{Z}$*-bilinear form* $\widehat{b}_J : \mathbb{Z}^J \times \mathbb{Z}^J \to \mathbb{Z}$ given by

$$\widehat{b}_J(x,y) = \sum_{a_i \in J} x_i y_i + \sum_{a_j < a_i \in \check{J}} x_i y_j - \sum_{a_p \in \max J} \sum_{a_i < a_p} x_i y_p = x \cdot \widehat{C}_J \cdot y^{\mathrm{tr}}, \tag{26}$$

where $\max J$ is the set of all maximal elements in the poset J and $\check{J} := J \setminus \max J$. We call $\widehat{q}_J(x) := \widehat{b}_J(x,x) = x \cdot \widehat{C}_J \cdot x^{\mathrm{tr}}$ the *Tits quadratic form* of J.

A homological interpretation of the $\mathbb{Z}$-bilinear forms $\widehat{b}_J(x,y) = x \cdot \widehat{C}_J \cdot y^{\mathrm{tr}}$ and $\overline{b}_J(x,y) = x \cdot \overline{C}_J \cdot y^{\mathrm{tr}}$ is given in [1, (1.3)]. For a geometric interpretation of the Tits form $\widehat{q}_I$ of a one-peak poset I, the reader is referred to Drozd [32] and Simson [26].

Note that, given a one-peak poset I of the form $I = \{1, 2, \ldots, n, * = n+1\}$, with a unique maximal element $* = n+1$, we have

$$\widehat{C}_I = \left[\begin{array}{c|c} C_T^{\mathrm{tr}} & -u \\ \hline 0 & 1 \end{array}\right] \in \mathbb{M}_{n+1}(\mathbb{Z}), \quad \text{with } u = \begin{bmatrix} 1 \\ \vdots \\ 1 \end{bmatrix}, \tag{27}$$

where $C_T \in \mathbb{M}_T(\mathbb{Z}) = \mathbb{M}_n(\mathbb{Z})$ is the incidence matrix of the poset $T = I \setminus \{*\} = \{1, 2, \ldots, n\}$; see [26]. Note that $\widehat{q}_I(x) = x \cdot \widehat{C}_I \cdot x^{\mathrm{tr}}$.

Now, we show that, in the Coxeter spectral study of finite posets J, we can use the Coxeter spectral technique introduced in [2, 4], for the edge-bipartite graphs (signed graphs [5]), and developed in [2, 34, 40] for the matrix morsifications of unit quadratic forms.

Following [3, 4, 24], by an *edge-bipartite graph* (bigraph, in short), we mean a pair $\Delta = (\Delta_0, \Delta_1)$, where Δ_0 is a finite nonempty set of vertices and Δ_1 is a finite set of edges equipped with a bipartition $\Delta_1 = \Delta_1^- \cup \Delta_1^+$ such that the set $\Delta_1(i,j) = \Delta_1^-(i,j) \cup \Delta_1^+(i,j)$ of edges connecting the vertices i and j does not contain edges lying in $\Delta_1^-(i,j) \cap \Delta_1^+(i,j)$, for each pair of vertices $i, j \in \Delta_0$, and either $\Delta_1(i,j) = \Delta_1^-(i,j)$ or $\Delta_1(i,j) = \Delta_1^+(i,j)$. Note that the edge-bipartite graphs can be viewed as signed multigraphs satisfying a separation property; see [4, 5].

We visualize Δ as a graph in a Euclidean space $\mathbb{R}^m$, $m \geq 2$, with the vertices numbered by the integers $1, \ldots, n$; usually, we simply write $\Delta_0 = \{1, \ldots, n\}$. An edge in $\Delta_1^-(i,j)$ is visualised as a continuous one $\bullet_i\!\!-\!\!\bullet_j$, and an edge in $\Delta_1^+(i,j)$ is visualised as a dotted one $\bullet_i \text{---} \bullet_j$. A bigraph Δ is said to be *loop-free* if it has no loops.

We view any finite graph $\Delta = (\Delta_0, \Delta_1)$ as an edge-bipartite one by setting $\Delta_1^-(i,j) = \Delta_1(i,j)$ and $\Delta_1^+(i,j) = \emptyset$, for each pair of vertices $i, j \in \Delta_0$.

To any loop-free edge-bipartite graph $\Delta = (\Delta_0, \Delta_1)$, with a fixed numbering $\Delta_0 = \{a_1, \ldots, a_m\}$ of its vertices, we associate the upper-triangular nonsymmetric *Gram matrix* $\check{G}_\Delta = E + [d_{ij}^\Delta] \in \mathbb{M}_m(\mathbb{Z})$ of the form (20), with $m := n+1$, where $d_{ij}^\Delta = -|\Delta_1^-(i,j)|$, if there is an edge $\bullet_i\!\!-\!\!\bullet_j$ and $i \leq j$, $d_{ij}^\Delta = |\Delta_1^+(i,j)|$, if there is an edge $\bullet_i \text{- - -} \bullet_j$ and $i \leq j$. We set $d_{ij}^\Delta = 0$, if $\Delta_1(i,j)$ is empty or $j < i$. Since Δ is loop-free, we have $d_{11}^\Delta = \cdots = d_{mm}^\Delta = 0$ and the main diagonal of $\check{G}_\Delta$ consists of unities.

Following [4], we call $\Delta = (\Delta_0, \Delta_1)$ *positive* (resp., *nonnegative*), if the symmetric Gram matrix

$$G_\Delta := (1/2)\left(\check{G}_\Delta + \check{G}_\Delta^{\mathrm{tr}}\right) \tag{28}$$

of Δ is positive definite (resp., positive semidefinite).

Following [4], we associate to any loop-free edge-bipartite graph Δ, with $|\Delta_0| = n \geq 2$, the *Coxeter spectrum* $\mathbf{specc}_\Delta \subseteq \mathbb{C}$ defined to be the spectrum of the Coxeter (-Gram) matrix

$$\mathrm{Cox}_\Delta := -\check{G}_\Delta \cdot \check{G}_\Delta^{-\mathrm{tr}} \in \mathbb{M}_n(\mathbb{Z}), \tag{29}$$

the Coxeter polynomial

$$\mathrm{cox}_\Delta(t) := \det\left(t \cdot E - \mathrm{Cox}_\Delta\right) \in \mathbb{Z}[t], \tag{30}$$

the Coxeter transformation $\Phi_\Delta : \mathbb{Z}^n \to \mathbb{Z}^n$, given by $x \mapsto \Phi_\Delta(x) := x \cdot \mathrm{Cox}_\Delta$, the Coxeter number $\mathbf{c}_\Delta$ (the order of Φ_Δ in the automorphism group of $\mathbb{Z}^n$, i.e., the minimal integer $r \geq 1$ such that $\Phi_\Delta^r = E$), the $\mathbb{Z}$-bilinear Gram form $b_\Delta : \mathbb{Z}^n \times \mathbb{Z}^n \to \mathbb{Z}$ of Δ given by $b_\Delta(x,y) := x \cdot \check{G}_\Delta \cdot y^{\mathrm{tr}}$, and the integral unit quadratic form

$$q_\Delta(x) := b_\Delta(x,x) = x_1^2 + \cdots + x_n^2 + \sum_{i<j} d_{ij}^\Delta x_i x_j = x \cdot G_\Delta \cdot x^{\mathrm{tr}}. \tag{31}$$

Conversely, following Ovsienko [24], to any integral unit form

$$q(x) = x_1^2 + \cdots + x_n^2 + \sum_{i<j} q_{ij} x_i x_j, \quad \text{with } q_{ij} \in \mathbb{Z}, \tag{32}$$

we associate the loop-free bigraph $\mathbf{bigr}(q)$ of q as follows (see also [34, 41]):

(a) the vertices of $\mathbf{bigr}(q)$ are the integers $1, \ldots, n$,

(b) two vertices $i \neq j$ are joined by $-q_{ij}$ continuous edges of the form $\bullet_i\!\!-\!\!\bullet_j$ if q_{ij} is negative, and by q_{ij} dotted edges of the form $\bullet_i \text{- - -} \bullet_j$, if q_{ij} is positive,

(c) there is no edge between i and j, if $q_{ij} = 0$, or $i = j$.

To any poset $J \equiv (J, \preceq)$, with a fixed numbering $J = \{a_1, \ldots, a_m\}$ of its points, we associate the following three edge-bipartite graphs:

$$\Delta_J := \mathbf{bigr}(q_J), \qquad \widehat{\Delta}_J := \mathbf{bigr}(\widehat{q}_J), \qquad \overline{\Delta}_J := \mathbf{bigr}(\overline{q}_J), \tag{33}$$

where $\mathbf{bigr}(q_J)$, $\mathbf{bigr}(\widehat{q}_J)$, and $\mathbf{bigr}(\overline{q}_J)$ are the bigraphs of the quadratic forms q_J, $\widehat{q}_J$, and $\overline{q}_J$, respectively; see (7). More precisely, the bigraphs (33) are defined as follows.

(i) The set of vertices of each of the bigraphs Δ_J, $\widehat{\Delta}_J$, and $\overline{\Delta}_J$ is the enumerated set $J = \{a_1, \ldots, a_m\}$.

(ii) There is an edge a_i- - -a_j in Δ_J, if $a_i \prec a_j$ or $a_j \prec a_i$ holds in J.

(iii) There is an edge a_i- - -a_j in $\widehat{\Delta}_J$, if a_i and a_j are not maximal in J and $a_i \prec a_j$ or $a_j \prec a_i$ holds in J. There is an edge a_i—a_j in $\widehat{\Delta}_J$, if $a_i \prec a_j$ holds and a_j is maximal in J.

(iv) Let $\overline{C}_J = C_J^{-1} = [\overline{c}_{ij}] \in \mathbb{M}_m(\mathbb{Z})$ be the Euler matrix of J. There is an edge a_i- - -a_j (resp., a_i—a_j) in $\overline{\Delta}_J$, if $\overline{c}_{ij} > 0$ or $\overline{c}_{ji} > 0$ (resp., $\overline{c}_{ij} < 0$ or $\overline{c}_{ji} < 0$).

We call Δ_J, $\widehat{\Delta}_J$, and $\overline{\Delta}_J$ the *incidence bigraph of* Δ, the *Tits bigraph of* Δ, and the *Euler bigraph of* Δ, respectively, (with respect to the numbering $J = \{a_1, \ldots, a_m\}$).

The following simple lemma is of importance.

Lemma 3. *Assume that J is a finite poset with a fixed numbering $J = \{a_1, \ldots, a_m\}$, and let Δ_J, $\widehat{\Delta}_J$, $\overline{\Delta}_J$ be the loop-free edge-bipartite graphs associated with J in* (33).

(a) *The symmetric Gram matrices G_J, $\widehat{G}_J$, $\overline{G}_J$ are $\mathbb{Z}$-congruent to the symmetric Gram matrices G_{Δ_J}, $G_{\widehat{\Delta}_J}$, $G_{\overline{\Delta}_J}$, respectively. The rank of each of the symmetric Gram matrices G_{Δ_J}, $G_{\widehat{\Delta}_J}$, $G_{\overline{\Delta}_J}$ does not depend of the numbering $J = \{a_1, \ldots, a_m\}$ and coincides with the common rank* $\mathrm{rk}\, G_{\Delta_J} = \mathrm{rk}\, G_{\widehat{\Delta}_J} = \mathrm{rk}\, G_{\overline{\Delta}_J}$.

(b) $P_J(t) = P_{\Delta_J}(t) = P_{\widehat{\Delta}_J}(t)$.

(c) *The poset J is positive (resp., nonnegative) if and only if the bigraph Δ_J (and $\widehat{\Delta}_J$, $\overline{\Delta}_J$) is positive (resp., nonnegative).*

(d) *The poset J is principal if and only if the bigraph Δ_J (and $\widehat{\Delta}_J$, $\overline{\Delta}_J$) is principal.*

Proof. For the proof of (a), we recall that the Gram matrices $G_J, \widehat{G}_J, \overline{G}_J, G_{\Delta_J}, G_{\widehat{\Delta}_J}, G_{\overline{\Delta}_J}$ are invariant, up to $\mathbb{Z}$-congruency, under permutations of the elements $\{a_1, \ldots, a_m\}$. Since J admits an upper-triangular numbering $J' = \{a_1', \ldots, a_m'\}$ and $G_{\Delta_{J'}} = G_{J'}$, then (a) follows. The proof of remaining statements is left to the reader.

Following the terminology used in [2–4, 34], we introduce the following definition.

Definition 4. Let J be a finite poset, with a fixed numbering $J = \{a_1, \ldots, a_m\}$.

(a) We associate with J the following three Coxeter matrices:

(a1) the (incidence) Coxeter matrix $\mathrm{Cox}_J := -C_J \cdot C_J^{-\mathrm{tr}} \in \mathbb{M}_m(\mathbb{Z})$;

(a2) the Coxeter-Tits matrix $\widehat{\mathrm{Cox}}_J := -\widehat{C}_J \cdot \widehat{C}_J^{-\mathrm{tr}} \in \mathbb{M}_m(\mathbb{Z})$;

(a3) the Coxeter-Euler matrix $\overline{\mathrm{Cox}}_J := -C_J^{-1} \cdot C_J^{\mathrm{tr}} \in \mathbb{M}_m(\mathbb{Z})$.

Moreover, we define the following three *Coxeter transformations*:

(a4) the (incidence) Coxeter transformation $\Phi_m : \mathbb{Z}^m \to \mathbb{Z}^m$ of J;

(a5) the Coxeter-Tits transformation $\widehat{\Phi}_J : \mathbb{Z}^m \to \mathbb{Z}^m$ of J;

(a6) the Coxeter-Euler transformation $\overline{\Phi}_J : \mathbb{Z}^m \to \mathbb{Z}^m$ of J, by the following formulae:

$$\Phi_J(x) = x \cdot \mathrm{Cox}_J, \qquad \widehat{\Phi}_J(x) = x \cdot \widehat{\mathrm{Cox}}_J, \qquad \text{and } \overline{\Phi}_J(x) = x \cdot \overline{\mathrm{Cox}}_J. \tag{34}$$

(b) The integral polynomial

$$\mathrm{cox}_J(t) = \det(t \cdot E - \mathrm{Cox}_J) = \det\left(t \cdot E - \widehat{\mathrm{Cox}}_J\right) = \det\left(t \cdot E - \overline{\mathrm{Cox}}_J\right) \in \mathbb{Z}[t] \tag{35}$$

is called the *Coxeter polynomial of the poset* J.

(c) The *Coxeter spectrum* of J is the set $\mathbf{specc}_J \subseteq \mathbb{C}$ of all $m = |J|$ eigenvalues of the matrix Cox_J, or, equivalently, the set $\mathbf{specc}_J$ of all $m = |J|$ roots of the Coxeter polynomial $\mathrm{cox}_J(t)$.

(d) The order $\mathbf{c}_J := \mathrm{ord}(\Phi_J)$ of the Coxeter transformation $\Phi_J : \mathbb{Z}^J \to \mathbb{Z}^J$ is called the *Coxeter number* of the poset J. In other words, $\mathbf{c}_J$ is the minimal integer $r \geq 1$ such that $\Phi_J^r = id$. We set $\mathbf{c}_J = \infty$, if $\Phi_J^r \neq id$, for any $r \geq 1$.

(e) Assume that J is nonnegative. The *Coxeter type* of J is defined to be the pair $\mathbf{Ctype}_J := (\mathbf{specc}_J, \mathbf{c}_J)$ if J is positive, and the triple $\mathbf{Ctype}_J := (\mathbf{specc}_J, \mathbf{c}_J, \check{\mathbf{c}}_J)$ if J is not positive, where $\check{\mathbf{c}}_J$ is the reduced Coxeter number of J in the sense of Theorems 10 and 18.

The following proposition shows that equality (35) holds.

Proposition 5. *Let J be a finite poset, with a fixed numbering $J = \{a_1, \ldots, a_m\}$, let $q_J, \widehat{q}_J, \overline{q}_J : \mathbb{Z}^m \to \mathbb{Z}$ be the incidence, Tits, and Euler quadratic form of J, and let $\Phi_J, \widehat{\Phi}_J, \overline{\Phi}_J : \mathbb{Z}^m \to \mathbb{Z}^m$ be the corresponding Coxeter transformations.*

(a) *The following equalities hold* $\widehat{C}_J = B \cdot \overline{C}_J \cdot B^{tr}$ *and* $C_J^{tr} = B' \cdot \overline{C}_J \cdot B'^{tr}$, *and the following diagrams are commutative*

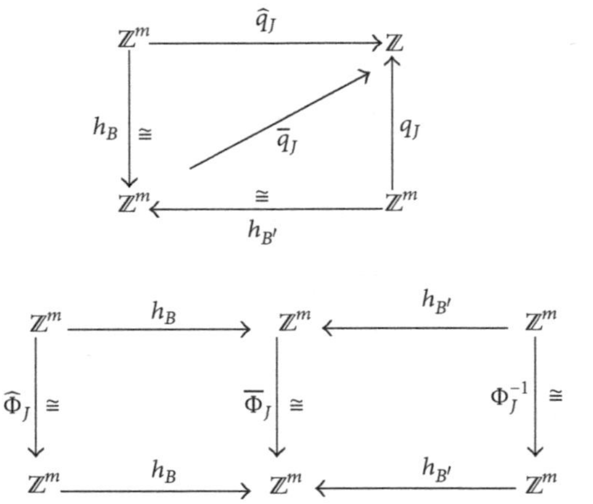

(36)

where $B' = C_{\check{J}}^{tr}$, $B = \left[\begin{array}{c|c} C_{\check{J}} & 0 \\ \hline 0 & E \end{array}\right]$, $\check{J} = J \setminus \max J$, *and* $h_B, h_{B'} : \mathbb{Z}^m \to \mathbb{Z}^m$ *are the group isomorphisms defined by the formulae* $h_B(x) = x \cdot B$ *and* $h_{B'}(x) = x \cdot B'$, *for* $x \in \mathbb{Z}^m$.

(b) $\widehat{Cox}_J = B \cdot \overline{Cox}_J \cdot B^{-1}$, $Cox_J^{-1} = Cox_{J^{op}} = B' \cdot \overline{Cox}_J \cdot B'^{-1}$, *and* $\Phi_{J^{op}} = \Phi_J^{-1}$.

(c) *The Coxeter number* $\mathbf{c}_J = ord(\Phi_J)$ *of the poset J coincides with the Coxeter number of* J^{op}. *Moreover,* $\mathbf{c}_J = ord(\widehat{\Phi}_J) = ord(\overline{\Phi}_J)$ *and* $cox_{J^{op}}(t) = cox_J(t)$.

(d) *Assume that J is connected and nonnegative.*

(d1) *If the numbering* $J = \{a_1, \ldots, a_m\}$ *is upper-triangular and* Δ_J *is the bigraph* (33) *associated to J, then* $Cox_{\Delta_J} = Cox_J$ *and* $cox_{\Delta_J}(t) = cox_J(t)$.

(d2) *The Coxeter type* $\mathbf{Ctype}_J := (\mathbf{specc}_J, \mathbf{c}_J, \check{\mathbf{c}}_J)$ *of J does not depend on the numbering* $J = \{a_1, \ldots, a_m\}$.

(d3) *The Coxeter spectrum* $\mathbf{specc}_J$ *is a subset of a unit circle* $\mathcal{S}^1 = \{z \in \mathbb{C}; |z| = 1\}$, *and any* $z \in \mathbf{specc}_J$ *is a root of unity.*

(d4) *The poset J is positive if and only if* $1 \notin \mathbf{specc}_J$.

Proof. The first equality $\widehat{C}_J = B \cdot \overline{C}_J \cdot B^{tr}$ is obvious, and the second one $C_J^{tr} = B' \cdot \overline{C}_J \cdot B'^{tr}$ follows by a direct calculation. Hence, (b) follows and, consequently, the diagrams (36) are commutative; see [1, Proposition 3.13]. Hence, the statement (c) follows from the commutativity of the diagrams (36).

(d1) We recall from Section 1 that, given two numberings $J = \{a_1, \ldots, a_m\}$ and $J' = \{a_1', \ldots, a_m'\}$ of elements in J, we have $C_{J'} = \widehat{\sigma}^{-1} \cdot C_J \cdot \widehat{\sigma}$, where $\widehat{\sigma} \in \mathrm{Gl}(m, \mathbb{Z})$ is the permutation matrix of a permutation $\sigma \in \mathbf{S}_m$. Hence, (d1) easily follows.

(d2) It is sufficient to note that the incidence matrix C_J is upper triangular. Hence, $C_J = \check{G}_{\Delta_J}$ and $\mathrm{Cox}_{\Delta_J} = \mathrm{Cox}_J$.

To prove (d3) and (d4), we recall from [2] and [3, Proposition 2.6] that the Coxeter spectrum $\mathbf{specc}_A$ of any matrix morsification $A \in \mathbf{Mor}_\Delta$ of a nonnegative bigraph Δ is a subset of the unit circle $\mathcal{S}^1$ and any $z \in \mathbf{specc}_A$ is a root of unity (see also [41, 42]). Moreover, Δ is positive iff $1 \notin \mathbf{specc}_A$. Assume that J is connected and nonnegative. Then, the bigraph Δ_J (33) associated to J is nonnegative, $A := \check{G}_{\Delta_J} = \nabla(C_J)$ is a morsification of Δ_J, and $\mathbf{specc}_J = \mathbf{specc}_A$, because the incidence matrix C_J is quasitriangular and [4, Proposition 2.2] applies. This completes the proof.

Corollary 6. *For any poset J, equality* (35) *holds.*

Proof. Apply Proposition 5(b).

The following example shows that the correspondence $J \mapsto \Delta_J$ defined in (33) does not preserve the Coxeter types of J and Δ_J. In particular, it shows that the equality $\mathrm{cox}_J(t) = \mathrm{cox}_{\Delta_J}(t)$ does not hold in general and the Coxeter polynomial $\mathrm{cox}_{\Delta_J}(t)$ depends on the numbering of J, whereas the Coxeter polynomial $\mathrm{cox}_J(t)$ does not depend on the numbering of J.

Example 7. Consider the poset J such that its Hasse quiver has the form

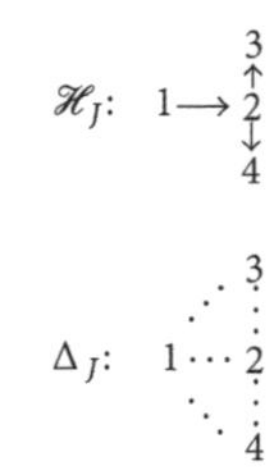

$\mathrm{cox}_J(t) = t^4 + t^3 + t + 1$

$\mathrm{cox}_{\Delta_J}(t) = t^4 + t^3 + t + 1$ (37)

By a permutation of the elements in J, we get

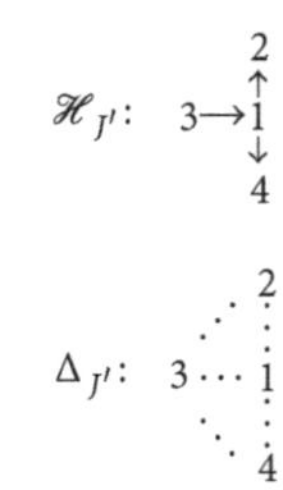

$\mathrm{cox}_{J'}(t) = t^4 + t^3 + t + 1$

$\mathrm{cox}_{\Delta_{J'}}(t) = t^4 + 2t^2 + 1$ (38)

Note that the first numbering is upper-triangular, whereas the second one is not upper-triangular.

3. Principal Posets

We recall that a poset J is *principal* if its incidence unit form q_J is principal in the sense of [34, Definition 2.1]; that is, $q_J : \mathbb{Z}^J \to \mathbb{Z}$ is nonnegative and not positive, and the kernel $\mathrm{Ker}\, q_J := \{v \in \mathbb{Z}^J; q_J(v) = 0\}$ is an infinite cyclic subgroup of $\mathbb{Z}^J$.

We start with the following useful observation.

Lemma 8. *Assume that J is a connected principal poset.*

(a) *The Coxeter number $\mathbf{c}_J$ of J is infinite.*

(b) *The Coxeter spectrum $\mathbf{specc}_J$ is a subset of a unit circle $\mathcal{S}^1 = \{z \in \mathbb{C}; |z| = 1\}$, $1 \in \mathbf{specc}_J$, and any $z \in \mathbf{specc}_J$ is a root of unity.*

(c) *If $\operatorname{Ker} q_J = \mathbb{Z} \cdot \mathbf{h}$, then $\operatorname{Ker} \widehat{q}_J = \mathbb{Z} \cdot \widehat{\mathbf{h}}$ and $\operatorname{Ker} \overline{q}_J = \mathbb{Z} \cdot \overline{\mathbf{h}}$, where*

(i) $\overline{\mathbf{h}} = \mathbf{h} \cdot B'$, $\overline{\mathbf{h}} = \widehat{\mathbf{h}} \cdot B$, $\widehat{\mathbf{h}} = \mathbf{h} \cdot B' \cdot B^{-1}$,

(ii) $B' = C_J^{tr}$, $B = \left[\begin{array}{c|c} C_{\check{J}} & 0 \\ \hline 0 & E \end{array}\right]$, *and* $\check{J} = J \setminus \max J$

are as in Proposition 5.

Proof. (a) By Proposition 5 (d2), $\mathbf{c}_J$ is independent of the numbering of J. Then, without loss of generality, we may suppose that the numbering of J is upper-triangular. Then, by Lemma 3(d) and Proposition 5(d1), the Coxeter number $\mathbf{c}_J$ coincides with the Coxeter number of the principal edge-bipartite graph Δ_J associated with J in (33). Then, (a) is a consequence of [3, Proposition 3.12].

The statements (b) and (c) follow by applying Proposition 5 and the commutative diagram (36).

Proposition 9. *Let J be a connected poset, $m = |J| \geq 2$, and let $G_J, \widehat{G}_J, \overline{G}_J, \in \mathbb{M}_J(\mathbb{Q})$ be the symmetric incidence Gram matrix of J, the symmetric Tits-Gram matrix of J, and the symmetric Euler-Gram matrix of J, respectively. The following five conditions are equivalent.*

(a) *The poset J is principal.*

(b) *The Gram matrix G_J is positive indefinite of rank $m-1$.*

(c) *The Tits quadratic form $\widehat{q}_J$ of J is nonnegative and $\operatorname{Ker} \widehat{q}_J = \mathbb{Z} \cdot \widehat{\mathbf{h}}$, for some nonzero vector $\widehat{\mathbf{h}} \in \mathbb{Z}^J$.*

(d) *The Euler quadratic form $\overline{q}_J$ of J is nonnegative and $\operatorname{Ker} \overline{q}_J = \mathbb{Z} \cdot \overline{\mathbf{h}}$, for some nonzero vector $\overline{\mathbf{h}} \in \mathbb{Z}^J$.*

(e) *If G is any of the symmetric Gram matrices $G_J, \widehat{G}_J, \overline{G}_J, \in \mathbb{M}_J(\mathbb{Q})$ of J, then there exists a simply laced Euclidean diagram $DJ \in \{\widetilde{\mathbb{A}}_s, s \geq 3, \widetilde{\mathbb{D}}_n, n \geq 4, \widetilde{\mathbb{E}}_6, \widetilde{\mathbb{E}}_7, \widetilde{\mathbb{E}}_8\}$ (uniquely determined by J) such that the matrix G is $\mathbb{Z}$-congruent to the symmetric Gram matrix $G_{DJ} \in \mathbb{M}_{DJ}(\mathbb{Q})$ of the Euclidean diagram DJ; that is, there is a $\mathbb{Z}$-invertible matrix $B \in Gl(m, \mathbb{Z})$ such that $G_{DJ} = B^{tr} \cdot G \cdot B$.*

Proof. (a)⇔(b) If $m = |J|$ and

$$Dq_J : \mathbb{Z}^m \longrightarrow \mathbb{Z}^m,$$
$$v \longmapsto Dq_J(v) = \left(\frac{\partial q_J}{\partial x_1}(v), \ldots, \frac{\partial q_J}{\partial x_m}(v)\right), \tag{39}$$

is the gradient group homomorphism of q_J, then $\operatorname{Ker} q_J = \operatorname{Ker}[Dq_J : \mathbb{Z}^m \to \mathbb{Z}^m]$ and the subgroup $\operatorname{Ker} q_J$ of $\mathbb{Z}^m$ is of rank $m - \operatorname{rk} G_J$ and consists of all integral solutions of the system $2 \cdot G_J \cdot x^{tr} = 0$ of linear equations with integral coefficients; see [34, Proposition 2.8]. Then, (a)⇔(b) follows.

The equivalences (a)⇔(c)⇔(d) follow from Proposition 5 (a) and the commutativity of the diagram (36).

(e)⇒(a) Assume that there exist a simply laced Euclidean diagram $DJ \in \{\widetilde{\mathbb{A}}_s, s \geq 3, \widetilde{\mathbb{D}}_n, n \geq 4, \widetilde{\mathbb{E}}_6, \widetilde{\mathbb{E}}_7, \widetilde{\mathbb{E}}_8\}$ and a $\mathbb{Z}$-invertible matrix $B \in \mathrm{Gl}(m, \mathbb{Z})$ such that $G_{DJ} = B^{tr} \cdot G \cdot B$. It follows that the quadratic form $q_{DJ}(x) = x \cdot G_{DJ} \cdot x^{tr}$ is $\mathbb{Z}$-congruent to q_J and $q_J = q_{DJ} \circ h_B$. Then, (a) is a consequence of [36, Lemma VII.4.2].

(a)⇒(e) Let $\overline{\Delta}_J$ be the Euler edge-bipartite graph defined in (33) of J. By (a) and Lemma 3 (d), $\overline{\Delta}_J$ is principal and the inflation algorithm defined in [4, 21] applies to Δ_J. Consequently, there exists a simply laced Euclidean diagram $DJ \in \{\widetilde{\mathbb{A}}_s, s \geq 3, \widetilde{\mathbb{D}}_n, n \geq 4, \widetilde{\mathbb{E}}_6, \widetilde{\mathbb{E}}_7, \widetilde{\mathbb{E}}_8\}$ and a $\mathbb{Z}$-invertible matrix $B \in \mathrm{Gl}(n, \mathbb{Z})$ defining the congruence $\overline{\Delta}_J \approx_{\mathbb{Z}} DJ$; that is, the equality $G_{DJ} = B^{tr} \cdot G_{\overline{\Delta}_J} \cdot B = B^{tr} \cdot G_J \cdot B$ holds. Then, in view of Proposition 5, the implication (a)⇒(e) follows from Lemma 3 (d); see also Section 6.

Theorem 10. *Let J be a finite principal poset, with a numbering $\{a_1, \ldots, a_m\}$ of elements of J. Fix a nonzero vector $\mathbf{h}_J \in \mathbb{Z}^J \equiv \mathbb{Z}^m$ such that $\operatorname{Ker} q_J = \mathbb{Z} \cdot \mathbf{h}_J$.*

(a) *There exist a minimal integer $\check{\mathbf{c}}_J \geq 2$ (called the reduced Coxeter number of J) and a group homomorphism $\partial_J : \mathbb{Z}^J \to \mathbb{Z}$ (called the incidence defect of J) such that*

$$\Phi_J^{\check{\mathbf{c}}_J}(v) = v + \partial_J(v) \cdot \mathbf{h}_J,$$
$$\partial_J(\Phi_J(v)) = \partial_J(v), \quad \forall v \in \mathbb{Z}^J, \tag{40}$$
$$\partial_J(\mathbf{h}_J) = 0.$$

(b) *Assume that $\check{\mathbf{c}}_J \geq 1$ and $\partial_J : \mathbb{Z}^J \to \mathbb{Z}$ are as in (a), and one sets $\overline{\mathbf{h}}_J = \mathbf{h}_J \cdot B'$, $\widehat{\mathbf{h}}_J = \mathbf{h}_J \cdot B' \cdot B^{-1}$, where $B', B \in \mathbb{M}_J(\mathbb{Z})$ are as in Proposition 5.*

(b1) *There exists a group homomorphism $\overline{\partial}_J : \mathbb{Z}^J \to \mathbb{Z}$ (called the Euler defect of J) such that*

$$\overline{\Phi}_J^{\check{\mathbf{c}}_J}(v) = v + \overline{\partial}_J(v) \cdot \overline{\mathbf{h}}_J, \quad \forall v \in \mathbb{Z}^J,$$
$$\overline{\partial}_J \circ \Phi_J = \overline{\partial}_J, \tag{41}$$
$$\overline{\partial}_J \circ h_{B'} = \partial_J, \quad \overline{\partial}_J\left(\overline{\mathbf{h}}_J\right) = 0.$$

(b2) *There exists a group homomorphism $\widehat{\partial}_J : \mathbb{Z}^J \to \mathbb{Z}$ (called the Tits defect of J) such that*

$$\widehat{\Phi}_J^{\check{\mathbf{c}}_J}(v) = v + \widehat{\partial}_J(v) \cdot \widehat{\mathbf{h}}_J, \quad \forall v \in \mathbb{Z}^J,$$
$$\widehat{\partial}_J = \overline{\partial}_J \circ h_B = \partial_J \circ h_{B'}^{-1} \circ h_B, \tag{42}$$
$$\widehat{\partial}_J = \widehat{\partial}_J \circ \Phi_J, \quad \widehat{\partial}_J\left(\widehat{\mathbf{h}}_J\right) = 0.$$

(c) *The Coxeter number* $\mathbf{c}_J$ *of* J *is infinite, and the incidence defect* $\partial_J : \mathbb{Z}^J \to \mathbb{Z}$ *is nonzero.*

(d) *Given* $v \in \mathbb{Z}^J$, *the order* $\mathbf{s}_v := |\mathcal{O}(v)|$ *of the* Φ_J*-orbit* $\mathcal{O}(v)$ *is finite if and only if* $\partial_J(v) = 0$. *If* $\mathbf{s}_v = |\mathcal{O}(v)|$ *is finite, then* $\mathbf{s}_v$ *divides* $\check{\mathbf{c}}_J$ *and there is a unique integer* m_v *such that*

$$m_v \cdot \mathbf{h} = v + \Phi_J(v) + \Phi_J^2(v) + \cdots + \Phi_J^{\mathbf{s}_v-1}(v)$$
$$= v + \Phi_J^{-1}(v) + \Phi_J^{-2}(v) + \cdots + \Phi_J^{-\mathbf{s}_v+1}(v). \quad (43)$$

Proof. We recall from the proof of Proposition 9 that

$$\mathbb{Z} \cdot \mathbf{h}_J = \operatorname{Ker} q_J = \operatorname{Ker}\left[Dq_J : \mathbb{Z}^m \longrightarrow \mathbb{Z}^m\right], \quad (44)$$

where $m = |J| \geq 2$ and $Dq_J : \mathbb{Z}^m \to \mathbb{Z}^m$, $v \mapsto Dq_J(v) = ((\partial q_J/\partial x_1)(v), \ldots, (\partial q_J/\partial x_m)(v))$, is the gradient group homomorphism. It follows that $\mathbb{Z}^m/\mathbb{Z}\cdot\mathbf{h}_J \cong \operatorname{Im} Dq_J \cong \mathbb{Z}^{m-1}$. Denote by $\phi : \mathbb{Z}^m \to \mathbb{Z}^{m-1}$ the composite quotient epimorphism. Then, the form q_J induces the form $\widetilde{q}_J : \mathbb{Z}^{m-1} \to \mathbb{Z}$ such that $\widetilde{q}_J(\phi(x)) = q_J(x)$, for all $x \in \mathbb{Z}^m$. Moreover, the Coxeter transformation $\Phi_J : \mathbb{Z}^m \to \mathbb{Z}^m$ induces a group automorphism $\widetilde{\Phi}_J : \mathbb{Z}^{m-1} \to \mathbb{Z}^{m-1}$ such that

$$\widetilde{\Phi}_J \circ \phi = \phi \circ \Phi_J, \qquad \widetilde{q}_J\left(\widetilde{\Phi}_J(y)\right) = \widetilde{q}_J(y), \quad \forall y \in \mathbb{Z}^{m-1}. \quad (45)$$

It follows that $\widetilde{q}_J$ is positive definite and there exists a minimal integer $\check{\mathbf{c}}_J \geq 1$ such that $\widetilde{\Phi}_J^{\check{\mathbf{c}}_J}$ is the identity map on $\mathbb{Z}^{m-1}$. Hence, (a) follows, because the equalities $\partial_J(\mathbf{h}_J) = 0$ and $\partial_J(\Phi_J(v)) = \partial_J(v)$, for all $v \in \mathbb{Z}^J$, are almost obvious; see [34, Theorem 4.7].

In view of Proposition 5, the statements (b)–(d) are a consequence of (a) and Lemma 8(a). The reader is referred to [34, Theorem 4.7, Corollary 4.15] for more details.

Corollary 11. *(a) If* J *is a principal connected poset with at most 13 elements, then its Coxeter spectrum* $\mathbf{specc}_J$ *is a subset of a unit circle* $\mathcal{S}^1 = \{z \in \mathbb{C}; |z| = 1\}$, $1 \in \mathbf{specc}_J$, *and any* $z \in \mathbf{specc}_J$ *is an mth root of unity, where* $m \leq \check{\mathbf{c}}_J$ *and* $\check{\mathbf{c}}_J$ *is the reduced Coxeter number of* J.

(b) If I *and* J *are one-peak principal posets with at most 13 elements and* DI, DJ *are the associated Euclidean diagrams, then the following conditions are equivalent:*

(b1) $DI = DJ$,

(b2) $\mathbf{specc}_I = \mathbf{specc}_J$,

(b3) $\check{\mathbf{c}}_I = \check{\mathbf{c}}_J$ *and* $|I| = |J|$,

(b4) *the incidence matrix* $C_J \in \mathbb{M}_J(\mathbb{Z})$ *is* $\mathbb{Z}$*-congruent to the incidence matrix* $C_I \in \mathbb{M}_I(\mathbb{Z})$; *that is, there is a* $\mathbb{Z}$*-invertible matrix* B *such that* $C_I = B^{tr} \cdot C_J \cdot B$.

Proof. (a) By Lemma 8, $\mathbf{specc}_J \subseteq \mathcal{S}^1$ and $1 \in \mathbf{specc}_J$. Assume that DJ is the associated Euclidean diagram of J, as in Proposition 9. By a computer search (using the results of [43] and the inflation algorithm given in [4, 21]), we show that

$$\operatorname{cox}_J(t) = \operatorname{cox}_{DJ}(t), \quad \check{\mathbf{c}}_J = \check{\mathbf{c}}_{DJ}, \quad (46)$$

for any poset J, with at most 13 elements. Hence, in view of [4, Proposition 2.17], we have

$$\operatorname{cox}_J(t) = \operatorname{cox}_{DJ}(t) = F_{DJ}(t), \quad (47)$$

where

$$\begin{aligned}
F_{DJ}(t) = {} & \\
& t^{n+1} + t^n - t^{n-1} - t^{n-2} - t^3 - t^2 + t + 1 \\
& = (t-1)^2 \mathfrak{v}_2^2(t)\, \mathfrak{v}_{n-2}(t), \quad \text{for } DJ = \widetilde{\mathbb{D}}_n, \\
& t^7 + t^6 - 2t^4 - 2t^3 + t + 1 \\
& = (t-1)^2 \mathfrak{v}_2(t)\, \mathfrak{v}_3^2(t), \quad \text{for } DJ = \widetilde{\mathbb{E}}_6, \\
& t^8 + t^7 - t^5 - 2t^4 - t^3 + t + 1 \\
& = (t-1)^2 \mathfrak{v}_2(t)\, \mathfrak{v}_3(t)\, \mathfrak{v}_4(t), \quad \text{for } DJ = \widetilde{\mathbb{E}}_7, \\
& t^9 + t^8 - t^6 - t^5 - t^4 - t^3 + t + 1 \\
& = (t-1)^2 \mathfrak{v}_2(t)\, \mathfrak{v}_3(t)\, \mathfrak{v}_5(t), \quad \text{for } DJ = \widetilde{\mathbb{E}}_8,
\end{aligned} \quad (48)$$

where $\mathfrak{v}_m(t) = t^{m-1} + t^{m-2} + t^{m-3} + \cdots + t^2 + t + 1$. For $DJ \in \{\widetilde{\mathbb{D}}_4, \widetilde{\mathbb{D}}_5\}$, we have

$$F_{DJ}(t) = \begin{cases} t^5 + t^4 - 2t^3 - 2t^2 + t + 1, & \text{for } DJ = \widetilde{\mathbb{D}}_4, \\ t^6 + t^5 - t^4 - 2t^3 - t^2 + t + 1, & \text{for } DJ = \widetilde{\mathbb{D}}_5. \end{cases} \quad (49)$$

Then, (a) follows by applying [38, Lemma XIII.1.3]. Hence, we also easily conclude that the statements (b1)–(b3) are equivalent.

To finish the proof of (b), we note that the equality $C_I = B^{\text{tr}} \cdot C_J \cdot B$ in (b4) implies that the matrices Cox_I and Cox_J are conjugate, and, hence, we get $\mathbf{specc}_I = \mathbf{specc}_J$; that is, the implication (b4)⇒(b2) holds. To prove the inverse implication (b2)⇒(b4), we apply the technique used in [18, Section 6]. On this way, given a principal poset J, with at most 13 elements and the associated Euclidean diagram DJ, we construct (by a computer search) a $\mathbb{Z}$-invertible matrix B_J such that $\check{G}_{DJ} = B_J^{\text{tr}} \cdot C_J \cdot B_J$ (compare with [17, 18, 33, 43]). Hence, (b4) follows, and the proof is complete.

If J is a principal poset, then the sets

$$\begin{aligned}
\mathcal{R}_{q_J} &= \{v \in \mathbb{Z}^m;\ q_J(v) = 1\}, \\
\mathcal{R}_{\widehat{q}_J} &= \{v \in \mathbb{Z}^m;\ \widehat{q}_J(v) = 1\}, \\
\mathcal{R}_{\overline{q}_J} &= \left\{v \in \mathbb{Z}^m;\ \overline{q}_J(v) = 1\right\}
\end{aligned} \quad (50)$$

of roots of the unit forms q_J, $\widehat{q}_J$, and $\overline{q}_J$ have the disjoint union decompositions

$$\begin{aligned}
\mathcal{R}_{q_J} &= \partial_J^- \mathcal{R}_{q_J} \cup \partial_J^+ \mathcal{R}_{q_J} \cup \partial_J^0 \mathcal{R}_{q_J}, \\
\mathcal{R}_{\widehat{q}_J} &= \widehat{\partial}_J^- \mathcal{R}_{\widehat{q}_J} \cup \widehat{\partial}_J^+ \mathcal{R}_{\widehat{q}_J} \cup \widehat{\partial}_J^0 \mathcal{R}_{\widehat{q}_J}, \\
\mathcal{R}_{\overline{q}_J} &= \overline{\partial}_J^- \mathcal{R}_{\overline{q}_J} \cup \overline{\partial}_J^+ \mathcal{R}_{\overline{q}_J} \cup \overline{\partial}_J^0 \mathcal{R}_{\overline{q}_J},
\end{aligned} \tag{51}$$

where

$$\begin{aligned}
\partial_J^- \mathcal{R}_{q_J} &= \left\{ v \in \mathcal{R}_{q_J};\ \partial_J(v) < 0 \right\}, \\
\partial_J^+ \mathcal{R}_{q_J} &= \left\{ v \in \mathcal{R}_{q_J};\ \partial_J(v) > 0 \right\}, \\
\partial_J^0 \mathcal{R}_{q_J} &= \left\{ v \in \mathcal{R}_{q_J};\ \partial_J(v) = 0 \right\}; \\
\widehat{\partial}_J^- \mathcal{R}_{\widehat{q}_J} &= \left\{ v \in \mathcal{R}_{\widehat{q}_J};\ \widehat{\partial}_J(v) < 0 \right\}, \\
\widehat{\partial}_J^+ \mathcal{R}_{\widehat{q}_J} &= \left\{ v \in \mathcal{R}_{\widehat{q}_J};\ \widehat{\partial}_J(v) > 0 \right\}, \\
\widehat{\partial}_J^0 \mathcal{R}_{\widehat{q}_J} &= \left\{ v \in \mathcal{R}_{\widehat{q}_J};\ \widehat{\partial}_J(v) = 0 \right\}; \\
\overline{\partial}_J^- \mathcal{R}_{\overline{q}_J} &= \left\{ v \in \mathcal{R}_{\overline{q}_J};\ \overline{\partial}_J(v) < 0 \right\}, \\
\overline{\partial}_J^+ \mathcal{R}_{\overline{q}_J} &= \left\{ v \in \mathcal{R}_{\overline{q}_J};\ \overline{\partial}_J(v) > 0 \right\}, \\
\overline{\partial}_J^0 \mathcal{R}_{\overline{q}_J} &= \left\{ v \in \mathcal{R}_{\overline{q}_J};\ \overline{\partial}_J(v) = 0 \right\}.
\end{aligned} \tag{52}$$

Note that the group isomorphism $\mathbb{Z}^J \to \mathbb{Z}^J$, $v \mapsto \widehat{v} := -v$, restricts to the bijections

$$\begin{gathered}
\partial_J^- \mathcal{R}_{q_J} \xrightarrow[1-1]{\simeq} \partial_J^+ \mathcal{R}_{q_J}, \qquad \widehat{\partial}_J^- \mathcal{R}_{\widehat{q}_J} \xrightarrow[1-1]{\simeq} \widehat{\partial}_J^+ \mathcal{R}_{\widehat{q}_J}, \\
\overline{\partial}_J^- \mathcal{R}_{\overline{q}_J} \xrightarrow[1-1]{\simeq} \overline{\partial}_J^+ \mathcal{R}_{\overline{q}_J}.
\end{gathered} \tag{53}$$

Example 12. We compute the reduced Coxeter number, the Coxeter polynomial, and the Euler defect of the following principal two-peak poset

J:
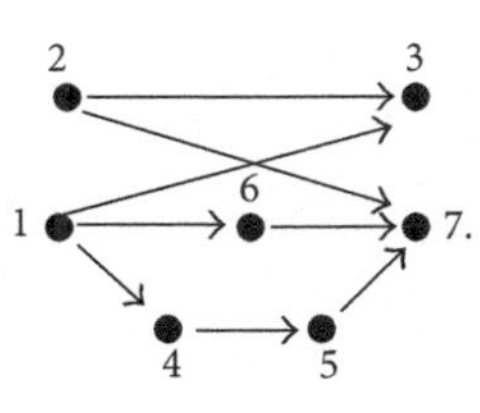

(54)

Note that J is principal, because

$$\begin{aligned}
\overline{q}_J(x) &= x_1^2 + x_2^2 + x_3^2 + x_4^2 + x_5^2 + x_6^2 \\
&\quad + x_7^2 - (x_1 + x_2) x_3 - (x_1 + x_5) x_4 \\
&\quad - x_1 x_6 + (x_1 - x_2 - x_5 - x_6) x_7 \\
&= \left(x_1 - \frac{1}{2}x_4 - \frac{1}{2}x_5 - \frac{1}{2}x_6 + \frac{1}{2}x_7 \right)^2 \\
&\quad + \left(x_2 - \frac{1}{2}x_3 - \frac{1}{2}x_7 \right)^2 \\
&\quad + \frac{5}{12}\left(x_3 - \frac{2}{5}x_5 - \frac{4}{5}x_6 + \frac{1}{5}x_7 \right)^2 \\
&\quad + \frac{3}{4}\left(-\frac{1}{3}x_3 + x_4 - \frac{2}{3}x_5 - \frac{1}{3}x_6 + \frac{1}{3}x_7 \right)^2 \\
&\quad + \frac{3}{5}\left(x_5 - \frac{1}{2}x_6 - \frac{1}{2}x_7 \right)^2 + \frac{1}{4}(x_6 - x_7)^2.
\end{aligned} \tag{55}$$

It follows that $\overline{q}_J$ is nonnegative and $\operatorname{Ker} \overline{q}_J = \mathbb{Z} \cdot \mathbf{h}$, where $\mathbf{h} = (1, 1, 1, 1, 1, 1, 1)$; $\overline{q}_J$ is critical in the sense of Ovsienko [24]; see also [38, 44]. Note that the Euler matrix $\overline{C}_J = C_J^{-1}$ of J and the inverse of the Coxeter-Euler matrix $\overline{\mathrm{Cox}}_J := -C_J^{-1} \cdot C_J^{tr}$ have the forms

$$\begin{gathered}
\overline{C}_J = C_J^{-1} = \begin{bmatrix} 1 & 0 & -1 & -1 & 0 & -1 & 1 \\ 0 & 1 & -1 & 0 & 0 & 0 & -1 \\ 0 & 0 & 1 & 0 & 0 & 0 & 0 \\ 0 & 0 & 0 & 1 & -1 & 0 & 0 \\ 0 & 0 & 0 & 0 & 1 & 0 & -1 \\ 0 & 0 & 0 & 0 & 0 & 1 & -1 \\ 0 & 0 & 0 & 0 & 0 & 0 & 1 \end{bmatrix}, \\
\overline{\mathrm{Cox}}_J^{-1} = \begin{bmatrix} -1 & 0 & -1 & -1 & -1 & -1 & -1 \\ 0 & -1 & -1 & 0 & 0 & 0 & -1 \\ 1 & 1 & 1 & 1 & 1 & 1 & 2 \\ 1 & 0 & 1 & 0 & 0 & 1 & 0 \\ 0 & 0 & 0 & 1 & 0 & 0 & 0 \\ 1 & 0 & 1 & 1 & 1 & 0 & 0 \\ -1 & 1 & 0 & -1 & 0 & 0 & 1 \end{bmatrix}.
\end{gathered} \tag{56}$$

Moreover, we have $G_{\widetilde{\mathbb{E}}_6} = B^{tr} \cdot \overline{G}_J \cdot B$, and the matrix $A := B^{tr} \cdot \overline{C}_J \cdot B$ is a morsification of the Euclidean diagram $\widetilde{\mathbb{E}}_6$ (see [34, 40]), where

$$\begin{gathered}
B = \begin{bmatrix} -2 & 0 & 1 & 0 & 0 & 0 & 0 \\ 0 & 0 & 0 & 0 & 0 & 0 & 1 \\ -1 & 0 & 0 & 0 & 0 & 1 & 0 \\ -1 & 0 & 0 & 0 & 1 & 0 & 0 \\ 0 & 0 & 0 & 1 & 0 & 0 & 0 \\ -1 & 1 & 0 & 0 & 0 & 0 & 0 \\ 1 & 0 & 0 & 0 & 0 & 0 & 0 \end{bmatrix}, \\
A = \begin{bmatrix} 1 & 1 & -2 & 1 & 1 & 1 & 0 \\ -2 & 1 & 0 & 0 & 0 & 0 & 0 \\ 2 & -1 & 1 & 0 & -1 & -1 & 0 \\ -1 & 0 & 0 & 1 & 0 & 0 & 0 \\ -1 & 0 & 0 & -1 & 1 & 0 & 0 \\ -1 & 0 & 0 & 0 & 0 & 1 & 0 \\ 0 & 0 & 0 & 0 & 0 & -1 & 1 \end{bmatrix}.
\end{gathered} \tag{57}$$

Hence, in view of [2, Proposition 2.8], we get the following:

(i) the Euclidean type DJ of J is the diagram $\widetilde{\mathbb{E}}_6$, and the Coxeter polynomial of the poset J has the form

$$\operatorname{cox}_J(t) = t^7 - t^5 - t^2 + 1 = \operatorname{cox}_A(t); \tag{58}$$

that is, $\operatorname{cox}_J(t)$ is the Coxeter polynomial $F_\Delta^{(2)}(t)$ (21), of the Euclidean diagram $\Delta = \widetilde{\mathbb{A}}_6$ (with a particular numbering of vertices), and is the Coxeter polynomial $\operatorname{cox}_A(t)$ of the morsification $A \in \mathbb{M}_7(\mathbb{Z})$ of the diagram $\widetilde{\mathbb{E}}_6$,

(ii) the Coxeter number $\mathbf{c}_J$ is infinite and the reduced Coxeter number $\check{\mathbf{c}}_J$ equals 10,

(iii) the Euler defect has the form $\overline{\partial}_J(x) = 3(x_1 + x_2 - x_3 - x_7)$,

(iv) the $\overline{\Phi}_J$-orbit of any vector of defect zero in $\overline{\partial}_J^0\mathcal{R}_{\widehat{q}_J}$ is of length 2 or of length 5. It is shown in [1, Remark 4.5] and [34, Example 5.14] that they lie on one sand-glass tube $\mathcal{T}_{1,2}$ of rank 2 and on six sand-glass tubes of rank five.

4. Nonnegative Posets of Positive Corank

In the study of nonnegative posets, the following extensions of [34, Definition 2.2] are of importance.

Definition 13. Assume that $m \geq 2$, $r \geq 0$, and $q : \mathbb{Z}^m \to \mathbb{Z}$ is a unit quadratic form.

(a) The form q is defined to be *nonnegative of corank* $r \geq 0$, if q is nonnegative and the $\mathbb{Q}$-rank $\operatorname{rk}_\mathbb{Q} G_q$ of the rational Gram matrix $G_q \in \mathbb{M}_m(\mathbb{Q})$ equals $\operatorname{rk}_\mathbb{Q} G_q = m - r$.

(b) The form q is defined to be *nonnegative critical of corank* $r \geq 1$, if q is nonnegative of corank $r \geq 1$ and each of the nonnegative quadratic forms $q^{(1)}, \ldots, q^{(m)} : \mathbb{Z}^{m-1} \to \mathbb{Z}$ is of corank at most $r - 1 \geq 0$, where

$$\begin{aligned} q^{(j)}&\left(x_1, \ldots, x_{j-1}, x_{j+1}, \ldots, x_m\right) \\ &= q\left(x_1, \ldots, x_{j-1}, 0, x_{j+1}, \ldots, x_m\right). \end{aligned} \tag{59}$$

Lemma 14. *Assume that $m \geq 2$, $r \geq 0$, and $q : \mathbb{Z}^m \to \mathbb{Z}$ is an integral quadratic form.*

(a) *q is nonnegative of corank $r \geq 0$ if and only if q is nonnegative and the subgroup $\operatorname{Ker} q$ of the abelian group $\mathbb{Z}^m$ is free of rank r.*

(b) *q is nonnegative of corank $r = 0$ if and only if q is positive, and q is nonnegative of corank one if and only if q is principal.*

(c) *q is nonnegative critical of corank $r \geq 1$ if and only if q is nonnegative and, for any $j \in \{1, \ldots, m\}$, the abelian subgroup $\mathbb{Z}^{m,j} \cap \operatorname{Ker} q$ of $\mathbb{Z}^m$ is free of rank at most $r - 1$, where*

$$\begin{aligned} \mathbb{Z}^{m,j} &:= \mathbb{Z}^{m-j-1} \times \{0\} \times \mathbb{Z}^{j-1} \\ &= \left\{v = (v_1, \ldots, v_m) \in \mathbb{Z}^m;\ v_j = 0\right\} \subseteq \mathbb{Z}^m \end{aligned} \tag{60}$$

is viewed as a subgroup of $\mathbb{Z}^m$.

(d) *q is nonnegative critical of corank $r = 1$ if and only if q is P-critical in the sense of [34, Definition 2.2] and [44].*

Proof. The proof of (a) follows by applying the arguments used in the proof of the equivalence (a)⇔(b) in Proposition 9. The statements (b) and (c) follow from (a).

(c) First, we note that the quadratic forms $q^{(1)}, \ldots, q^{(m)} : \mathbb{Z}^{m-1} \to \mathbb{Z}$ are nonnegative, if q is nonnegative. Then, (c) is a consequence of the group isomorphism

$$\begin{aligned} \operatorname{Ker} q^{(j)} &\xrightarrow{\simeq} \mathbb{Z}^{m,j} \cap \operatorname{Ker} q, \\ w \longmapsto \widehat{w}^{(j)} &:= \left(w_1, \ldots, w_{j-1}, 0, w_{j+1}, \ldots, w_m\right). \end{aligned} \tag{61}$$

Since (d) is a consequence of (c), the proof is complete.

Definition 15. Assume that J is a connected poset and $q_J, \widehat{q}_J : \mathbb{Z}^J \to \mathbb{Z}$ are its incidence and Tits quadratic forms (6), respectively.

(a) J is defined to be *nonnegative of corank* $r \geq 0$ if its incidence quadratic form $q_J : \mathbb{Z}^J \to \mathbb{Z}$ (resp., one of the forms $\widehat{q}_J$ and $\overline{q}_J$) is nonnegative and the free abelian subgroup $\operatorname{Ker} q_J$ of $\mathbb{Z}^J$ is of $\mathbb{Z}$-rank r (resp., $\operatorname{Ker} \widehat{q}_J \cong \operatorname{Ker} \overline{q}_J \cong \operatorname{Ker} q_J$ is of $\mathbb{Z}$-rank r); see (36).

(b) J is defined to be *nonprincipal critical* if the incidence quadratic form $q_J : \mathbb{Z}^J \to \mathbb{Z}$ is nonnegative and not positive, J is not principal, and the quadratic form $q_{J'} : \mathbb{Z}^{J'} \to \mathbb{Z}$ is principal or positive, for every proper subposet J' of J.

(c) A one-peak poset I, with $\max I = \{*\}$, is defined to be *nonprincipal Tits-critical* if the Tits quadratic form $\widehat{q}_I : \mathbb{Z}^I \to \mathbb{Z}$ is nonnegative and not positive, I is not principal, and the Tits quadratic form $\widehat{q}_{I'} : \mathbb{Z}^{I'} \to \mathbb{Z}$ is principal or positive, for every proper subposet I' of

I containing the peak $*$. We call a nonprincipal Tits-critical poset I *exceptional*, if the subposet $T = I \setminus \{*\}$ is nonprincipal Tits-critical; see [33, 34].

(d) A poset J is defined to be *P-hypercritical* if J is not nonnegative and each of its proper subposet is nonnegative; see [34, Definition 2.2].

Remark 16. Assume that T is a poset and $T^* = T \cup \{*\}$ is its one-peak enlargement.

(a) If T^* is P-hypercritical, then T is NP-critical in the sense of [14], but not conversely.

(b) By [43], many of the NP-critical posets T listed in [14, Table 2] are of corank at most two.

(c) A Coxeter spectral classification of one-peak positive (resp., almost Tits P-critical) posets is given in [17, 18] (resp., in [33]).

We frequently use the following important characterisation.

Theorem 17. *Assume that J is a connected poset and $q_J, \widehat{q}_J : \mathbb{Z}^J \to \mathbb{Z}$ are the incidence and the Tits quadratic forms of J (7), respectively.*

(a) *If J is nonnegative of corank two, then J contains at least 6 elements, and $|J| = 6$ if and only if J is the garland*

$\mathcal{G}_3$: 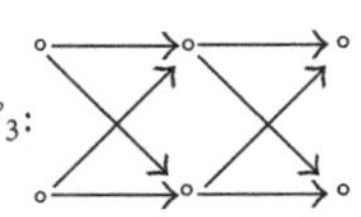

(62)

with $\mathrm{cox}_{\mathcal{G}_3}(t) = t^6 + 2t^5 - t^4 - 4t^3 - t^2 + 2t + 1$

and $\mathrm{Ker}\, q_{\mathcal{G}_3} := \mathbb{Z} \cdot \mathbf{h}^{(1)} \oplus \mathbb{Z} \cdot \mathbf{h}^{(2)}$, *where* $\mathbf{h}^{(1)} = (1, 1, -1, -1, 0, 0)$ *and* $\mathbf{h}^{(2)} = (1, 1, 0, 0, -1, -1)$. *The garland $\mathcal{G}_3$ is nonprincipal critical.*

(b) *The following four conditions are equivalent.*

(b1) *The poset J is nonprincipal critical.*

(b2) *$|J| \geq 6$ and the form $q_J : \mathbb{Z}^J \to \mathbb{Z}$ is nonnegative critical of corank two.*

(b3) *$|J| \geq 6$ and $q_J : \mathbb{Z}^J \to \mathbb{Z}$ is nonnegative, the group* $\mathrm{Ker}\, q_J$ *is of $\mathbb{Z}$-rank two, and for any $j \in J$, the subposet $J^{(j)} := J \setminus \{j\}$ of J is principal or positive.*

(b4) *$|J| \geq 6$ and $q_J : \mathbb{Z}^J \to \mathbb{Z}$ is nonnegative, and the group* $\mathrm{Ker}\, q_J$ *has a $\mathbb{Z}$-basis $\mathbf{h}, \mathbf{h}'$ such that there is no $j \in J$, with $h_j = h'_j = 0$.*

(c) *Let I be a one-peak poset I, with $\max I = \{*\}$. The following three conditions are equivalent.*

(c1) *I is nonprincipal Tits-critical.*

(c2) *$|I| \geq 7$, the Tits form $\widehat{q}_I : \mathbb{Z}^I \to \mathbb{Z}$ is nonnegative, the group* $\mathrm{Ker}\, \widehat{q}_I$ *is of $\mathbb{Z}$-rank two, and for any $j \in I \setminus \{*\}$, the one-peak subposet $I^{(j)} := I \setminus \{j\}$ of I is principal or positive.*

(c3) *$|I| \geq 7$ and $\widehat{q}_I : \mathbb{Z}^I \to \mathbb{Z}$ is nonnegative, and the group* $\mathrm{Ker}\, \widehat{q}_I$ *has a $\mathbb{Z}$-basis $\mathbf{h}, \mathbf{h}'$ such that there is no $j \in I \setminus \{*\}$, with $h_j = h'_j = 0$.*

(d) *A nonprincipal Tits-critical one-peak poset I, with $\max I = \{*\}$ and $|I| = 7$, is exceptional if and only if I is the one-peak garland*

$\mathcal{G}_3^*$: 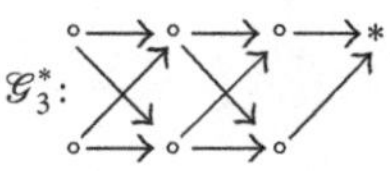

(63)

with $\mathrm{cox}_{\mathcal{G}_3^*}(t) = t^7 + t^6 - t^5 - t^4 - t^3 - t^2 + t + 1$

and $\mathrm{Ker}\, \widehat{q}_{\mathcal{G}_3^*} := \mathbb{Z} \cdot \widehat{\mathbf{h}}^{(1)} \oplus \mathbb{Z} \cdot \widehat{\mathbf{h}}^{(2)}$, *where* $\widehat{\mathbf{h}}^{(1)} = (1, 1, -1, -1, 0, 0, 0)$, $\widehat{\mathbf{h}}^{(2)} = (1, 1, 0, 0, -1, -1, 0)$.

Proof. (a) It is easy to check that any poset J with at most 5 elements is either positive or principal. Moreover, if J is nonnegative of corank two and $|J| = 6$, then J is the garland $\mathcal{G}_3$. Since

$$q_{\mathcal{G}_3}(x) = x_1^2 + x_2^2 + x_3^2 + x_4^2 + x_5^2 + x_6^2 + (x_1 + x_2)(x_3 + x_4 + x_5 + x_6) + (x_3 + x_4)(x_5 + x_6), \tag{64}$$

the Lagrange's algorithm yields

$$q_{\mathcal{G}_3}(x) = \left(x_1 + \frac{1}{2}x_3 + \frac{1}{2}x_4 + \frac{1}{2}x_5 + \frac{1}{2}x_6\right)^2 + \left(x_2 + \frac{1}{2}x_3 + \frac{1}{2}x_4 + \frac{1}{2}x_5 + \frac{1}{2}x_6\right)^2 + \frac{1}{2}(x_3 - x_4)^2 + \frac{1}{2}(x_5 - x_6)^2. \tag{65}$$

It follows that $q_{\mathcal{G}_3} : \mathbb{Z}^6 \to \mathbb{Z}$ is nonnegative and its kernel is a rank-two free abelian group of the form shown in (a). Hence, (a) follows.

(b) We show by a computer search that there is no nonprincipal critical poset J such that $|J| \leq 5$. Then, in view of Lemma 14, the equivalences (b1)⇔(b2)⇔(b3)⇔(b4) easily follow.

(c) We show by a computer search that there is no one-peak nonprincipal Tits-critical poset I such that $|I| \leq 6$. Then, in view of Lemma 14, the equivalences (c1)⇔(c2)⇔(c3) easily follow.

(d) Note that

$$\begin{aligned}\widehat{q}_{\mathscr{G}_3^*}(x) &= x_1^2+x_2^2+x_3^2+x_4^2+x_5^2+x_6^2+x_7^2\\&+(x_1+x_2)(x_3+x_4+x_5+x_6)\\&+(x_3+x_4)(x_5+x_6)\\&-(x_1+x_2+x_3+x_4+x_5+x_6)x_7,\end{aligned}\tag{66}$$

and the Lagrange's algorithm yields

$$\begin{aligned}\widehat{q}_{\mathscr{G}_3^*}(x) &= \left(x_1+\frac{1}{2}x_3+\frac{1}{2}x_4+\frac{1}{2}x_5+\frac{1}{2}x_6-\frac{1}{2}x_7\right)^2\\&+\left(x_2+\frac{1}{2}x_3+\frac{1}{2}x_4+\frac{1}{2}x_5+\frac{1}{2}x_6-\frac{1}{2}x_7\right)^2\\&+\frac{1}{2}(x_3-x_4)^2+\frac{1}{2}(x_5-x_6)^2+\frac{1}{2}x_7^2.\end{aligned}\tag{67}$$

It follows that $\widehat{q}_{\mathscr{G}_3^*}:\mathbb{Z}^7\to\mathbb{Z}$ is nonnegative and its kernel is a rank-two free abelian group of the form shown in (d). Hence, the one-peak garland $\mathscr{G}_3^*$ is nonprincipal Tits-critical and exceptional. On the other hand, one shows by a computer search that $\mathscr{G}_3^*$ is the only one-peak poset that is nonprincipal Tits-critical and exceptional. This finishes the proof.

Following [34, Section 4], we will study nonnegative posets J of corank $r\geq 2$ by means of the spectrum $\mathbf{specc}_J$, the reduced Coxeter number $\check{\mathbf{c}}_J$, and the rank $r\geq 2$ defects

$$\begin{aligned}&\partial_J=\left(\partial_J^{(1)},\ldots,\partial_J^{(r)}\right),\qquad \overline{\partial}_J=\left(\overline{\partial}_J^{(1)},\ldots,\overline{\partial}_J^{(r)}\right),\\&\widehat{\partial}_J=\left(\widehat{\partial}_J^{(1)},\ldots,\widehat{\partial}_J^{(r)}\right):\mathbb{Z}^J\longrightarrow\mathbb{Z}^r\end{aligned}\tag{68}$$

defined in the following extension of Theorem 10.

Theorem 18. *Let J be a finite nonnegative poset of corank $r\geq 2$, and let $m=|J|\geq 2$. One fixes nonzero vectors $\mathbf{h}_J^{(1)},\ldots,\mathbf{h}_J^{(r)}\in\mathbb{Z}^J$ such that $\operatorname{Ker}q_J=\mathbb{Z}\cdot\mathbf{h}_J^{(1)}\oplus\cdots\oplus\mathbb{Z}\cdot\mathbf{h}_J^{(r)}\cong\mathbb{Z}^r$, and one sets $\mathbf{h}_J=(\mathbf{h}_J^{(1)},\ldots,\mathbf{h}_J^{(r)})$.*

(a) *There exist a minimal integer $\check{\mathbf{c}}_J\geq 1$ (called the reduced Coxeter number of J) and a group homomorphism $\partial_J=(\partial_J^{(1)},\ldots,\partial_J^{(r)}):\mathbb{Z}^J\to\mathbb{Z}^r\cong\operatorname{Ker}q_J$ (called the incidence defect of J) such that*

$$\begin{aligned}\Phi_J^{\check{\mathbf{c}}_J}(v)&=v+\partial_J(v)\bullet\mathbf{h}_J\\&=v+\partial_J^{(1)}(v)\cdot\mathbf{h}_J^{(1)}+\cdots+\partial_J^{(r)}(v)\cdot\mathbf{h}_J^{(r)},\\ \partial_J(\Phi_J(v))&=\partial_J(v),\quad\forall v\in\mathbb{Z}^J,\end{aligned}\tag{69}$$

and $\partial_J(\mathbf{h})=0$, for all $\mathbf{h}\in\operatorname{Ker}q_J$, where one sets

$$\begin{aligned}\partial_J(v)&=\left(\partial_J^{(1)}(v),\ldots,\partial_J^{(r)}(v)\right),\\ \partial_J(v)\bullet\mathbf{h}_J&:=\partial_J^{(1)}(v)\cdot\mathbf{h}_J^{(1)}+\cdots+\partial_J^{(r)}(v)\cdot\mathbf{h}_J^{(r)}.\end{aligned}\tag{70}$$

(b) *Assume that $\check{\mathbf{c}}_J\geq 1$ and $\partial_J:\mathbb{Z}^J\to\mathbb{Z}^r$ are as in (a), and one sets*

$$\begin{aligned}\overline{\mathbf{h}}_J^{(1)}&=\mathbf{h}_J^{(1)}\cdot B',\ldots,\overline{\mathbf{h}}_J^{(r)}=\mathbf{h}_J^{(r)}\cdot B',\\ \widehat{\mathbf{h}}_J^{(1)}&=\mathbf{h}_J^{(1)}\cdot B'\cdot B^{-1},\ldots,\widehat{\mathbf{h}}_J^{(r)}=\mathbf{h}_J^{(r)}\cdot B'\cdot B^{-1},\end{aligned}\tag{71}$$

where $B',B\in\mathbb{M}_J(\mathbb{Z})$ are as in Proposition 5.

(b1) *There exists a group homomorphism $\overline{\partial}_J=(\overline{\partial}_J^{(1)},\ldots,\overline{\partial}_J^{(r)}):\mathbb{Z}^J\to\mathbb{Z}^r\cong\operatorname{Ker}\overline{q}_J$ (called the Euler defect of J) such that*

$$\begin{aligned}\overline{\Phi}_J^{\check{\mathbf{c}}_J}(v)&=v+\overline{\partial}_J(v)\bullet\overline{\mathbf{h}}_J\\&=v+\overline{\partial}_J^{(1)}(v)\cdot\overline{\mathbf{h}}_J^{(1)}+\cdots+\overline{\partial}_J^{(r)}(v)\cdot\overline{\mathbf{h}}_J^{(r)},\\&\qquad\forall v\in\mathbb{Z}^J,\end{aligned}\tag{72}$$

$\overline{\partial}_J\circ\overline{\Phi}_J=\overline{\partial}_J$, $\overline{\partial}_J=\overline{\partial}_J\circ h_{B'}$, and $\overline{\partial}_J(\mathbf{h})=0$, for all $\mathbf{h}\in\operatorname{Ker}\overline{q}_J$, where one sets

$$\begin{aligned}\overline{\mathbf{h}}_J&=\left(\overline{\mathbf{h}}_J^{(1)},\ldots,\overline{\mathbf{h}}_J^{(r)}\right),\\ \overline{\partial}_J(v)\bullet\overline{\mathbf{h}}_J&:=\overline{\partial}_J^{(1)}(v)\cdot\overline{\mathbf{h}}_J^{(1)}+\cdots+\overline{\partial}_J^{(r)}(v)\cdot\overline{\mathbf{h}}_J^{(r)}.\end{aligned}\tag{73}$$

(b2) *There exists a group homomorphism $\widehat{\partial}_J=(\widehat{\partial}_J^{(1)},\ldots,\widehat{\partial}_J^{(r)}):\mathbb{Z}^J\to\mathbb{Z}^r\cong\operatorname{Ker}\widehat{q}_J$ (called the Tits defect of J) such that*

$$\begin{aligned}\widehat{\Phi}_J^{\check{\mathbf{c}}_J}(v)&=v+\widehat{\partial}_J(v)\bullet\widehat{\mathbf{h}}_J\\&=v+\overline{\partial}_J^{(1)}(v)\cdot\widehat{\mathbf{h}}_J^{(1)}+\cdots+\widehat{\partial}_J^{(r)}(v)\cdot\widehat{\mathbf{h}}_J^{(r)},\\&\qquad\forall v\in\mathbb{Z}^J,\end{aligned}\tag{74}$$

$\widehat{\partial}_J\circ\widehat{\Phi}_J=\widehat{\partial}_J$, $\widehat{\partial}_J=\overline{\partial}_J\circ h_B=\partial_J\circ h_{B'}^{-1}\circ h_B$, and $\widehat{\partial}_J(\mathbf{h})=0$, for all $\mathbf{h}\in\operatorname{Ker}\widehat{q}_J$, where one sets

$$\begin{aligned}\widehat{\mathbf{h}}_J&=\left(\widehat{\mathbf{h}}_J^{(1)},\ldots,\widehat{\mathbf{h}}_J^{(r)}\right),\\ \widehat{\partial}_J(v)\bullet\widehat{\mathbf{h}}_J&:=\widehat{\partial}_J^{(1)}(v)\cdot\widehat{\mathbf{h}}_J^{(1)}+\cdots+\widehat{\partial}_J^{(r)}(v)\cdot\widehat{\mathbf{h}}_J^{(r)}.\end{aligned}\tag{75}$$

(c) *The Coxeter number $\mathbf{c}_J$ of J is finite if and only if the incidence defect $\partial_J:\mathbb{Z}^J\to\mathbb{Z}^r$ is zero. In this case, $\check{\mathbf{c}}_J=\mathbf{c}_J$.*

(d) *Given $v\in\mathbb{Z}^m\equiv\mathbb{Z}^J$, the order $\mathbf{s}_v:=|\mathcal{O}(v)|$ of the Φ_J-orbit $\mathcal{O}(v)$ is finite if and only if $\partial_J(v)=0$. If $\mathbf{s}_v=|\mathcal{O}(v)|$ is finite, then $\mathbf{s}_v$ divides $\check{\mathbf{c}}_J$ and there is a unique integer m_v such that*

$$\begin{aligned}m_v\cdot\mathbf{h}&=v+\Phi_J(v)+\Phi_J^2(v)+\cdots+\Phi_J^{\mathbf{s}_v-1}(v)\\&=v+\Phi_J^{-1}(v)+\Phi_J^{-2}(v)+\cdots+\Phi_J^{-\mathbf{s}_v+1}(v).\end{aligned}\tag{76}$$

(e) *The statement (d) holds with Φ_J and $\widehat{\Phi}_J$ (resp., $\overline{\Phi}_J$) interchanged.*

Proof. For simplicity of presentation, we assume that $r = 2$. We recall from the proof of Proposition 9 that $\mathbb{Z} \cdot \mathbf{h}_J^{(1)} \oplus \mathbb{Z} \cdot \mathbf{h}_J^{(2)} = \operatorname{Ker} q_J = \operatorname{Ker}[Dq_J : \mathbb{Z}^m \to \mathbb{Z}^m]$, where $m = |J| \geq 2$ and

$$Dq_J : \mathbb{Z}^m \longrightarrow \mathbb{Z}^m,$$
$$v \longmapsto Dq_J(v) = \left(\frac{\partial q_J}{\partial x_1}(v), \ldots, \frac{\partial q_J}{\partial x_m}(v)\right), \tag{77}$$

is the gradient group homomorphism. It follows that

$$\frac{\mathbb{Z}^m}{\mathbb{Z} \cdot \mathbf{h}_J^{(1)} \oplus \mathbb{Z} \cdot \mathbf{h}_J^{(2)}} = \frac{\mathbb{Z}^m}{\operatorname{Ker} q_J} \cong \operatorname{Im} Dq_J \cong \mathbb{Z}^{m-2}. \tag{78}$$

Denote by $\phi : \mathbb{Z}^m \to \mathbb{Z}^{m-2}$ the composite quotient epimorphism. Then, the form q_J induces the form $\widetilde{q}_J : \mathbb{Z}^{m-2} \to \mathbb{Z}$ such that $\widetilde{q}_J(\phi(x)) = q_J(x)$, for all $x \in \mathbb{Z}^m$. Moreover, the Coxeter transformation $\Phi_J : \mathbb{Z}^m \to \mathbb{Z}^m$ induces a group automorphism $\widetilde{\Phi}_J : \mathbb{Z}^{m-2} \to \mathbb{Z}^{m-2}$ such that

$$\widetilde{\Phi}_J \circ \phi = \phi \circ \Phi_J, \qquad \widetilde{q}_J\left(\widetilde{\Phi}_J(y)\right) = \widetilde{q}_J(y), \tag{79}$$

for all $y \in \mathbb{Z}^{m-2}$. It follows that $\widetilde{q}_J$ is positive definite, and there exists a minimal integer $\check{\mathbf{c}}_J \geq 1$ such that $\widetilde{\Phi}_J^{\check{\mathbf{c}}_J}$ is the identity map on $\mathbb{Z}^{m-2}$. Hence, given $v \in \mathbb{Z}^m$, the element $\Phi_J^{\check{\mathbf{c}}_J}(v) - v$ lies in the kernel of q_J; that is, it has the form

$$\Phi_J^{\check{\mathbf{c}}_J}(v) - v = \partial_J^{(1)}(v) \cdot \mathbf{h}_J^{(1)} + \partial_J^{(2)}(v) \cdot \mathbf{h}_J^{(2)}, \tag{80}$$

where $\partial_J^{(1)}(v), \partial_J^{(2)}(v)$ are integers uniquely determined by v. Since Φ_J is a group homomorphism, then

$$\partial_J^{(1)}\left(v + v'\right) = \partial_J^{(1)}(v) + \partial_J^{(1)}\left(v'\right),$$
$$\partial_J^{(2)}\left(v + v'\right) = \partial_J^{(2)}(v) + \partial_J^{(2)}\left(v'\right); \tag{81}$$

that is, we have defined a pair of group homomorphisms $\partial_J^{(1)}, \partial_J^{(2)} : \mathbb{Z}^J \to \mathbb{Z}$; hence, $\partial_J = (\partial_J^{(1)}, \partial_J^{(2)}) : \mathbb{Z}^J \to \mathbb{Z}^2$ is a group homomorphism. It is easy to see that ∂_J has the properties required in (a), and (a) follows.

In view of Proposition 5, the statements (b)–(e) are a consequence of (a). The reader is referred to [34, Theorem 4.17] for more details and a generalization.

Corollary 19. *Assume that J is a finite nonnegative poset of corank $r \geq 2$.*

(a) *The Coxeter number $\mathbf{c}_J$ of J is infinite if and only if the defect homomorphism $\partial_J : \mathbb{Z}^J \to \mathbb{Z}^r$ is nonzero, or, equivalently, if and only if the Φ_J-orbit $\mathcal{O}(e_j)$ of some basis vector $e_j \in \mathbb{Z}^J$ is infinite.*

(b) *The Coxeter transformation Φ_J is weakly periodic in the sense of Sato [42]; that is, $\Phi_J^s - id$ is nilpotent, for some $s \geq 1$.*

Proof. The statement (a) follows immediately from Theorem 18. To prove (b), we check that $(\Phi_J^{\check{\mathbf{c}}_J} - id)^2 = 0$.

Remark 20. (a) It was shown in [34, Example 5.18] that, for the one-peak garland $I = \mathscr{G}_3^*$ of Theorem 17(d), we have

(i) $\partial_I = \widehat{\partial}_I = \overline{\partial}_I = 0$ and $\mathbf{c}_I = \check{\mathbf{c}}_I = 4$,

(ii) the set $\mathscr{R}_{\widehat{q}_I}$ of Tits roots of I lies on 22 sand-glass tubes; six of them are of rank two, and each of the remaining fourteen tubes is of rank four; see [34, pp. 459–461] for details.

(b) By Lemma 8(a), the Coxeter number $\mathbf{c}_J$ is infinite, for every principal poset J.

(c) By Theorem 17, there is no nonnegative connected poset J of corank 2, with $|J| \leq 5$. Moreover, a minimal such a poset is the garland

$\mathscr{G}_3$: 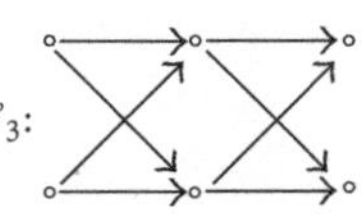 (82)

(d) We show in [43] that most of the nonnegative connected posets J of corank 2, with at most 15 elements, are of zero defect. We also show there that a smallest nonnegative connected poset J with nonzero defect has 8 elements and is one of the following two posets:

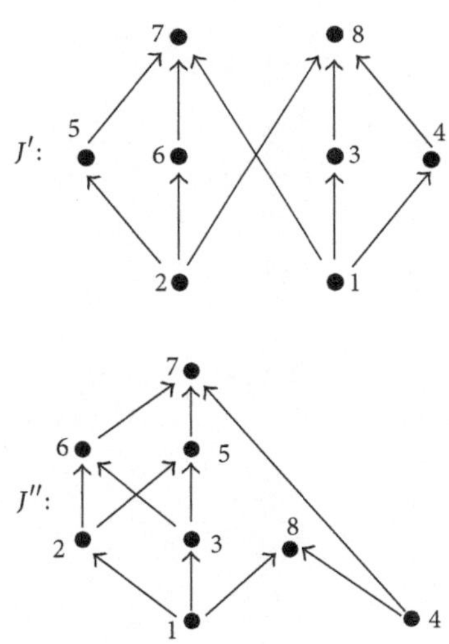

(83)

It is easy to check that

(i) $\check{\mathbf{c}}_{J'} = \check{\mathbf{c}}_{J''} = 2$,

(ii) $\operatorname{cox}_{J'}(t) = \operatorname{cox}_{J''}(t) = t^8 - 4t^6 + 6t^4 - 4t^2 + 1$,

(iii) the coordinates of the Tits defect $\widehat{\partial}_{J'} = (\widehat{\partial}_{J'}^{(1)}, \widehat{\partial}_{J'}^{(2)}) : \mathbb{Z}^8 \to \mathbb{Z}^2$ of J', with respect to the $\mathbb{Z}$-basis

$$\widehat{\mathbf{h}}_{J'}^{(1)} = (2, 0, -1, -1, 1, 1, 2, 0),$$
$$\widehat{\mathbf{h}}_{J'}^{(2)} = (0, 2, 1, 1, -1, -1, 0, 2) \tag{84}$$

of $\operatorname{Ker}\widehat{q}_{J'}$, are given by the formulae

$$\widehat{\partial}_{J'}^{(1)}(x) = x_2 - \frac{1}{2}x_3 - \frac{1}{2}x_4 + \frac{1}{2}x_5 + \frac{1}{2}x_6 - x_7,$$
$$\widehat{\partial}_{J'}^{(2)}(x) = x_1 + \frac{1}{2}x_3 + \frac{1}{2}x_4 - \frac{1}{2}x_5 - \frac{1}{2}x_6 - x_8, \quad (85)$$

(iv) the coordinates of the Tits defect $\widehat{\partial}_{J''} = (\widehat{\partial}_{J''}^{(1)}, \widehat{\partial}_{J''}^{(2)}) : \mathbb{Z}^8 \to \mathbb{Z}^2$ of J'', with respect to the $\mathbb{Z}$-basis

$$\widehat{\mathbf{h}}_{J''}^{(1)} = (0,-1,-1,0,1,1,0,0),$$
$$\widehat{\mathbf{h}}_{J''}^{(2)} = (1,0,0,1,0,0,1,1) \quad (86)$$

of $\operatorname{Ker}\widehat{q}_{J''}$, are given by

$$\widehat{\partial}_{J'}^{(1)}(x) = x_1 + x_2 + x_3 + x_4 + x_5 + x_6 - x_7 - x_8,$$
$$\widehat{\partial}_{J''}^{(2)}(x) = 2x_1 + x_2 + x_3 + 2x_4 + x_5 + x_6 - 2x_7 - 2x_8. \quad (87)$$

5. An Example

In this section, we illustrate the results of Section 3 by an example of a principal one-peak poset I of the Euclidean type $DI = \widetilde{\mathbb{D}}_4$. We give a description of the set $\mathscr{R}_{\widehat{q}_I}$ of roots of $\widehat{q}_I$ and the mesh translation quiver $\Gamma(\mathscr{R}_{\widehat{q}_I}, \widehat{\Phi}_I)$ together with the decomposition (see (51))

$$\Gamma(\mathscr{R}_{\widehat{q}_I}, \widehat{\Phi}_I) = \Gamma(\widehat{\partial}_I^{-}\mathscr{R}_{\widehat{q}_I}, \widehat{\Phi}_I) \cup \Gamma(\widehat{\partial}_I^{+}\mathscr{R}_{\widehat{q}_I}, \widehat{\Phi}_I)$$
$$\cup \Gamma(\widehat{\partial}_I^{0}\mathscr{R}_{\widehat{q}_I}, \widehat{\Phi}_I). \quad (88)$$

Let I be the one-peak garland

$I = \mathscr{G}_2^*$: 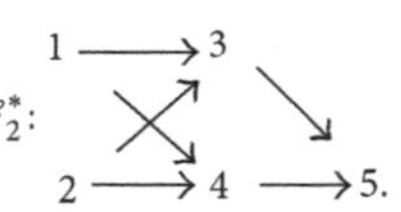 (89)

The incidence matrix C_I, the Tits matrix $\widehat{C}_I$, and the Coxeter-Tits matrix $\widehat{\operatorname{Cox}}_I = -\widehat{C}_I \cdot \widehat{C}_I^{-\mathrm{tr}}$ of I are the following:

$$C_I = \begin{bmatrix} 1 & 0 & 1 & 1 & 1 \\ 0 & 1 & 1 & 1 & 1 \\ 0 & 0 & 1 & 0 & 1 \\ 0 & 0 & 0 & 1 & 1 \\ 0 & 0 & 0 & 0 & 1 \end{bmatrix},$$

$$\widehat{C}_I = \left[\begin{array}{cccc|c} 1 & 0 & 0 & 0 & -1 \\ 0 & 1 & 0 & 0 & -1 \\ 1 & 1 & 1 & 0 & -1 \\ 1 & 1 & 0 & 1 & -1 \\ \hline 0 & 0 & 0 & 0 & 1 \end{array}\right], \quad (90)$$

$$\widehat{\operatorname{Cox}}_I = \begin{bmatrix} 0 & 1 & 0 & 0 & 1 \\ 1 & 0 & 0 & 0 & 1 \\ 0 & 0 & 0 & 1 & 1 \\ 0 & 0 & 1 & 0 & 1 \\ -1 & -1 & 1 & 1 & -1 \end{bmatrix}.$$

The Coxeter polynomial $\operatorname{cox}_I(t)$, the Tits quadratic form $\widehat{q}_I : \mathbb{Z}^5 \to \mathbb{Z}$, and the Coxeter-Tits transformation $\widehat{\Phi}_I : \mathbb{Z}^5 \to \mathbb{Z}^5$ of I are

$$\operatorname{cox}_I(t) = t^5 + t^4 - 2t^3 - 2t^2 + t + 1 = F_{\widetilde{\mathbb{D}}_4}(t),$$
$$\widehat{q}_I(x) = x \cdot \widehat{C}_I \cdot x^{\mathrm{tr}} = x_1^2 + x_2^2 + x_3^2 + x_4^2 + x_5^2$$
$$+ (x_1 + x_2)(x_3 + x_4) - (x_1 + x_2 + x_3 + x_4)x_5,$$
$$\widehat{\Phi}_I(x) = x \cdot \widehat{\operatorname{Cox}}_I = (x_2 - x_5, x_1 - x_5, x_4 + x_5,$$
$$x_3 + x_5, x_1 + x_2 + x_3 + x_4 - x_5), \quad (91)$$

for $x = (x_1, x_2, x_3, x_4, x_5) \in \mathbb{Z}^I \equiv \mathbb{Z}^5$. Note that the $\widehat{\Phi}_I$-orbit $\mathscr{O}(e_5)$ of e_5 consists of two vectors e_5 and $\widehat{\Phi}_I(e_5) = (-1,-1,1,1,-1) = \widehat{\Phi}_I^{-1}(e_5)$. Since

$$\widehat{q}_I(x) = \left(x_1 + \frac{1}{2}x_3 + \frac{1}{2}x_4 - \frac{1}{2}x_5\right)^2$$
$$+ \left(x_2 + \frac{1}{2}x_3 + \frac{1}{2}x_4 - \frac{1}{2}x_5\right)^2 \quad (92)$$
$$+ \frac{1}{2}(x_3 - x_4)^2 + \frac{1}{2}x_5^2,$$

then the form $\widehat{q}_I : \mathbb{Z}^5 \to \mathbb{Z}$ is positive semidefinite, is not positive definite,

$$\operatorname{Ker}\widehat{q}_I = \mathbb{Z} \cdot \widehat{\mathbf{h}}_I, \quad \text{where } \widehat{\mathbf{h}}_I = (1,1,-1,-1,0), \quad (93)$$

and $\widehat{\Phi}_I(\widehat{\mathbf{h}}_I) = \widehat{\mathbf{h}}_I$. This means that $\widehat{q}_I$ is principal, but not P-critical; see [44]. One easily shows that the reduced Coxeter number of I equals $\check{\mathbf{c}}_I = 2$ and the Tits defect $\widehat{\partial}_I : \mathbb{Z}^5 \to \mathbb{Z}$ of I is given by $\widehat{\partial}_I(x) = -(x_1 + x_2 + x_3 + x_4)$, because $\widehat{\Phi}_I \neq id$ and $\widehat{\Phi}_I^2(v) = v + \widehat{\partial}_I(v) \cdot \widehat{\mathbf{h}}_I$, for any $v \in \mathbb{Z}^5$. The set $\mathscr{R}_{\widehat{q}_I}$ of roots of $\widehat{q}_I$ has the disjoint union decomposition (see (51))

$$\mathscr{R}_{\widehat{q}_I} = \widehat{\partial}_I^{-}\mathscr{R}_{\widehat{q}_I} \cup \widehat{\partial}_I^{+}\mathscr{R}_{\widehat{q}_I} \cup \widehat{\partial}_I^{0}\mathscr{R}_{\widehat{q}_I}, \quad (94)$$

and $\widehat{\partial}_I^{-}\mathscr{R}_{\widehat{q}_I}$, $\widehat{\partial}_I^{+}\mathscr{R}_{\widehat{q}_I}$, $\widehat{\partial}_I^{0}\mathscr{R}_{\widehat{q}_I}$ are $\widehat{\Phi}_I$-invariant subsets of $\mathscr{R}_{\widehat{q}_I}$. Obviously, the $\widehat{\Phi}_I$-orbit $\mathscr{O}(v)$ of any $v \in \widehat{\partial}_I^{0}\mathscr{R}_{\widehat{q}_I}$ is of length two, whereas the $\widehat{\Phi}_I$-orbit $\mathscr{O}(w)$ of any vector $w \in \widehat{\partial}_I^{-}\mathscr{R}_{\widehat{q}_I} \cup \widehat{\partial}_I^{+}\mathscr{R}_{\widehat{q}_I}$ is infinite. By (92), a vector $v = (v_1, v_2, v_3, v_4, v_5) \in \mathbb{Z}^5$ is a root of $\widehat{q}_I : \mathbb{Z}^5 \to \mathbb{Z}$ if and only if $(2v_1 + v_3 + v_4 - v_5)^2 + (2v_2 + v_3 + v_4 - v_5)^2 + 2(v_3 - v_4)^2 + 2v_5^2 = 4$. Hence, looking at all possible decompositions $4 = a_1^2 + a_2^2 + 2a_3^2 + 2a_4^2$, with $a_1, a_2, a_3, , a_4, \in \mathbb{Z}$, we show that $v = (v_1, v_2, v_3, v_4, v_5) \in \mathbb{Z}^5$ is a root of $\widehat{q}_I : \mathbb{Z}^5 \to \mathbb{Z}$ if and only if v or $\widetilde{v} := -v$ is one of the vectors listed in Table 1 or in Table 2.

(1) *The $\widehat{\Phi}_I$-orbits in $\widehat{\mathscr{P}}_I := \widehat{\partial}_I^{-}\mathscr{R}_{\widehat{q}_I}$.* Since $\widehat{\partial}_I(u) < 0$, if $u \in \{e_1, e_2, e_3, e_4\}$ or u is the vector $\mathbf{p}_{12} = (1,1,0,0,1)$, then the $\widehat{\Phi}_I$-orbits of the vectors $e_1, e_2, e_3, e_4, \mathbf{p}_{12}$ lie in $\widehat{\mathscr{P}}_I := \widehat{\partial}_I^{-}\mathscr{R}_{\widehat{q}_I}$, because $\widehat{\mathscr{P}}_I$ is a $\widehat{\Phi}_I$-invariant subset of $\mathscr{R}_{\widehat{q}_I}$. It is easy to see that the $\widehat{\Phi}_I$-orbits consist of the vectors listed in Table 1.

TABLE 1

j	$\widehat{\Phi}_I^j(e_1)$	$\widehat{\Phi}_I^j(e_2)$	$\widehat{\Phi}_I^j(\mathbf{p}_{12})$	$\widehat{\Phi}_I^j(e_3)$	$\widehat{\Phi}_I^j(e_4)$
$\vdots$	$\vdots$	$\vdots$	$\vdots$	$\vdots$	$\vdots$
$j=7$	$(-3,-2,3,3,1)$	$(-2,-3,3,3,1)$	$(-6,-6,7,7,1)$	$(-3,-3,3,4,1)$	$(-3,-3,4,3,1)$
$j=6$	$(-2,-3,3,3,0)$	$(-3,-2,3,3,0)$	$(-5,-5,6,6,1)$	$(-3,-3,4,3,0)$	$(-3,-3,3,4,0)$
$j=5$	$(-2,-1,2,2,1)$	$(-1,-2,2,2,1)$	$(-4,-4,5,5,1)$	$(-2,-2,2,3,1)$	$(-2,-2,3,2,1)$
$j=4$	$(-1,-2,2,2,0)$	$(-2,-1,2,2,0)$	$(-3,-3,4,4,1)$	$(-2,-2,3,2,0)$	$(-2,-2,2,3,0)$
$j=3$	$(-1,0,1,1,1)$	$(0,-1,1,1,1)$	$(-2,-2,3,3,1)$	$(-1,-1,1,2,1)$	$(-1,-1,2,1,1)$
$j=2$	$(0,-1,1,1,0)$	$(-1,0,1,1,0)$	$(-1,-1,2,2,1)$	$(-1,-1,2,1,0)$	$(-1,-1,1,2,0)$
$j=1$	$(0,1,0,1,0)$	$(1,0,0,0,1)$	$(0,0,1,1,1)$	$(0,0,0,1,1)$	$(0,0,1,0,1)$
$j=0$	$(1,0,0,0,0)$	$(0,1,0,0,0)$	$(1,1,0,0,1)$	$(0,0,1,0,0)$	$(0,0,0,1,0)$
$j=-1$	$(1,2,-1,-1,1)$	$(2,1,-1,-1,1)$	$(2,2,-1,-1,1)$	$(1,1,-1,0,1)$	$(1,1,0,-1,1)$
$j=-2$	$(2,1,-1,-1,0)$	$(1,2,-1,-1,0)$	$(3,3,-2,-2,1)$	$(1,1,0,-1,0)$	$(1,1,-1,0,0)$
$j=-3$	$(2,3,-2,-2,1)$	$(3,2,-2,-2,1)$	$(4,4,-3,-3,1)$	$(2,2,-2,-1,1)$	$(2,2,-1,-2,1)$
$j=-4$	$(3,2,-2,-2,0)$	$(2,3,-2,-2,0)$	$(5,5,-4,-4,1)$	$(2,2,-1,-2,0)$	$(2,2,-2,-1,0)$
$j=-5$	$(3,4,-3,-3,1)$	$(4,3,-3,-3,1)$	$(6,6,-5,-5,1)$	$(3,3,-3,-2,1)$	$(3,3,-2,-3,1)$
$j=-6$	$(4,3,-3,-3,0)$	$(3,4,-3,-3,0)$	$(7,7,-6,-6,1)$	$(3,3,-2,-3,0)$	$(3,3,-3,-2,0)$
$j=-7$	$(4,5,-4,-4,1)$	$(5,4,-4,-4,1)$	$(8,8,-7,-7,1)$	$(4,4,-4,-3,1)$	$(4,4,-3,-4,1)$
$\vdots$	$\vdots$	$\vdots$	$\vdots$	$\vdots$	$\vdots$

TABLE 2

j	$u^{(j)}$	$u_+^{(j)}$	$w^{(j)}$	$w_+^{(j)}$	$v^{(j)}$	$v_+^{(j)}$
$j=0$	$-e_5=(0,0,0,0,-1)$	$(1,1,-1,-1,1)$	$(1,0,0,-1,0)$	$(0,1,-1,0,0)$	$(0,1,0,-1,0)$	$(1,0,-1,0,0,)$
$j=1$	$(2,2,-2,-2,1)$	$(1,1,-1,-1,-1)$	$(1,2,-2,-1,0)$	$(2,1,-1,-2,0)$	$(2,1,-2,-1,0)$	$(1,2,-1,-2,0)$
$j=2$	$(2,2,-2,-2,-1)$	$(3,3,-3,-3,1)$	$(3,2,-2,-3,0)$	$(2,3,-3,-2,0)$	$(2,3,-2,-3,0)$	$(3,2,-3,-2,0)$
$j=3$	$(4,4,-4,-4,1)$	$(3,3,-3,-3,-1)$	$(3,4,-4,-3,0)$	$(4,3,-3,-4,0)$	$(4,3,-4,-3,0)$	$(3,4,-3,-4,0)$
$j=4$	$(4,4,-4,-4,-1)$	$(5,5,-5,-5,1)$	$(5,4,-4,-5,0)$	$(4,5,-5,-4,0)$	$(4,5,-4,-5,0)$	$(5,4,-5,-4,0)$
$j=5$	$(6,6,-6,-6,1)$	$(5,5,-5,-5,-1)$	$(5,6,-6,-5,0)$	$(6,5,-5,-6,0)$	$(6,5,-6,-5,0)$	$(5,6,-5,-6,0)$
$j=6$	$(6,6,-6,-6,-1)$	$(7,7,-7,-7,1)$	$(7,6,-6,-7,0)$	$(6,7,-7,-6,0)$	$(6,7,-6,-7,0)$	$(7,6,-7,-6,0)$
$j=7$	$(8,8,-8,-8,1)$	$(7,7,-7,-7,-1)$	$(7,8,-8,-7,0)$	$(8,7,-7,-8,0)$	$(8,7,-8,-7,0)$	$(7,8,-7,-8,0)$
$\vdots$	$\vdots$	$\vdots$	$\vdots$	$\vdots$	$\vdots$	$\vdots$

Throughout this section, we freely use the $\widehat{\Phi}_I$-mesh terminology and notation introduced in [2, 34, 40].

(2) *$\widehat{\Phi}_I$-mesh quiver* $\Gamma(\widehat{\mathscr{P}}_I, \widehat{\Phi}_I) = \Gamma(\widehat{\partial}_I^- \mathscr{R}_{\widehat{q}_I}, \widehat{\Phi}_I)$. It follows from our earlier remarks that the set $\widehat{\mathscr{P}}_I := \widehat{\partial}_I^- \mathscr{R}_{\widehat{q}_I}$ of the negative defect roots of $\widehat{q}_I$ splits into the five $\widehat{\Phi}_I$-orbits $\mathscr{O}(e_1)$, $\mathscr{O}(e_2)$, $\mathscr{O}(e_3)$, $\mathscr{O}(e_4)$, $\mathscr{O}(\mathbf{p}_{12})$. By applying the mesh toroidal algorithm defined in [2, 34], one constructs the following infinite $\widehat{\Phi}_I$-mesh translation quiver of the negative defect roots of $\widehat{q}_I$; see Figure 2, where we set $\widehat{a} := -a$ for any positive integer $a \geq 1$.

(3) *$\widehat{\Phi}_I$-mesh quiver* $\Gamma(\widehat{\mathscr{Q}}_I, \widehat{\Phi}_I) = \Gamma(\widehat{\partial}_I^+ \mathscr{R}_{\widehat{q}_I}, \widehat{\Phi}_I)$. Since the group isomorphism $\mathbb{Z}^I \to \mathbb{Z}^I$, $v \mapsto -v$, carries roots to roots, $\widehat{\Phi}_I$-meshes to $\widehat{\Phi}_I$-meshes, and $\widehat{\Phi}_I$-orbits to $\widehat{\Phi}_I$-orbits, then it defines the bijections $\widehat{\partial}_I^- \mathscr{R}_{\widehat{q}_I} \to \widehat{\partial}_I^+ \mathscr{R}_{\widehat{q}_I}$ and $\widehat{\partial}_I^0 \mathscr{R}_{\widehat{q}_I} \to \widehat{\partial}_I^0 \mathscr{R}_{\widehat{q}_I}$, because $\widehat{\partial}_I(-v) = -\widehat{\partial}_I(v)$. It follows that the set $\widehat{\mathscr{Q}}_I := \widehat{\partial}_I^+ \mathscr{R}_{\widehat{q}_I}$ of the positive defect roots of $\widehat{q}_I$ splits into the five $\widehat{\Phi}_I$-orbits $\mathscr{O}(\widehat{e}_1)$, $\mathscr{O}(\widehat{e}_2)$, $\mathscr{O}(\widehat{e}_3)$, $\mathscr{O}(\widehat{e}_4)$, $\mathscr{O}(\widehat{\mathbf{p}}_{12})$, and one constructs the infinite $\widehat{\Phi}_I$-mesh translation quiver

$$\Gamma\left(\widehat{\mathscr{Q}}_I, \widehat{\Phi}_I\right) = \Gamma\left(\widehat{\partial}_I^+ \mathscr{R}_{\widehat{q}_I}, \widehat{\Phi}_I\right) \tag{95}$$

of the positive defect roots of $\widehat{q}_I$ by interchanging any vector v in $\Gamma(\widehat{\mathscr{P}}_I, \widehat{\Phi}_I) = \Gamma(\widehat{\partial}_I^- \mathscr{R}_{\widehat{q}_I}, \widehat{\Phi}_I)$ with its negative $\widehat{v} := -v$.

(4) $\widehat{\Phi}_I$-*mesh quiver* $\Gamma(\widehat{\partial}_I^0\mathscr{R}_{\widehat{q}_I}, \widehat{\Phi}_I)$. By the equality $\widehat{\Phi}_I^2(v) = v - \widehat{\partial}_I(v)\cdot\widehat{\mathbf{h}}_I$, the $\widehat{\Phi}_I$-orbit of any $v \in \widehat{\partial}_I^0\mathscr{R}_{\widehat{q}_I}$ consists of two vectors v and $\widehat{\Phi}_I(v)$. Now, we show that the $\widehat{\Phi}_I$-orbits in $\widehat{\partial}_I^0\mathscr{R}_{\widehat{q}_I}$ form a $\widehat{\Phi}_I$-mesh translation quiver $\Gamma(\widehat{\partial}_I^0\mathscr{R}_{\widehat{q}_I}, \widehat{\Phi}_I)$.

Note that $\widehat{\Phi}_I(e_5) = (-1,-1,1,1,-1)$, $\widehat{\Phi}_I^2(e_5) = e_5$, $\widehat{\partial}_I(e_5) = 0$, and $\widehat{\partial}_I(\widehat{\Phi}_I(e_5)) = 0$. It follows that the two-element $\widehat{\Phi}_I$-orbits of e_5 and $-e_5$ lie in $\widehat{\partial}_I^0\mathscr{R}_{\widehat{q}_I}$. Moreover, the vectors

$$\begin{aligned} u_+^{(1)} &= (1,1,-1,-1,-1), \quad -u_+^{(1)}, \\ w^{(0)} &= (1,0,0,-1,0), \\ w_+^{(0)} &= \widehat{\Phi}_I\left(w^{(0)}\right) = (0,1,-1,0,0), \\ v^{(0)} &= (0,1,0,-1,0), \\ v_+^{(0)} &= \widehat{\Phi}_I\left(v^{(0)}\right) = (1,0,-1,0,0) \\ w_+^{(1)} &= (2,1,-1,-2,0), \\ v_+^{(1)} &= (1,2,-1,-2,0) \end{aligned} \tag{96}$$

belong to $\widehat{\partial}_I^0\mathscr{R}_{\widehat{q}_I}$. It is easy to see that we have the following $\widehat{\Phi}_I^2$-mesh quivers of vectors in $\widehat{\partial}_I^0\mathscr{R}_{\widehat{q}_I}$:

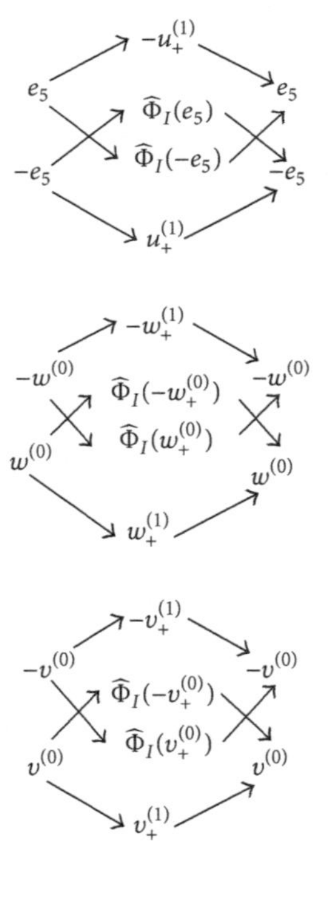

(97)

Note that the $\widehat{\Phi}_I$-orbit of $u_+^{(1)}$ consists of the following two vectors:

$$u^{(1)} = (2,-2,-2,-2,1), \qquad u_+^{(1)} = \widehat{\Phi}_I\left(u^{(1)}\right). \tag{98}$$

By (92), a vector $v = (v_1, v_2, v_3, v_4, v_5) \in \mathbb{Z}^5$ is a root of $\widehat{q}_I : \mathbb{Z}^5 \to \mathbb{Z}$ of defect zero if and only if

$$\begin{gathered} (2v_1 + v_3 + v_4 - v_5)^2 + (2v_2 + v_3 + v_4 - v_5)^2 \\ + 2(v_3 - v_4)^2 + 2v_5^2 = 4, \\ v_1 + v_2 + v_3 + v_4 = 0. \end{gathered} \tag{99}$$

It follows that v or $-v$ belongs to any of the six series of roots presented in Table 2.

Hence, we conclude that the $\widehat{\Phi}_I$-orbits in the set $\widehat{\partial}_I^0\mathscr{R}_{\widehat{q}_I}$ form three $\widehat{\Phi}_I$-mesh quivers $\mathscr{T}_u$, $\mathscr{T}_w$, $\mathscr{T}_v$, and each of them has the form of infinite two-surface tube of rank 2:

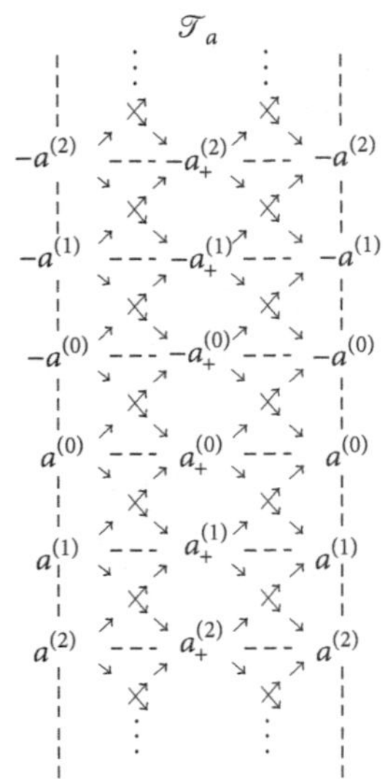

(100)

where a is one of the vectors

$$\begin{aligned} u &= u^{(0)} = (0,0,0,0,-1), \\ w &= w^{(0)} = (1,0,0,-1,0), \\ v &= v^{(0)} = (0,1,0,-1,0). \end{aligned} \tag{101}$$

(5) $\widehat{\Phi}_I$-*mesh quiver* $\Gamma(\widehat{\partial}_I^0\mathscr{R}_{\widehat{q}_I} \cup \operatorname{Ker}\widehat{q}_I, \widehat{\Phi}_I)$. We recall that $\operatorname{Ker}\widehat{q}_I = \mathbb{Z}\cdot\widehat{\mathbf{h}}_I$, where $\widehat{\mathbf{h}}_I = (1,1,-1,-1,0)$. Note that

$$\widehat{\Phi}_I\left(\widehat{\mathbf{h}}_I\right) = \widehat{\mathbf{h}}_I, \quad \widehat{\Phi}_I\left(m\cdot\widehat{\mathbf{h}}_I\right) = m\cdot\widehat{\mathbf{h}}_I, \text{ for any } m \in \mathbb{Z}. \tag{102}$$

Obviously, the vectors lying in $\operatorname{Ker}\widehat{q}_I$ form the $\widehat{\Phi}$-mesh translation quiver $\widehat{\mathscr{T}}_{\widehat{\mathbf{h}}_I}$ presented in (108).

Now, we construct from the $\widehat{\Phi}_I$-orbits in the set $\widehat{\partial}_I^0\mathscr{R}_{\widehat{q}_I} \cup \operatorname{Ker}\widehat{q}_I$ an infinite $\widehat{\Phi}_I$-mesh translation quiver. For this purpose, we note that the following six vectors

$$-e_5, e_5, \widehat{\Phi}_I(-e_5), \widehat{\Phi}_I(e_5), \quad -\widehat{\mathbf{h}}_I, \widehat{\mathbf{h}}_I \tag{103}$$

form two $\widehat{\Phi}_I$-meshes of width 1. If we complete them by the three vectors

$$0, u_+^{(1)} := (1,1,-1,-1,-1), \quad -u_+^{(1)}, \tag{104}$$

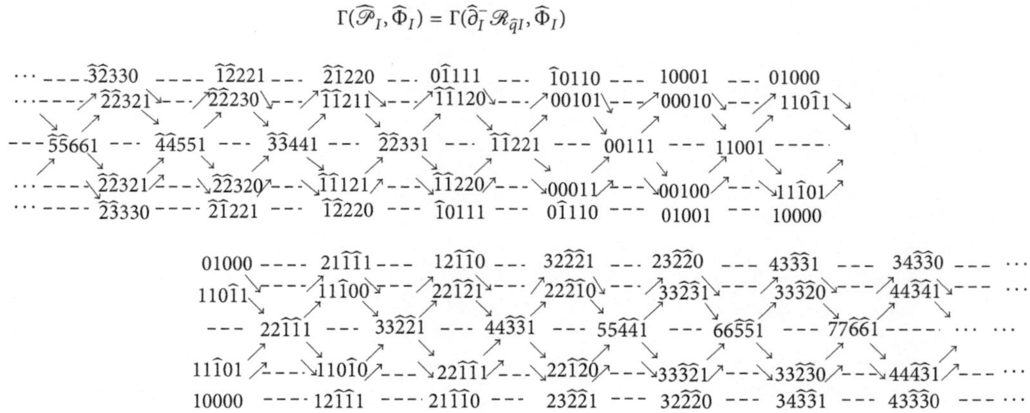

Figure 2

we get the $\widehat{\Phi}_I$-mesh quiver

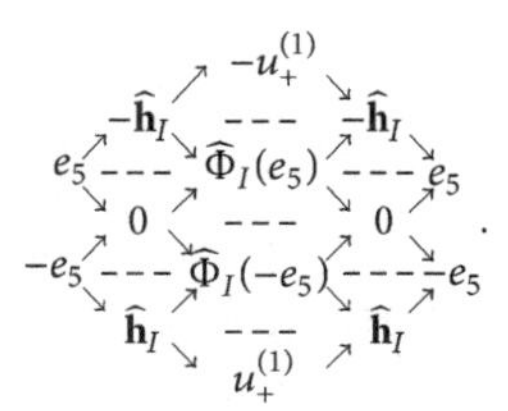 (105)

Analogously, we construct the following two $\widehat{\Phi}_I$-mesh quivers

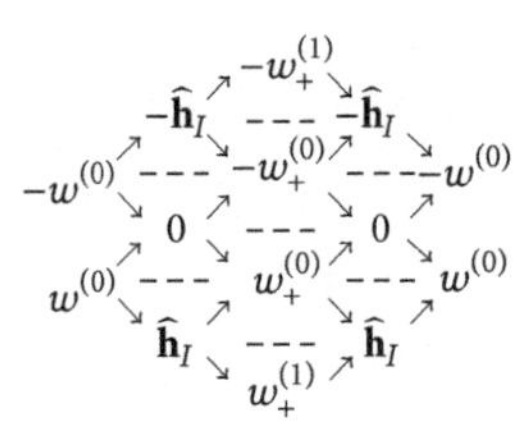

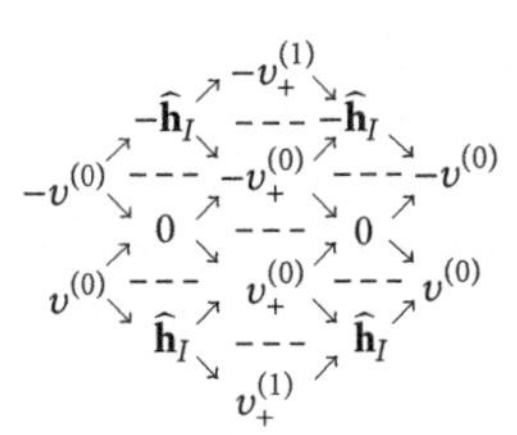 (106)

where

$$u^{(1)} = (2, -2, -2, -2, 1),$$

$$u_+^{(1)} = \widehat{\Phi}_I\left(u^{(1)}\right),$$

$$w^{(0)} = (1, 0, 0, -1, 0),$$

$$w_+^{(0)} = \widehat{\Phi}_I\left(w^{(0)}\right) = (0, 1, -1, 0, 0),$$

$$v^{(0)} = (0, 1, 0, -1, 0),$$

$$v_+^{(0)} = \widehat{\Phi}_I\left(v^{(0)}\right) = (1, 0, -1, 0, 0),$$

$$w_+^{(1)} = (2, 1, -1, -2, 0),$$

$$v_+^{(1)} = (1, 2, -1, -2, 0). \tag{107}$$

We recall that if $v \in \widehat{\partial}_I^0 \mathscr{R}_{\widehat{q}_I}$, then v or $-v$ is one of the vectors presented in Table 2. It follows that the $\widehat{\Phi}_I$-orbits in $\widehat{\partial}_I^0 \mathscr{R}_{\widehat{q}_I} \cup \mathbb{Z} \cdot \widehat{\mathbf{h}}_I$ form three infinite $\widehat{\Phi}_I$-mesh sand-glass tubes $\widehat{\mathscr{T}}_u$, $\widehat{\mathscr{T}}_w$, $\widehat{\mathscr{T}}_v$ of rank (2, 1), and each of them has the shape presented in (109)

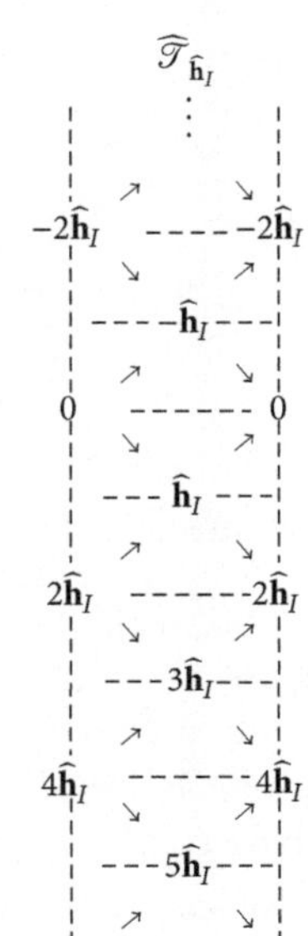 (108)

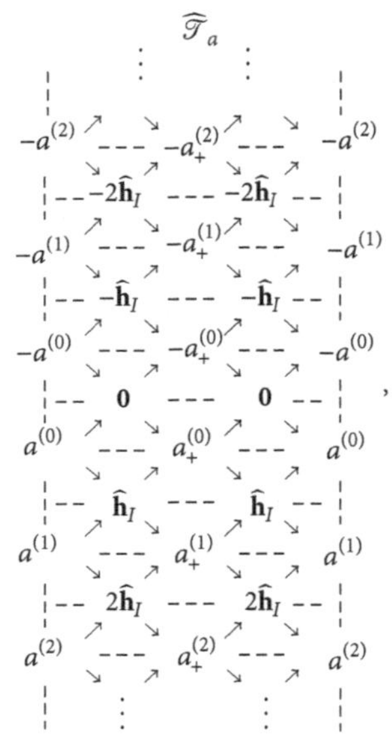

(109)

where a is one of the vectors

$$u = u^{(0)} = (0,0,0,0,-1),$$
$$w = w^{(0)} = (1,0,0,-1,0), \quad (110)$$
$$v = v^{(0)} = (0,1,0,-1,0).$$

Construct the disjoint union $\widehat{\mathcal{T}}_u \cup \widehat{\mathcal{T}}_w \cup \widehat{\mathcal{T}}_v$ of the tubes $\widehat{\mathcal{T}}_u$, $\widehat{\mathcal{T}}_w$, $\widehat{\mathcal{T}}_v$, and note that each of them contains the tube $\widehat{\mathcal{T}}_{\widehat{\mathbf{h}}_I}$. By making the identification of the vectors $m \cdot \widehat{\mathbf{h}}_I$, with $m \in \mathbb{Z}$, lying in the corresponding $\widehat{\Phi}_I$-orbits, we get the quotient $\widehat{\Phi}_I$-mesh translation quiver

$$\Gamma\left(\widehat{\partial}_I^0 \mathcal{R}_{\widehat{q}_I} \cup \operatorname{Ker} \widehat{q}_I, \widehat{\Phi}_I\right) = \frac{\widehat{\mathcal{T}}_u \cup \widehat{\mathcal{T}}_w \cup \widehat{\mathcal{T}}_v}{\simeq} \quad (111)$$

that has a shape of a threefold sand-glass tube of rank (2, 2, 2, 1) in the sense of [40]. It is obtained from the disjoint union of three copies of the onefold sand-glass tube of rank (2, 1) presented in Figure 3 (see also [34, Figure 5.8]) by making an obvious identification of their waist vectors.

(6) *A $\mathbb{Z}$-congruence of the bigraph $\widehat{\Delta}_I$ with the Euclidean diagram $\widetilde{\mathbb{D}}_4$.* Since we have $\operatorname{cox}_I(t) = F_{\widetilde{\mathbb{D}}_4}(t) = t^5 + t^4 - 2t^3 - 2t^2 + t + 1$ and $\mathbf{specc}_I = \mathbf{specc}_{\widetilde{\mathbb{D}}_4}$, the Euclidean diagram $\widetilde{\mathbb{D}}_4$ is the diagram DI associated to I. A technique developed in [2, 17, 18, 34, 40] allows us to construct a $\mathbb{Z}$-invertible matrix $B \in \mathrm{Gl}(5, \mathbb{Z})$ such that the following diagrams are commutative:

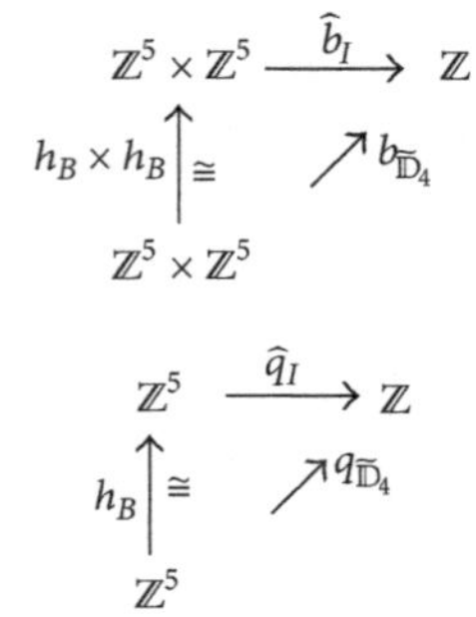

(112)

where $b_{\widetilde{\mathbb{D}}_4}$ and $q_{\widetilde{\mathbb{D}}_4}$ are the forms of the Euclidean diagram

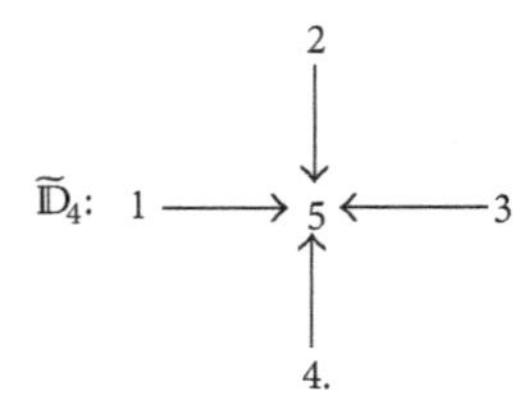

(113)

defined by the formulae $b_{\widetilde{\mathbb{D}}_4}(x, y) = x_1y_1 + x_2y_2 + x_3y_3 + x_4y_4 + x_5y_5 - (x_1 + x_2 + x_3 + x_4)y_5 = x \cdot \check{G}_{\widetilde{\mathbb{D}}_4} \cdot y^{\mathrm{tr}}$, $q_{\widetilde{\mathbb{D}}_4}(x) = b_{\widetilde{\mathbb{D}}_4}(x, x)$, for $x, y \in \mathbb{Z}^5$, $h_B : \mathbb{Z}^5 \to \mathbb{Z}^5$ is the group automorphism defined by the formula $h_B(x) = x \cdot B$, and

$$B = \begin{bmatrix} 1 & 0 & -1 & -1 & 0 \\ 0 & 0 & -1 & 0 & -1 \\ 0 & 0 & 0 & -1 & -1 \\ 0 & 1 & -1 & -1 & 0 \\ 0 & 0 & 1 & 1 & 1 \end{bmatrix},$$
$$\check{G}_{\widetilde{\mathbb{D}}_4} = \begin{bmatrix} 1 & 0 & 0 & 0 & -1 \\ 0 & 1 & 0 & 0 & -1 \\ 0 & 0 & 1 & 0 & -1 \\ 0 & 0 & 0 & 1 & -1 \\ 0 & 0 & 0 & 0 & 1 \end{bmatrix}. \quad (114)$$

It is easy to check that the equality $\check{G}_{\widetilde{\mathbb{D}}_4} = B \cdot \widehat{C}_I \cdot B^{\mathrm{tr}}$ holds, and therefore the diagrams (112) are commutative. Furthermore, by the same technique, we construct another matrix

$$B_1 = \begin{bmatrix} -1 & 0 & 0 & 0 & -1 \\ 0 & 0 & 0 & -1 & 0 \\ 0 & 0 & -1 & 0 & 0 \\ 0 & -1 & 0 & 0 & -1 \\ 1 & 1 & 0 & 0 & 1 \end{bmatrix} \quad (115)$$

such that the equality $\check{G}_{\widetilde{\mathbb{D}}_4} = B_1 \cdot \widehat{C}_I \cdot B_1^{\mathrm{tr}}$ holds.

6. Concluding Remarks

6.1 It follows from Lemma 3 and the results obtained recently in [3, 4] that for any connected positive (resp.,

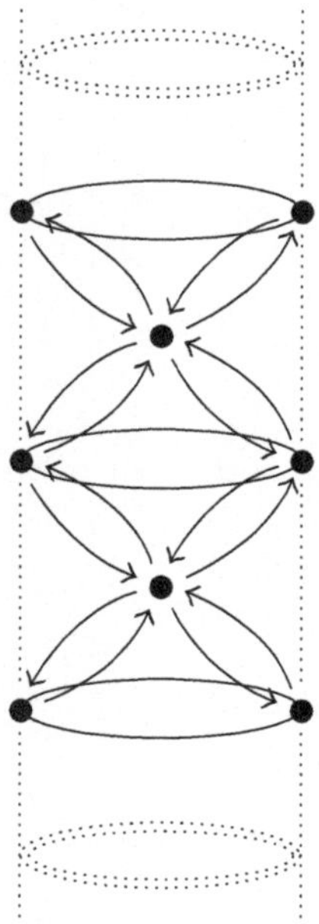

FIGURE 3: Sand-glass tube of rank (2, 1).

principal) poset J, there exists a simply laced Dynkin diagram $DJ \in \{\mathbb{A}_m, \mathbb{D}_m, \mathbb{E}_6, \mathbb{E}_7, \mathbb{E}_8\}$ (resp., a simply laced Euclidean diagram DJ), uniquely determined by J, such that the symmetric Gram matrices G_J, G_{DJ} are $\mathbb{Z}$-congruent.

Analogous Coxeter spectral classification of one-peak posets I, with almost P-critical Tits form $\widehat{q}_I : \mathbb{Z}^I \to \mathbb{Z}$, is obtained in [33] by a reduction to computer calculations.

6.2. Although the Coxeter spectral classification problem for arbitrary finite posets remains unsolved, we have a solution for positive one-peak posets. Indeed, it follows from the results in [17] that for any one-peak positive poset J, there exists a simply laced Dynkin diagram $DJ \in \{\mathbb{A}_s, \mathbb{D}_n, \mathbb{E}_6, \mathbb{E}_7, \mathbb{E}_8\}$ (uniquely determined by J) such that $\mathbf{specc}_J = \mathbf{specc}_{DJ}$, the nonsymmetric Gram matrices $\check{G}_J$, $\check{G}_{DJ}$ are $\mathbb{Z}$-congruent, and the symmetric Gram matrices G_J, G_{DJ} are $\mathbb{Z}$-congruent.

6.3. We can determine the diagram DJ as follows. Fix an upper-triangular numbering $\{a_1, \ldots, a_m\}$ of elements of J. Then, the incidence matrix $C_J \in \mathbb{M}_m(\mathbb{Z})$ is upper-triangular, and the Euler matrix $\overline{C}_J := C_J^{-1}$ is also upper triangular. Then, the Euler edge-bipartite graph $\overline{\Delta}_J$ (33) is loop-free, and we have $C_{\overline{\Delta}_J} = \overline{C}_J$. Hence, the symmetric Gram matrices $G_{\overline{\Delta}_J}, \overline{G}_J$ coincide, and, by Lemma 3, the poset J is positive (resp., principal) if and only if the bigraph Δ_J is positive (resp., principal). By applying to $\overline{\Delta}_J$ the inflation algorithm constructed in [4, 21] (see also [45]), we get (in a finite number of steps) an edge-bipartite graph $D\Delta_J$ such that the symmetric Gram matrix $G_{\overline{\Delta}_J} = \overline{G}_J$ is $\mathbb{Z}$-congruent with the symmetric Gram matrix $G_{D\Delta_J}$, and the edge-bipartite graph $D\Delta_J$ has no dotted edges; that is, $D\Delta_J$ is a (multi) graph. We set $DJ := D\Delta_J$. It follows from the results in [3, 4] that DJ is a simply laced Dynkin diagram, if J is positive, and DJ is a simply laced Euclidean diagram, if J is principal. Moreover, the matrix $\overline{G}_J$ is $\mathbb{Z}$-congruent with G_{DJ}. Since the incidence Gram matrix G_J of J is $\mathbb{Z}$-congruent with the matrix $\overline{G}_J$ (by Proposition 5), then the matrices G_J and G_{DJ} are $\mathbb{Z}$-congruent.

6.4. Although we can apply in 6.3 the inflation algorithm to the incidence edge-bipartite graph Δ_J, we use in the construction of DJ the Euler edge-bipartite graph $\overline{\Delta}_J$, because the number of nonzero entries in the Euler matrix $\overline{C}_J := C_J^{-1}$ does not increase the number for the matrix C_J; see [28, Proposition 2.12]. It follows that the number of the dotted edges in $\overline{\Delta}_J$ does not increase the number of the dotted edges in Δ_J, and the use in 6.3 the bigraph $\overline{\Delta}_J$ reduces the time of calculation in the procedure $\overline{\Delta}_J \mapsto DJ$.

Acknowledgment

The research is supported by Polish Research Grant NCN 2011/03/B/ST1/00824.

References

[1] D. Simson, "Integral bilinear forms, Coxeter transformations and Coxeter polynomials of finite posets," *Linear Algebra and Its Applications*, vol. 433, no. 4, pp. 699–717, 2010.

[2] D. Simson, "Algorithms determining matrix morsifications, Weyl orbits, Coxeter polynomials and mesh geometries of roots for Dynkin diagrams," *Fundamenta Informatica*, vol. 123, 2013.

[3] D. Simson, "A framework for Coxeter spectral analysis of loop-free edge-bipartite graphs, their rational morsifications and mesh geometries of root orbits," *Fundamenta Informaticae*, vol. 125, 2013.

[4] D. Simson, "A Coxeter-Gram classification of simply-laced edge bipartite graphs," *SIAM Journal on Discrete Mathematics*, vol. 27, 2013.

[5] T. Zaslavsky, "Signed graphs," *Discrete Applied Mathematics*, vol. 4, no. 1, pp. 47–74, 1982.

[6] G. R. Belitskii and V. V. Sergeichuk, "Complexity of matrix problems," *Linear Algebra and its Applications*, vol. 361, pp. 203–222, 2003.

[7] V. M. Bondarenko, "Representations of bundles of semichained sets and their applications," *Algebra i Analiz*, vol. 3, no. 5, pp. 38–61, 1991 (Russian).

[8] V. M. Bondarenko, "Representations of bundles of semichained sets and their applications," *St. Petersburg Mathematical Journal*, vol. 3, pp. 973–996, 1992.

[9] V. M. Bondarenko, "Minimax isomorphism algorithm and primitive posets," *Algebra and Discrete Mathematics*, vol. 12, no. 2, pp. 31–37, 2011.

[10] V. M. Bondarenko, T. G. Gerasimova, and V. V. Sergeichuk, "Pairs of mutually annihilating operators," *Linear Algebra and its Applications*, vol. 430, no. 1, pp. 86–105, 2009.

[11] V. M. Bondarenko and M. V. Styopochkina, "On posets of width two with positive Tits form," *Algebra and Discrete Mathematics*, vol. 2, no. 2, pp. 20–35, 2005.

[12] V. M. Bondarenko and M. V. Styopochkina, "On finite posets of *P*-finite type and their Tits forms," *Algebra and Discrete Mathematics*, vol. 2, pp. 17–21, 2006.

[13] V. M. Bondarenko and M. V. Stepochkina, "(Min, max)-equivalence of posets, and nonnegative Tits forms," *Ukrains'kyi Matematychnyi Zhurnal*, vol. 60, no. 9, pp. 1157–1167, 2008.

[14] V. M. Bondarenko and M. V. Stepochkina, "Description of partially ordered sets that are critical with respect to the nonnegativity of the quadratic Tits form," *Ukrains'kyi Matematychnyi Zhurnal*, vol. 61, no. 5, pp. 611–624, 2009.

[15] R. A. Brualdi, "Spectra of digraphs," *Linear Algebra and its Applications*, vol. 432, no. 9, pp. 2181–2213, 2010.

[16] P. Gabriel and A. V. Roiter, *Representations of Finite Dimensional Algebras, Algebra VIII*, vol. 73 of *Encyclopaedia of Mathematical Sciences*, Springer, Berlin, Germany, 1992.

[17] M. Gąsiorek and D. Simson, "One-peak posets with positive quadratic Tits form, their mesh translation quivers of roots, and programming in Maple and Python," *Linear Algebra and its Applications*, vol. 436, no. 7, pp. 2240–2272, 2012.

[18] M. Gąsiorek and D. Simson, "A computation of positive one-peak posets that are Tits-sincere," *Colloquium Mathematicum*, vol. 127, pp. 83–103, 2012.

[19] A. Kisielewicz and M. Szykuła, "Rainbow induced subgraphs in proper vertex colorings," *Fundamenta Informaticae*, vol. 111, no. 4, pp. 437–451, 2011.

[20] J. Kosakowska, "A classification of two-peak sincere posets of finite prinjective type and their sincere prinjective representations," *Colloquium Mathematicum*, vol. 87, no. 1, pp. 27–77, 2001.

[21] J. Kosakowska, "Inflation algorithms for positive and principal edge-bipartite graphs and unit quadratic forms," *Fundamenta Informaticae*, vol. 119, pp. 149–162, 2012.

[22] S. Ladkani, "On the periodicity of Coxeter transformations and the non-negativity of their Euler forms," *Linear Algebra and its Applications*, vol. 428, no. 4, pp. 742–753, 2008.

[23] P. Lakatos, "Additive functions on trees," *Colloquium Mathematicum*, vol. 89, no. 1, pp. 135–145, 2001.

[24] S. A. Ovsienko, "Integral weakly positive forms," in *Schur Matrix Problems and Quadratic Forms*, pp. 3–17, Akademiia Nauk SSSR, 1978.

[25] V. V. Sergeichuk, "Canonical matrices for linear matrix problems," *Linear Algebra and its Applications*, vol. 317, no. 1–3, pp. 53–102, 2000.

[26] D. Simson, *Linear Representations of Partially Ordered Sets and Vector Space Categories, Algebra, Logic and Applications*, vol. 4, Gordon and Breach Science, New York, NY, USA, 1992.

[27] D. Simson, "Posets of finite prinjective type and a class of orders," *Journal of Pure and Applied Algebra*, vol. 90, no. 1, pp. 77–103, 1993.

[28] D. Simson, "Incidence coalgebras of intervally finite posets, their integral quadratic forms and comodule categories," *Colloquium Mathematicum*, vol. 115, no. 2, pp. 259–295, 2009.

[29] A. P. Wojda, "A condition for a graph to contain *k*-matching," *Discrete Mathematics*, vol. 276, no. 1–3, pp. 375–378, 2004.

[30] A. P. Wojda, M. Woźniak, and I. A. Zioło, "On self-complementary cyclic packing of forests," *Electronic Journal of Combinatorics*, vol. 14, no. 1, 11 pages, 2007.

[31] Y. B. Zhang, "Eigenvalues of Coxeter transformations and the structure of regular components of an Auslander-Reiten quiver," *Communications in Algebra*, vol. 17, no. 10, pp. 2347–2362, 1989.

[32] Ju. A. Drozd, "Coxeter transformations and representations of partially ordered sets," *Functional Analysis and its Applications*, vol. 8, no. 3, pp. 219–225, 1974.

[33] A. Polak and D. Simson, "One-peak posets with almost *P*-critical Tits form and a spectral Coxeter classification using computer algebra tools," *European Journal of Combinatorics*. To appear.

[34] D. Simson, "Mesh algorithms for solving principal Diophantine equations, sand-glass tubes and tori of roots," *Fundamenta Informaticae*, vol. 109, no. 4, pp. 425–462, 2011.

[35] R. A. Horn and V. V. Sergeichuk, "Congruences of a square matrix and its transpose," *Linear Algebra and its Applications*, vol. 389, pp. 347–353, 2004.

[36] I. Assem, D. Simson, and A. Skowroński, *Elements of the Representation Theory of Associative Algebras, Volume 1*, Techniques of Representation Theory, London Mathematical Society Student Texts 65, Cambridge University Press, Cambridge, UK, 2006.

[37] L. F. Barannyk, "Finite-dimensional twisted group algebras of semi-wild representation type," *Colloquium Mathematicum*, vol. 120, no. 2, pp. 277–298, 2010.

[38] D. Simson and A. Skowroński, *Elements of the Representation Theory of Associative Algebras, Volume 2*, Tubes and Concealed Algebras of Euclidean Type, London Mathematical Society Student Texts 71, Cambridge University Press, Cambridge, UK, 2007.

[39] D. Simson and M. Wojewódzki, "An algorithmic solution of a Birkhoff type problem," *Fundamenta Informaticae*, vol. 83, no. 4, pp. 389–410, 2008.

[40] D. Simson, "Mesh geometries of root orbits of integral quadratic forms," *Journal of Pure and Applied Algebra*, vol. 215, no. 1, pp. 13–34, 2011.

[41] M. Barot and J. A. de la Peña, "The Dynkin type of a non-negative unit form," *Expositiones Mathematicae*, vol. 17, no. 4, pp. 339–348, 1999.

[42] M. Sato, "Periodic Coxeter matrices and their associated quadratic forms," *Linear Algebra and its Applications*, vol. 406, pp. 99–108, 2005.

[43] M. Gąsiorek, D. Simson, and K. Zając, "Algorithmic computation of principal posets using Maple and Python," *Algebra and Discrete Mathematics*. To appear.

[44] G. Marczak, A. Polak, and D. Simson, "P-critical integral quadratic forms and positive unit forms: an algorithmic approach," *Linear Algebra and its Applications*, vol. 433, no. 11-12, pp. 1873–1888, 2010.

[45] D. Simson and K. Zając, "An inflation algorithm and a toroidal mesh algorithm for edge-bipartite graphs," *Electronic Notes in Discrete Mathematics*. To appear.

Some Nonunique Common Fixed Point Theorems in Symmetric Spaces through $\mathrm{CLR}_{(S,T)}$ Property

E. Karapınar,[1] D. K. Patel,[2] M. Imdad,[3] and D. Gopal[2]

[1] *Department of Mathematics, Atilim University, Incek, 06836 Ankara, Turkey*
[2] *Department of Mathematics and Humanities, S. V. National Institute of Technology, Surat, Gujarat 395 007, India*
[3] *Department of Mathematics, Aligarh Muslim University, Aligarh 202 002, India*

Correspondence should be addressed to E. Karapınar; erdalkarapinar@yahoo.com

Academic Editor: N. Hussain

We introduce a new class of mappings satisfying the "common limit range property" in symmetric spaces and utilize the same to establish common fixed point theorems for such mappings in symmetric spaces. Our results generalize and improve some recent results contained in the literature of metric fixed point theory. Some illustrative examples to highlight the realized improvements are also furnished.

1. Introduction

In 1986, Jungck [1] generalized the idea of weakly commuting pair of mappings due to Sessa [2] by introducing the notion of compatible pair of mappings and also showed that compatible pair of mappings commute on the set of coincidence points of the involved mappings. Recall that a point $x \in X$ is called a coincidence point of the pair of self-mappings (f, g) defined on X if $fx = gx(= w)$ while the point w is then called a point of coincidence for the pair (f, g). In the recent past and even now, the concept of compatible mappings is frequently used to prove results on the existence of common fixed points. The study of common fixed points of noncompatible pairs is also equally natural and fascinating. Pant [3] initiated the study of noncompatible pairs employing the idea of pointwise R-weakly commuting pairs. Pant [4] proved an interesting fixed point theorem for maps satisfying Lipschitz type conditions. In recent years, the result of Pant [4] was generalized and improved by Sastry and Murthy [5] (see also [6]) by introducing the idea of tangential maps (or the property (E.A)) and g-continuity. In continuation of this, Imdad and Soliman [7] and Soliman et al. [8] extended the results of Sastry and Murthy [5] as well as Pant [4] to symmetric space utilizing the idea of weakly compatible pair together with common property (E.A) (a notion due to Liu et al. [9]). For more references on the recent development of common fixed point theory in symmetric spaces, we refer readers to [10–14]. Most recently, Gopal et al. [15] improved these results by utilizing the idea of absorbing pair which is essentially due to Gopal et al. [16].

In this paper, we introduce a new notion called the common limit range property and show that this new notion buys a typically required condition up to a pair of mappings along with the notion of absorbing property in proving common fixed point theorems for Lipschitz type mappings in symmetric spaces. Consequently, the relevant recent fixed point theorems due to Soliman et al. [8] and Gopal et al. [15] are generalized and improved.

2. Preliminaries

A symmetric d on a nonempty set X is a function $d : X\times X \to [0,\infty)$ which satisfies $d(x, y) = d(y, x)$ and $d(x, y) = 0 \Leftrightarrow x = y$ (for all $x, y \in X$). If d is a symmetric on a set X, then for $x \in X$ and $\epsilon > 0$, we write $B(x,\epsilon) = \{y \in X : d(x, y) < \epsilon\}$. A topology $\tau(d)$ on X is given by the sets U (along with empty set) in which for each $x \in U$, one can find some $\epsilon > 0$ such that $B(x,\epsilon) \subset U$. A set $S \subset X$ is a neighbourhood of $x \in X$ if and only if there is a U containing x such that $x \in$

$U \subset S$. A symmetric d is said to be a semimetric if for each $x \in X$ and for each $\epsilon > 0$, $B(x, \epsilon)$ is a neighbourhood of x in the topology $\tau(d)$. Thus a symmetric (resp. a semimetric) space X is a topological space whose topology $\tau(d)$ on X is induced by a symmetric (resp. a semimetric) d. Notice that $\lim_{n\to\infty} d(x_n, x) = 0$ if and only if $x_n \to x$ in the topology $\tau(d)$. The distinction between a symmetric and a semimetric is apparent as one can easily construct a semimetric d such that $B(x, \epsilon)$ need not be a neighbourhood of x in $\tau(d)$.

Since symmetric spaces are not necessarily Hausdorff and the symmetric d is not generally continuous, in the course of proving fixed point theorems, some additional axioms are required. The following axioms are relevant to this note which are available in the papers of Aliouche [17], Galvin and Shore [18], Hicks and Rhoades [19], and Wilson [20].

(W_3) [20] Given $\{x_n\}$, x and y in X with $d(x_n, x) \to 0$ and $d(x_n, y) \to 0$ imply $x = y$.

(W_4) [20] Given $\{x_n\}, \{y_n\}$ and an x in X with $d(x_n, x) \to 0$ and $d(x_n, y_n) \to 0$ imply $d(y_n, x) \to 0$.

(HE) [17] Given $\{x_n\}, \{y_n\}$ and an x in X with $d(x_n, x) \to 0$ and $d(y_n, x) \to 0$ imply $d(x_n, y_n) \to 0$.

$(1C)$ [18] A symmetric d is said to be 1-continuous if $\lim_{n\to\infty} d(x_n, x) = 0$ implies $\lim_{n\to\infty} d(x_n, y) = d(x, y)$.

(CC) [18] A symmetric d is said to be continuous if $\lim_{n\to\infty} d(x_n, x) = 0$ and $\lim_{n\to\infty} d(y_n, y) = 0$ imply $\lim_{n\to\infty} d(x_n, y_n) = d(x, y)$ where x_n, y_n are sequences in X and $x, y \in X$.

Clearly, (CC) implies $(1C)$ but not conversely. Also (W_4) implies (W_3) and $(1C)$ implies (W_3) but converse implications are not true. All other possible implications amongst (W_3), $(1C)$, and (HE) are not true in general. A nice illustration via demonstrative examples is given by Cho et al. [21]. However, (CC) implies all the remaining four conditions namely: (W_3), (W_4), (HE), and $(1C)$.

Recall that a sequence $\{x_n\}$ in a semimetric space (X, d) is said to be d-Cauchy if it satisfies the usual metric condition. Here, one needs to notice that in a semimetric space, Cauchy convergence criterion is not a necessary condition for the convergence of a sequence but this criterion becomes a necessary condition if semimetric is suitably restricted (see Wilson [20]). In [22], Burke furnished an illustrative example to show that a convergent sequence in semimetric spaces need not admit a Cauchy subsequence. He was able to formulate an equivalent condition under which every convergent sequence in semimetric space admits a Cauchy subsequence. There are several concept of completeness in semimetric space for example, S-completeness, d-Cauchy completeness, strong and weak completeness (see Wilson [20]). We omit the details of these notions which are not relevant to this paper.

Let (f, g) be a pair of self-mappings defined on a nonempty set X equipped with a symmetric (semimetric) d. Then for the pair (f, g), we recall some relevant concepts as follows.

A pair (f, g) of self-mappings is said to be

(i) compatible (cf. [1]) if $\lim_{n\to\infty} d(fgx_n, gfx_n) = 0$ whenever $\{x_n\}$ is a sequence such that $\lim_{n\to\infty} fx_n = \lim_{n\to\infty} gx_n = t$ for some t in X,

(ii) noncompatible (cf. [4, 23]) if there exists at least one sequence $\{x_n\}$ such that $\lim_{n\to\infty} fx_n = \lim_{n\to\infty} gx_n = t$ for some t in X but $\lim_{n\to\infty}(fgx_n, gfx_n)$ is either nonzero or nonexistent,

(iii) tangential (or satisfying the property (E.A)) (cf. [5, 24]) if there exists a sequence $\{x_n\}$ in X such that $\lim_{n\to\infty} fx_n = \lim_{n\to\infty} gx_n = t$ for some $t \in X$.

Let Y be an arbitrary set and X be a nonempty set equipped with symmetric (semimetric) d. Then the pairs (f, S) and (g, T) of mappings from Y into X are said to have,

(iv) (cf. [9]) the common property (E.A) if there exist two sequences $\{x_n\}$ and $\{y_n\}$ in Y such that

$$\lim_{n\to\infty} fx_n = \lim_{n\to\infty} Sx_n = \lim_{n\to\infty} gy_n = \lim_{n\to\infty} Ty_n = t \quad \text{for some } t \in X, \tag{1}$$

while the pair (g, T) is said to have

(v) the common limit range property with respect to the map g (denoted by (CLRg)) (cf. [25–29]) if there exists a sequence $\{x_n\}$ in X such that $\lim_{n\to\infty} Tx_n = \lim_{n\to\infty} gx_n = gu$ for some $u \in X$,

(vi) let Y be an arbitrary set and X be a nonempty set equipped with symmetric (semimetric) d. Then f is said to be g-continuous (cf. [5]) if $gx_n \to gx \Rightarrow fx_n \to fx$ whenever $\{x_n\}$ is a sequence in Y and $x \in Y$,

(vii) a pair (f, g) of self-mappings defined on a set X is said to be weakly compatible (or partially commuting or coincidentally commuting (cf. [5, 30])) if the pair commutes on the set of coincidence points that is, $fx = gx$ (for $x \in X$) implies that $fgx = gfx$,

(viii) let f and $g (f \neq g)$ be two self-mappings defined on a symmetric (or semimetric) space (X, d), then f is called g-absorbing if there exists some real number $R > 0$ such that $d(gx, gfx) \leq Rd(fx, gx)$ for all x in X. Analogously, g will be called f-absorbing (cf. [16]) if there exists some real number $R > 0$ such that $d(fx, fgx) \leq Rd(fx, gx)$ for all x in X. The pair of self maps (f, g) will be called absorbing if it is both g-absorbing as well as f-absorbing,

(ix) let f and g $(f \neq g)$ be two self-mappings defined on a symmetric (or semimetric) space (X, d), then f is called pointwise g-absorbing if for given x in X, there exists some $R > 0$ such that $d(gx, gfx) \leq Rd(fx, gx)$,

On similar lines, we can define pointwise f-absorbing map. If we take $g = I$, the identity map on X, then f is trivially I-absorbing. Similarly, I is f-absorbing in respect of any f. It has been shown in [16] that a pair of compatible or R-weakly commuting pair need not be g-absorbing or

f-absorbing. Also absorbing pairs are neither a subclass of compatible pairs nor a subclass of noncompatible pairs as the absorbing pairs need not commute at their coincidence points. For other properties and related results for absorbing pair of maps, one can consult [16].

For the sake of completeness, we state below some theorems contained in Soliman et al. [8] and Gopal et al. [15].

Theorem 1 (see cf. [8]). *Let Y be an arbitrary nonempty set while X be another nonempty set equipped with a symmetric (semimetric) d which enjoys (W_3) (Hausdorffness of $\tau(d)$) and (HE). Let $f, g, S, T : Y \rightarrow X$ be four mappings which satisfy the following conditions:*

(i) *f is S-continuous and g is T-continuous,*

(ii) *the pairs (f, S) and (g, T) share the common property (E.A),*

(iii) *SX and TX are d-closed ($\tau(d)$-closed) subset of X (resp., $fX \subset TX$ and $gX \subset SX$).*

Then there exist $u, w \in X$ such that $fu = Su = Tw = gw$. Moreover, if $Y = X$ along with

(iv) *the pairs (f, S) and (g, T) are weakly compatible and*

(v) *$(fx, gfx) \neq \max\{d(Sx, Tfx), d(gfx, Tfx), d(fx, Tfx), d(fx, Sx), d(gfx, Sx)\}$, whenever the right hand side is nonzero.*

Then, f, g, S, and T have a common fixed point in X.

Theorem 2 (see cf. [15]). *Let Y be an arbitrary nonempty set while X be another nonempty set equipped with a symmetric (semimetric) d which enjoys (W_3) (Hausdorffness of $\tau(d)$) and (HE). Let $f, g, S, T : Y \rightarrow X$ be four mappings which satisfy the following conditions:*

(i) *f is S-continuous and g is T-continuous,*

(ii) *the pairs (f, S) and (g, T) share the common property (E.A),*

(iii) *TY is a d-closed ($\tau(d)$-closed) subset of X and $gY \subset SY$ (resp., SY is a d-closed ($\tau(d)$-closed) subset of X and $fY \subset TY$).*

Then, there exist $u, w \in Y$ such that $fu = Su = Tw = gw$.

Moreover, if $Y = X$, then f, g, S and T have a common fixed point provided the pairs (f, S) and (g, T) are pointwise absorbing.

Theorem 3 (see cf. [15]). *Let Y be an arbitrary set while (X, d) be a symmetric (semimetric) space equipped with a symmetric (semimetric) d which enjoys (W_3) (Hausdorffness of $\tau(d)$) and (HE). Let $f, g, S, T : Y \rightarrow X$ be four mappings which satisfy the following conditions:*

(i) *the pair (g, T) satisfies the property (E.A) (resp., (f, S) satisfies the property (E.A)),*

(ii) *TY is a d-closed ($\tau(d)$-closed) subset of X and $gY \subset SY$ (resp., SY is a d-closed ($\tau(d)$-closed) subset of X and $fY \subset TY$) and*

(iii) *$d(fx, gy) \leq km(x, y)$ for any $x, y \in X$ where $k \geq 0$ and $m(x, y) = \max\{d(Sx, Ty), \min\{d(fx, Sx), d(gy, Ty)\}, \min\{d(fx, Ty), d(gy, Sx)\}\}$.*

Then, there exist $u, w \in Y$ such that $fu = Su = Tw = gw$.

Moreover, if $Y = X$, then f, g, S and T have a common fixed point provided the pairs (f, S) and (g, T) are pointwise absorbing.

In this paper, we provide a unified approach to certain theorems in symmetric (semimetric) spaces using a blend of common limit range property along with absorbing pair property and obtain generalizations of various results due to Gopal et al. [15], Soliman et al. [8], Pant [31], Sastry and Murthy [5], Imdad et al. [7], Cho et al. [21], and some others.

3. Main Results

We start to section with the following definition.

Definition 4. Let f, g, S, and T be four self-mappings defined on a symmetric space (X, d). Then the pairs (f, S) and (g, T) are said to have the common limit range property (with respect to S and T), often denoted by $\mathrm{CLR}_{(S,T)}$, if there exist two sequences $\{x_n\}$ and $\{y_n\}$ in X such that

$$\lim_{n\to\infty} fx_n = \lim_{n\to\infty} Sx_n = \lim_{n\to\infty} gy_n = \lim_{n\to\infty} Ty_n = t \qquad (2)$$

with $t = Su = Tw$, for some $t, u, w \in X$.

If $f = g$ and $S = T$, then the above definition implies (CLRg) property due to Sintunavarat and Kumam [28]. Also notice that the preceeding definition implies the common property (E.A) but the converse implication is not true in general. The following example substantiates this fact.

Example 5. Consider $X = [2, 20]$ equipped with the symmetric defined by $d(x, y) = (x - y)^2$ for all $x, y \in X$ which satisfies (W_3) and (HE). Define self mappings f, g, S and T on X as

$$fx = \begin{cases} 2 & \text{if } x = 2, \\ 7 & \text{if } 2 < x \leq 5, \\ \dfrac{2x+5}{3} & \text{if } x > 5, \end{cases}$$
$$Sx = \begin{cases} 3 & \text{if } x = 2, \\ 2 & \text{if } 2 < x \leq 5, \\ \dfrac{x+5}{2} & \text{if } x > 5, \end{cases}$$
$$gx = \begin{cases} 4 & \text{if } x = 2, \\ \dfrac{4x+7}{3} & \text{if } 2 < x \leq 5, \\ 3 & \text{if } x > 5, \end{cases} \qquad (3)$$
$$Tx = \begin{cases} 6 & \text{if } x = 2, \\ \dfrac{3x+4}{2} & \text{if } 2 < x \leq 5, \\ 4 & \text{if } x > 5. \end{cases}$$

For sequences $x_n = 5 + (1/n)$ and $y_n = 2 + (1/n)$, we have

$$\lim_{n\to\infty} fx_n = \lim_{n\to\infty} Sx_n = \lim_{n\to\infty} gy_n = \lim_{n\to\infty} Ty_n = 5 (= t), \quad (4)$$

which shows that the pairs (f, S) and (g, T) share the common property (E.A). However, there does not exist points u and w in X for which $t = Su = Tw$.

In view of the preceeding example, the following proposition is predictable.

Proposition 6. *If the pairs (f, S) and (g, T) share the common property (E.A) and $S(X)$ as well as $T(X)$ are closed subsets of X, then the pairs also share the $CLR_{(S,T)}$ property.*

We now prove our first result employing S-continuity of f and T-continuity of g instead of utilizing some Lipschitz or contractive type condition.

Theorem 7. *Let Y be an arbitrary nonempty set while X be a nonempty set equipped with a symmetric (semimetric) d which enjoys (W_3) (Hausdorffness of $\tau(d)$) and (HE). If $f, g, S, T : Y \to X$ are four mappings which satisfy the following conditions:*

(i) *f is S-continuous and g is T-continuous,*

(ii) *the pairs (f, S) and (g, T) satisfy the $CLR_{(S,T)}$ property,*

then, (f, S) and (g, T) have a coincidence point. Moreover, if $Y = X$, then f, g, S, and T have a common fixed point provided the pairs (f, S) and (g, T) are pointwise absorbing.

Proof. Since the pairs (f, S) and (g, T) satisfy the $CLR_{(S,T)}$ property, therefore there exist two sequences $\{x_n\}$ and $\{y_n\}$ in X such that

$$\lim_{n\to\infty} fx_n = \lim_{n\to\infty} Sx_n = \lim_{n\to\infty} gy_n = \lim_{n\to\infty} Ty_n = t \quad (5)$$

with $t = Su = Tw$, for some $t, u, w \in X$.

On using S-continuity of f along with the condition (W_3), we get $fu = Su$ which shows that u is a coincidence point of the mappings f and S. Similarly, using the T-continuity of g along with the condition (W_3), we obtain $gw = Tw$ which shows that w is a coincidence point of g and T. Owing to $CLR_{(S,T)}$ property, we have $fu = Su = gw = Tw = t$.

As the pairs (f, S) and (g, T) are pointwise absorbing, we can write

$$fu = fSu, \quad Su = Sfu, \quad gw = gTw, \quad Tw = Tgw$$
$$\Longrightarrow fu = Sfu, \quad fu = ffu, \quad gw = Tgw,$$
$$gw = ggw, \quad (6)$$

which show that $fu(fu = gw)$ is a common fixed point of f, g, S and T. This concludes the proof.

With a view to demonstrate the utility of Theorem 7 over Theorem 1 and Theorem 2, we adopt the following example.

Example 8. Consider $X = Y = (-1, 1] \cup \{2, 3, 4\}$ equipped with the symmetric defined by $d(x, y) = (x - y)^2$ for all $x, y \in X$ which satisfies (W_3) and (HE). Define self mappings f, g, S, and T on X as

$$fx = \begin{cases} \frac{3}{5} & \text{if } -1 < x \le \frac{-1}{2}, \\ \frac{x}{4} & \text{if } \frac{-1}{2} < x < \frac{1}{2}, \\ \frac{3}{5} & \text{if } \frac{1}{2} \le x < 1, \\ 3 & \text{if } x = 1, 4, \\ 2 & \text{if } x = 2, 3, \end{cases}$$

$$gx = \begin{cases} \frac{3}{5} & \text{if } -1 < x \le \frac{-1}{2}, \\ \frac{-x}{4} & \text{if } \frac{-1}{2} < x < \frac{1}{2}, \\ \frac{3}{5} & \text{if } \frac{1}{2} \le x < 1, \\ 3 & \text{if } x = 1, 4, \\ 2 & \text{if } x = 2, 3, \end{cases} \quad (7)$$

$$Sx = \begin{cases} \frac{3}{4} & \text{if } -1 < x \le \frac{-1}{2} \\ \frac{x}{2} & \text{if } \frac{-1}{2} < x < \frac{1}{2}, \\ \frac{-3}{4} & \text{if } \frac{1}{2} \le x < 1, \\ 2 & \text{if } x = 1, 2, 3, 4, \end{cases}$$

$$Tx = \begin{cases} \frac{-3}{4} & \text{if } -1 < x \le \frac{-1}{2}, \\ \frac{-x}{2} & \text{if } \frac{-1}{2} < x < \frac{1}{2}, \\ \frac{3}{4} & \text{if } \frac{1}{2} \le x < 1, \\ 2 & \text{if } x = 1, 2, 3, 4. \end{cases} \quad (8)$$

Consider sequences $\{x_n\} = \{1/(n+2)\}$ and $\{y_n\} = \{-1/(n+2)\}$ in X. Clearly,

$$\lim_{n\to\infty} fx_n = \lim_{n\to\infty} Sx_n = \lim_{n\to\infty} gy_n = \lim_{n\to\infty} Ty_n = 0 \quad (9)$$

with $0 = S(0) = T(0)$, and

$$\lim_{n\to\infty} Sx_n = 0 = S(0) \Longrightarrow \lim_{n\to\infty} fx_n = 0 = f(0),$$
$$\lim_{n\to\infty} Ty_n = 0 = T(0) \Longrightarrow \lim_{n\to\infty} gy_n = 0 = g(0), \quad (10)$$

which show that the pairs (f, S) and (g, T) share the $CLR_{(S,T)}$ property while the map f is S-continuous and the map g is T-continuous. Further $f(X) = (-1/8, 1/8) \cup \{3/5, 2, 3\} \not\subseteq T(X) = (-1/4, 1/4) \cup \{-3/4, 3/4, 2\}$ and $g(X) = (-1/8, 1/8) \cup \{3/5, 2, 3\} \not\subseteq S(X) = (-1/4, 1/4) \cup \{-3/4, 3/4, 2\}$ and evidently none of the involved subspaces are closed. Also, by a routine calculation, one can easily verify that the pairs (f, S) and (g, T) are pointwise absorbing. Thus, the involved pairs of maps (f, S) and (g, T) satisfy all the conditions of Theorem 7 and have two common fixed points namely: $x = 0$ and $x = 2$.

Notice that at $x = 1$, the involved maps do not satisfy the condition

$$d(fx, gfx) \neq \max\{d(Sx, Tfx), d(gfx, Tfx), d(fx, Tfx), d(fx, Sx), d(gfx, Sx)\}, \quad (11)$$

whenever the right hand side is nonzero. Moreover, it can also be verified that at points $x = 1$ and $y = 2$, the involved maps do not satisfy the Lipschitz type condition employed in [4]. Thus, this example substantiates the fact that Theorem 7 is genuine extension of Theorems 1 and 2.

By restricting f, g, S, and T suitably, one can derive corollaries involving two as well as three mappings. Here, it may be pointed out that any result involving three maps is itself a new result. For the sake of brevity, we opt to mention just one such corollary by restricting Theorem 7 to three mappings f, S, and T which is still new and presents yet another sharpened form of a relevant theorem contained in [15] besides admitting a nonself setting upto coincidence points.

Corollary 9. *Let Y be an arbitrary set while (X, d) be a symmetric (semimetric) space equipped with a symmetric (semimetric) d which enjoys (W_3) (Hausdorffness of $\tau(d)$) and (HE). If $f, S, T : Y \to X$ are three mappings which satisfy the following conditions:*

(i) *f is S-continuous and f is T-continuous,*

(ii) *the pairs (f, S) and (f, T) satisfy the $CLR_{(S,T)}$ property,*

then, there exist $u, w \in Y$ such that $fu = Su = Tw$. Moreover, if $Y = X$, then f, S and T have a common fixed point provided the pairs (f, S) and (f, T) are pointwise absorbing.

The following example illustrates the preceding corollary involving a pair of two self-mappings.

Example 10. Consider $X = Y = [2, 23)$ equipped with the symmetric defined by $d(x, y) = e^{|x-y|} - 1$, for all $x, y \in X$ which satisfies (W_3) and (HE). Define self mappings $f, S : X \to X$ as

$$fx = \begin{cases} 2 & \text{if } x \in \{2\} \cup (5,7) \cup (7,10) \cup (10,11) \cup (11,12) \cup (12,13) \cup (13,21) \cup (21,23), \\ \dfrac{x+5}{2} & \text{if } 2 < x \le 5, \\ 7 & \text{if } x = 7, \\ 12 & \text{if } x = 10, \\ 11 & \text{if } x = 11, 13, \\ 11.5 & \text{if } x = 12, \\ 10 & \text{if } x = 21, \end{cases}$$

$$Sx = \begin{cases} 2 & \text{if } x \in \{2\} \cup [7,10) \cup (10,11) \cup (11,12) \cup (12,13) \cup (13,21) \cup (21,22) \cup (22,23), \\ 6 & \text{if } 2 < x \le 5, \\ \dfrac{x+1}{3} & \text{if } x \in (5,7), \\ 11 & \text{if } x = 10, 11, 13, 22, \\ 11.6 & \text{if } x = 12, \\ 10 & \text{if } x = 21. \end{cases} \quad (12)$$

By routine calculations, one can easily verify that the maps in the pair (f, S) satisfies all the conditions of Corollary 9 and have two common fixed points, namely: 2 and 11. Also, the present example does not satisfy the Lipschitz type condition utilized in [4]. To view this claim, consider $x = 13$ and $y = 22$, then we have $e^9 - 1 \le k\,0 = 0$, which is a contradiction. Also, observe that at $x = 21$, the involved maps do not satisfy the condition:

$$d(fx, ffx) \neq \max\{d(Sx, Sfx), d(ffx, Sfx), d(fx, Sfx), d(fx, Sx), d(ffx, Sx)\}, \quad (13)$$

whenever the right hand side is nonzero. Here, it is worth noting that none of the earlier relevant theorems for example, Imdad and Soliman [7], Soliman et al. [8] and Gopal et al. [15] can be used in the context of this example as Corollary 9 does not require conditions on containment and closedness amongst the ranges of the involved mappings.

Our next theorem is essentially inspired by Theorem 3 due to Gopal et al. [15].

Theorem 11. *Let Y be an arbitrary set while (X, d) be a symmetric (semimetric) space equipped with a symmetric (semimetric) d which enjoys (W_3) (or Hausdorffness of $\tau(d)$) and (HE). If $f, g, S, T : Y \to X$ are four mappings which satisfy the following conditions:*

(i) *the pairs (f, S) and (g, T) satisfy the $CLR_{(S,T)}$ property,*

(ii) *$d(fx, gy) \le km(x, y)$, for any $x, y \in X$, where $k \ge 0$ and $m(x, y) = \max\{d(Sx,Ty), \min\{d(fx, Sx), d(gy, Ty)\}, \min\{d(fx, Ty), d(gy, Sx)\}\}$,*

then, there exist $u, w \in Y$ such that $fu = Su = Tw = gw$. Moreover, if $Y = X$, then f, g, S, and T have a common fixed point provided the pairs (f, S) and (g, T) are pointwise absorbing.

Proof. Since the pairs (f, S) and (g, T) share the $CLR_{(S,T)}$ property, therefore there exist two sequences $\{x_n\}$ and $\{y_n\}$ in X such that

$$\lim_{n\to\infty} fx_n = \lim_{n\to\infty} Sx_n = \lim_{n\to\infty} gy_n = \lim_{n\to\infty} Ty_n = t \quad (14)$$

with $t = Su = Tw$, for some $t, u, w \in X$.

On using condition (ii), we have

$$d(fu, gy_n) \leq k \max\{d(Su, Ty_n), \min\{d(fu, Su), d(gy_n, Ty_n)\}, \min\{d(fu, Ty_n), d(gy_n, Su)\}\} \tag{15}$$

which on letting $n \to \infty$, gives rise $\lim_{n\to\infty} d(fu, gy_n) = 0$. Now appealing to (W_3), we get $fu = Su$ so that $fu = Su = Tw$.

Next, we show that $Tw = gw$. To accomplish this, using (ii), we have

$$\begin{aligned} d(fu, gw) &\leq k \max\{d(Su, Tw), \min\{d(fu, Su), d(gw, Tw)\}, \min\{d(fu, Tw), d(gw, Su)\}\} \\ &= k \max\{d(Tw, Tw), \min\{d(fu, fu), d(gw, Tw)\}, \min\{d(fu, fu), d(gw, gw)\}\} \\ &= 0 \end{aligned} \tag{16}$$

so that $fu = gw$ and hence in all $fu = Su = gw = Tw$ which shows that both the pairs have a point of coincidence.

On using pointwise absorbing property of the pairs (f, S) and (g, T), we have

$$\begin{aligned} &fu = fSu, \quad Su = Sfu, \quad gw = gTw, \quad Tw = Tgw, \\ &\Longrightarrow fu = Sfu, \quad fu = ffu, \quad gw = Tgw, \\ &gw = ggw, \end{aligned} \tag{17}$$

which show that fu $(fu = gw)$ is a common fixed point of f, g, S, and T.

The following example demonstrates Theorem 11.

Example 12. Consider $X = Y = [0, 20)$ equipped with the symmetric $d(x, y) = (x - y)^2$ for all $x, y \in X$ which satisfies (W_3) and (HE). Set $f = g$ and $S = T$. Define $f, S : X \to X$ as follows:

$$fx = \begin{cases} 2 & \text{if } 0 \leq x \leq 2,\ x \geq \dfrac{11}{2}, \\ 6 & \text{if } 2 < x \leq 5, \\ \dfrac{x+3}{4} & \text{if } 5 < x < \dfrac{11}{2}, \\ 10 & \text{if } x = 10, \end{cases}$$
$$Sx = \begin{cases} 2 & \text{if } 0 \leq x \leq 2,\ x \geq \dfrac{11}{2}, \\ 4 & \text{if } 2 < x \leq 5, \\ \dfrac{x+1}{3} & \text{if } 5 < x < \dfrac{11}{2}, \\ 10 & \text{if } x = 10. \end{cases} \tag{18}$$

Then, by a routine calculation, it can be easily verified that f and S satisfy condition (ii) (of Theorem 11) for $k = 4.271$. Also, the mappings f and S satisfies the $CLR_{(S,T)}$ property with the sequence $x_n = 5 + 1/n$. The verification of the pointwise absorbing property of the pair (f, S) is straight forward. Thus f and S satisfy all the conditions of Theorem 11 and have two common fixed points, namely: $x = 2$ and $x = 10$.

Observe that $f(X) = [2, 17/8) \cup \{6, 10\} \not\subseteq S(X) = [2, 13/6) \cup \{4, 10\}$ and none of $f(X)$ and $S(X)$ is closed. Further, it is also worth noting that for all x with $2 < x \leq 5$ and with $f = g$ and $S = T$, the involved pair (f, S) does not satisfy the condition

$$d(fx, gfx) \neq \max\{d(Sx, Tfx), d(gfx, Tfx), d(fx, Tfx), d(fx, Sx), d(gfx, Sx)\}, \tag{19}$$

whenever the right hand side is nonzero. Thus, this example also establishes the utility of Theorem 11 over corresponding results proved in Soliman et al. [8] and Gopal et al. [15].

Remark 13. Choosing $k = 1$ in Theorem 11, we can derive a slightly sharpened form of a theorem due to Cho et al. [21] as conditions on the ranges of involved mappings are completely relaxed.

By restricting f, g, S, and T suitably, one can derive corollaries for two as well as three mappings. For the sake of brevity, we derive just one corollary by restricting Theorem 11 to three mappings which is yet another sharpened and unified form of a theorem due to Gopal et al. [15] in symmetric spaces and also remains relevant to some results in Pant [4] and Pant [31].

Corollary 14. *Suppose that (in the setting of Theorem 11) d satisfies (W_3) and (HE). If $f, S, T : Y \to X$ are three mappings which satisfy the following conditions:*

(i) *the pairs (f, S) and (f, T) satisfy the $CLR_{(S,T)}$ property,*

(ii) *$d(fx, fy) \leq k m_2(x, y)$, for any $x, y \in X$, where $k \geq 0$ and $m_2(x, y) = \max\{d(Sx, Ty), \min\{d(fx, Sx), d(fy, Ty)\}, \min\{d(fx, Ty), d(fy, Sx)\}\}$,*

then, there exist $u, w \in Y$ such that $fu = Su = Tw$. Moreover, if $Y = X$, then f, S, and T have a common fixed point provided the pair (f, S) is pointwise S-absorbing while the pair (f, T) is pointwise T-absorbing.

Corollary 15. *Let (X, d) be symmetric (semimetric) space wherein d satisfies (W_3) (Hausdoffness of $\tau(d)$) and (HE). If $f, g, S, T : X \to X$ are four self mappings of X which satisfy the following conditions:*

(i) *the pairs (f, S) and (g, T) satisfy the $CLR_{(S,T)}$ property,*

(ii) *$d(fx, gy) < m(x, y)$, where $m(x, y) = \max\{d(Sx,Ty), \min\{d(fx, Sx), d(gy, Ty)\}, \min\{d(fx, Ty), d(gy, Sx)\}\}$*

then there exist $u, w \in X$ such that $fu = Su = Tw = gw$.

Moreover, if $Y = X$, then f, g, S, and T have a unique common fixed point provided the pair (f, S) is pointwise S-absorbing whereas the pair (g, T) is pointwise T-absorbing.

Proof. Proof follows from Theorem 11 by setting $k = 1$.

Our next theorem is essentially inspired by a Lipschitzian condition utilized by Cho et al. [21] as well as Gopal et al. [15].

Theorem 16. *Theorem 11 remains true if (W_3) is replaced by (1C) while condition (ii) (of Theorem 11) is replaced by the following condition (ii′) besides retaining rest of the hypotheses:*

(ii′) $d(fx, gy) \leq km_1(x, y)$, *for any* $x, y \in X$, *where* $k \geq 0$ *together with* $k\alpha < 1$, *and wherein* $m_1(x, y) = \max\{d(Sx, Ty), \alpha[d(fx, Sx) + d(gy, Ty)], \alpha[d(fx,Ty) + d(gy,Sx)]\}$.

Proof. The proof can be completed on the lines of proof of Theorem 11, hence details are not included.

By restricting f, g, S, and T suitably, one can derive corollaries for two as well as three mappings. For the sake of brevity, we derive just one corollary by restricting Theorem 16 to three mappings which is yet another sharpened form of a theorem contained in [15] which also remains relevant to some results in Pant [4] and Pant [31].

Corollary 17. *Suppose that (in the setting of Theorem 16) d satisfies (IC) and (HE). If $f, S, T : Y \to X$ are three mappings which satisfy the following conditions:*

(i) *the pairs (f, S) and (f, T) satisfy the $CLR_{(S,T)}$ property,*

(ii) $d(fx, fy) \leq km_3(x, y)$, *for any* $x, y \in X$, *where* $k \geq 0$ *together with* $k\alpha < 1$, *and* $m_3(x, y) = \max\{d(Sx, Ty), \alpha[d(fx, Sx)+ d(fy, Ty)], \alpha[d(fx, Ty) + d(fy, Sx)]\}$

then, there exist $u, w \in Y$ such that $fu = Su = Tw$. Moreover, if $Y = X$, then f, S, and T have a common fixed point provided the pair (f, S) is pointwise S-absorbing while the pair (f, T) is pointwise T-absorbing.

Corollary 18. *Let (X, d) be symmetric (semimetric) space wherein d satisfies (IC) and (HE). If f, g, S, and T are four self mappings of X which satisfy the following conditions:*

(i) *the mappings satisfy the $CLR_{(S,T)}$ property,*

(ii) $d(fx, gy) < m_1(x, y)$, *where* $m_1(x, y) = \max\{d(Sx, Ty), \alpha[d(fx, Sx) + d(gy, Ty)], \alpha[d(fx, Ty) + d(gy, Sx)]\}$ *with* $0 < \alpha < 1$,

then, there exist $u, w \in X$ such that $fu = Su = Tw = gw$. Moreover, if $Y = X$, then f, g, S, and T have a unique common fixed point provided the pair (f, S) is pointwise S-absorbing while the pair (f, T) is pointwise T-absorbing.

Proof. The proof can be completed on the lines of proof of Theorem 11.

Acknowledgments

The authors are grateful to two anonymous referees for their helpful comments and suggestions.

References

[1] G. Jungck, "Compatible mappings and common fixed points," *International Journal of Mathematics and Mathematical Sciences*, vol. 9, no. 4, pp. 771–779, 1986.

[2] S. Sessa, "On a weak commutativity condition of mappings in fixed point considerations," *Institut Mathématique. Publications*, vol. 32, pp. 149–153, 1982.

[3] R. P. Pant, "Common fixed points of noncommuting mappings," *Journal of Mathematical Analysis and Applications*, vol. 188, no. 2, pp. 436–440, 1994.

[4] R. P. Pant, "Common fixed points of Lipschitz type mapping pairs," *Journal of Mathematical Analysis and Applications*, vol. 240, no. 1, pp. 280–283, 1999.

[5] K. P. R. Sastry and I. S. R. Krishna Murthy, "Common fixed points of two partially commuting tangential selfmaps on a metric space," *Journal of Mathematical Analysis and Applications*, vol. 250, no. 2, pp. 731–734, 2000.

[6] S. L. Singh and A. Kumar, "Fixed point theorems for Lipschitz type maps," *Rivista di Matematica della Università di Parma. Serie 7*, vol. 3, pp. 25–34, 2004.

[7] M. Imdad and A. H. Soliman, "Some common fixed point theorems for a pair of tangential mappings in symmetric spaces," *Applied Mathematics Letters*, vol. 23, no. 4, pp. 351–355, 2010.

[8] A. H. Soliman, M. Imdad, and M. Hasan, "Proving unified common fixed point theorems via common property (E-A) in symmetric spaces," *Communications of the Korean Mathematical Society*, vol. 25, no. 4, pp. 629–645, 2010.

[9] Y. Liu, J. Wu, and Z. Li, "Common fixed points of single-valued and multivalued maps," *International Journal of Mathematics and Mathematical Sciences*, no. 19, pp. 3045–3055, 2005.

[10] M. Abbas and A. R. Khan, "Common fixed points of generalized contractive hybrid pairs in symmetric spaces," *Fixed Point Theory and Applications*, vol. 2009, Article ID 869407, 11 pages, 2009.

[11] D. Gopal, M. Imdad, and C. Vetro, "Common fixed point theorems for mappings satisfying common property (E.A.) in symmetric spaces," *Filomat*, vol. 25, no. 2, pp. 59–78, 2011.

[12] N. Hussain, M. A. Khamsi, and A. Latif, "Common fixed points for *JH*-operators and occasionally weakly biased pairs under relaxed conditions," *Nonlinear Analysis*, vol. 74, no. 6, pp. 2133–2140, 2011.

[13] N. Hussain and M. Abbas, "Common fixed point results for two new classes of hybrid pairs in symmetric spaces," *Applied Mathematics and Computation*, vol. 218, no. 2, pp. 542–547, 2011.

[14] M. Imdad, J. Ali, and L. Khan, "Coincidence and fixed points in symmetric spaces under strict contractions," *Journal of Mathematical Analysis and Applications*, vol. 320, no. 1, pp. 352–360, 2006.

[15] D. Gopal, M. Hasan, and M. Imdad, "Absorbing pairs facilitating common fixed point theorems for Lipschitzian type mappings in symmetric spaces," *Communications of the Korean Mathematical Society*, vol. 27, no. 2, pp. 385–397, 2012.

[16] D. Gopal, R. P. Pant, and A. S. Ranadive, "Common fixed point of absorbing maps," *Bulletin of the Marathwada Mathematical Society*, vol. 9, no. 1, pp. 43–48, 2008.

[17] A. Aliouche, "A common fixed point theorem for weakly compatible mappings in symmetric spaces satisfying a contractive condition of integral type," *Journal of Mathematical Analysis and Applications*, vol. 322, no. 2, pp. 796–802, 2006.

[18] F. Galvin and S. D. Shore, "Completeness in semimetric spaces," *Pacific Journal of Mathematics*, vol. 113, no. 1, pp. 67–75, 1984.

[19] T. L. Hicks and B. E. Rhoades, "Fixed point theory in symmetric spaces with applications to probabilistic spaces," *Nonlinear Analysis*, vol. 36, pp. 331–344, 1999.

[20] W. A. Wilson, "On semi-metric spaces," *American Journal of Mathematics*, vol. 53, no. 2, pp. 361–373, 1931.

[21] S.-H. Cho, G.-Y. Lee, and J.-S. Bae, "On coincidence and fixed-point theorems in symmetric spaces," *Fixed Point Theory and Applications*, vol. 2008, Article ID 562130, 9 pages, 2008.

[22] D. K. Burke, "Cauchy sequences in semimetric spaces," *Proceedings of the American Mathematical Society*, vol. 33, pp. 161–164, 1972.

[23] R. P. Pant and V. Pant, "Common fixed points under strict contractive conditions," *Journal of Mathematical Analysis and Applications*, vol. 248, no. 1, pp. 327–332, 2000.

[24] M. Aamri and D. El Moutawakil, "Some new common fixed point theorems under strict contractive conditions," *Journal of Mathematical Analysis and Applications*, vol. 270, no. 1, pp. 181–188, 2002.

[25] S. Chauhan, W. Sintunavarat, and P. Kumam, "Common fixed point theorems for weakly compatible mappings in fuzzy metric spaces using (JCLR) property," *Applied Mathematics*, vol. 3, no. 9, pp. 976–982, 2012.

[26] M. Imdad, B. D. Pant, and S. Chauhan, "Fixed point theorems in menger spaces using the (CLRST) property and applications," *Journal of Nonlinear Analysis and Optimization*, vol. 3, no. 2, pp. 215–223, 2012.

[27] S. L. Singh, B. D. Pant, and S. Chauhan, "Fixed point theorems in non-archimedean menger PM-spaces," *Journal of Nonlinear Analysis and Optimization*, vol. 3, no. 2, pp. 153–160, 2012.

[28] W. Sintunavarat and P. Kumam, "Common fixed point theorems for a pair of weakly compatible mappings in fuzzy metric spaces," *Journal of Applied Mathematics*, vol. 2011, Article ID 637958, 14 pages, 2011.

[29] W. Sintunavarat and P. Kumam, "Common fixed points for R-weakly commuting mappings in fuzzy metric spaces," *Annali dell'Universita' di Ferrara*, vol. 58, no. 2, pp. 389–406, 2012.

[30] G. Jungck, "Common fixed points for noncontinuous nonself maps on nonmetric spaces," *Far East Journal of Mathematical Sciences*, vol. 4, no. 2, pp. 199–215, 1996.

[31] V. Pant, "Common fixed points under Lipschitz type condition," *Bulletin of the Korean Mathematical Society*, vol. 45, no. 3, pp. 467–475, 2008.

Univalence of a New General Integral Operator Associated with the q-Hypergeometric Function

Huda Aldweby and Maslina Darus

School of Mathematical Sciences, Faculty of Science and Technology, Universiti Kebangsaan Malaysia, 43600 Bangi, Selangor, Malaysia

Correspondence should be addressed to Maslina Darus; maslina@ukm.my

Academic Editor: Shyam Kalla

Motivated by the familiar q-hypergeometric functions, we introduce a new family of integral operators and obtain new sufficient conditions of univalence criteria. Several corollaries and consequences of the main results are also pointed out.

1. Introduction

Let $\mathcal{A}$ denote the class of functions of the form

$$f(z) = z + \sum_{n=2}^{\infty} c_n z^n, \quad c_n \geq 0, \tag{1}$$

which are analytic in the open unit disk $\mathcal{U} = \{z \in \mathbb{C} : |z| < 1\}$, and $\mathcal{S}$ the class of functions $f \in \mathcal{A}$ which are univalent in $\mathcal{U}$.

Let $f, g \in \mathcal{A}$, where f is defined by (1) and g is given by

$$g(z) = z + \sum_{n=2}^{\infty} b_n z^n, \quad b_n \geq 0. \tag{2}$$

Then the Hadamard product (or convolution) $f * g$ of the functions f and g is defined by

$$(f * g)(z) = z + \sum_{n=2}^{\infty} c_n b_n z^n. \tag{3}$$

For complex parameters a_i, b_j, and q ($i = 1, \ldots, r$, $j = 1, \ldots, s$, $b_j \in \mathbb{C} \setminus \{0, -1, -2, \ldots\}$, $|q| < 1$), we define the q-hypergeometric function ${}_r\Phi_s(a_1, \ldots, a_r; b_1, \ldots, b_s; q, z)$ by

$${}_r\Phi_s\left(a_i; b_j; q, z\right) = \sum_{n=0}^{\infty} \frac{(a_1, q)_n \cdots (a_r, q)_n}{(q, q)_n (b_1, q)_n \cdots (b_s, q)_n} z^n \tag{4}$$

($r = s + 1$; $r, s \in \mathbb{N}_0 = \mathbb{N} \cup \{0\}$; $z \in \mathcal{U}$), where $\mathbb{N}$ denotes the set of positive integers and $(a, q)_n$ is the q-shifted factorial defined by

$$(a, q)_n = \begin{cases} 1, & n = 0; \\ (1 - a)(1 - aq)\left(1 - aq^2\right) \cdots \left(1 - aq^{n-1}\right), & n \in \mathbb{N}. \end{cases} \tag{5}$$

By using the ratio test, we should note that, if $|q| < 1$, the series (4) converges absolutely for $|z| < 1$ if $r = s + 1$. For more mathematical background of these functions, one may refer to [1].

Corresponding to the function defined by (4), consider

$${}_r\mathcal{G}_s\left(a_i; b_j; q, z\right) = z \quad {}_r\Phi_s\left(a_i; b_j; q, z\right). \tag{6}$$

Recently, the authors [2] defined the linear operator $\mathcal{M}(a_i, b_j; q) f : \mathcal{A} \to \mathcal{A}$ by

$$\begin{aligned} \mathcal{M}\left(a_i, b_j; q\right) f(z) &= {}_r\mathcal{G}_s\left(a_i; b_j; q, z\right) * f(z) \\ &= z + \sum_{n=2}^{\infty} \Upsilon_n c_n z^n, \end{aligned} \tag{7}$$

where

$$\Upsilon_n = \frac{(a_1, q)_{n-1} \cdots (a_r, q)_{n-1}}{(q, q)_{n-1} (b_1, q)_{n-1} \cdots (b_s, q)_{n-1}}, \quad (|q| < 1). \tag{8}$$

It should be remarked that the linear operator (7) is a generalization of many operators considered earlier. For $a_i = q^{\alpha_i}$, $b_j = q^{\beta_j}$, α_i, $\beta_j \in \mathbb{C}$, $\beta_j \neq 0, -1, -2, \ldots, (i = 1, \ldots, r,\ j = 1, \ldots, s)$, and $q \to 1$, we obtain the Dziok-Srivastava linear operator [3] (for $r = s + 1$), so that it includes (as its special cases) various other linear operators introduced and studied by Ruscheweyh [4], Carlson and Shaffer [5] and the Bernardi-Libera-Livingston operators [6–8].

The q-difference operator is defined by

$$d_q h(z) = \frac{h(qz) - h(z)}{(q-1)z}, \quad q \neq 1,\ z \neq 0, \qquad \lim_{q \to 1} d_q h(z) = h'(z), \tag{9}$$

where $h'(z)$ is the ordinary derivative. For more properties of d_q see [9, 10].

Lemma 1 (see [2]). *Let $f \in \mathcal{A}$; then*

(i) *for $r = 1$, $s = 0$, and $a_1 = q$, one has $\mathcal{M}(q, -; q) f(z) = f(z)$.*

(ii) *For $r = 1$, $s = 0$, and $a_1 = q^2$, one has $\mathcal{M}(q^2, -; q) f(z) = z d_q f(z)$ and $\lim_{q \to 1} \mathcal{M}(q^2, -; q) f(z) = z f'(z)$, where d_q is the q-derivative defined by (9).*

Definition 2. A function $f \in \mathcal{A}$ is said to be in the class $\mathfrak{B}_s^r(a_i, b_j; q; \mu)$ if it is satisfying the condition

$$\left| \frac{z^2 \left(\mathcal{M}(a_i, b_j; q) f(z) \right)'}{\left[\mathcal{M}(a_i, b_j; q) f(z) \right]^2} - 1 \right| < 1 - \mu \quad (z \in \mathcal{U};\ 0 \leq \mu < 1), \tag{10}$$

where $\mathcal{M}(a_i, b_j; q) f$ is the operator defined by (7).

Note that $\mathfrak{B}_0^1(q, -; q; \mu) = \mathfrak{B}(\mu)$, where the class $\mathfrak{B}(\mu)$ of analytic and univalent functions was introduced and studied by Frasin and Darus [11].

Using the operator $\mathcal{M}(a_i, b_j; q) f(z) f$, we now introduce the following new general integral operator.

For $m \in \mathbb{N} \cup \{0\}$, $\gamma_1, \gamma_2, \ldots, \gamma_m$, $\delta \in \mathbb{C} \setminus \{0, -1, -2, \ldots\}$, and $|q| < 1$, we define the integral operator $I_{\gamma_k, \delta}(a_i, b_j; q; z) : \mathcal{A}^n \to \mathcal{A}^n$ by

$$I_{\gamma_k, \delta}(a_i, b_j; q; z) = \left(\delta \int_0^z t^{\delta - 1} \prod_{k=1}^m \left(\frac{\mathcal{M}(a_i, b_j; q) f(z) f_k(t)}{t} \right)^{1/\gamma_k} dt \right)^{1/\delta}, \tag{11}$$

where $f_k \in \mathcal{A}$.

Remark 3. It is interesting to note that the integral operator $I_{\gamma_k, \delta}(a_i, b_j; q; z)$ generalizes many operators introduced and studied by several authors, for example,

(1) for $r = s + 1$, $a_i = q^{\alpha_i}$, $b_j = q^{\beta_j}$, $i = 1, \ldots, r$, $j = 1, \ldots, s$, $q \to 1$, $\gamma_k = 1/(\alpha - 1)$, and $\delta = 1 + m(\alpha - 1)$, where $\alpha \in \mathbb{C}$ and $\Re(\alpha) > 0$, we obtain the following integral operator introduced and studied by Selvaraj and Karthikeyan [12]:

$$F_\alpha(\alpha_1, \beta_1; z) = \left(1 + m(\alpha - 1) \int_0^z \left(H_s^r(\alpha_1, \beta_1) f_1(t) \right)^{\alpha - 1} \cdots \left(H_s^r(\alpha_1, \beta_1) f_m(t) \right)^{\alpha - 1} dt \right)^{1/(1 + m(\alpha - 1))}, \tag{12}$$

where for convenience $H_s^r(\alpha_1, \beta_1) f := H(\alpha_1, \ldots, \alpha_r; \beta_1, \ldots, \beta_s; z) f(z)$, and $H_s^r(\alpha_1, \beta_1) f(z) = z + \sum_{n=2}^{\infty} ((\alpha_1)_{n-1} \cdots (\alpha_r)_{n-1} / (\beta_1)_{n-1} \cdots (\beta_s)_{n-1} (n-1)!) a_n z^n$ is the Dziok-Srivastava operator [3].

(2) For $r = 1$, $s = 0$, $a_1 = q$, $\gamma_k = 1/(\alpha - 1)$, and $\delta = 1 + m(\alpha - 1)$, we obtain the integral operator

$$F_{m,\alpha}(z) = \left(1 + m(\alpha - 1) \times \int_0^z \left(f_1(t) \right)^{\alpha - 1} \cdots \left(f_m(t) \right)^{\alpha - 1} dt \right)^{1/(1 + m(\alpha - 1))} \tag{13}$$

studied recently by Breaz et al. [13].

(3) For $r = 1$, $s = 0$, $a_1 = q$, $\gamma_k = 1/\alpha_k$, and $\delta = 1$, we obtain the integral operator

$$F_\alpha(z) = \int_0^z \left(\frac{f_1(t)}{t} \right)^{\alpha_1} \cdots \left(\frac{f_m(t)}{t} \right)^{\alpha_m} dt \tag{14}$$

introduced and studied by D. Breaz and N. Breaz [14].

(4) For $r = 1$, $s = 0$, $a_1 = q^2$, $\gamma_k = 1/(\alpha - 1)$, and $\delta = 1 + m(\alpha - 1)$, we obtain the integral operator

$$G_\alpha(z) = \left(1 + m(\alpha - 1) \times \int_0^z t^{m(\alpha - 1)} \left(f_1'(t) \right)^{\alpha - 1} \cdots \left(f_m'(t) \right)^{\alpha - 1} dt \right)^{1/(1 + m(\alpha - 1))} \tag{15}$$

introduced by Selvaraj and Karthikeyan [12].

(5) For $r = 1$, $s = 0$, $a_1 = q^2$, $\gamma_k = 1/\alpha$, and $\delta = 1$, we obtain the integral operator

$$G_\alpha(z) = \int_0^z \left(f_1'(t) \right)^\alpha \cdots \left(f_m'(t) \right)^\alpha dt, \tag{16}$$

recently introduced and studied by Breaz and Güney [15].

(6) For $r = 1$, $s = 0$, $a_1 = q$, $f_1 = \cdots = f_m = f \in \mathcal{A}$, $\gamma_k = 1/(\alpha - 1)$, and $\delta = \alpha$, where $\alpha \in \mathbb{C}$ and $\Re(\alpha) > 0$, we obtain the integral operator

$$G_\alpha(z) = \left(\alpha \int_0^z \left(f(t) \right)^{\alpha - 1} \right)^{1/\alpha} dt, \tag{17}$$

introduced and studied by Pescar [16].

In order to derive our main results, we have to recall the following univalence criteria.

Lemma 4 (see [17, 18]). *Let $\delta \in \mathbb{C}$ with $\operatorname{Re}(\delta) > 0$. If $f \in \mathcal{A}$ satisfies*

$$\frac{1-|z|^{2\operatorname{Re}(\delta)}}{\operatorname{Re}(\delta)}\left|\frac{zf''(z)}{f'(z)}\right| \le 1, \tag{18}$$

for all $z \in \mathcal{U}$, then the integral operator

$$F_\delta(z) = \left\{\delta \int_0^z t^{\delta-1} f'(t)\,dt\right\}^{1/\delta} \tag{19}$$

is in the class $\mathcal{S}$.

Lemma 5 (see [16]). *Let $\delta \in \mathbb{C}$ with $\operatorname{Re}(\delta) > 0$, $c \in \mathbb{C}$, with $|c| \le 1$, $c \ne -1$. If $f \in \mathcal{A}$ satisfies*

$$\left|c|z|^{2\delta} + \left(1-|z|^{2\delta}\right)\frac{zf''(z)}{\delta f'(z)}\right| \le 1, \tag{20}$$

for all $z \in \mathcal{U}$ then the integral operator

$$F_\delta(z) = \left\{\delta \int_0^z t^{\delta-1} f'(t)\,dt\right\}^{1/\delta} \tag{21}$$

is in the class $\mathcal{S}$.

Lemma 6 (Generalized Schwarz Lemma, see [19]). *(Generalized Schwarz Lemma) Let the function f be analytic in the disk $\mathcal{U}_R = \{z : |z| < R\}$, with $|f(z)| < M$ for fixed M. If $f(z)$ has one zero with multiplicity order bigger that m for $z = 0$, then*

$$|f(z)| \le \frac{M}{R^m}|z|^m, \quad (z \in \mathcal{U}_R). \tag{22}$$

Equality can hold only if

$$f(z) = e^{i\theta}\left(\frac{M}{R^m}\right)z^m, \tag{23}$$

where θ is constant.

2. Univalence Conditions for $I_{\gamma_k,\delta}(a_i,b_j;q;z)$

Theorem 7. *Let $f_k \in \mathcal{A}$ for all $k = 1,\dots,m$, $\gamma_k \in \mathcal{C}$, and $M \ge 1$ with*

$$\frac{1}{\operatorname{Re}(\delta)}\sum_{k=1}^{m}\frac{[(2-\mu_k)M+1]}{|\gamma_k|} \le 1. \tag{24}$$

If for all $k = 1,\dots,m$, $f_k \in \mathfrak{B}_s^r(a_i,b_j,q,\mu_k)$, $0 \le \mu_k < 1$, and

$$\left|\mathcal{M}\left(a_i,b_j;q\right)f(z)f_k(z)\right| \le M, \quad (z \in \mathcal{U}) \tag{25}$$

then the integral operator $I_{\gamma_k,\delta}(a_i,b_j;q;z)$ defined by (11) is analytic and univalent in $\mathcal{U}$.

Proof. From the definition of the operator $\mathcal{M}(a_i,b_j;q)f(z)f$ it can be observed that

$$\frac{\mathcal{M}\left(a_i,b_j;q\right)f(z)}{z} \ne 0, \quad (z \in \mathcal{U}), \tag{26}$$

and for $z = 0$, we have

$$\left(\frac{\mathcal{M}\left(a_i,b_j;q\right)f(z)f_1(z)}{z}\right)^{1/\gamma_1} \cdots \left(\frac{\mathcal{M}\left(a_i,b_j;q\right)f(z)f_m(z)}{z}\right)^{1/\gamma_m} = 1. \tag{27}$$

We define the function $h(z)$ by the form

$$h(z) = \int_0^z \prod_{k=1}^{m}\left(\frac{\mathcal{M}\left(a_i,b_j;q\right)f(z)f_k(t)}{t}\right)^{1/\gamma_k} dt. \tag{28}$$

Therefore

$$h'(z) = \prod_{k=1}^{m}\left(\frac{\mathcal{M}\left(a_i,b_j;q\right)f(z)f_k(z)}{z}\right)^{1/\gamma_k}. \tag{29}$$

Differentiating logarithmically and multiplying by z on both sides of (29)

$$\frac{zh''(z)}{h'(z)} = \sum_{k=1}^{m}\frac{1}{\gamma_k}\left(\frac{z\left(\mathcal{M}\left(a_i,b_j;q\right)f(z)f_k(z)\right)'}{\mathcal{M}\left(a_i,b_j;q\right)f(z)f_k(z)} - 1\right). \tag{30}$$

Thus we have

$$\left|\frac{zh''(z)}{h'(z)}\right| \le \sum_{k=1}^{m}\frac{1}{|\gamma_k|}\left|\frac{z\left(\mathcal{M}\left(a_i,b_j;q\right)f(z)f_k(z)\right)'}{\mathcal{M}\left(a_i,b_j;q\right)f(z)f_k(z)} - 1\right|. \tag{31}$$

So

$$\begin{aligned}
&\frac{1-|z|^{2\operatorname{Re}(\delta)}}{\operatorname{Re}(\delta)}\left|\frac{zh''(z)}{h'(z)}\right| \\
&\le \frac{1-|z|^{2\operatorname{Re}(\delta)}}{\operatorname{Re}(\delta)} \times \left[\sum_{k=1}^{m}\frac{1}{|\gamma_k|}\left(\left|\frac{z\left(\mathcal{M}\left(a_i,b_j;q\right)f(z)f_k(z)\right)'}{\mathcal{M}\left(a_i,b_j;q\right)f(z)f_k(z)}\right| + 1\right)\right] \\
&\le \frac{1-|z|^{2\operatorname{Re}(\delta)}}{\operatorname{Re}(\delta)} \times \left[\sum_{k=1}^{m}\frac{1}{|\gamma_k|}\left(\left|\frac{z^2\left(\mathcal{M}\left(a_i,b_j;q\right)f(z)f_k(z)\right)'}{\left[\mathcal{M}\left(a_i,b_j;q\right)f(z)f_k(z)\right]^2}\right| \times \left|\frac{\mathcal{M}\left(a_i,b_j;q\right)f(z)f_k(z)}{z}\right| + 1\right)\right].
\end{aligned} \tag{32}$$

Since $|\mathcal{M}(a_i,b_j;q)f(z)f_k(z)| \le M$, $(z \in \mathcal{U}, k = 1,\ldots,m)$, and $f_k \in \mathfrak{B}_s^r(a_i,b_j,q,\mu_k)$ for all $k = 1,\ldots,m$, then from the Schwarz Lemma and (10), we obtain

$$\begin{aligned}
&\frac{1-|z|^{2\operatorname{Re}(\delta)}}{\operatorname{Re}(\delta)}\left|\frac{zh''(z)}{h'(z)}\right| \\
&\le \frac{1-|z|^{2\operatorname{Re}(\delta)}}{\operatorname{Re}(\delta)} \\
&\quad\times\left[\left(\sum_{k=1}^{m}\frac{1}{|\gamma_k|}\left|\frac{z^2\left(\mathcal{M}\left(a_i,b_j;q\right)f(z)f_k(z)\right)'}{\left[\mathcal{M}\left(a_i,b_j;q\right)f(z)f_k(z)\right]^2}\right|M\right.\right. \\
&\qquad\left.\left.+M+1\right)\right] \\
&\le \frac{1}{\operatorname{Re}(\delta)}\sum_{k=1}^{m}\frac{1}{|\gamma_k|}\left[(2-\mu_k)M+1\right], \quad (z\in\mathcal{U})
\end{aligned} \tag{33}$$

which, in the light of the hypothesis (24), yields

$$\frac{1-|z|^{2\operatorname{Re}(\delta)}}{\operatorname{Re}(\delta)}\left|\frac{zh''(z)}{h'(z)}\right| \le 1, \quad (z\in\mathcal{U}). \tag{34}$$

Applying Lemma (1) for the function $h(z)$ we obtain that $I_{\gamma_k,\delta}(a_i,b_j;q;z)$ is univalent.

Taking $\mu_k = 0$ (for all $k = 1,\ldots,m$), $M = 1$, $a_i = q^{\alpha_i}$, $b_j = q^{\beta_j}$, $q \to 1$, and $\gamma_k = 1/(\alpha-1)$, $\delta = 1+m(\alpha-1)$ in Theorem 7, we have the following.

Corollary 8 (see [12]). *Let $f_k \in \mathcal{A}$ for all $k = 1,\ldots,m$ and $\alpha \in \mathbb{C}$ with*

$$|\alpha-1| \le \frac{\operatorname{Re}(\alpha)}{3m}. \tag{35}$$

If

$$\left|\frac{z^2\left(H_s^r(\alpha_1,\beta_1)f_k(z)\right)'}{\left(H_s^r(\alpha_1,\beta_1)f_k(z)\right)^2}-1\right| < 1, \quad (z\in\mathcal{U}) \tag{36}$$

and for all $k = 1,\ldots,m$, then the integral operator $F_\alpha(\alpha_1,\beta_1;z)$ defined by (12) is analytic and univalent in $\mathcal{U}$.

Taking $\mu_k = 0$ (for all $k = 1,\ldots,m$), $M = 1$, $r = 1$, $s = 0$, $a_1 = q$, and $\gamma_k = 1/(\alpha-1)$, $\delta = 1+m(\alpha-1)$ in Theorem 7, we have the following.

Corollary 9. *Let $f_k \in \mathcal{A}$ for all $k = 1,\ldots,m$ and $\alpha \in \mathbb{C}$ with*

$$|\alpha-1| \le \frac{\operatorname{Re}(\alpha)}{3m}. \tag{37}$$

If

$$\left|\frac{z^2 f_k'(z)}{\left(f_k(z)\right)^2}-1\right| < 1, \quad (z\in\mathcal{U}) \tag{38}$$

and for all $k = 1,\ldots,m$, then the integral operator $F_{m,\alpha}(z)$ defined by (13) is analytic and univalent in $\mathcal{U}$.

Theorem 10. *Let $f_k \in \mathcal{A}$ for all $k = 1,\ldots,m$, δ, $\gamma_k \in \mathbb{C}$, and $M \ge 1$ with*

$$|c| \le 1-\frac{1}{\delta}\sum_{k=1}^{m}\frac{[(2-\mu_k)M+1]}{|\gamma_k|}, \quad c\in\mathbb{C}. \tag{39}$$

If for all $k = 1,\ldots,m$, $f_k \in \mathfrak{B}_s^r(a_i,b_j,q,\mu_k)$, $0 \le \mu_k < 1$, and

$$\left|\mathcal{M}\left(a_i,b_j;q\right)f(z)f_k(z)\right| \le M, \quad (z\in\mathcal{U}), \tag{40}$$

then the integral operator $I_{\gamma_k,\delta}(a_i,b_j;q)$ defined by (11) is analytic and univalent in $\mathcal{U}$.

Proof. From the proof of Theorem 7, we have

$$\frac{zh''(z)}{h'(z)} = \sum_{k=1}^{m}\frac{1}{\gamma_k}\left(\frac{z\left(\mathcal{M}\left(a_i,b_j;q\right)f(z)f_k(z)\right)'}{\mathcal{M}\left(a_i,b_j;q\right)f(z)f_k(z)}-1\right). \tag{41}$$

Thus we have

$$\left|c|z|^{2\delta}+\left(1-|z|^{2\delta}\right)\frac{zh''(z)}{\delta h'(z)}\right| \le |c|+\left|\left(1-|z|^{2\delta}\right)\frac{zh''(z)}{\delta h'(z)}\right|. \tag{42}$$

From this result and using the proof of Theorem 7 we obtain

$$\begin{aligned}
&\left|c|z|^{2\delta}+\left(1-|z|^{2\delta}\right)\frac{zh''(z)}{\delta h'(z)}\right| \\
&\quad\le |c|+\frac{1}{\delta}\sum_{k=1}^{m}\frac{1}{|\gamma_k|}\left[(2-\mu_k)M+1\right].
\end{aligned} \tag{43}$$

Since $|c| \le 1-(1/\delta)\sum_{k=1}^{m}(1/\gamma_k)[(2-\mu_k)M+1]$, then we have

$$\left|c|z|^{2\delta}+\left(1-|z|^{2\delta}\right)\frac{zh''(z)}{\delta h'(z)}\right| \le 1, \quad (z\in\mathcal{U}). \tag{44}$$

Applying Lemma (4) for the function $h(z)$ we obtain that $I_{\gamma_k,\delta}(a_i,b_j;q;z)$ is univalent.

Taking $\mu_k = 0$ (for all $k = 1,\ldots,m$), $r = 1$, $s = 0$, $a_1 = q$, and $\gamma_k = 1/(\alpha-1)$, $\delta = 1+m(\alpha-1)(\alpha \in \mathbf{R})$ in Theorem 10, we have the following.

Corollary 11. *Let $f_k \in \mathcal{A}$ for all $k = 1,\ldots,m$; $c \in \mathbb{C}$, $\alpha \in \mathbf{R}$, and $M \ge 1$ with*

$$\begin{aligned}
&|c| \le 1+\left(\frac{1-\alpha}{1+m(\alpha-1)}\right)(2M+1)m, \\
&\alpha \in \left[1,\frac{2Mm+1}{2Mm}\right].
\end{aligned} \tag{45}$$

If for all $k = 1,\ldots,m$

$$\begin{aligned}
&\left|\frac{z^2 f_k'(z)}{f_k^2(z)}-1\right| < 1, \quad (z\in\mathcal{U}), \\
&\left|f_k(z)\right| \le M, \quad (z\in\mathcal{U}; k = 1,\ldots,m),
\end{aligned} \tag{46}$$

then the integral operator $F_{m,\alpha}(z)$ defined by (13) is analytic and univalent in $\mathcal{U}$.

Letting $m = 1$, $M = 1$, and $f_1 = f$ in Corollary 11, we have the following.

Corollary 12. *Let $f \in \mathscr{A}$, $c \in \mathbb{C}$ and $\alpha \in \mathbf{R}$ with*

$$|c| \leq \frac{3-2\alpha}{\alpha}, \quad (c \neq -1), \qquad \alpha \in \left[1, \frac{3}{2}\right]. \tag{47}$$

If

$$\left|\frac{z^2 f'(z)}{f^2(z)} - 1\right| < 1, \quad (z \in \mathscr{U}), \qquad |f(z)| \leq 1, \quad (z \in \mathscr{U}), \tag{48}$$

then the integral operator $G_\alpha(z)$ defined by (17) *is analytic and univalent in $\mathscr{U}$.*

Remark 13. Many other interesting corollaries and results can be obtained by specializing the parameters in Theorem 10; for example, see [13, 20, 21].

Acknowledgments

The work presented here was partially supported by GUP-2012-023 and UKM-DLP-2011-050.

References

[1] G. Gasper and M. Rahman, *Basic Hypergeometric Series*, vol. 35 of *Encyclopedia of Mathematics and its Applications*, Cambridge University Press, Cambridge, UK, 1990.

[2] A. Mohammed and M. Darus, "A generalized operator involving the q-hypergeometric function," *Mathematici Vesnik*, Available online 10.06.2012, 12 pages.

[3] J. Dziok and H. M. Srivastava, "Classes of analytic functions associated with the generalized hypergeometric function," *Applied Mathematics and Computation*, vol. 103, no. 1, pp. 1–13, 1999.

[4] S. Ruscheweyh, "New criteria for univalent functions," *Proceedings of the American Mathematical Society*, vol. 49, pp. 109–115, 1975.

[5] B. C. Carlson and D. B. Shaffer, "Starlike and prestarlike hypergeometric functions," *SIAM Journal on Mathematical Analysis*, vol. 15, no. 4, pp. 737–745, 1984.

[6] S. D. Bernardi, "Convex and starlike univalent functions," *Transactions of the American Mathematical Society*, vol. 135, pp. 429–446, 1969.

[7] R. J. Libera, "Some classes of regular univalent functions," *Proceedings of the American Mathematical Society*, vol. 16, pp. 755–758, 1965.

[8] A. E. Livingston, "On the radius of univalence of certain analytic functions," *Proceedings of the American Mathematical Society*, vol. 17, pp. 352–357, 1966.

[9] H. Exton, *q-Hypergeometric Functions and Applications*, Ellis Horwood, Chichester, UK, 1983.

[10] H. A. Ghany, "q-derivative of basic hypergeometric series with respect to parameters," *International Journal of Mathematical Analysis*, vol. 3, no. 33-36, pp. 1617–1632, 2009.

[11] B. A. Frasin and M. Darus, "On certain analytic univalent functions," *International Journal of Mathematics and Mathematical Sciences*, vol. 25, no. 5, pp. 305–310, 2001.

[12] C. Selvaraj and K. R. Karthikeyan, "Sufficient conditions for univalence of a general integral operator," *Acta Universitatis Apulensis*, no. 17, pp. 87–94, 2009.

[13] D. Breaz, N. Breaz, and H. M. Srivastava, "An extension of the univalent condition for a family of integral operators," *Applied Mathematics Letters*, vol. 22, no. 1, pp. 41–44, 2009.

[14] D. Breaz and N. Breaz, "Two integral operators," *Studia Universitatis Babes-Bolyai, Mathematica, Cluj-Napoca*, vol. 47, no. 3, pp. 13–19, 2002.

[15] D. Breaz and H. Ö. Güney, "The integral operator on the classes $S^*_\alpha(b)$ and $C_\alpha(b)$," *Journal of Mathematical Inequalities*, vol. 2, no. 1, pp. 97–100, 2008.

[16] V. Pescar, "A new generalization of Ahlfors's and Becker's criterion of univalence," *Malaysian Mathematical Society Bulletin*, vol. 19, no. 2, pp. 53–54, 1996.

[17] N. N. Pascu, "On a univalence criterion. II," in *Itinerant Seminar on Functional Equations, Approximation and Convexity (Cluj-Napoca, 1985)*, vol. 85, pp. 153–154, Babeş-Bolyai University, Cluj-Napoca, Romania, 1985.

[18] N. N. Pascu, "An improvement of Becker's univalence criterion," in *Proceedings of the Commemorative Session: Simion Stoïlow*, pp. 43–48, University of Braşov, Braşov, Romania, 1987.

[19] Z. Nehari, *Conformal Mapping*, Dover, New York, NY, USA, 1975.

[20] D. Breaz and H. Ö. Güney, "On the univalence criterion of a general integral operator," *Journal of Inequalities and Applications*, vol. 2008, Article ID 702715, 8 pages, 2008.

[21] G. I. Oros, G. Oros, and D. Breaz, "Sufficient conditions for univalence of an integral operator," *Journal of Inequalities and Applications*, vol. 2008, Article ID 127645, 7 pages, 2008.

An Enhanced Matrix-Free Secant Method via Predictor-Corrector Modified Line Search Strategies for Solving Systems of Nonlinear Equations

M. Y. Waziri[1,2] and Z. A. Majid[1,3]

[1] *Department of Mathematics, Faculty of Science, Universiti Putra Malaysia, 43400 Serdang, Selangor, Malaysia*
[2] *Department of Mathematical Sciences, Faculty of Science, Bayero University Kano, Kano, Nigeria*
[3] *Institute for Mathematical Research, Universiti Putra Malaysia, 43400 Serdang, Selangor, Malaysia*

Correspondence should be addressed to M. Y. Waziri; mywaziri@gmail.com

Academic Editor: Paolo Ricci

Diagonal updating scheme is among the cheapest Newton-like methods for solving system of nonlinear equations. Nevertheless, the method has some shortcomings. In this paper, we proposed an improved matrix-free secant updating scheme via line search strategies, by using the steps of backtracking in the Armijo-type line search as a step length predictor and Wolfe-Like condition as corrector. Our approach aims at improving the overall performance of diagonal secant updating scheme. Under mild assumptions, the global convergence results have been presented. Numerical experiments verify that the proposed approach is very promising.

1. Introduction

Consider the problem

$$F(x) = 0, \tag{1}$$

with $F : R^n \rightarrow R^n$.

The mapping F is assumed to satisfy the following standard assumptions:

(1) F is continuously differentiable in an open convex set Φ;

(2) there exists a solution x^* of (1) in Φ such that $F(x^*) = 0$ and $F'(x^*) \neq 0$;

(3) the Jacobian $F'(x)$ is local Lipschitz continuous at x^*.

The famous scheme for finding the solution to (1) is the Newton method. The method is simple to implement, and it produces an iterative sequence $\{x_k\}$ from any given initial guess x_0 in the neighborhood of x^* via

$$x_{k+1} = x_k - \left(F'(x_k)\right)^{-1} F(x_k), \tag{2}$$

where $k = 0, 1, 2 \ldots n$.

The attractive features of this method are that it is easy to implement and converges rapidly [1]. However, the major difficulty of Newton-type method is matrix storage requirements especially when handling large systems of nonlinear equations [1]. To overcome such difficulty, the simple modification on the Newton method is the fixed Newton method. Fixed Newton method for the determination of solution x^* is given by

$$x_{k+1} = x_k - \left(F'(x_0)\right)^{-1} F(x_k), \quad k = 0, 1, 2, \ldots. \tag{3}$$

This method avoids computation and storing the Jacobian in each iterations (except at $k = 0$). However, it still requires to solve systems of n linear equations and consumes more CPU time as the system's dimension increases [2, 3].

Quasi-Newton method is another variant of Newton-type methods. It replaces the Jacobian or its inverse with an approximation which can be updated at each iteration and is given as

$$x_{k+1} = x_k - B_k^{-1} F(x_k), \tag{4}$$

where the matrix B_k is the approximation of the Jacobian at x_k. The main idea behind quasi-Newton's method is to eliminate the evaluation cost of the Jacobian matrix [4, 5]. However, the most critical part of such quasi-Newton method is storing the full matrix of the approximate Jacobian, which can be a very expensive task as the dimension of systems increases [6]. In contrast, this paper presents an improved variant of quasi-Newton update via the steps of backtracking in the Armijo-type line search and Wolfe-like condition and incorporating restarting strategy whenever the updating matrix is singular or nearly singular. The anticipation has been to improve the overall performance of the diagonal quasi-Newton method.

We organized the rest of this paper as follows. In the next section, we present the proposed method. Section 3 presents convergence results. Numerical experiments are reported in Section 4, and finally, conclusion is given in Section 5.

2. The Improved Secant Diagonal Updating

It is well known that it is not always feasible to compute the full elements of the Jacobian matrix of the given nonlinear function or it may be very expensive. We often have to approximate the Jacobian matrix by some other approach, and the famous method of doing so is quasi-Newton's method [4]. The basic idea underlining this approach has been to reduce the evaluation cost of the Jacobian matrix. This new approach generates a sequence of points $\{x_k\}$ via

$$x_{k+1} = x_k - \alpha_k B_k F(x_k), \tag{5}$$

where α_k is a step length, and B_k is a diagonal approximation of the inverse Jacobian matrix which can be updated in each iterations. To achieve this, we incorporate some new line search strategies, via the steps of backtracking in the Armijo-type line search as predictor and then improved via Wolfe-like condition as a corrector. The Armijo rule is among the inexact line search methods which guarantees a sufficient degree of accuracy to ensure the algorithm convergence. Nevertheless, the scheme needs high floating points operations and function call. In this paper, we present a simple line search strategy, which is less computational cost, floating points operations and CPU time consumptions respectively compared to classical Armijor line search. The approach is as follows.

Given $\epsilon \in (0,1)$ and $\sigma < 1$, the proposed approach finds the appropriate α, such that

$$\|F(x_k + \alpha_k d_k)\| \leq \sigma \|F(x_k)\|. \tag{6}$$

In addition, the new strategy is implemented in an iterative way using a fixed initial value of α as follows.

Algorithm 1 (Armijo-Like).

Step 0. Set $K = 0$, $\alpha_0 > 0$, and $\sigma < 1$.

Step 1. $\|F(x_{k+1})\| \leq \sigma\|F(x_k)\|$. Choose α_k as the step size; stop. Otherwise $\alpha_{k+1} = \alpha_k/2$, $k = k + 1$.

Waziri et al. [7] have set the step length ($\alpha_k = 1$, for all k); this approach is mostly used in many Newton-like methods. Here, we continue in the spirit of diagonal updating, using a new line search strategy to obtain a good step length (α_k) in every iteration, anticipating to produce a more accurate approximation of the Jacobian inverse matrix and then employing restating strategy whenever the updating matrix is undefined. To this end, B_{k+1} would be obtained almost similar to the diagonal updating scheme presented in [7] in which $s_k = -\alpha_k B_k F(x_k)$ instead of $s_k = -B_k F(x_k)$. Now, let the deviation between B_{k+1} and B_k denoted as $\Phi_k = B_{k+1} - B_k$ be minimized under some norms; the optimal solution is given as

$$\Phi_k = \frac{\left(y_k^T s_k - y_k^T B_k y_k\right)}{\operatorname{tr}\left(\Psi_k^2\right)} \Psi_k, \tag{7}$$

where $\Psi_k = \operatorname{diag}((y_k^{(1)})^2, (y_k^{(2)})^2, \ldots, (y_k^{(n)})^2)$, $\sum_{i=1}^n (y_k^{(i)})^4 = \operatorname{tr}(\Psi_k^2)$, $y_k = F(x_{k+1}) - F(x_k)$, $s_k = x_{k+1} - x_k$, and tr is the trace operation, respectively. The updated formula for the proposed matrix B_k is given as [7]

$$B_{k+1} = B_k + \frac{\left(y_k^T s_k - y_k^T B_k y_k\right)}{\operatorname{tr}\left(\Psi_k^2\right)} \Psi_k. \tag{8}$$

To safeguard on the possibilities of generating undefined B_{k+1}, we let $B_{k+1} = B_0$, whenever this situation happens

$$B_{k+1} = \begin{cases} B_k + \dfrac{\left(y_k^T s_k - y_k^T B_k y_k\right)}{\operatorname{tr}\left(\Psi_k^2\right)} \Psi_k, & \operatorname{tr}\left(\Psi_k^2\right) \neq 0, \\ B_0, & \text{otherwise.} \end{cases} \tag{9}$$

Now, we can describe the algorithm for our proposed approach as follows.

Algorithm 2 (EMFM).

Step 1. Choose an initial guess x_0, $\sigma \in (0,1)$, $B_0 = I_n$, $\alpha_0 > 0$ and let $k := 0$.

Step 2. Compute $F(x_k)$, and if $\|F(x_k)\| \leq 10^{-4}$, stop.

Step 3. Compute $d = -F(x_k)B_k$.

Step 4. If $\|F(x_k + \alpha_k d_k)\| \leq \sigma\|F(x_k)\|$, retain α_k and go to 5. Otherwise, set $\alpha_{k+1} = \alpha_k/2$ and repeat 4.

Step 5. If $\|F(x_k + \alpha_k d_k) - F(x_k)\| \geq \|F(x_k + \alpha_k d_k)\| - \|F(x_k)\|$, retain α_k and go to 6. Otherwise, set $\alpha_{k+1} = \alpha_k \times 1.1$ and repeat 5.

Step 6. Let $x_{k+1} = x_k + \alpha_k d_k$.

Step 7. If $\|x_{k+1} - x_k\|_2 + \|F(x_k)\|_2 \leq 10^{-4}$, stop. Otherwise go to Step 8.

Step 8. If $\|\Delta F_k\|_2 \geq \epsilon_1$, where $\epsilon_1 = 10^{-4}$, compute B_{k+1}, and if not, $B_{k+1} = B_0$.

Step 9. Set $k := k + 1$ and go to 2.

3. Convergence Analysis

We present the convergence result of EMFM method by proving the existence of the step length $\alpha_k > 0$. We will make the following assumptions on nonlinear system F.

Assumption 3. (i) F is differentiable in an open convex set E in $\mathfrak{R}^n$.

(ii) There exists $x^* \in E$ such that $(x^*) = 0$, and $F'(x)$ is continuous for all x.

(iii) $F(x)$ satisfies Lipschitz condition of order one; that is, there exists a positive constant μ such that

$$\|F(x) - F(y)\| \leq \mu \|x - y\|, \tag{10}$$

for all $x, y \in \mathfrak{R}^n$.

(iv) There exist constants $c_1 \leq c_2$ such that $c_1\|\omega\|^2 \leq \omega^T F'(x)\omega \leq c_2\|\omega\|^2$ for all $x \in E$ and $\omega \in \mathfrak{R}^n$.

To this end, we proceed by given the following result for the step length generated by the proposed strategies.

Theorem 4. *Assume that F is a strictly convex function. Suppose that the new strategies are employed with $d_k \neq 0$ and positive α_k exist, for all sufficiently large k. Then, the iterates $\{x_k\}$ generated by the line search algorithm have the property that*

$$\lim \|F(x_k + \alpha_k d_k)\| = 0, \quad \text{as } k \longrightarrow \infty. \tag{11}$$

Proof. From condition (iii) of Assumption 3 and the fact that $s_k = x_{k+1} - x_k$, we have

$$\begin{aligned}\|F(x_k) - F(x_k + \alpha_k d_k)\|_F &\leq \mu\|s_k\|_F \\ &= \mu\|\alpha d_k\|_F \\ &\leq \mu|\alpha|\,\|d_k\|_F.\end{aligned} \tag{12}$$

Recall that

$$\|F(x) - F(x_k + \alpha_k d_k)\|_F \leq \|F(x)\|_F - \|F(x_k + \alpha_k d_k)\|_F. \tag{13}$$

Equation (13) gives

$$\begin{aligned}\|F(x_k)\|_F - \|F(x_k + \alpha_k d_k)\|_F &\leq \alpha\mu\|d_k\|_F \\ &= \mu\|\alpha d_k\|_F \\ &\leq \mu|\alpha|\,\|d_k\|_F.\end{aligned} \tag{14}$$

Since $\|d_k\| \neq 0$ and $\|F(x_k + \alpha_k d_k)\| \leq \sigma\|F(x_k)\|$, hence, it follows that

$$\begin{aligned}\|F(x_k)\|_F - \sigma\|F(x_k)\|_F &\leq \alpha\mu\|d_k\|_F, \\ (1-\sigma)\|F(x_k)\|_F &\leq \alpha\mu\|d_k\|_F.\end{aligned} \tag{15}$$

After little simplifications, we obtain

$$\alpha \geq \frac{(1-\sigma)\|F(x_k)\|_F}{\mu\|d_k\|_F} > 0. \tag{16}$$

Hence, α_k exist and are positive.

We continue to show the convergence of the iterates $\{x_k\}$ by recalling that

$$\|F(x_k + \alpha_k d_k)\| \leq \sigma\|F(x_k)\|. \tag{17}$$

Then, we have

$$\begin{aligned}\|F(x_k + \alpha_k d_k)\| &\leq \sigma^2\|F(x_k)\| \\ &\leq \sigma^3\|F(x_k)\|\end{aligned} \tag{18}$$

for a finite k, and (18) yields

$$\|F(x_k + \alpha_k d_k)\| \leq \sigma^k\|F(x_k)\|. \tag{19}$$

Since the algorithm terminates at $\|F(x_k)\| = 0$, it follows from (19) that

$$\frac{\|F(x_k + \alpha_k d_k)\|}{\|F(x_k)\|} \leq \sigma^k. \tag{20}$$

We have

$$\lim \frac{\|F(x_k + \alpha_k d_k)\|}{\|F(x_k)\|} \leq \lim \sigma^k, \quad \text{as } k \longrightarrow \infty. \tag{21}$$

Therefore, due to $\sigma < 1$, it implies that

$$\lim \sigma^k \longrightarrow 0, \quad \text{as } k \longrightarrow \infty. \tag{22}$$

hence

$$\lim \|F(x_k + \alpha_k d_k)\| = 0, \quad \text{as } k \longrightarrow \infty. \tag{23}$$

To show the convergence results of EMFM method, we require to show that the updating matrix B_k is bounded above and below by some positive constants. Hence, we can state the following result on the boundedness of $\{\|\Phi_k\|_F\}$ by assuming that, without loss of generality, the updating matrix (8) is always used; then, we have the following.

Theorem 5. *Let F satisfy Assumption 3, and let $\|y_k\| \neq 0$ for all finite k. Let $\{B_k\}$ be the sequence generated by (8). If the given nonsingular B_0 satisfies*

$$\theta \leq B_0^i \leq \omega, \quad i = 1, 2, \ldots, n, \tag{24}$$

for some constants θ and ω, then the sequence $\{\|B_k\|_F\}$ is bounded for all finite k.

Proof. Since $B_{k+1} = B_k + \Phi_k$, it follows that

$$\|B_{k+1}\|_F \leq \|B_k\|_F + \|\Phi_k\|_F. \tag{25}$$

For $k = 0$ and assuming $B_0 = I$, we have

$$\begin{aligned}\left|\Phi_0^{(i)}\right| &= \left|\frac{y_0^T s_0 - y_0^T B_0 y_0}{\operatorname{tr}(\Psi_0^2)}\left(y_0^{(i)}\right)^2\right| \\ &\leq \frac{\left|y_0^T s_0 - y_0^T B_0 y_0\right|}{\operatorname{tr}(\Psi_0^2)}\left(y_0^{(\max)}\right)^2,\end{aligned} \tag{26}$$

where $(y_0^{(\max)})^2$ is the largest element among $(y_0^{(i)})^2$, $i = 1, 2, \ldots, n$.

After multiplying (26) by $(y_0^{(\max)})^2/(y_0^{(\max)})^2$ and substituting $\operatorname{tr}(\Psi_0^2) = \sum_{i=1}^n (y_0^{(i)})^4$, we have

$$\left|\Phi_0^{(i)}\right| \leq \frac{\left|y_0^T s_0 - y_0^T B_0 y_0\right|}{\left(y_0^{(\max)}\right)^2 \sum_{i=1}^n \left(y_0^{(i)}\right)^4} \left(y_0^{(\max)}\right)^4. \quad (27)$$

Since $(y_0^{(\max)})^4 / \sum_{i=1}^n (y_0^{(i)})^4 \leq 1$, then (27) turns into

$$\left|\Phi_0^{(i)}\right| \leq \frac{\left|y_0^T F'(x)\ y_0 - y_0^T B_0 y_0\right|}{\left(y_0^{(\max)}\right)^2}. \quad (28)$$

From Assumption 3 and $B_0 = I$, (28) becomes

$$\left|\Phi_0^{(i)}\right| \leq \frac{|c-1|\left(y_0^T y_0\right)}{\left(y_0^{(\max)}\right)^2}, \quad (29)$$

where $c = \max\{|c_1|, |c_2|\}$.

Since $(y_0^{(i)})^2 \leq (y_0^{(\max)})^2$ for $i = 1, \ldots, n$, it follows that

$$\left|\Phi_0^{(i)}\right| \leq \frac{n|c-1|\left(y_0^{(\max)}\right)^2}{\left(y_0^{(\max)}\right)^2}. \quad (30)$$

Hence, we obtain

$$\left\|\Phi_0\right\|_F \leq n^{3/2}|c-1|. \quad (31)$$

Suppose that $\eta = n^{3/2}|c-1|$; then,

$$\left\|\Phi_0\right\|_F \leq \eta. \quad (32)$$

From the fact that $\|B_0\|_F = \sqrt{n}$, it follows that

$$\left\|B_1\right\|_F \leq \beta, \quad (33)$$

where $\beta = \sqrt{n} + \eta > 0$.

4. Numerical Results

In this section, we consider some benchmark problems to illustrate the performance of the method proposed in this paper for solving large-scale systems of nonlinear equations when compared to some Newton-like methods. The computations are performed in MATLAB 7.0 using double precision computer, and the stopping rule used is

$$\left\|s_k\right\| + \left\|F(x_k)\right\| \leq 10^{-4}. \quad (34)$$

The identity matrix has been chosen as an initial approximate Jacobian inverse. We further design the codes to terminate whenever one of the following happens:

(i) the number of iteration is at least 250, but no point of x_k that satisfies (34) is obtained;

(ii) CPU time in seconds reaches 250;

(iii) There is insufficient memory to initiate the run.

The performances of these methods are be compared in terms of number of iterations and CPU time in seconds. In the following, some details on the benchmarks test problems are presented.

Problem 1. System of n nonlinear equations is as follows:

$$f_i(x) = x_i - 3x_i\left(\frac{\sin x_i}{3} - 0.66\right) + 2, \quad i = 1, 2, \ldots, n, \quad x_0 = (3, 3, \ldots, 3). \quad (35)$$

Problem 2. Extended Trigonometric function of Spedicator [8] is as follows:

$$f_i(x) = n - \sum_{j=1}^n \cos x_j + i\left(1 - \cos x_i\right) - \sin x_i + \exp^{(1-\cos x_i)} - \sum_{j=1}^n \left(x_j + 1\right), \quad i = 1, \ldots, n, \quad x_0 = \left(\frac{1}{n}, \frac{1}{n}, \ldots, \frac{1}{n}\right). \quad (36)$$

Problem 3. System of n nonlinear equations is as follows:

$$f_i(x) = x_i - 0.1x_{i+1}^2, \qquad f_n(x) = x_n - 0.1x_1^2, \quad i = 1, \ldots, n-1, \quad x_0 = (7, 7, \ldots, 7). \quad (37)$$

(A1) Nonadiabatic Stirred Tank Reactors. The first application is the model of [9]. The model deals with two continuous nonadiabatic stirred tank reactors. The reactors are in a series, at steady state having a recycle component, and with an exothermic first-order irreversible reaction. By eliminating certain variables, the model results into two nonlinear systems of equations as follows:

$$f_1 = (1-\lambda)\left[\frac{D}{10(1+\beta_1)} - x_1\right]\exp\left(\frac{10x_1}{1+(10x_1/\gamma)}\right) - x_1,$$
$$f_2 = x_1 - (1+\beta_2)x_2 + (1-\lambda) \times \left[\frac{D}{10} - \beta_1 x_1 - (1+\beta_2 x_2)\right]\exp\left(\frac{10x_2}{1+(10x_2/\gamma)}\right). \quad (38)$$

The dimensionless temperatures of the two reactors are represented by x_1 and x_2. The parameters λ, γ, β_1, β_2, and D are given as 1, 1000, 2, 2, and 22, respectively.

(A2) Navigation by Range Measurements. Consider two beacons determining position by measuring the distances $r1$ and $r2$, where $(p1, p2)$ is the position of beacon 1, $(q1, q2)$ is the position of beacon 2, and (u, v) is an unknown point (see Figure 1). This can be modeled into a two nonlinear systems of equations as follows:

$$f_1 = \sqrt{(p_1-u)^2 + (q_1-v)^2} - r_1 = 0,$$
$$f_1 = \sqrt{(p_2-u)^2 + (q_2-v)^2} - r_2 = 0. \quad (39)$$

TABLE 1: Numerical comparison of NM, FN, BM, I-VDN, and EMFM methods.

Prob	Dim	NM		FN		BM		I-VDN		EMFM	
		NI	CPU	NI	CPU	NI	CPU	NI	CPU	NI	CPU
1	25	4	0.031	11	0.016	8	0.015	8	0.006	6	0.001
2	25	4	0.062	11	0.031	8	0.016	8	0.002	7	0.001
3	25	6	0.031	—	—	10	0.031	10	0.001	7	0.001
1	50	4	0.046	11	0.031	8	0.046	8	0.011	6	0.008
2	50	4	0.109	11	0.062	8	0.034	8	0.010	7	0.005
3	50	6	0.156	—	—	10	0.124	10	0.008	7	0.004
1	100	4	0.064	11	0.032	8	0.048	8	0.014	6	0.010
2	100	4	0.125	11	0.046	8	0.032	8	0.015	7	0.011
3	100	6	0.508	—	—	10	0.187	10	0.031	7	0.015
1	1000	4	1.919	12	0.7332	9	2.122	9	0.033	6	0.018
2	1000	4	2.277	12	0.998	8	1.778	8	0.049	5	0.031
3	1000	6	88.031	—	—	10	13.057	10	0.041	7	0.024

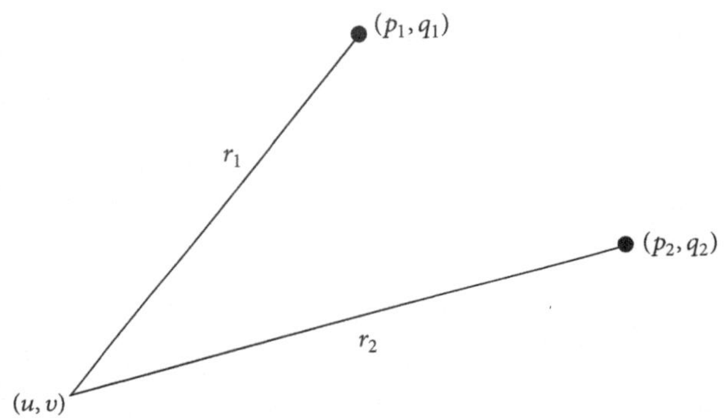

FIGURE 1

The parameters (p_1, q_1), (p_2, q_2), r_1, and r_2 are given as $(10, 10)$, $(10, -10)$, 14, and 16, respectively.

Table 1 shows the number of iterations and CPU time for these five methods, respectively. In Table 1, the value "Dim" denotes the dimension of the systems, "NI" denotes number of iterations, while "CPU" is the CPU time in seconds respectively. We analyze the performance of each method via execution time, floating points operations, and storage locations. One can observe that EMFM has the smallest number of iterations compared to the classical diagonal updating (I-VDN) proposed by Waziri et al. [7]. This shows that the line search strategies presented in this paper have increased the convergence speed of the classical diagonal updating method.

If we compare the performance of all methods, in terms of CPU time, it is clear that EMFM method consumes less CPU time than the others and still keeping memory requirement and CPU time in seconds to only $O(n)$. All five methods are able to obtain the solution ($x^* = -1.1243, 1.5001$) of A2, but EMFM method consumes less CPU time in second (0.001) compared to the other 4 methods. Moreover, for A1, still proposed method has shown a promising performance with less storage locations (2 locations) whereas NM, FN, and BM, respectively, required 4 locations for each.

5. Conclusion

From the fact that there is a rapid development in research on solving nonlinear systems, nevertheless, the dimension of the nonlinear system is most of the times so large that it requires a very costly arithmetic operations when using some other Newton-like methods; so we usually employ cheap iterative approach, and the good candidate is classical diagonal updating. Notwithstanding, the classical updating scheme has some lapses. One is that it usually needs high number of iterations. The other is that the method generally has slow convergence behavior. This paper proposes an enhanced diagonal secant updating scheme based on the steps of backtracking in the Armijo-type line search and then improved via Wolfe-like condition. Our approach aims at improving the overall performance of diagonal secant updating scheme. The algorithm only requires to store a row vector while ignoring all the off and low diagonal elements and therefore largely reduces memory locations. In addition, as it uses two line search strategies (predictor and corrector) to obtain a new iterates point, the spectral properties of the diagonal updating scheme is improved, and rapid convergence property is gained.

Computational experiment suggests that it is very vital for diagonal updating scheme to use line search strategy. EMFM method has very good solving speed and the best performance among the Newton-like methods. Finally, it can be concluded that this approach would certainly be quite useful for solving large-scale systems of nonlinear equations.

References

[1] J. E. Dennis, Jr. and R. B. Schnabel, *Numerical Methods for Unconstrained Optimization and Nonlinear Equations*, Prentice Hall, Englewood Cliffs, NJ, USA, 1983.

[2] M. Y. Waziri, W. J. Leong, M. A. Hassan, and M. Monsi, "A New Newton method with diagonal Jacobian approximation for

systems of Non-Linear equations," *Journal of Mathematics and Statistics*, vol. 6, no. 3, pp. 246–252, 2010.

[3] M. Y. Waziri, W. J. Leong, and M. Mamat, "A two-step matrix-free secant method for solving large-scale systems of nonlinear equations," *Journal of Applied Mathematics*, vol. 2012, Article ID 348654, 9 pages, 2012.

[4] C. G. Broyden, "A class of methods for solving nonlinear simultaneous equations," *Mathematics of Computation*, vol. 19, pp. 577–593, 1965.

[5] K. Natasa and L. Zorna, "Newton-like method with modification of the right-hand vector," *Mathematics of Computation*, vol. 71, pp. 237–250, 2001.

[6] C. T. Kelley, *Iterative Methods for Linear and Nonlinear Equations*, vol. 16, SIAM, Philadelphia, Pa, USA, 1995.

[7] M. Y. Waziri, W. J. Leong, and M. A. Hassan, "Diagonal Broyden-like method for large-scale systems of nonlinear equations," *Malaysian Journal of Mathematical Sciences*, vol. 6, no. 1, pp. 59–73, 2012.

[8] E. Spedicator, "Computational experience with quas-Newton algorithms for minimization problems of moderately large size," Tech. Rep. CISE-N-175, Segrate, Milano, Italy, 1975.

[9] J. Sinkule, "Multiplicity and stability in a sequence of two nonadiabatic non-isothermal CSTR," *Chemical Engineering Sceinces*, vol. 35, pp. 987–996, 1980.

Further Results on Colombeau Product of Distributions

Biljana Jolevska-Tuneska[1] **and Tatjana Atanasova-Pacemska**[2]

[1] *Faculty of Electrical Engineering and Informational Technologies, Saints Cyril and Methodius University of Skopje, Karpos II bb, 1000 Skopje, Macedonia*
[2] *Faculty of Computer Sciences, Goce Delčev University of Štip, Krste Misirkov, Br 10A, 2000 Štip, Macedonia*

Correspondence should be addressed to Biljana Jolevska-Tuneska; biljanaj@feit.ukim.edu.mk

Academic Editor: Shyam Kalla

Results on Colombeau product of distributions $x_+^{-r-1/2}$ and $x_-^{-r-1/2}$ are derived. They are obtained in Colombeau differential algebra $\mathscr{G}(\mathbf{R})$ of generalized functions that contains the space $\mathscr{D}'(\mathbf{R})$ of Schwartz distributions as a subspace and has a notion of "association" that is a faithful generalization of the weak equality in $\mathscr{D}'(\mathbf{R})$.

1. Introduction

In quantum physics one finds the need to evaluate δ^2, when calculating the transition rates of certain particle interactions; see [1]. The problem of defining products of distributions is also closely connected with the problem of renormalization in quantum field theory. Due to the large use of distributions in the natural sciences and other mathematical fields, the problem of the product of distributions [2] is an objective of many research studies. Starting with the historically first construction of distributional multiplication by König [3], and the sequential approach developed in [4] by Antosik et al., there have been numerous attempts to define products of distributions, see [5–7], or rather to enlarge the number of existing products.

Having a look at the theory of distributions [8, 9] we realize that there are two complementary points of view.

(1) A distribution $f \in \mathscr{D}'(\mathbf{R}^n)$ is a continuous functional on the space $\mathscr{D}(\mathbf{R}^n)$ of space functions (compactly supported smooth functions). Here we have a linear action

$$\varphi \longrightarrow \langle f, \varphi \rangle \tag{1}$$

of f on a test function ϕ.

(2) Letting $\{\varphi_n\}$ be a sequence of smooth functions converging to the Dirac measure, a family of regularization $\{f_n\}$ can be produced by convolution,

$$f_n(x) = f * \varphi_n = \langle f(y), \varphi_n(x-y) \rangle, \tag{2}$$

which converges weakly to the original distribution $f \in \mathscr{D}'(\mathbf{R}^n)$. Identifying two sequences $\{\varphi_n\}$ and $\{\overline{\varphi}\}_n$ if they have the same limit, we obtain a sequential representation of the space of distributions. Other authors use the equivalence classes of nets of regularization. The delta-net $\{\varphi_\varepsilon\}_{\varepsilon>0}$ is defined by

$$\varphi_\varepsilon(x) = \varepsilon^{-n}\varphi\left(\frac{x}{\varepsilon}\right). \tag{3}$$

When we work with regularization the nonlinear structure is lost by identifying sequences (nets) with the same limit. So we have to get nonlinear theory of generalized functions that will work with regularization, but identify less. The actual construction of such algebras enjoying these optimal properties is due to Colombeau [10]. His theory of algebras of generalized functions offers the possibility of applying large classes of nonlinear operations to distributional objects. Some of them are used on solving differential equations with nonstandard coefficients [11]. The "association process" in

Colombeau algebra providing a faithful generalization of the equality of distributions in $\mathscr{D}(\mathbf{R}^n)$ enables us to obtain results in terms of distributions, the so-called Colombeau products. In fact, we evaluate particular product of distributions with coinciding singularities, as embedded in Colombeau algebra, in terms of associated distributions; see [12–14]. Therefore, the results obtained can be reformulated as regularized model products in the classical distribution theory.

In 1966, Mikusinski published his famous result [15]:

$$x^{-1} \cdot x^{-1} - \pi^2 \delta(x)\delta(x) = x^{-2}, \tag{4}$$

where neither of the products on the left-hand side here exists, but their difference still has a correct meaning in distribution space $\mathscr{D}'(\mathbf{R})$. Formula including balanced products of distributions with coinciding singularities can be found in mathematical and physical literature. In this paper results on product of distributions $x_+^{-r-1/2}$ and $x_-^{-r-1/2}$ are derived.

2. Colombeau Algebra

The basic idea underlying Colombeau's theory in its simplest form is that of embedding the space of distributions into a factor algebra of $\mathscr{C}^\infty(\mathbf{R}^n)^I$, $I = (0,1)$ with regularization by convolution with a fixed "mollifier" φ.

The space of test function is

$$\mathscr{A}_0(\mathbf{R}^n) = \left\{\varphi \in \mathscr{D}(\mathbf{R}^n) : \int \varphi(x)\,dx = 1\right\}. \tag{5}$$

Functional act on test functions and points is

$$u : \mathscr{A}_0(\mathbf{R}^n) \times \mathbf{R}^n \longrightarrow \mathbf{C}, \quad (\varphi, x) \longrightarrow u(\varphi, x), \tag{6}$$

where $u(\varphi, x)$ is required to belong to $\mathscr{C}^\infty(\mathbf{R}^n)$ with respect to the second variable $x \in \mathbf{R}^n$. Now let $\varphi_\varepsilon(x) = \varepsilon^{-n}\varphi(x/\varepsilon)$ for $\varphi \in \mathscr{A}_0(\mathbf{R}^n)$. The sequence $(u * \varphi_\varepsilon)_{\varepsilon\in I}$ converges to u in $\mathscr{D}'(\mathbf{R}^n)$. Taking this sequence as a representative of u we obtain an embedding of $\mathscr{D}'(\mathbf{R}^n)$ into the algebra $\mathscr{C}^\infty(\mathbf{R}^n)$. However, embedding $\mathscr{C}^\infty(\mathbf{R}^n) \subset \mathscr{D}'(\mathbf{R}^n)$ into this algebra via convolution as above will not yield a subalgebra since of course $(f * \varphi_\varepsilon)(g * \varphi_\varepsilon) \neq (fg) * \varphi_\varepsilon$ in general. The idea, therefore, is to factor out an ideal $\mathscr{N}(\mathbf{R}^n)$ such that this difference vanishes in the resulting quotient. In order to construct $\mathscr{N}(\mathbf{R}^n)$ it is obviously sufficient to find an ideal containing all differences $(f * \varphi_\varepsilon) - (f)_{\varepsilon\in I}$. Taylor expansion of $(f * \varphi_\varepsilon) - f$ shows that this term will vanish faster than any power of ε, uniformly on compact sets, in all derivatives. We use the following space:

$$\mathscr{A}_q(\mathbf{R}^n) = \left\{\varphi \in \mathscr{A}_0(\mathbf{R}^n) : \int x^\alpha \varphi(x)\,dx = 0 \text{ for } 1 \le |\alpha| \le q\right\}; \tag{7}$$

moderate functionals, denoted by $\mathscr{E}_M(\mathbf{R}^n)$, are defined by the property:

$$\forall K \subset\subset \mathbf{R}^n \ \forall \alpha \in \mathbf{N}_0^n \ \exists p \ge 0 \quad \text{such that } \forall \varphi \in \mathscr{A}_q(\mathbf{R}^n), \quad \sup_{x\in K} |\partial^\alpha u(\varphi_\varepsilon, x)| = \mathscr{O}(\varepsilon^{-p}) \quad \text{as } \varepsilon \longrightarrow 0. \tag{8}$$

Null functionals, denoted by $\mathscr{N}(\mathbf{R}^n)$, are defined by the property:

$$\forall K \subset\subset \mathbf{R}^n \ \forall \alpha \in \mathbf{N}_0^n \ \exists p \ge 0 \quad \text{such that } \forall q \ge p, \ \forall \varphi \in \mathscr{A}_q(\mathbf{R}^n), \quad \sup_{x\in K} |\partial^\alpha u(\varphi_\varepsilon, x)| = \mathscr{O}(\varepsilon^{q-p}) \quad \text{as } \varepsilon \longrightarrow 0. \tag{9}$$

In other words, moderate functionals satisfy a locally uniform polynomial estimate as $\varepsilon \to 0$ when acting on φ_ε, together with all derivatives, while null functionals vanish faster than any power of ε in the same situation. The null functionals form a differential ideal in the collection of moderate functionals.

We define space of functions $u : \mathscr{A}_0 \to \mathscr{C}$ and denote by $\mathscr{E}_0(\mathbf{R}^n)$. It is a subalgebra in $\mathscr{E}(\mathbf{R}^n)$ in sense of natural identification.

Moderate functionals, denoted by $\mathscr{E}_{0M}(\mathbf{R}^n)$, are defined by the property:

$$\exists p \ge 0 \quad \text{such that } \forall \varphi \in \mathscr{A}_p(\mathbf{R}^n), \quad |u(\varphi_\varepsilon)| = \mathscr{O}(\varepsilon^{-p}) \quad \text{as } \varepsilon \longrightarrow 0. \tag{10}$$

Null functionals, denoted by $\mathscr{N}_0(\mathbf{R}^n)$, are defined by the property:

$$\exists p \ge 0 \quad \text{such that } \forall q \ge p, \ \forall \varphi \in \mathscr{A}_q(\mathbf{R}^n), \quad |u(\varphi_\varepsilon)| = \mathscr{O}(\varepsilon^{q-p}) \quad \text{as } \varepsilon \longrightarrow 0. \tag{11}$$

Definition 1. Space of generalized functions $\mathscr{G}(\mathbf{R}^n)$, generalized complex numbers, and generalized real numbers is the factor algebra defined as

$$\mathscr{G}(\mathbf{R}^n) = \frac{\mathscr{E}_M(\mathbf{R}^n)}{\mathscr{N}(\mathbf{R}^n)}, \qquad \overline{\mathbf{C}} = \frac{\mathscr{E}_{0M}(\mathbf{C})}{\mathscr{N}_0(\mathbf{C})}, \qquad \overline{\mathbf{R}} = \frac{\mathscr{E}_{0M}(\mathbf{R})}{\mathscr{N}_0(\mathbf{R})}. \tag{12}$$

The space of distributions is imbedded by convolution:

$$i : \mathscr{D}'(\mathbf{R}^n) \longrightarrow \mathscr{G}(\mathbf{R}^n), \quad i(u) = \text{class of } [(\varphi, x) \longrightarrow u * \varphi(x)]. \tag{13}$$

Equivalence classes of sequences $(u_\varepsilon)_{\varepsilon\in I}$ in $\mathscr{G}(\mathbf{R}^n)$ will be denoted by $U = \text{class } [(u_\varepsilon)_{\varepsilon\in I}]$. If G is a generalized function with compact support $K \subset\subset \mathbf{R}^n$ and $G_\varepsilon(x)$ is a representative of G, then its integral is defined by

$$\int G\,dx = \text{class}\left[\int \varphi(x)\,G_\varepsilon(x)\,dx\right]. \tag{14}$$

Let $F, G \in \mathscr{G}(\mathscr{R}^n)$. Then

(i) they are equal in the distribution sense, $G \overset{\mathcal{D}'}{=} F$ if

$$\int (G - F)\,\varphi\, dx = 0 \in \overline{\mathscr{C}}, \quad \text{for any } \varphi \in \mathscr{D}(\mathbf{R}^n); \tag{15}$$

(ii) they are associated $G \approx F$ if there exist a representative G_ε and F_ε of G and F, respectively, such that

$$\lim_{\varepsilon \to 0} \int (G_\varepsilon(x) - F_\varepsilon(x))\,\varphi(x)\,dx = 0, \quad \text{for any } \varphi \in \mathscr{D}(\mathbf{R}^n). \tag{16}$$

3. Results on Some Products of Distributions

It was proved in [12] that for any $a \in \mathbf{R}/\mathbf{Z}$ the product of the generalized functions $\widetilde{x_+^a}$ and $\widetilde{x_-^{-a-1}}$ in $\mathscr{G}(\mathbf{R})$ admits associated distributions and it holds

$$\begin{aligned}\widetilde{x_+^a} \cdot \widetilde{x_-^{-a-1}} = \widetilde{x_-^{-a-1}} \cdot \widetilde{x_+^a} &\approx \frac{\Gamma(1+a)\,\Gamma(-a)}{2}\delta(x) \\ &= -\frac{\pi}{2}\csc(\pi a)\,\delta(x),\end{aligned} \tag{17}$$

with the particular cases

$$\widetilde{x_+^{-r-1/2}} \cdot \widetilde{x_-^{r-1/2}} = \frac{(-1)^r \pi}{2}\delta(x), \tag{18}$$

for $r = 0, \pm 1, \pm 2, \ldots$ and

$$\widetilde{x_+^{-1/2}} \cdot \widetilde{x_-^{-1/2}} = \frac{\pi}{2}\delta(x). \tag{19}$$

Here we will make a generalization of (19). In order to prove the main theorem we need the following lemmas, easily proved by induction.

Lemma 2. *For $p < r$ one has*

$$\int_t^d s^p \varphi^{(r)}(s)\,ds = \sum_{k=0}^{p} (-1)^k \frac{p!}{(p-k)!} t^{p-k} \varphi^{(r-k-1)} \times (t)\,dt, \quad \text{for p < r.} \tag{20}$$

Lemma 3. *For $p = r$ one has*

$$\int_t^d s^r \varphi^{(r)}(s)\,ds = \sum_{k=0}^{r} (-1)^k \frac{r!}{(r-k)!} t^{r-k} \varphi^{(r-k-1)}(t)\,dt, \tag{21}$$

and $\varphi^{(-1)}(t)$ stands for $\int_t^d \varphi(s)ds$.

Theorem 4. *The product of the generalized functions $\widetilde{x_+^{-r-1/2}}$ and $\widetilde{x_-^{-r-1/2}}$ for $r = 0, 1, 2, \ldots$ in $\mathscr{G}(\mathbf{R})$ admits associated distributions and it holds*

$$\widetilde{x_+^{-r-1/2}} \cdot \widetilde{x_-^{-r-1/2}} \approx \frac{\pi}{2\,(2r)!}\delta^{(2r)}(x). \tag{22}$$

Proof. For given $\varphi \in A_0(\mathbf{R})$ we suppose that $\operatorname{supp}\varphi(x) \subseteq [c, d]$, without loss of generality. Then using the embedding rule and the substitution $t = (y - x)/\varepsilon$ we have the representatives of the distribution $x_+^{-r-1/2}$ in Colombeau algebra:

$$\begin{aligned}\widetilde{x_+^{-r-1/2}}(\varphi_\varepsilon, x) &= (-1)^r \frac{2^r}{(2r-1)!!} \frac{1}{\varepsilon} \frac{\partial^r}{\partial x} \int_0^\infty y^{-1/2} \varphi\left(\frac{y-x}{\varepsilon}\right) dy \\ &= \frac{2^r}{(2r-1)!!} \frac{1}{\varepsilon^{r+1}} \int_0^\infty y^{-1/2} \varphi^{(r)}\left(\frac{y-x}{\varepsilon}\right) dy \\ &= \frac{2^r}{(2r-1)!!} \frac{1}{\varepsilon^r} \int_{-x/\varepsilon}^d (x + \varepsilon t)^{-1/2} \varphi^{(r)}(t)\,dt.\end{aligned} \tag{23}$$

Similar, using the embedding rule and the substitution $s = (y - x)/\varepsilon$ we have the representatives of the distribution $x_-^{-r-1/2}$ in Colombeau algebra:

$$\begin{aligned}\widetilde{x_-^{-r-1/2}}(\varphi_\varepsilon, x) &= (-1)^r \frac{2^r}{(2r-1)!!} \frac{1}{\varepsilon} \frac{\partial^r}{\partial x} \times \int_{-\infty}^0 (-y)^{-1/2} \varphi\left(\frac{y-x}{\varepsilon}\right) dy \\ &= \frac{2^r}{(2r-1)!!} \frac{1}{\varepsilon^{r+1}} \times \int_{-\infty}^0 (-y)^{-1/2} \varphi^{(r)}\left(\frac{y-x}{\varepsilon}\right) dy \\ &= \frac{2^r}{(2r-1)!!} \frac{1}{\varepsilon^r} \int_c^{-x/\varepsilon} (-x - \varepsilon s)^{-1/2} \varphi^{(r)}(s)\,ds.\end{aligned} \tag{24}$$

Then, for any $\psi(x) \in \mathscr{D}(\mathbf{R})$ we have

$$\begin{aligned}&\left\langle \widetilde{x_+^{-r-1/2}}(\varphi_\varepsilon, x) \cdot \widetilde{x_-^{-r-1/2}}(\varphi_\varepsilon, x), \psi(x) \right\rangle \\ &= \int_{-\infty}^{\infty} \widetilde{x_+^{-r-1/2}}(\varphi_\varepsilon, x)\, \widetilde{x_-^{-r-1/2}}(\varphi_\varepsilon, x)\, \psi(x)\,dx \\ &= \frac{2^{2r}}{((2r-1)!!)^2} \frac{1}{\varepsilon^{2r}} \int_{-d\varepsilon}^{c\varepsilon} \psi(x) \int_{-x/\varepsilon}^d \varphi^{(r)}(t) \times \int_c^{-x/\varepsilon} (x + \varepsilon t)^{-1/2} (-x - \varepsilon s)^{-1/2} \varphi^{(r)}(s)\,ds\,dt\,dx \\ &= \frac{2^{2r}}{((2r-1)!!)^2} \frac{1}{\varepsilon^{2r}} \int_c^d \psi(-\varepsilon\omega) \int_\omega^d \varphi^{(r)}(t) \times \int_c^\omega (t - \omega)^{-1/2} (\omega - s)^{-1/2} \varphi^{(r)}(s)\,ds\,dt\,d\omega,\end{aligned} \tag{25}$$

using the substitution $\omega = -x/\varepsilon$.

By the Taylor theorem we have that

$$\psi(-\varepsilon\omega) = \sum_{k=0}^{2r} \frac{\psi^{(k)}(0)}{k!} (-\varepsilon\omega)^k + \frac{\psi^{(2r+1)}(\eta\omega)}{(2r+1)!} (-\varepsilon\eta)^{2r+1}, \tag{26}$$

for $\eta \in (0,1)$. Using this for (25) we have

$$\left\langle \widetilde{x_+^{-r-1/2}}(\varphi_\varepsilon, x)\cdot \widetilde{x_-^{-r-1/2}}(\varphi_\varepsilon, x), \psi(x)\right\rangle = \frac{2^{2r}}{((2r-1)!!)^2}\sum_{k=0}^{2r}\frac{(-1)^k\psi^{(k)}(0)}{k!\varepsilon^{2r-k}}I_k + O(\varepsilon), \tag{27}$$

where

$$I_k = \int_c^d \varphi^{(r)}(t)\int_t^d \varphi^{(r)}(s) \times \int_s^t (t-\omega)^{-1/2}(\omega-s)^{-1/2}\omega^k d\omega\, ds\, dt, \tag{28}$$

for $k = 0,1,\ldots,2r$ and we have changed the order of integration.

Putting $\omega - s = (t-s)v$ we have

$$\begin{aligned}\int_s^t (t-\omega)^{-1/2}(\omega-s)^{-1/2}\omega^k d\omega &= \int_0^1 v^{-1/2}(1-v)^{-1/2}[sv+(1-v)t]^k dv\\ &= \sum_{p=0}^k \frac{k!}{p!(k-p)!}\int_0^1 v^{p-1/2}(1-v)^{k-p-1/2}s^p t^{k-p}dv\\ &= \sum_{p=0}^k \frac{k!}{p!(k-p)!}B\left(p+\frac{1}{2}, k-p+\frac{1}{2}\right)s^p t^{k-p}.\end{aligned} \tag{29}$$

Thus

$$\begin{aligned}I_k &= \sum_{p=0}^k \frac{k!}{p!(k-p)!}B\left(p+\frac{1}{2}, k-p+\frac{1}{2}\right) \times \int_c^d t^{k-p}\varphi^{(r)}(t)\int_t^d s^p\varphi^{(r)}(s)\,ds\,dt\\ &= \sum_{p=0}^k \frac{k!}{p!(k-p)!}B\left(p+\frac{1}{2}, k-p+\frac{1}{2}\right)J_{k,p}.\end{aligned} \tag{30}$$

Next, suppose that k is even, less than $2r$ and that p is less than r. It follows from Lemma 2 that $\int_t^d s^p\varphi^{(r)}(s)ds$ is an even or odd function accordingly as $r+p$ is odd or even. We thus have that $t^{k-p}\varphi^{(r)}(t)\int_t^d s^p\varphi^{(r)}(s)ds$ is an odd function and $J_{k,p} = 0$. If $p \geq r$ then $k-p < r$ and by changing the order of integration we can prove that again $J_{k,p} = 0$. If we suppose that k is odd and less than $2r$ we can prove in a similar manner that $J_{k,p} = 0$.

For the case $k = 2r$ if $p \neq r$ again we have $J_{k,p} = 0$. For $k = 2r$ and $p = r$ using Lemma 3 and changing the order of integration we have

$$\begin{aligned}J_{2r,r} &= \int_c^d t^r\varphi^{(r)}(t)\int_t^d s^r\varphi^{(r)}(s)\,ds\,dt\\ &= \sum_{i=0}^r \frac{(-1)^i r!}{(r-i)!}\int_c^d t^{2r-i}\varphi^{(r)}(t)\varphi^{(r-i-1)}(t)\,dt\\ &= (-1)^r r!\int_c^d t^r\varphi^{(r)}(t)\int_t^d \varphi(s)\,ds\,dt\\ &= (-1)^r r!\int_c^d \varphi(s)\int_c^s t^r\varphi^{(r)}(t)\,dt\,ds\\ &= (-1)^r r!\sum_{i=0}^r \frac{(-1)^i r!}{(r-i)!}\int_c^d s^{r-i}\varphi^{(r-i-1)}(s)\varphi(s)\,ds\\ &= (r!)^2\int_c^d \varphi(s)\left(\int_c^s \varphi(t)\,dt\right)ds.\end{aligned} \tag{31}$$

Further,

$$\begin{aligned}\int_c^d \varphi(s)\left(\int_c^s \varphi(t)\,dt\right)ds &= \int_c^d \left(\int_c^s \varphi(t)\,dt\right)d\left(\int_c^s \varphi(t)\,dt\right)\\ &= \frac{1}{2}\left(\int_c^s \varphi(t)\,dt\right)^2\Bigg|_c^d = \frac{1}{2},\end{aligned} \tag{32}$$

and $J_{2r,r} = (r!)^2/2$. So, $I_k = 0$ for $k = 0,1,\ldots,2r-1$ and

$$I_{2r} = \frac{(2r)!}{2}B\left(r+\frac{1}{2}, r+\frac{1}{2}\right) = \frac{((2r-1)!!)^2\pi}{2^{2r+1}}. \tag{33}$$

Finally we have

$$\begin{aligned}&\left\langle \widetilde{x_+^{-r-1/2}}(\varphi_\varepsilon, x)\cdot \widetilde{x_-^{-r-1/2}}(\varphi_\varepsilon, x), \psi(x)\right\rangle\\ &\quad = \frac{\pi}{2}\frac{\psi^{(2r)}(0)}{(2r)!} + O(\varepsilon)\\ &\quad = \frac{\pi}{2(2r)!}\left\langle \delta^{(2r)}(x), \psi(x)\right\rangle + O(\varepsilon).\end{aligned} \tag{34}$$

Therefore passing to the limit, as $\varepsilon \to 0$, we obtain (22) proving the theorem.

References

[1] S. Gasiorawics, *Elementary Particle Physics*, John Wiley & Sons, New York, NY, USA, 1966.

[2] M. Oberguggenberger, *Multiplication of Distributions and Application to Partial Differential Equations*, vol. 259 of *Pitman Research Notes in Mathematics Series*, Longman, 1992.

[3] H. König, "Neue Begründung der theorie der distributions," *Mathematische Nachrichten*, vol. 9, pp. 129–148, 1953.

[4] P. Antosik, J. Mikusinski, and R. Sigorski, *theory of Distributions. The Sequential Approach*, Elsevier, Amsterdam, The Netherlands, 1973.

[5] B. Fisher, "The product of distributions," *The Quarterly Journal of Mathematics*, vol. 22, pp. 291–298, 1971.

[6] B. Fisher, "A noncommutative neutrix product of distributions," *Mathematische Nachrichten*, vol. 108, pp. 117–127, 1982.

[7] L. Z. Cheng and B. Fisher, "Several products of distributions on $\mathbf{R}^m$," *Proceedings of the Royal Society London A*, vol. 426, no. 1871, pp. 425–439, 1989.

[8] A. H. Zemanian, *Distribution Theory and Transform Analysis*, Dover, 1965.

[9] I. M. Gel'fand and G. E. Shilov, *Generalized Functions*, vol. 1, chapter 1, Academic Press, New York, NY, USA, 1964.

[10] J.-F. Colombeau, *New Generalized Functions and Multiplication of Distributions*, vol. 84 of *North-Holland Mathematics Studies*, North-Holland, Amsterdam, The Netherlands, 1984.

[11] B. Jolevska-Tuneska, A. Takači, and E. Ozcag, "On differential equations with nonstandard coefficients," *Applicable Analysis and Discrete Mathematics*, vol. 1, no. 1, pp. 276–283, 2007.

[12] B. Damyanov, "Results on Colombeau product of distributions," *Commentationes Mathematicae Universitatis Carolinae*, vol. 38, no. 4, pp. 627–634, 1997.

[13] B. Damyanov, "Some distributional products of Mikusiński type in the Colombeau algebra $G(R^m)$," *Journal of Analysis and its Applications*, vol. 20, no. 3, pp. 777–785, 2001.

[14] B. Damyanov, "Modelling and products of singularities in Colombeau's algebra $G(R)$," *Journal of Applied Analysis*, vol. 14, no. 1, pp. 89–102, 2008.

[15] J. Mikusiński, "On the square of the Dirac delta-distribution," *Bulletin de l'Académie Polonaise des Sciences*, vol. 14, pp. 511–513, 1966.

22

On the Set of Fixed Points and Periodic Points of Continuously Differentiable Functions

Aliasghar Alikhani-Koopaei

Department of Mathematics, Berks College, Pennsylvania State University, Tulpehocken Road, P.O. Box 7009, Reading, PA 19610-6009, USA

Correspondence should be addressed to Aliasghar Alikhani-Koopaei; axa12@psu.edu

Academic Editor: Paolo Ricci

In recent years, researchers have studied the size of different sets related to the dynamics of self-maps of an interval. In this note we investigate the sets of fixed points and periodic points of continuously differentiable functions and show that typically such functions have a finite set of fixed points and a countable set of periodic points.

1. Introduction and Notation

The set of periodic points of self-maps of intervals has been studied for different reasons. The functions with smaller sets of periodic points are more likely not to share a periodic point. Of course, one has to decide what "big" or "small" means and how to describe this notion. In this direction one would be interested in studying the size of the sets of periodic points of self-maps of an interval, in particular, and other sets arising in dynamical systems in general (see [1–5]). For example, typically continuous functions have a first category set of periodic points (see [1, 5]). This result was generalized in [2] for the set of chain recurrent points. At times, even the smallness of these sets in some sense could be useful. For example, in [6] we showed that two commuting continuous self-maps of an interval share a periodic point if one has a countable set of periodic points. Schwartz (see [7]) was able to show that if one of the two commuting continuous functions is also continuously differentiable, then it would necessarily follow that the functions share a periodic point. Schwartz's result along with the results given in [6] may suggest that continuously differentiable functions have a countable set of periodic points. This is not true in general. However, in this note we show that typically such functions have a finite set of fixed points and a countable set of periodic points. Here $F_1(f) = \{x : f(x) = x\}$ denotes the set of fixed points of f.

For f and $n \in \mathbb{N}$ we define $f^n(x)$ by induction:

$$f^n(x) = f\left(f^{n-1}(x)\right) \quad \text{with } f^0(x) = x, \qquad f^1(x) = f(x). \tag{1}$$

The orbit of x under f is given by the sequence $\operatorname{orb}(x, f) = \{f^i(x)\}_{i=0}^{\infty}$. For $n \in \mathbb{N}$, let $F_n(f) = \{x : f^n(x) = x\}$ and P_f^n be the set of periodic points of order n; that is,

$$P_f^n = F_n(f) \setminus \cup_{k<n} F_k(f). \tag{2}$$

Two functions on a given interval are said to be of the same monotone type if both are either strictly increasing or strictly decreasing on that interval. Here, for a partition T of the interval $[a, b]$, $\|T\|$ is the length of the largest subinterval of T, $B(f, \epsilon)$ is the open ball about f with radius ϵ, A° denotes the set of interior points of A, and $\lambda(I)$ is the length of the interval I.

2. Continuously Differentiable Functions

For $I = [a, b]$, consider $C(I, I)$ and $C^1(I, I)$ to be the family of all continuous maps and continuously differentiable maps from I into itself, respectively. Recall that the usual metrics ρ_0 and ρ_1 on $C(I,I)$ and $C^1(I,I)$, respectively, are given by

$$\rho_0(f,g)=\sup_{x\in I}|f(x)-g(x)| \quad \text{for } f,g\in C(I,I),$$
$$\rho_1(f,g)=\rho_0(f,g)+\rho_0(f',g') \quad \text{for } f,g\in C^1(I,I). \tag{3}$$

It is well known that the metric spaces $(C(I,I);\rho_0)$ and $(C^1(I,I);\rho_1)$ are complete and hence Baire's category theorem holds in these spaces. We say that a typical function in $C(I,I)$ or $C^1(I,I)$ has a certain property if the set of those functions which does not have this property is of first category in $C(I,I)$ or in $C^1(I,I)$. (Some authors prefer using the term generic instead of typical.) It is known that typically continuous self-maps of an interval have σ-perfect, measure zero sets of periodic points (see [2]). Here we show that typically members of $(C^1(I,I),\rho_1)$ have a finite set of fixed points and a countable set of periodic points.

Lemma 1. *Let $n\ge 1$ and $f\in C^1(I,I)$ so that $F_n(f)$ is not finite. Then there exists $x_0\in I$ so that $f^n(x_0)=x_0$ and $(f^n)'(x_0)=1$.*

Proof. Suppose $F_n(f)$ is not finite, and then we can choose a strictly monotone sequence in $F_n(f)$. Without loss of generality, assume $\{x_i\}_{i=1}^{\infty}\subset F_n(f)$ so that $x_i<x_{i+1}$ for each $i\ge 1$. Let $h(x)=f^n(x)-x$, and then, for each i, $h(x_i)=h(x_{i+1})=0$. Thus by Rolle's Theorem we have y_i, $x_i<y_i<x_{i+1}$ so that $h'(y_i)=0$. Let $x_0=\lim_{i\to\infty}x_i$, then $x_0\in I$, and $\lim_{i\to\infty}y_i=x_0$, and from the continuity of f^n and $(f^n)'$ it follows that $f^n(x_0)=x_0$ and $(f^n)'(x_0)=1$.

Lemma 2. *For each $f\in C^1(I,I)$ and $0<\epsilon<1$ there is a polynomial $p:I\to I^o$ with $\rho_1(f,p)<\epsilon$.*

Proof. Let $M=\sup_{x\in I}\{|f'(x)|,|a|,|b|,1\}$ where $I=[a,b]$. Take $\beta=\epsilon/8M$, $\alpha=\beta(a+b)/2$, and $g(x)=\alpha+(1-\beta)f(x)$. It is easy to see that $0<\beta<1$, $g(I)\subset I^o$, and $\rho_1(f,g)<\epsilon/4$. Let

$$\epsilon_1=\frac{1}{2}\inf_{x\in[a,b]}\left\{g(x)-a,b-g(x),\frac{\epsilon}{8(b-a)},\frac{\epsilon}{8}\right\} \tag{4}$$

and $s(t)$ be a polynomial with $\rho_0(g',s)<\epsilon_1/(b-a+1)$. Then $p(x)=g(a)+\int_a^x s(t)dt$ is a polynomial and $g(x)-p(x)=\int_a^x(g'(t)-s(t))dt$. Thus we have

$$\rho_1(g,p)=\rho_0(g',s)+\rho_0(g,p)\le\frac{\epsilon_1}{(b-a)+1}+\frac{\epsilon_1(b-a)}{(b-a)+1}$$
$$=\epsilon_1<\frac{\epsilon}{4}, \tag{5}$$

hence $\rho_1(f,p)\le\rho_1(f,g)+\rho_1(g,p)<\epsilon/2$. From $g(I)\subset[a+2\epsilon_1,b-2\epsilon_1]$ and $\rho_1(g,p)\le\epsilon_1$ it follows that $p(I)\subset I^o$.

Lemma 3. *Let $n>2$ be a positive integer and $f\in C^1(I,I)$. If x_0 is a periodic point of order n with $(f^n)'(x_0)=1$, then $orb(x_0,f)\cap\{a,b\}=\emptyset$.*

Proof. We show that $a\notin \mathrm{orb}(x_0,f)$. The case $b\notin\mathrm{orb}(x_0,f)$ follows similarly. Let $\mathrm{orb}(x_0,f)=\{a=x_0,x_1,\dots,x_n=x_0\}$ with $x_i=f(x_{i-1})$ for $1\le i\le n$. Since $n>2$, there exist $x_i\in\mathrm{orb}(x_0,f)\cap(a,b)$ and $1\le s\le n-1$ so that $f^s(x_i)=a$. Since $f^s\in C^1(I,I)$ and $(f^s)'(x_0)\neq 0$, there exists $\delta_1>0$ so that $a<x_i-\delta_1<x_i<x_i+\delta_1<b$ and f^s is strictly increasing or strictly decreasing on $J=[x_i-\delta_1,x_i+\delta_1]$. Let f^s be strictly increasing on J, and then for $x_i-\delta_1\le x\le x_i$ we have $f^s(x)\le f^s(x_i)=a$; hence, $f^s|_{[x_i-\delta_1,x_i]}=a$, implying that $(f^s)'(x_i)=0$, a contradiction to $(f^n)'(x_0)=1$. On the other hand when f^s is strictly decreasing on J, for $x_i\le x\le x_i+\delta_1$ we have $f^s(x)\le f^s(x_i)=a$; hence, $f^s|_{[x_i,x_i+\delta_1]}=a$ implying that $(f^s)'(x_i)=0$, a contradiction to $(f^n)'(x_0)=1$.

Lemma 4. *For $k\ge 1$, the set*

$$\mathbb{E}_k=\left\{f\in C^1(I,I):F_k(f)\cap\left\{x:\left(f^k\right)'(x)=1\right\}\neq\emptyset\right\} \tag{6}$$

is closed in $C^1(I,I)$.

Proof. Let $g_n\in\mathbb{E}_k$ and $\rho_1(g_n,g)\to 0$. Then there exists $\{t_n\}_{n=1}^{\infty}\subset I$ such that $g_n^k(t_n)=t_n$ and $(g_n^k)'(t_n)=1$. Without loss of generality, we may assume that $\lim_{n\to\infty}t_n=t_0$, then $t_0\in I$. Let $\epsilon>0$ be arbitrary. Due to the uniform continuity of g^k on I there exists a positive integer n_1 such that, for $n\ge n_1$, $|g^k(t_n)-g^k(t_0)|<\epsilon$. Since $\rho_0(g_n,g)\to 0$, there exists a positive integer n_2 such that, for $n\ge n_2$, $\rho_0(g_n^k,g^k)<\epsilon$. Choose the positive integer n_3 so that $|t_n-t_0|<\epsilon$ for $n\ge n_3$, and let $m_1=\max(n_1,n_2,n_3)$. Then for $m\ge m_1$ we have

$$\begin{aligned}&\left|g^k(t_0)-t_0\right|\\&\le\left|g^k(t_0)-g^k(t_m)\right|+\left|g^k(t_m)-g_m^k(t_m)\right|\\&\quad+\left|g_m^k(t_m)-t_0\right|\\&\le\epsilon+\rho_0\left(g^k,g_m^k\right)+\left|t_m-t_0\right|\le 3\epsilon,\end{aligned} \tag{7}$$

Implying that $g^k(t_0)=t_0$. We also have

$$\left|\left(g_n^k\right)'(t_n)-\left(g^k\right)'(t_n)\right|\le\lim_{n\to\infty}\rho_1\left(g_n^k,g^k\right)=0. \tag{8}$$

Thus we have $\lim_{n\to\infty}|1-(g^k)'(t_n)|=|1-(g^k)'(t_0)|=0$ and $g\in\mathbb{E}_k$.

Theorem 5. *There exists a residual subset M_1 of $C^1(I,I)$ such that, for every $f\in M_1$, $F_1(f)$ is finite.*

Proof. If $F_1(f)$ has infinitely many elements, then from Lemma 1 it follows that there exists a point $x_0\in F_1(f)$ so that $f'(x_0)=1$. Let

$$\mathbb{E}_1=\left\{f\in C^1(I,I):\exists x_0\in I\cap F_1(f),\ f'(x_0)=1\right\}. \tag{9}$$

From Lemma 4 the set $\mathbb{E}_1$ is closed in $C^1(I,I)$. To show that $\mathbb{E}_1$ has no interior point, let $f\in\mathbb{E}_1$ and $\epsilon>0$, and then, by

Lemma 2, there exists a polynomial p such that $\rho_1(f,p) < \epsilon/2$ and $p(I) \subset [a+\epsilon_1, b-\epsilon_1]$ for some $0 < \epsilon_1 < \epsilon/2$. Let

$$M(p) = \{x : p'(x) = 1\} = \{t_1, t_2, \dots, t_m\}. \quad (10)$$

Choose $0 < \epsilon_2 < \epsilon_1$ so that $p(x) - x \neq \epsilon_2$ for all $x \in M(p)$ and take $h = p - \epsilon_2$. Then $h \in C^1(I,I)$, $\rho_1(h,f) < \epsilon$, $M(h) = \{x : h'(x) = 1\} = M(p)$, and $h(x) \neq x$ on $M(h)$, implying that $\mathbb{E}_1$ is of first category and $M_1 = C^1(I,I) \setminus \mathbb{E}_1$ is residual.

Theorem 6. *The set of functions $f \in C^1(I,I)$ with $f(a) = b$, $f(b) = a$, and $(f^2)'(a) = 1$ is a nowhere dense subset of $C^1(I,I)$.*

Proof. Let $A = \{f \in C^1(I,I) : f(a) = b, f(b) = a, \text{ and } (f^2)'(a) = 1\}$. It is easy to see that A is a closed subset of $C^1(I,I)$. To show that A is nowhere dense, let $f \in C^1(I,I)$ and $\epsilon > 0$. Choose $0 < \gamma < \min\{\epsilon/(4(b-a)+4), 1\}$ and $a < c < b$ such that $|f(t) - f(a)| < (b-a)[1-\epsilon/4]$ for $t \in [a,c]$. Let $h(x) = f(x) + \int_a^x k(t)dt$, where

$$k(t) = \begin{cases} -\gamma, & t = a, \\ 0, & t = a + \dfrac{c-a}{4}, \\ \dfrac{\gamma}{3}, & t = a + \dfrac{(c-a)}{2}, \\ 0, & t = c,\ t = b, \\ \text{linear}, & \text{otherwise.} \end{cases} \quad (11)$$

It is clear that $h(a) = f(a) = b$, $h(b) = f(b) = a$, $h'(a) = f'(a) - \gamma$, and $h'(b) = f'(b)$. It is easy to see that $h \in C^1(I,I)$, $(h^2)'(a) = (h^2)'(b) \neq 1$, and $\rho_1(f,h) < \epsilon/2$.

Theorem 7. *Let $\epsilon > 0$, $f \in C^1(I,I)$, and $A_n = \{x : (f^n)'(x) = 1\} \cap F_n(f)$ be a finite set, and $\mathrm{orb}(x,f) \subset (a,b)$ for $x \in A_n$. Then there exists a function $g \in C^1(I,I)$ with $\rho_1(f,g) < \epsilon/4$ such that $\{x : (g^n)'(x) = 1\} \cap F_n(g) = \emptyset$.*

Proof. Let $A_n = \{x : (f^n)'(x) = 1\} \cap F_n(f) = \{z_j\}_{j=1}^l$. Let $B = \{t_j\}_{j=1}^k$ be the elements of A_n with distinct orbits, that is, $\mathrm{orb}(t_i,f) \cap \mathrm{orb}(t_j,f) = \emptyset$ for $i \neq j$ and $\cup_{j=1}^k \mathrm{orb}(t_j,f) \supseteq A_n$. It is clear that each $t_j \in B$ is a periodic point of f with some period m where m is a factor of n. This suggests that if one can construct a function g that is sufficiently close to f, and either $t_j \notin P_f^m$ or $(f^m)'(t_j) \neq 1$, then

$$\{x : (g^n)'(x) = 1\} \cap F_n(g) \subseteq A_n \setminus \mathrm{orb}(t_j, f). \quad (12)$$

By repeating this process for each $t_j \in B$ in a finite number of steps we can construct the desired function g. Thus for convenience, we assume that, for $1 \le j \le k$, $z_j \in P_f^n$. Consider the partition T of $[a,b]$ obtained by $\{f^i(t_j), 1 \le i \le n, 1 \le j \le k\}$, $\epsilon_1 = (1/4)\min\{\epsilon, \| T \|\}$, and choose a positive δ less than ϵ_1 such that, for each i, $1 \le i \le n$ and each j, $1 \le j \le k$, and if $|t - t_j| < \delta$, then $|f^i(t) - f^i(t_j)| < \epsilon_1$. Given that, for $x \in A_n$, $\mathrm{orb}(x,f) \subset (a,b)$, $(f^n)'$ is continuous, and $(f^n)'(x) = 1$ on B, we may choose the nondegenerate closed intervals H_j, $1 \le j \le k$ of length less than δ and the positive numbers γ_j such that

(i) $t_j \in (H_j)^\circ$ is the midpoint of H_j and $f^i(H_j) \cap \mathrm{orb}(t_j,f) = \{f^i(t_j)\}$ for $0 \le i \le n$,

(ii) f is strictly monotone on each $H_{j,i} = f^i(H_j)$ for $0 \le i \le n-1$, $1 \le j \le k$, where $H_{j,0} = H_j$,

(iii) the intervals $H_{j,i} = f^i(H_j)$ are mutually disjoint for $1 \le i \le n$, $1 \le j \le k$,

(iv) $a < \min(H_{j,i}) - \gamma_j \lambda(H_{j,i}) < \max(H_{j,i}) + \gamma_j \lambda(H_{j,i}) < b$, for $0 \le i \le n-1$,

(v) $\max_{1 \le j \le k}\{\gamma_j, (b-a)\gamma_j\} < (1/8)[\min_{0 \le i \le n,\, 1 \le j \le k}\{\epsilon/k, \lambda(f^i(H_j)), \beta_{j,i}\}]$, where $\beta_{j,i} = \inf\{|f'(x)| : x \in f^i(H_j)\}$.

Let $x_0 = t_i$ for some i, γ, and $J = [c,d]$ be the associated γ_i and H_i, respectively. Define

$$g_{J,x_0,\gamma}(x) = \begin{cases} -\dfrac{\gamma}{3}, & \text{if } x = c + \dfrac{x_0 - c}{2}, \\ 0, & \text{if } x = c + \dfrac{3(x_0 - c)}{4}, \\ \gamma, & \text{if } x = x_0, \\ 0, & \text{if } x = x_0 + \dfrac{d - x_0}{4}, \\ -\dfrac{\gamma}{3}, & \text{if } x = x_0 + \dfrac{d - x_0}{2}, \\ 0, & \text{if } x \in [a,b] \setminus (c,d), \\ \text{linear}, & \text{otherwise,} \end{cases}$$

$$r_{J,x_0,\gamma}(x) = \begin{cases} \gamma, & \text{if } x = c + \dfrac{x_0 - c}{4}, \\ 0, & \text{if } x = x_0, \\ \dfrac{-\gamma(x_0 - c)}{d - x_0} & \text{if } x = x_0 + \dfrac{3(d - x_0)}{4}, \\ 0 & \text{if } x \in [a,b] \setminus (c,d), \\ \text{linear}, & \text{otherwise,} \end{cases} \quad (13)$$

$$h_1(x) = \int_a^x g_{J,x_0,\gamma}(t)\,dt,$$

$$h_2(x) = \int_a^x r_{J,x_0,\gamma}(t)\,dt.$$

We have $h_m(c) = h_m(d) = h_m'(c) = h_m'(d) = 0$ for $m = 1, 2$, $h_1(x_0) = 0$, $h_1'(x_0) = \gamma$, $h_2(x_0) = (x_0 - c)\gamma/2$, and $h_2'(x_0) = 0$. By considering four different cases, we construct a function $h \in C^1(I,I) \cap B(f,\epsilon)$ so that, for Cases A and B, $F_n(h) \cap \{x : (h^n)'(x) = 1\} \cap J = \emptyset$ and $F_n(h) \cap \{x : (h^n)'(x) = 1\} \cap J = \{x_0\}$ for Cases C and D. From conditions (iv) and (v) we have $h(I) \subset I$, and from condition (v) it follows that h is of the same monotone type as f on each $f^s(J)$, $0 \le s < n$.

Case A. Let $f^n(x) < x$ for $J \setminus \{x_0\}$ and $f^n(x_0) = x_0$.

(i) If f is strictly increasing on J, then by taking $h(x) = f(x) - h_2(x)$ we have $h(x) < f(x)$ on J^o

and $h(x) = f(x)$ on $I \setminus J^o$ and the function h^{n-1} is also strictly increasing on $f(J)$, and as a result $h^n(x) < f^n(x) \le x$ on J^o. Thus $F_n(h) \cap J = \emptyset$.

(ii) If f is strictly decreasing on J, then by taking $h(x) = f(x) + h_2(x)$ we have $h(x) > f(x)$ on J^o and $h(x) = f(x)$ on $I \setminus J^o$ and the function h^{n-1} is also strictly decreasing on $f(J)$ and as a result, $h^n(x) < f^n(x) \le x$ on J^o. Thus $F_n(h) \cap J = \emptyset$.

Case B. Let $f^n(x) > x$ for $J \setminus \{x_0\}$ and $f^n(x_0) = x_0$.

If f is strictly increasing on J, take $h(x) = f(x) + h_2(x)$, and if f is strictly decreasing on J, take $h(x) = f(x) - h_2(x)$. Then similar to Case A, we may show that $h^n(x) > f^n(x) \ge x$ for $x \in J^o$; hence, $F_n(h) \cap J = \emptyset$.

Case C. Let $f^n(x) < x$ for $c < x < x_0$, $f^n(x) > x$ for $x_0 < x < d$, and $f^n(x_0) = x_0$.

(i) If f is strictly increasing on J, then by taking $h(x) = f(x) + h_1(x)$ we have $h(x) < f(x)$ for $c < x < x_0$ and $h(x) > f(x)$ for $x_0 < x < d$, and the function h^{n-1} is also strictly increasing on $f(J)$. Thus $h^n(x) < f^n(x) < x$ for $c < x < x_0$ and $h^n(x) > f^n(x) > x$ for $x_0 < x < d$; hence, $F_n(h) \cap J = \{x_0\}$.

(ii) If f is strictly decreasing on J, then by taking $h(x) = f(x) - h_1(x)$ we have $h(x) > f(x)$ for $c < x < x_0$ and $h(x) < f(x)$ for $x_0 < x < d$, and the function h^{n-1} is also decreasing on $f(J)$. Thus $h^n(x) < f^n(x) < x$ for $c < x < x_0$ and $h^n(x) > f^n(x) > x$ for $x_0 < x < d$; hence, $F_n(h) \cap J = \{x_0\}$.

Case D. Let $f^n(x) > x$ for $c < x < x_0$, $f^n(x) < x$ for $x_0 < x < d$, and $f^n(x_0) = x_0$.

If f is strictly increasing on J, take $h(x) = f(x) - h_1(x)$, and if f is strictly decreasing on J, take $h(x) = f(x) + h_1(x)$. Then similar to Case C, we may show that $h^n(x) > f^n(x) > x$ for $c < x < x_0$ and $h^n(x) < f^n(x) < x$ for $x_0 < x < d$. Thus $F_n(h) \cap J = \{x_0\}$.

Note that h^n is strictly increasing on $J \cup f^n(J)$, so we have

$$F_n(h) \cap [(J \setminus f^n(J)) \cup (f^n(J) \setminus J)] = \emptyset. \tag{14}$$

Thus for $x \in I$ either $\operatorname{orb}(x,h) = \operatorname{orb}(x,f)$ or $\operatorname{orb}(x,h) \cap J \ne \emptyset$, of which in such case $F_n(h) \cap J = \{x_0\}$, $h'(x_0) = f'(x_0) \pm \gamma$, and $h'(h^i(x_0)) = f'(f^i(x_0))$ for $1 \le i \le n-1$. Thus

$$\begin{aligned}(h^n)'(x_0) &= h'(h^{n-1}(x_0))\, h'(h^{n-2}(x_0)) \cdots h'(h(x_0))\, h'(x_0) \\ &= f'(f^{n-1}(x_0))\, f'(f^{n-2}(x_0)) \\ &\quad \times f'(f^{n-3}(x_0)) \cdots f'(f(x_0))\, h'(x_0) \\ &= \frac{(f^n)'(x_0)\, h'(x_0)}{f'(x_0)} = \frac{f'(x_0) \pm \gamma}{f'(x_0)} \\ &= \left[1 \pm \frac{\gamma}{f'(x_0)}\right] \ne 1.\end{aligned} \tag{15}$$

We have

$$\begin{aligned}F_n(h) \cap \{x : (h^n)'(x) = 1\} \\ = [F_n(f) \cap \{x : (f^n)'(x) = 1\}] \setminus \{x_0\}.\end{aligned} \tag{16}$$

Also

$$\begin{aligned}\rho_0(f', h') &\le \sup_{x \in J} \{|g(x)|, |r(x)|\} \le \frac{\epsilon}{8k}, \\ \rho_0(f, h) &\le \sup_{x \in J} \{|h_1(x)|, |h_2(x)|\} \le \frac{\epsilon}{8k}.\end{aligned} \tag{17}$$

Hence $\rho_1(f,h) \le \rho_0(f,h) + \rho_0(f',h') \le \epsilon/4k$.

Doing this recursively for each t_j, $1 \le j \le k$, we get a function $g \in B(f, \epsilon/4)$ such that $F_n(g) \cap \{x : (g^n)'(x) = 1\} = \emptyset$.

Theorem 8. *Typically continuously differentiable self-maps of intervals have a countable set of periodic points.*

Proof. Take

$$\begin{aligned} H_n &= \{f \in C^1(I,I) : F_n(f) \text{ is not finite}\}, \\ B_1 &= \{f \in C^1(I,I) : a \in P_f^2 \cap \{x : (f^2)'(x) = 1\} \\ &\qquad \text{with } f(a) = b\}, \\ F_1 &= \{f \in C^1(I,I) : \exists x_0 \in I \text{ such that} \\ &\qquad x_0 \in F_1(f) \text{ and } f'(x_0) = 1\}, \\ \mathbb{E}_n &= \{f \in C^1(I,I) : F_n(f) \cap \{x : (f^n)'(x) = 1\} \ne \emptyset\}. \end{aligned} \tag{18}$$

The sets F_1 and B_1 are first category sets (see Theorems 5 and 6). For $n \ge 1$ from Lemmas 4 and 1 we have that $\mathbb{E}_n$ is closed and $H_n \subseteq \mathbb{E}_n$. Without loss of generality we may assume that $f \in C^1(I,I)$ is neither a constant function nor a polynomial of first degree, and then by Lemma 2 we may choose a polynomial $g \in C^1(I,I)$ of degree $n \ge 2$ such that $g(I) \subset I^o$ and $\rho_1(f,g) < \epsilon/4$. Using Theorem 7 we can construct $h \in C^1(I,I)$ so that $\rho_1(g,h) < \epsilon/4$ and the set $A_n = F_n(h) \cap \{x : (h^n)'(x) = 1\} = \emptyset$, thus $\rho_1(f,h) < \epsilon$ and $h \notin \mathbb{E}_n$, hence $\mathbb{E}_n^o = \emptyset$. This implies that for each $k \ge 2$ the set $\mathbb{E}_k$ is nowhere dense. Thus $F = (\cup_{k=2}^{\infty} \mathbb{E}_k) \cup F_1 \cup B_1$ is a first category set, so $M = C^1(I,I) \setminus F$ is a residual set, and, for $f \in M$, $P(f) = \cup_{n=1}^{\infty} F_n(f)$ is countable.

Acknowledgments

The author would like to thank professor David Preiss for his valuable comments and suggestions as well as the anonymous referees for their comments.

References

[1] S. J. Agronsky, A. M. Bruckner, and M. Laczkovich, "Dynamics of typical continuous functions," *Journal of the London Mathematical Society*, vol. 40, no. 2, pp. 227–243, 1989.

[2] A. A. Alikhani-Koopaei, "On the size of the sets of chain recurrent points, periodic points, and fixed points of functions," *Global Journal of Pure and Applied Mathematics*, vol. 4, no. 2, pp. 113–121, 2008.

[3] A. A. Alikhani-Koopaei, "On periodic and recurrent points of continuous functions," in *Proceedings of the 28th Summer Symposium Conference*, Real Analysis Exchange, pp. 37–40, 2004.

[4] A. Alikhani-Koopaei, "On common fixed points, periodic points, and recurrent points of continuous functions," *International Journal of Mathematics and Mathematical Sciences*, vol. 2003, no. 39, pp. 2465–2473, 2003.

[5] T. H. Steele, "A note on periodic points and commuting functions," *Real Analysis Exchange*, vol. 24, no. 2, pp. 781–789, 1998-1999.

[6] A. Alikhani-Koopaei, "On common fixed and periodic points of commuting functions," *International Journal of Mathematics and Mathematical Sciences*, vol. 21, no. 2, pp. 269–276, 1998.

[7] A. J. Schwartz, "Common periodic points of commuting functions," *The Michigan Mathematical Journal*, vol. 12, pp. 353–355, 1965.

23

Perpendicularity in an Abelian Group

Pentti Haukkanen,[1] Mika Mattila,[1] Jorma K. Merikoski,[1] and Timo Tossavainen[2]

[1] *School of Information Sciences, University of Tampere, 33014, Finland*
[2] *School of Applied Educational Science and Teacher Education, University of Eastern Finland, P.O. Box 86, 57101 Savonlinna, Finland*

Correspondence should be addressed to Pentti Haukkanen; pentti.haukkanen@uta.fi

Academic Editor: Petru Jebelean

We give a set of axioms to establish a perpendicularity relation in an Abelian group and then study the existence of perpendicularities in $(\mathbb{Z}_n, +)$ and $(\mathbb{Q}_+, \cdot)$ and in certain other groups. Our approach provides a justification for the use of the symbol $\perp$ denoting relative primeness in number theory and extends the domain of this convention to some degree. Related to that, we also consider parallelism from an axiomatic perspective.

1. Introduction

In [1, page 115], Graham et al. made the following suggestion:

When $\gcd(m, n) = 1$, the integers m and n have no prime factors in common and we say that they are *relatively prime*.

This concept is so important in practice, we ought to have a special notation for it; but alas, number theorists have not agreed on a very good one yet. Therefore we cry: HEAR US, O MATHEMATICIANS OF THE WORLD! LET US NOT WAIT ANY LONGER! WE CAN MAKE MANY FORMULAS CLEARER BY ADOPTING A NEW NOTATION NOW! LET US AGREE TO WRITE "$m \perp n$", AND TO SAY "m IS PRIME TO n," IF m AND n ARE RELATIVELY PRIME. Like perpendicular lines do not have a common direction, perpendicular numbers do not have common factors.

In fact, this cry had been answered even before it was made. Namely, in studying l-groups (i.e., groups with a lattice structure), Birkhoff [2, page 295] defines that two positive elements a and b of an l-group are disjoint if $a \wedge b = 0$ and uses the notation $a \perp b$ for disjoint elements. He also remarks that disjointness specializes to relative primeness in the l-group of positive integers.

A motivation for the present paper is to study how justified ultimately it is to use the symbol of perpendicularity to denote relative primeness. Does this practice rely only on the analogy between having no common direction and having no common factor or is there a deeper linkage to entitle this convention? This question leads us to ask which properties essentially establish the notion of perpendicularity in the algebraic context and what the most suitable algebraic context for the axiomatization of perpendicularity actually is; we have recently studied the axioms of perpendicularity from an elementary geometric point of view [3].

In an inner product space, perpendicularity obviously traces back to the inner product being zero. However, certain features of this perpendicularity can be shifted down to simpler algebraic structures. We will define perpendicularity in an Abelian group and examine it in Section 2. In Section 3, we will focus on perpendicularity in $(\mathbb{Z}_n, +)$. Davis [4] defined perpendicularity in an Abelian group differently. In Section 4, we will introduce his approach and compare it with ours. Thereafter, we will consider divisibility in $(\mathbb{Q}_+, \cdot)$ in Section 5 and parallelism in an Abelian group in Section 6. We will conclude our paper with a brief discussion and a supplement to the suggestion cited previously.

2. Axioms and Properties of Perpendicularity

Throughout this paper, $G = (G, +)$ is an Abelian group so that $G \neq \{0\}$. Unless otherwise stated, $\perp$ is a binary relation in G satisfying

(A1) $\forall a \in G : \exists b \in G : a \perp b$,

(A2) $\forall a \in G \setminus \{0\} : a \not\perp a$,

(A3) $\forall a, b \in G : a \perp b \Rightarrow b \perp a$,

(A4) $\forall a,b,c \in G : a \perp b \wedge a \perp c \Rightarrow a \perp (b+c)$,

(A5) $\forall a,b \in G : a \perp b \Rightarrow a \perp -b$.

We call $\perp$ a *perpendicularity* in G. This concept can be defined also in weaker structures by changing these axioms appropriately. For example, if G is an Abelian monoid, then we simply omit (A5). Since the *trivial perpendicularity*

$$x \perp y \Longleftrightarrow x = 0 \vee y = 0 \tag{1}$$

always exists, we are mainly interested in nontrivial perpendicularities.

We call $\perp$ *maximal* if it is not a subrelation of any other perpendicularity in G. There always exists a maximal perpendicularity. This is obvious if G is finite and, otherwise, it follows from Zorn's lemma.

Proposition 1 records some elementary properties of perpendicularity; we leave the proof for the reader.

Proposition 1. *Perpendicularity $\perp$ has the following properties:*

(a) $\forall a \in G:\ a \perp 0$,

(b) $\forall a \in G \setminus \{0\}:\ a \not\perp -a$,

(c) $\forall a, b_1, \ldots, b_k \in G, \gamma_1, \ldots, \gamma_k \in \mathbb{Z} : a \perp b_1, \ldots, b_k \Rightarrow a \perp (\gamma_1 b_1 + \cdots + \gamma_k b_k)$,

(d) $\forall a, b \in G, \mu, \nu \in \mathbb{Z} : a \perp b \Rightarrow \mu a \perp \nu b$.

The following characterization is useful in proving that a given relation is perpendicularity.

Proposition 2. *A binary relation $\perp$ in G is perpendicularity if and only if it satisfies (A1) and (A2) and*

(A6) $\forall a,b,c \in G : a \perp b \wedge a \perp c \Rightarrow (b-c) \perp a$.

Proof. The "only if"-part is trivial. To prove the "if"-part, we first show that our assumptions imply Proposition 1(a). Let $a \in G$. By (A1), there is $b \in G$ such that $a \perp b$. Putting $c := b$ in (A6) implies $0 \perp a$; in particular $0 \perp 0$. Further, (A6) with $a := 0$, $b := a$ and $c := 0$ gives $(a-0) \perp 0$, that is, $a \perp 0$. Now we can verify the remaining axioms.

(A3) Assume $a \perp b$. Apply (A6) with $c := 0$; then $b \perp a$.

(A5) Assume $a \perp b$. Apply (A6) with $b := 0$ and $c := b$. Then $(-b) \perp a$, and so, by (A3), $a \perp -b$.

(A4) Assume $a \perp b$ and $a \perp c$; then $a \perp -c$ by (A5). Now (A6) with $c := -c$ implies $(b-(-c)) \perp a$, that is, $(b+c) \perp a$. Hence, by (A3), $a \perp (b+c)$.

Is there a simple condition under which (A5) follows from (A1)–(A4)? The answer is positive.

Proposition 3. *If all elements of G have finite order and if $\perp$ satisfies (A1)–(A4), then it satisfies (A5). If G has at least one element of infinite order, then there exists a relation $\perp$ which satisfies (A1)–(A4) but not (A5).*

Proof. For the first part, assume that $a, b \in G$ satisfy $a \perp b$, and let the order of b be n. Then $a \perp (n-1)b$ by (A4). But $(n-1)b = -b$ and (A5) follows. For the second part, let $a \in G$ have infinite order. Then the subgroup $\{0, \pm a, \pm 2a, \ldots\}$ is isomorphic to $\mathbb{Z}$. The relation $\perp$ defined by

$$\begin{aligned} x \perp y \Longleftrightarrow & (\exists \mu, \nu \in \mathbb{Z} : x = \mu a \wedge y = \nu a \wedge \mu\nu < 0) \\ & \vee x = 0 \vee y = 0 \end{aligned} \tag{2}$$

satisfies (A1)–(A4) but not (A5).

If $\emptyset \neq A \subseteq G$, we define the *perpendicular complement* or *$\perp$-complement* of A as follows:

$$A^{\perp} = \{y \in G \mid y \perp A\} = \bigcup_{G \supseteq B \perp A} B. \tag{3}$$

Here $y \perp A$ means that $y \perp x$ for all $x \in A$, and $B \perp A$ means that $y \perp A$ for all $y \in B$. Thus $A^{\perp}$ is the maximal set perpendicular to A. In particular, $G^{\perp} = \{0\}$ and $\{0\}^{\perp} = G$. We also define $\emptyset^{\perp} = G$.

Proposition 4. *If $A \subseteq G$, then $A^{\perp}$ is a subgroup of G. If G is cyclic, then $A^{\perp}$ is cyclic.*

Proof. The first part follows by applying the subgroup test and Proposition 2. The second part follows from the fact that any subgroup of a cyclic group is cyclic.

The next theorem tells when G has a nontrivial perpendicularity.

Theorem 5. *The following conditions are equivalent:*

(a) *G has a nontrivial perpendicularity $\perp$,*

(b) *G has nontrivial cyclic subgroups H and K satisfying $H \cap K = \{0\}$,*

(c) *G has nontrivial subgroups H and K satisfying $H \cap K = \{0\}$.*

Proof. (a)⇒(b). Since $\perp$ is nontrivial, there exist $x, y \in G \setminus \{0\}$ such that $x \perp y$. Then $H = \langle x \rangle$ and $K = \langle y \rangle$ apply. Here $\langle a \rangle$ stands for the cyclic group generated by a.

(b)⇒(c). Trivial.

(c)⇒(a). Define $\perp$ by

$$\begin{aligned} x \perp y \Longleftrightarrow & (x \in H \wedge y \in K) \vee (x \in K \wedge y \in H) \\ & \vee x = 0 \vee y = 0. \end{aligned} \tag{4}$$

Next we consider the maximal perpendicularity in some examples of groups. In Examples 6–9, the group operation is addition.

Example 6. Let $G = \mathbb{Z}_6$. By Lagrange's theorem [5, page 130, Theorem 2], the smallest n such that $\mathbb{Z}_n$ has a nontrivial perpendicularity is $6 = 2 \cdot 3$ because n must have at least two different prime factors. The nontrivial subgroups of $\mathbb{Z}_6$ are $H = \langle 3 \rangle = \{0, 3\}$ and $K = \langle 2 \rangle = \{0, 2, 4\}$. Since $G = H \oplus K$, it has exactly one nontrivial perpendicularity, defined by $0 \perp 0, 1, 2, 3, 4, 5$ and $3 \perp 2, 4$ and vice versa. Consequently, this perpendicularity is maximal.

Example 7. Let $G = \mathbb{Z}_2 \times \mathbb{Z}_2$, the Klein four group. Denote $0 = (0, 0)$, $a = (0, 1)$, $b = (1, 0)$, $c = (1, 1)$. The nontrivial subgroups are $A = \{0, a\}$, $B = \{0, b\}$, $C = \{0, c\}$. So, there are three nontrivial perpendicularities obtained as follows: Choose two elements of a, b, and c. Define that they are perpendicular to each other and to 0. Define that the remaining element is perpendicular to 0 only. All perpendicularities arising in this way are clearly maximal. We also note that $G = A \oplus B = B \oplus C = C \oplus A$.

Example 8. Let $G = \mathbb{Z}$. For each $n \geq 2$, the subgroup $\langle n \rangle = n\mathbb{Z}$ is nontrivial and there are no other nontrivial subgroups than those found in this way. Because $mn \in \langle m \rangle \cap \langle n \rangle$, there is no pair of nontrivial subgroups with intersection $\{0\}$. Hence G has only the trivial perpendicularity.

Example 9. Let $G = \mathbb{R}$. Since $\mathbb{R}$ has infinitely many pairs of nontrivial subgroups with intersection $\{0\}$, it has infinitely many nontrivial perpendicularities. For example, let $H = \mathbb{Q}$ and $K = \{x\sqrt{2} \mid x \in \mathbb{Q}\}$ and define $\perp$ by (4). To see that this perpendicularity is not maximal, let $H_1 = \{x\sqrt{3} \mid x \in \mathbb{Q}\}$ and $K_1 = \{x\sqrt{5} \mid x \in \mathbb{Q}\}$ and define $\perp'$ by

$$\begin{aligned} x \perp' y \Longleftrightarrow & (x \in H \wedge y \in K) \vee (x \in K \wedge y \in H) \\ & \vee (x \in H_1 \wedge y \in K_1) \vee (x \in K_1 \wedge y \in H_1) \\ & \vee x = 0 \vee y = 0. \end{aligned} \tag{5}$$

Then $x \perp y \Rightarrow x \perp' y$.

Example 10. Let $G = (\mathbb{Q}_+, \cdot)$, where $\mathbb{Q}_+$ denotes the set of positive rational numbers.

Every $c \in \mathbb{Q}_+$ can be uniquely expressed as

$$c = \prod_{p \in \mathbb{P}} p^{\nu_p(c)}, \tag{6}$$

where $\nu_p(c) \in \mathbb{Z}$ for each $p \in \mathbb{P}$ and only a finite number of them are nonzero. The symbol $\mathbb{P}$ stands for the set of primes. For example, if $c = 8/25$, then $\nu_2(c) = 3$, $\nu_3(c) = 0$, $\nu_5(c) = -2$, $\nu_7(c) = \nu_{11}(c) = \cdots = 0$.

Assign now

$$a \perp b \Longleftrightarrow \forall p \in \mathbb{P} : \nu_p(a) = 0 \vee \nu_p(b) = 0. \tag{7}$$

In other words, if

$$\begin{gathered} a = \frac{m}{u}, \quad b = \frac{n}{v}, \quad m, u, n, v \in \mathbb{Z}_+, \\ \gcd(m, u) = \gcd(n, v) = 1, \end{gathered} \tag{8}$$

then

$$a \perp b \Longleftrightarrow \gcd(mu, nv) = 1. \tag{9}$$

Hence, for example, $8/9 \perp 7/5$. In particular, for $m, n \in \mathbb{Z}_+$, applying (9) to $m/1$ and $n/1$ yields that

$$m \perp n \Longleftrightarrow \gcd(m, n) = 1. \tag{10}$$

So, it seems that Graham et al. were prophetically quite right with their suggestion—and not forgetting Birkhoff either! We will discuss the perpendicularity of positive rational numbers in more detail in Section 5.

3. Perpendicularity in $\mathbb{Z}_n$

Studying perpendicularities requires that we know the structure of G. Next we take a more thorough look at perpendicularity in $\mathbb{Z}_n$. To that end, we begin by introducing a suitable notation to discuss the structure of $\mathbb{Z}_n$ and record two lemmas which are useful in the search for the maximal perpendicularity. We will also use the notations introduced in Theorem 11 and the following lemmas throughout the next sections.

Theorem 11. *If*

$$n = p_1^{\alpha_1} \cdots p_r^{\alpha_r}, \tag{11}$$

where $p_1, \ldots, p_r \in \mathbb{P}$ are distinct and $\alpha_1, \ldots, \alpha_r > 0$, then

$$\mathbb{Z}_n = H_1 \oplus \cdots \oplus H_r, \tag{12}$$

where

$$H_i = \langle e_i \rangle, \quad e_i = \frac{n}{p_i^{\alpha_i}}, \quad i = 1, \ldots, r. \tag{13}$$

The decomposition (12) is unique (up to the order of subgroups).

Proof. The claim (12) follows from [5, page 399, Corollary 1] and from the facts that $\mathbb{Z}_{p_i^{\alpha_i}} \cong H_i$ and $H_i \cap H_j = \{0\}$ for all $i, j = 1, \ldots, r$, $i \neq j$. Uniqueness follows from [5, page 399, Corollary 2].

Although we consider $\mathbb{Z}_n$ mainly as an Abelian group, it is now useful to work with $\mathbb{Z}_n$ as a ring.

Lemma 12. *For all $i, j = 1, \ldots, r$, $i \neq j$,*

$$e_i^2 \neq 0, \qquad e_i e_j = 0. \tag{14}$$

Proof. It is enough to consider $i = 1$, $j = 2$. Regarding e_1 and e_2 as integers, we have

$$\begin{aligned} e_1^2 &= \frac{n^2}{p_1^{2\alpha_1}} = p_2^{2\alpha_2} \cdots p_r^{2\alpha_r} \not\equiv 0 \pmod{n}, \\ e_1 e_2 &= \frac{n}{p_1^{\alpha_1}} \frac{n}{p_2^{\alpha_2}} \\ &= p_2^{\alpha_2} \cdots p_r^{\alpha_r} p_1^{\alpha_1} p_3^{\alpha_3} \cdots p_r^{\alpha_r} \\ &= p_3^{\alpha_3} \cdots p_r^{\alpha_r} n \equiv 0 \pmod{n}, \end{aligned} \tag{15}$$

and (14) follows.

Lemma 13. *Let $\perp$ be a perpendicularity in $\mathbb{Z}_n$. Then*

$$\forall a, b, c, d \in \mathbb{Z}_n : a \perp b \Longrightarrow ca \perp db. \quad (16)$$

Proof. Let $c = \gamma 1$, $d = \delta 1$, where γ and δ are integers with $0 \le \gamma,\ \delta < n$. Since $ca = (\gamma 1)a = \gamma(1a) = \gamma a$ and, similarly, $db = \delta b$, Proposition 1(d) implies (16).

Now we are ready to introduce a perpendicularity which turns out to be maximal in $\mathbb{Z}_n$. Let

$$x = x_1 + \cdots + x_r, \qquad y = y_1 + \cdots + y_r \in \mathbb{Z}_n, \quad (17)$$

where $x_i, y_i \in H_i$, $i = 1, \ldots, r$. The relation $\perp_0$, defined in $\mathbb{Z}_n$ by

$$x \perp_0 y \Longleftrightarrow \forall i \in \{1, \ldots, r\} : x_i = 0 \vee y_i = 0, \quad (18)$$

is clearly a perpendicularity.

Theorem 14. *The perpendicularity $\perp_0$ is maximal and every other perpendicularity in $\mathbb{Z}_n$ is contained in it.*

Proof. Let $\perp$ be another perpendicularity in $\mathbb{Z}_n$. Our claim is that $x \perp y \Rightarrow x \perp_0 y$. By (12), we can express

$$x = \xi_1 e_1 + \cdots + \xi_r e_r, \qquad y = \eta_1 e_1 + \cdots + \eta_r e_r, \quad (19)$$

where the integers $\xi_i, \eta_i \in \{0, \ldots, p_i^{\alpha_i} - 1\}$ and the residue class $e_i = n/p_i^{\alpha_i}$, $i = 1, \ldots, r$.

Suppose against the claim of theorem that there exist $x, y \in \mathbb{Z}_n$ such that $x \perp y$ but $x \not\perp_0 y$. Then $\xi_i, \eta_i \neq 0$ for some i. Reordering the indices so that $i = 1$ and applying (16), we have $xe_1 \perp ye_1$ which implies that

$$\xi_1 e_1^2 \perp \eta_1 e_1^2 \quad (20)$$

by (14). Hence, by Proposition 1(d),

$$\frac{\eta_1}{\gcd(\xi_1, \eta_1)} \xi_1 e_1^2 \perp \frac{\xi_1}{\gcd(\xi_1, \eta_1)} \eta_1 e_1^2, \quad (21)$$

that is,

$$\operatorname{lcm}(\xi_1, \eta_1) e_1^2 \perp \operatorname{lcm}(\xi_1, \eta_1) e_1^2. \quad (22)$$

Consequently, $\operatorname{lcm}(\xi_1, \eta_1) e_1^2 = 0$ by (A2). In other words, regarding also e_1 as an integer,

$$\begin{aligned} \operatorname{lcm}(\xi_1, \eta_1) e_1^2 &= \operatorname{lcm}(\xi_1, \eta_1) \frac{n^2}{p_1^{2\alpha_1}} \\ &= \operatorname{lcm}(\xi_1, \eta_1) p_2^{2\alpha_2} \cdots p_r^{2\alpha_r} \\ &\equiv 0 \pmod{n}, \end{aligned} \quad (23)$$

and $p_1^{\alpha_1}$ divides $\operatorname{lcm}(\xi_1, \eta_1)$. However, since it divides neither ξ_1 nor η_1, this is a contradiction. Hence, $x \perp_0 y$.

Considering the direct sum (12) external, we can identify x and y in (17) with vectors $(x_1, \ldots, x_r)$ and $(y_1, \ldots, y_r)$, respectively. So, it is natural to define their "inner product" by

$$\langle x, y \rangle = x_1 y_1 + \cdots + x_r y_r. \quad (24)$$

Proposition 15 shows that this operation coincides with the ordinary multiplication in $\mathbb{Z}_n$.

Proposition 15. *Given $x, y \in \mathbb{Z}_n$,*

$$\langle x, y \rangle = xy. \quad (25)$$

Proof. We have

$$xy = \left(\sum_{i=1}^{r} x_i \right) \left(\sum_{i=1}^{r} y_i \right) = \sum_{i=1}^{r} x_i y_i + \sum_{\substack{i,j=1 \\ i \neq j}}^{r} x_i y_j. \quad (26)$$

But, recalling (19) and (14),

$$\sum_{\substack{i,j=1 \\ i \neq j}}^{r} x_i y_j = \sum_{\substack{i,j=1 \\ i \neq j}}^{r} \xi_i \eta_j e_i e_j = 0. \quad (27)$$

The claim follows.

Theorem 16. *Let $\perp$ be a perpendicularity in $\mathbb{Z}_n$. Then*

$$\forall x, y \in \mathbb{Z}_n : x \perp y \Longrightarrow xy = 0. \quad (28)$$

Proof. If $x \perp y$, then $x \perp_0 y$ by Theorem 14. So, $xy = 0$ by (18) and (25).

Does the converse of Theorem 16 hold if $\perp = \perp_0$? And, related to Proposition 15, is $\langle x, y \rangle = xy$ a proper inner product? Namely, an inner product in a real vector space is symmetric and bilinear and it satisfies $\langle x, x \rangle = 0 \Rightarrow x = 0$. The operation $\langle x, y \rangle = xy$ in $\mathbb{Z}_n$ has clearly the first and second properties but what about the third one? The answers to both questions are contained in Theorem 17.

Theorem 17. *The following conditions are equivalent:*

(a) $\alpha_1 = \cdots = \alpha_r = 1$,

(b) $\forall x, y \in \mathbb{Z}_n : xy = 0 \Rightarrow x \perp_0 y$,

(c) $\forall x \in \mathbb{Z}_n : x^2 = 0 \Rightarrow x = 0$.

Proof. (a)$\Rightarrow$(b). Assume that $x \not\perp_0 y$. Express x and y as in (19). We can rearrange the indices so that, for some $s \in \{1, \ldots, r\}$,

$$\begin{aligned} &\xi_i, \eta_i \neq 0, \quad i = 1, \ldots, s, \\ &\xi_i = 0 \vee \eta_i = 0, \quad i = s+1, \ldots, r. \end{aligned} \quad (29)$$

By (25),

$$xy = \xi_1 \eta_1 e_1^2 + \cdots + \xi_s \eta_s e_s^2. \quad (30)$$

If $\xi_1 \eta_1 e_1^2 + \cdots + \xi_s \eta_s e_s^2 = 0$, then the integer $\xi_1 \eta_1 e_1^2 + \cdots + \xi_s \eta_s e_s^2 \equiv 0 \pmod{n}$, that is,

$$\xi_1 \eta_1 \frac{n^2}{p_1^2} + \cdots + \xi_s \eta_s \frac{n^2}{p_s^2} \equiv 0 \quad (\operatorname{mod} n = p_1 \cdots p_r). \quad (31)$$

However, this is impossible because none of $p_1, \dots, p_s$ divides the left-hand side. (Namely, p_i divides every other summand except the ith one.) Therefore, $xy \neq 0$ and our claim follows by contradiction.

(b)⇒(c). If $x^2 = 0$, then $x \perp_0 x$ by (b) and $x = 0$ by (A2).

(c)⇒(a). Suppose that (a) does not hold. Then, say, $\alpha_1 > 1$. Let $x = n/p_1$. Since the integer

$$x^2 = \frac{n^2}{p_1^2} = \frac{p_1^{2\alpha_1} \cdots p_r^{2\alpha_r}}{p_1^2} = p_1^{2\alpha_1 - 2} p_2^{2\alpha_2} \cdots p_r^{2\alpha_r} \equiv 0 \pmod{n}, \tag{32}$$

the residue class $x^2 = 0$. But $x \neq 0$ and hence (c) does not hold. Again, our claim follows now by contradiction.

Corollary 18. *If and only if the conditions of Theorem 17 are satisfied, then*

$$\forall x, y \in \mathbb{Z}_n : x \perp_0 y \Longleftrightarrow n \mid (xy), \tag{33}$$

where xy is the product of integers x and y.

Example 19. Let $G = \mathbb{Z}_{30}$. Since $30 = 2 \cdot 3 \cdot 5$, the decomposition (12) is

$$\mathbb{Z}_{30} = \left\langle \frac{30}{2} \right\rangle \oplus \left\langle \frac{30}{3} \right\rangle \oplus \left\langle \frac{30}{5} \right\rangle = \{0, 15\} \oplus \{0, 10, 20\} \oplus \{0, 6, 12, 18, 24\}. \tag{34}$$

For example, since $2 = 0 \cdot 15 + 2 \cdot 10 + 2 \cdot 6$ and $15 = 1 \cdot 15 + 0 \cdot 10 + 0 \cdot 6$, we have $2 \perp_0 15$. Generally, (33) implies that $x \perp_0 y$ if and only if the corresponding integers satisfy $30 \mid (xy)$.

Example 20. Let $G = \mathbb{Z}_{360}$. Since $360 = 2^3 \cdot 3^2 \cdot 5$, we have

$$\mathbb{Z}_{360} = \left\langle \frac{360}{2^3} \right\rangle \oplus \left\langle \frac{360}{3^2} \right\rangle \oplus \left\langle \frac{360}{5} \right\rangle = \{45, 90, \dots, 315\} \oplus \{40, 80, \dots, 320\} \oplus \{72, 144, \dots, 288\}. \tag{35}$$

For example, $5 \perp_0 72$ because $5 = 1 \cdot 45 + 8 \cdot 40 + 0 \cdot 72$ and $72 = 0 \cdot 45 + 0 \cdot 40 + 1 \cdot 72$. Now (33) is only necessary for $\perp_0$ but not sufficient. For example, $10 \not\perp_0 36$ due to the fact that $10 = 2 \cdot 45 + 7 \cdot 40 + 0 \cdot 72$ and $36 = 4 \cdot 45 + 0 \cdot 40 + 3 \cdot 72$. However, $360 \mid (10 \cdot 36)$.

4. Another Definition of Perpendicularity

Davis [4] defined perpendicularity as a binary relation $\perp$ in G satisfying

(D1) $\forall a, b \in G : a \perp b \Rightarrow b \perp a$,

(D2) $\forall a \in G : 0 \perp a$,

(D3) $\forall a \in G : a \perp a \Rightarrow a = 0$,

(D4) $\forall a, b, c \in G : b \perp a \wedge c \perp a \Rightarrow (b + c) \perp a$,

(D5) $\forall a, b \in G : a \perp b \Leftrightarrow \{a\}^{\perp\perp} \cap \{b\}^{\perp\perp} = \{0\}$.

He assumes that G is an Abelian group, but the definition applies more generally to an Abelian monoid, too. It is easy to see that (D1)–(D4) are equivalent to (A1)–(A4). Axiom (D5) arises from introducing the concept of "disjointness" on a vector lattice; see [2, page 295], [6]. In fact, $\Leftrightarrow$ can be replaced with $\Leftarrow$ in (D5) due to the following observation.

Proposition 21. *Assume that $\perp$ satisfies (D1)–(D3) (or, equivalently, (A1)–(A3)). Then*

$$\forall a, b \in G : a \perp b \Rightarrow \{a\}^{\perp\perp} \cap \{b\}^{\perp\perp} = \{0\}. \tag{36}$$

Proof. We show first that if $\emptyset \neq A \subseteq G$, then

$$A \cap A^{\perp} = \{0\}. \tag{37}$$

If $x \in A \cap A^{\perp}$, then $x \perp y$ for all $y \in A$. In particular, $x \perp x$, and hence $x = 0$ by (D3) and (37) follows.

Assume next that $a \perp b$ and let $x \in \{a\}^{\perp\perp} \cap \{b\}^{\perp\perp}$. Since $x \perp \{b\}^{\perp}$ and $a \in \{b\}^{\perp}$, we have $x \perp a$ implying that $x \in \{a\}^{\perp}$. Thus $x \in \{a\}^{\perp} \cap \{a\}^{\perp\perp}$. But (37) applied to $A = \{a\}^{\perp}$ implies that $\{a\}^{\perp} \cap \{a\}^{\perp\perp} = \{0\}$ and $x = 0$ follows.

How are these two perpendicularities related? We give a partial answer. Let us denote by A and D the axioms (A1)–(A5) and (D1)–(D5), respectively.

Proposition 22. *If all elements of G have finite order, then $D \Rightarrow A$. If G has at least one element of infinite order, then there exists a relation $\perp$ satisfying D but not A.*

Proof. The first claim follows from Proposition 3. Concerning the second one, $\perp$ defined by (2) establishes a relation satisfying D but not (A5).

Proposition 23. *Assume that G has elements $a_1, a_2, a_3, a_4 \neq 0$ such that $\langle a_i \rangle \cap \langle a_j \rangle = \{0\}$ whenever $i \neq j$. Then there exists a relation $\perp$ satisfying A but not D.*

Proof. The relation $\perp$ defined by (5) with $H = \langle a_1 \rangle$, $K = \langle a_2 \rangle$, $H_1 = \langle a_3 \rangle$, and $K_1 = \langle a_4 \rangle$ satisfies A. Since $\{a_1\}^{\perp\perp} \cap \{a_3\}^{\perp\perp} = \langle a_1 \rangle \cap \langle a_3 \rangle = \{0\}$ and $a_1 \not\perp a_3$, it does not satisfy (D5).

5. Divisibility in $\mathbb{Q}_+$

It will turn out that perpendicularity has got something to do also with divisibility in $\mathbb{Q}_+$. To that end, we begin by noticing that every $b \in \mathbb{Q}_+$ can be said to be a rational divisor of every $a \in \mathbb{Q}_+$ because $a = cb$ for some $c \in \mathbb{Q}_+$. So, this divisibility is trivial. In order to be able to discuss nontrivial divisibilities in $\mathbb{Q}_+$, we have to consider which properties essentially establish this relation. The following three ones seem quite obvious.

Let $\mid$ be a relation in $\mathbb{Q}_+$ satisfying

(i) $\forall a \in \mathbb{Q}_+ : a \mid a$,

(ii) $\forall a, b, c \in \mathbb{Q}_+ : c \mid a \wedge c \mid b \Rightarrow c \mid (ab)$,

(iii) $\forall a, b, c \in \mathbb{Q}_+ : c \mid b \wedge b \mid a \Rightarrow c \mid a$.

We call $\mid$ a *divisibility* in $\mathbb{Q}_+$. In other words, divisibility is a reflexive and transitive relation (i.e., a preorder) satisfying (ii).

If $b \mid a$, then we say that b is a *divisor* of a and that a is *divisible* by b. If $d \mid a$, $d \mid b$ and $c \mid a \wedge c \mid b \Rightarrow c \mid d$, then d is a *greatest common divisor* of a and b, denoted by $\gcd_{|}(a, b)$. All these notions are meaningful also in any Abelian monoid.

Let us recall that every $c \in \mathbb{Q}_+$ can be expressed as

$$c = \prod_{p\in\mathbb{P}} p^{\nu_p(c)}, \tag{38}$$

where $\nu_p(c) \in \mathbb{Z}$ for each $p \in \mathbb{P}$, and only a finite number of them are nonzero. If $\nu_p(c) \neq 0$, then p is a *prime factor* of c.

Consider the set S of all sequences $(n_2, n_3, \ldots, n_p, \ldots)$, where the index runs through $\mathbb{P}$, each $n_p \in \mathbb{Z}$, and only a finite number of them are nonzero. The mapping

$$f(c) = \left(\nu_2(c), \nu_3(c), \ldots, \nu_p(c), \ldots\right) \tag{39}$$

is an isomorphism from $(\mathbb{Q}_+, \cdot)$ onto $(S, +)$ where addition is defined termwise. For example,

$$\begin{aligned} f(45) + f\left(\frac{8}{25}\right) &= (0,2,1,0,0,\ldots) + (3,0,-2,0,0,\ldots) \\ &= (3,2,-1,0,0,\ldots), \\ f\left(45\cdot\frac{8}{25}\right) &= f\left(\frac{72}{5}\right) = f\left(2^3\cdot 3^2\cdot 5^{-1}\right) \\ &= (3,2,-1,0,0,\ldots). \end{aligned} \tag{40}$$

Given $a, b \in \mathbb{Q}_+$, we define their "inner product" being the Euclidean inner product of the vectors $f(a)$ and $f(b)$:

$$\langle a, b\rangle = \langle f(a), f(b)\rangle = \sum_{p\in\mathbb{P}} \nu_p(a)\,\nu_p(b). \tag{41}$$

Since only a finite number of summands are nonzero, this sum is finite. For example,

$$\left\langle 45, \frac{8}{25}\right\rangle = 0\cdot 3 + 2\cdot 0 + 1\cdot(-2) + 0 + 0 + \cdots = -2. \tag{42}$$

Next we define $|c|$ by setting $\nu_p(|c|) = |\nu_p(c)|$ for all $p \in \mathbb{P}$ or, equivalently, $|c| = f^{-1}((\nu_2(|c|), \nu_3(|c|), \nu_5(|c|), \ldots))$. For example, if $c = 40/63 = 2^3 \cdot 3^{-2} \cdot 5^1 \cdot 7^{-1}$, then $|c| = 2^3 \cdot 3^2 \cdot 5^1 \cdot 7^1 = 2520$. Letting $\perp_1$ be the same relation as the one defined by (7), it can be characterized now by

$$a \perp_1 b \Longleftrightarrow \langle |a|, |b|\rangle = 0. \tag{43}$$

Also the relation $\perp_2$ in $\mathbb{Q}_+$, defined by

$$a \perp_2 b \Longleftrightarrow \langle a, b\rangle = 0, \tag{44}$$

is a perpendicularity.

We will introduce one more nontrivial perpendicularity using divisibility. For that purpose, we first notice that the relation δ defined by

$$b\delta a \Longleftrightarrow \forall p \in \mathbb{P} : \nu_p(b) \le \nu_p(a) \tag{45}$$

is a divisibility, $\gcd_\delta(a, b)$ exists and is unique for all $a, b \in \mathbb{Q}_+$, and

$$\gcd_\delta(a, b) = \prod_{p\in\mathbb{P}} p^{\min(\nu_p(a), \nu_p(b))}. \tag{46}$$

Assume now that $m, n, u, v \in \mathbb{Z}_+$ so that $\gcd(m, u) = \gcd(n, v) = 1$. An alternative expression for (45) is

$$\frac{n}{v}\,\delta\,\frac{m}{u} \Longleftrightarrow n \mid m \wedge u \mid v, \tag{47}$$

and that for (46) is

$$\gcd_\delta\left(\frac{m}{u}, \frac{n}{v}\right) = \frac{\gcd(m, n)}{\operatorname{lcm}(u, v)}. \tag{48}$$

For example, if $a = 45/14 = 2^{-1}\cdot 3^2\cdot 5^1\cdot 7^{-1}$ and $b = 33/100 = 2^{-2}\cdot 3^1\cdot 5^{-2}\cdot 11^1$, then $\gcd_\delta(a, b) = 2^{-2}\cdot 3^1\cdot 5^{-2}\cdot 7^{-1} = 3/700$. Alternatively,

$$\gcd_\delta(a, b) = \frac{\gcd(45, 33)}{\operatorname{lcm}(14, 100)} = \frac{3}{700}. \tag{49}$$

Since $\gcd_\delta(|m/u|, |n/v|) = \gcd(mu, nv)$, we have by (9)

$$a \perp_1 b \Longleftrightarrow \gcd_\delta(|a|, |b|) = 1. \tag{50}$$

This relation generalizes (10) and answers the cry of Graham et al. in a slightly wider context than what they, perhaps, had thought.

Eugeni and Rizzi [7, Section 2] defined divisibility in $\mathbb{Q}_+$ by setting the relation γ so that

$$\frac{n}{v}\,\gamma\,\frac{m}{u} \Longleftrightarrow n \mid m \wedge v \mid u. \tag{51}$$

Then $\gcd_\gamma(a, b)$ always exists and is unique, and

$$\gcd_\gamma\left(\frac{m}{u}, \frac{n}{v}\right) = \frac{\gcd(m, n)}{\gcd(u, v)}. \tag{52}$$

For example,

$$\gcd_\gamma\left(\frac{45}{14}, \frac{33}{100}\right) = \frac{\gcd(45, 33)}{\gcd(14, 100)} = \frac{3}{2}. \tag{53}$$

We define now the corresponding perpendicularity by writing

$$\begin{aligned} a \perp_{\mathrm{ER}} b &\Longleftrightarrow \gcd_\gamma(a, b) = 1 \\ &\Longleftrightarrow \gcd(m, n) = \gcd(u, v) = 1. \end{aligned} \tag{54}$$

Summing up, we have at least three nontrivial perpendicularities in $\mathbb{Q}_+$. Let us see how they relate to one another.

$\perp_1$ *versus* $\perp_2$. Clearly $\perp_1 \Rightarrow \perp_2$ (i.e., $x \perp_1 y \Rightarrow x \perp_2 y$). The converse does not hold. For example, $6 \perp_2 2/3$ but $6 \not\perp_1 2/3$.

$\perp_1$ *versus* $\perp_{\mathrm{ER}}$. Clearly $\perp_1 \Rightarrow \perp_{\mathrm{ER}}$. The converse does not hold. For example, $2/3 \perp_{\mathrm{ER}} 3/2$ but $2/3 \not\perp_1 3/2$.

$\perp_2$ *versus* $\perp_{\mathrm{ER}}$. These perpendicularities are independent. For example, $6 \perp_2 2/3$ but $6 \not\perp_{\mathrm{ER}} 2/3$. On the other hand, $2/3 \perp_{\mathrm{ER}} 3/2$ but $2/3 \not\perp_2 3/2$.

However, regarding $(\mathbb{Z}_+, \cdot)$ as a submonoid of $(\mathbb{Q}_+, \cdot)$, it is obvious that $\perp_1 = \perp_2 = \perp_{\mathrm{ER}}$ in $\mathbb{Z}_+$. Moreover, in $\mathbb{Z}_+$, they yield the very perpendicularity proposed by Graham et al.

6. Parallelism

Parallelism is closely related to perpendicularity. Considering different geometric contexts we notice soon that, in general, parallelism does not have any other properties except those of equivalence. However, any equivalence relation cannot be said to stand for parallelism in any reasonable way. This leads us to ask whether it is possible or not to define parallelism in Abelian groups having a perpendicularity so that it makes sense.

Let G have a perpendicularity $\perp$ and let $a, b \in G$. We say that a and b are *parallel* and write $a \parallel b$ if $\{a\}^{\perp} = \{b\}^{\perp}$. The relation $\parallel$ is clearly an equivalence. If $a \neq 0$, then $a \nparallel 0$, since $\{0\}^{\perp} = G$ by Proposition 1(a) but $\{a\}^{\perp} \neq G$ by (A2). All nonzero elements are parallel if and only if $\perp$ is trivial.

If $G = \mathbb{Z}_n$ and $\perp = \perp_0$, then, recalling (19),

$$\begin{aligned} x \parallel y &\Longleftrightarrow (\forall i \in \{1, \dots, r\} : \xi_i = 0 \Longleftrightarrow \eta_i = 0) \\ &\Longleftrightarrow \{x\}^{\perp} = \{y\}^{\perp} = H_{i_1} \oplus \cdots \oplus H_{i_t}, \end{aligned} \tag{55}$$

where $\xi_i = \eta_i = 0 \Leftrightarrow i \in \{i_1, \dots, i_t\}$. For example, consider $\mathbb{Z}_{30}$ (see Example 19). Since $2 = 0 \cdot 15 + 2 \cdot 10 + 2 \cdot 6$ and $16 = 0 \cdot 15 + 1 \cdot 10 + 1 \cdot 6$, we have $\{2\}^{\perp} = \{16\}^{\perp} = \{0, 15\}$, and so $2 \parallel 16$.

Now, let $G = \mathbb{Q}_+$ and let $\perp_1$, $\perp_2$, and $\perp_{\mathrm{ER}}$ be as before. Denote the corresponding parallelisms by $\parallel_1$, $\parallel_2$, and $\parallel_{\mathrm{ER}}$, respectively. Then $a \parallel_1 b$ if and only if a and b have the same prime factors. Further, $m/u \parallel_{\mathrm{ER}} n/v$ if and only if m and n have the same prime factors and u and v have the same prime factors.

Let us study how these parallelisms relate to one another.

$\parallel_1$ *versus* $\parallel_2$. We show that $\parallel_2 \;\Rightarrow\; \parallel_1$. Assume first that $a \parallel_2 b$. If $a \nparallel_1 b$, then there exists $p_0 \in \mathbb{P}$ such that, say, $\nu_{p_0}(a) = 0$ and $\nu_{p_0}(b) \neq 0$. But now $p_0 \perp_2 a$ and $p_0 \not\perp_2 b$, and so $\{a\}^{\perp_2} \neq \{b\}^{\perp_2}$ contradicting the assumption. The converse does not hold. For example, let $a = 6$ and $b = 12$; then $a \parallel_1 b$. If $x = 2/3$, then $x \perp_2 a$ but $x \not\perp_2 b$, and hence $\{a\}^{\perp_2} \neq \{b\}^{\perp_2}$. In other words, $a \nparallel_2 b$.

$\parallel_1$ *versus* $\parallel_{\mathrm{ER}}$. Clearly $\parallel_{\mathrm{ER}} \;\Rightarrow\; \parallel_1$. The converse does not hold. For example, $2/3 \parallel_1 3/2$ but $2/3 \nparallel_{\mathrm{ER}} 3/2$.

$\parallel_2$ *versus* $\parallel_{\mathrm{ER}}$. We show that $\parallel_2 \;\Rightarrow\; \parallel_{\mathrm{ER}}$. Given $p_1, \dots, p_t \in \mathbb{P}$, denote by $N(p_1, \dots, p_t)$ the set of such positive integers that are not divisible by any p_i, $i = 1, \dots, t$. Let $a = m/u \in \mathbb{Q}_+$, $\gcd(m, u) = 1$. Factorize

$$m = p_1^{\alpha_1} \cdots p_h^{\alpha_h}, \qquad u = q_1^{\beta_1} \cdots q_k^{\beta_k}, \tag{56}$$

where $p_1, \dots, p_h, q_1, \dots, q_k \in \mathbb{P}$ are distinct and $\alpha_1, \dots, \alpha_h, \beta_1, \dots, \beta_k > 0$. (If $m = 1$ or $u = 1$, then the corresponding "empty product" is one.) Now

$$\begin{aligned} \{a\}^{\perp_2} = \Bigg\{ &\frac{p_1^{\xi_1} \cdots p_h^{\xi_h}}{q_1^{\eta_1} \cdots q_k^{\eta_k}} \frac{x}{y} \mid \alpha_1 \xi_1 + \cdots + \alpha_h \xi_h + \beta_1 \eta_1 \\ &+ \cdots + \beta_k \eta_k = 0, \\ &x, y \in N(p_1, \dots, p_h, q_1, \dots, q_k) \Bigg\}. \end{aligned} \tag{57}$$

(The "empty sum" is zero.) Assume that $b = n/v \in \mathbb{Q}_+$, $\gcd(n, v) = 1$, satisfies $a \parallel_2 b$, that is, $\{a\}^{\perp_2} = \{b\}^{\perp_2}$. Then, by (57), necessarily

$$n = p_1^{\rho_1} \cdots p_h^{\rho_h}, \qquad v = q_1^{\sigma_1} \cdots q_k^{\sigma_k}, \tag{58}$$

where $\rho_1, \dots, \rho_h, \sigma_1, \dots, \sigma_k > 0$. Hence $a \parallel_{\mathrm{ER}} b$, and the claim follows. The converse is not valid. For example, $2/3 \parallel_{\mathrm{ER}} 4/3$ but $2/3 \nparallel_2 4/3$.

7. Discussion

This paper began with a citation by three established mathematicians and computer scientists who showed a remarkable intuition by promoting the use of the symbol of perpendicularity in number theory. Indeed, we have previously seen how this notion settles comfortably in this setting and gains new meanings at a more general level in the context of Abelian group theory. We conclude this paper with the following supplement to their proposal.

Let perpendicularity and parallelism mean here $\perp_1$ and $\parallel_1$, respectively. Consider the "direction vector" of $c \in \mathbb{Q}_+$ by $(c(2), c(3), \dots, c(p), \dots)$, where $c(p) = 0$ if $\nu_p(c) = 0$ and $c(p) = 1$ otherwise. For example, the direction vectors of $45, 1$, and $8/25$ are, respectively $(0, 1, 1, 0, 0, \dots)$, $(0, 0, \dots)$, and $(1, 0, 1, 0, 0, \dots)$.

Now, like the directions of perpendicular lines are as different as possible, the prime factors of perpendicular (positive rational) numbers are as different as possible; that is, such numbers do not have common prime factors. In other words, the direction vectors of perpendicular numbers are as different as possible in the sense that they have no common element of value one. Like parallel lines have the same direction, parallel numbers have the same prime factors. In other words, their direction vectors are equal.

Finally, we note that perpendicularity can be axiomatized in a natural way also in many other algebraic structures. Davis [8] did that in a ring. In a vector space, perpendicularity is customarily defined based on an inner product. Another possible approach is to supplement (A1)–(A5) with suitable axioms concerning the multiplication of a vector by a scalar. It might be interesting to study under which additional conditions there exists an inner product inducing this perpendicularity.

Acknowledgment

The authors would like to thank the referees for carefully reading the paper and kind comments.

References

[1] R. L. Graham, D. E. Knuth, and O. Patashnik, *Concrete Mathematics: A Foundation for Computer Science*, Addison-Wesley, Reading, Mass, USA, 2nd edition, 1994.

[2] G. Birkhoff, *Lattice Theory*, American Mathematical Society, Providence, RI, USA, 3rd edition, 1993.

[3] P. Haukkanen, J. K. Merikoski, and T. Tossavainen, "Axiomatizing perpendicularity and parallelism," *Journal for Geometry and Graphics*, vol. 15, no. 2, pp. 129–139, 2011.

[4] G. Davis, "Orthogonality relations on abelian groups," *Journal of the Australian Mathematical Society. Series A*, vol. 19, pp. 173–179, 1975.

[5] W. K. Nicholson, *Introduction to Abstract Algebra*, John Wiley & Sons, New York, NY, USA, 2nd edition, 1999.

[6] A. I. Veksler, "Linear spaces with disjoint elements and their conversion into vector lattices," *Leningradskiĭ Gosudarstvennyĭ Pedagogičeskiĭ Institut imeni A. I. Gercena. Učenye Zapiski*, vol. 328, pp. 19–43, 1967 (Russian).

[7] F. Eugeni and B. Rizzi, "An incidence algebra on rational numbers," *Rendiconti di Matematica*, vol. 12, no. 3-4, pp. 557–576, 1979.

[8] G. Davis, "Rings with orthogonality relations," *Bulletin of the Australian Mathematical Society*, vol. 4, pp. 163–178, 1971.

On the Equivalence of B-Rigidity and C-Rigidity for Quasitoric Manifolds

Jin Hong Kim

Department of Mathematics Education, Chosun University, 309 Pilmundaero, Dong-gu, Gwangju 501-759, Republic of Korea

Correspondence should be addressed to Jin Hong Kim; jinhkim11@gmail.com

Academic Editor: Christian Corda

For quasitoric manifolds and moment-angle complexes which are central objects recently much studied in toric topology, there are several important notions of rigidity formulated in terms of cohomology rings. The aim of this paper is to show that, among other things, Buchstaber-rigidity (or B-rigidity) is equivalent to cohomological-rigidity (or C-rigidity) for simple convex polytopes supporting quasitoric manifolds.

1. Introduction and Main Results

In general, a cohomology ring of a given manifold is not enough to determine the manifold completely. However, there are some cases where we can characterize a given manifold in terms of a cohomology ring and which have recently attracted a great amount of attention in toric topology (see [1, 2]). For example, certain Bott manifolds and quasitoric manifolds, but not all of them, are such cases. The aim of this paper is, roughly speaking, to establish certain equivalence between two well-known notions of rigidity which essentially characterize quasitoric manifolds and also are formulated in terms of cohomology rings.

In order to describe our results more precisely, we first need to collect some definitions and notations. To do so, throughout this paper $\mathbf{k}$ will denote a field of characteristic zero. A *quasitoric manifold* M of dimension $2n$ is a closed $2n$-dimensional smooth manifold with a locally standard action of an n-torus $T^n := (S^1)^n$ whose orbit space is a simple convex polytope P. The combinatorial structure of P can be decoded from the equivariant cohomology ring $H^*_{T^n}(M;\mathbf{k})$ of M. The reason is that the equivariant cohomology ring of M is isomorphic to the Stanley-Reisner face ring $\mathbf{k}(K)$ of the dual $K = (\partial P)^*$ of the boundary ∂P of P and that the Stanley-Reisner face ring $\mathbf{k}(K)$ is in turn obtained by using certain combinatorial information of P (refer to, e.g., [2], Theorem 4.8). In a similar vein, it is also expected that one can possibly obtain some information on a simple convex polytope P from the usual cohomology ring of the manifold M. If we have a quasitoric manifold M over a simple convex polytope P, from now on we will say that P (or $K = (\partial P)^*$) *supports* the quasitoric manifold M, for simplicity.

From these contexts, it is natural to give Definition 1. In order to explain it, recall first that the faces of a convex polytope form a *face poset* (or *face lattice*) where the partial ordering is by set containment of faces. Two polytopes are defined to be *combinatorially isomorphic* or *combinatorially equivalent* if their face posets are isomorphic (refer to [1], Section 1.1). An analogous definition obviously applies to two simplicial complexes.

Definition 1. A simple convex polytope P is said to be *cohomologically rigid*, or simply *C-rigid*, if the following two conditions hold.

(i) There exists a quasitoric manifold M over P.

(ii) Let M' be another quasitoric manifold over a simple convex polytope P' such that as a ring

$$H^*(M;\mathbf{k}) \cong H^*\left(M';\mathbf{k}\right). \tag{1}$$

Then P' is combinatorially equivalent to P.

As mentioned above, the Stanley-Reisner face ring $\mathbf{k}(K)$ contains certain significant information of a simple convex polytope P supporting the quasitoric manifold, and the dimension of its Tor algebra gives rise to the bigraded Betti numbers $\beta_{i,2j}(P)$ which are purely combinatorial invariants of the polytope. It can be shown that the cohomology ring of the moment-angle complex $\mathcal{Z}_K$ of the simplicial complex $K = (\partial P)^*$ is isomorphic to $\mathrm{Tor}(\mathbf{k}(K); \mathbf{k})$ (refer to [1], Theorem 7.6, and see Section 2 for the definition of a moment-angle complex $\mathcal{Z}_K$). Hence, it will be also natural to consider the following notion of rigidity, introduced first by Buchstaber in [3].

Definition 2. A simplicial complex K is said to be *Buchstaber-rigid*, or simply *B-rigid*, if the following condition holds.

(i) Let K' be another simplicial complex such that as a ring

$$H^*(\mathcal{Z}_K; \mathbf{k}) \cong H^*(\mathcal{Z}_{K'}; \mathbf{k}). \quad (2)$$

Then K' is combinatorially equivalent to K.

Note that there are simple convex polytopes P supporting quasitoric manifolds which are not C-rigid such that their dual simplicial complexes $K = (\partial P)^*$ are not B-rigid, either (refer to, e.g., [4], Example 1.1). Nonetheless, the aim of this paper is to show that B-rigidity is equivalent to C-rigidity in case of simple convex polytopes supporting quasitoric manifolds. This affirmatively answers a question in [4], Section 8, as follows.

Theorem 3. *Let P be a simple convex polytope supporting a quasitoric manifold, and let $K = (\partial P)^*$ be the dual of the boundary of P. Then B-rigidity of K is equivalent to C-rigidity of P.*

We organize this paper as follows. In Section 2, we will collect some basic facts about the Stanley-Reisner face ring $\mathbf{k}(K)$ and its Tor-algebra $\mathrm{Tor}_{\mathbf{k}[\mathbf{v}]}(\mathbf{k}(K), \mathbf{k})$ of $\mathbf{k}(K)$ and recall some well-known terminology used in this paper. In Section 3, we give a proof of our main Theorem 3.

2. Stanley-Reisner Face Ring and Moment-Angle Complexes

The aim of this section is to set up some notations and briefly collect some basic material necessary for the proof of main Theorem 3 given in Section 3. In particular, we recall well-known facts about Stanley-Reisner face ring, following the works of [5, 6]. Refer to [1, 4] for more details and other notations used in this paper.

To do so, let $\mathbf{k}[\mathbf{v}] = \mathbf{k}[v_1, v_2, \ldots, v_m]$ be the polynomial algebra over $\mathbf{k}$ on m variables $v_1, v_2, \ldots, v_m$ of degree 2. Let $\mathbb{N}$ denote the set of nonnegative integers and $\mathbf{a} = (a_1, a_2, \ldots, a_m) \in \mathbb{N}^m$. Each monomial in $\mathbf{k}[\mathbf{v}]$ has the form $\mathbf{v}^{\mathbf{a}} = v_1^{a_1} v_2^{a_2} \cdots v_m^{a_m}$. Thus $\mathbf{k}[\mathbf{v}]$ is $\mathbb{N}^m$-graded, so that we have

$$\mathbf{k}[\mathbf{v}] = \oplus_{\mathbf{a}\in\mathbb{N}^m} \mathbf{k}[\mathbf{v}]_{\mathbf{a}}, \qquad \mathbf{k}[\mathbf{v}]_{\mathbf{a}} \cdot \mathbf{k}[\mathbf{v}]_{\mathbf{b}} = \mathbf{k}[\mathbf{v}]_{\mathbf{a}+\mathbf{b}}, \quad (3)$$

where $\mathbf{k}[\mathbf{v}]_{\mathbf{a}}$ denotes the vector space $\langle \mathbf{v}^{\mathbf{a}} \rangle$ over $\mathbf{k}$ spanned by $\mathbf{v}^{\mathbf{a}}$.

The *Stanley-Reisner ideal* I_K of a simplicial complex K is defined as

$$I_K = \langle \mathbf{v}^{\tau} \mid \tau \notin K \rangle, \quad (4)$$

and its quotient ring $\mathbf{k}(K) = \mathbf{k}[\mathbf{v}]/I_K$ is called the *Stanley-Reisner face ring* of K.

Since $\mathbf{k}(K)$ has finitely generated graded $\mathbf{k}[\mathbf{v}]$-module, there exists a free resolution of $\mathbf{k}(K)$ of length at most m. Moreover, it has an $\mathbb{N}^m$-graded minimal free resolution as follows:

$$0 \longleftarrow \mathbf{k}(K) \xleftarrow{\phi_0} F_0 \xleftarrow{\phi_1} F_1 \xleftarrow{\phi_2} \cdots \xleftarrow{\phi_{h-1}} F_{h-1} \xleftarrow{\phi_h} F_h \longleftarrow 0, \quad (5)$$

where each homomorphism ϕ_i is an $\mathbb{N}^m$-graded degree-preserving homomorphism.

Now, applying the functor $\otimes_{\mathbf{k}[\mathbf{v}]}\mathbf{k}$ to the sequence (5), we can obtain the following chain complex of $\mathbb{N}^m$-graded $\mathbf{k}[\mathbf{v}]$-modules:

$$\begin{aligned} &0 \longleftarrow F_0 \otimes_{\mathbf{k}[\mathbf{v}]} \mathbf{k} \xleftarrow{\phi_1'} F_1 \otimes_{\mathbf{k}[\mathbf{v}]} \mathbf{k} \\ &\xleftarrow{\phi_2'} \cdots \xleftarrow{\phi_{h-1}'} F_{h-1} \otimes_{\mathbf{k}[\mathbf{v}]} \mathbf{k} \xleftarrow{\phi_h'} F_h \otimes_{\mathbf{k}[\mathbf{v}]} \mathbf{k} \longleftarrow 0, \end{aligned} \quad (6)$$

where each homomorphism ϕ_i' is given by $\phi_i \otimes \mathrm{Id}_{\mathbf{k}}$. Since the free resolution (5) is minimal, the differentials ϕ_i' are actually zero homomorphisms. Hence the ith homology module of the above chain complex, denoted $\mathrm{Tor}^i_{\mathbf{k}[\mathbf{v}]}(\mathbf{k}(K), \mathbf{k})$, is given by $F_i \otimes_{\mathbf{k}[\mathbf{v}]} \mathbf{k}$. In particular, we have

$$\dim_{\mathbf{k}} \mathrm{Tor}^i_{\mathbf{k}[\mathbf{v}]}(\mathbf{k}(K), \mathbf{k}) = \mathrm{rank}_{\mathbf{k}[\mathbf{v}]} F_i. \quad (7)$$

With these notations, by definition we have

$$\begin{aligned} \beta_i(K) &= \dim_{\mathbf{k}} \mathrm{Tor}^i_{\mathbf{k}[\mathbf{v}]}(\mathbf{k}(K), \mathbf{k}), \\ \beta_{i,2j}(K) &= \dim_{\mathbf{k}} \mathrm{Tor}^{i,2j}_{\mathbf{k}[\mathbf{v}]}(\mathbf{k}(K), \mathbf{k}). \end{aligned} \quad (8)$$

In case of a simple convex polytope P, the bigraded Betti number $\beta_{i,2j}(P)$ is defined to be the bigraded Betti number $\beta_{i,2j}(K)$ of the simplicial complex $K = (\partial P)^*$ that is dual to the boundary ∂P of P.

Next, we recall the notion of a regular sequence of the Stanley-Reisner face ring $\mathbf{k}(K)$ (refer to, e.g., [2], Section 5). The *Krull dimension* of $\mathbf{k}(K)$ is defined to be the maximal number of algebraically independent elements of $\mathbf{k}(K)$. Suppose that the Krull dimension of $\mathbf{k}(K)$ is n. A sequence $(\lambda_1, \lambda_2, \ldots, \lambda_n)$ of homogeneous elements of $\mathbf{k}(K)$ is called a *homogeneous system of parameters* if the Krull dimension of $\mathbf{k}(K)/\langle \lambda_1, \lambda_2, \ldots, \lambda_n \rangle$ is zero. A homogeneous system of parameters $(\lambda_1, \lambda_2, \ldots, \lambda_n)$ of $\mathbf{k}(K)$ is called *regular* if λ_{i+1} is not a zero divisor in $\mathbf{k}(K)/\langle \lambda_1, \lambda_2, \ldots, \lambda_i \rangle$. Equivalently, a sequence $(\lambda_1, \lambda_2, \ldots, \lambda_n)$ is regular if $\lambda_1, \lambda_2, \ldots, \lambda_n$ are algebraically independent and if $\mathbf{k}(K)$ is a finite dimensional free $\mathbf{k}[\lambda_1, \lambda_2, \ldots, \lambda_n]$-module.

Finally, we close this section with reviewing the construction of a moment-angle complex associated with an abstract simplicial complex on a vertex set. To do so, let m be a positive integer and let us denote by $[m]$ the set $\{1, 2, \ldots, m\}$. Let K be an abstract simplicial complex on the vertex set $[m]$. For each simplex $\sigma \in K$, we set

$$B_\sigma \left(D^2, S^1\right) = \prod_{i=1}^{m} A_i, \tag{9}$$

where $D^2 = \{z \in \mathbb{C} \mid |z| \le 1\}$, $S^1 = \partial D^2$, and

$$A_i = \begin{cases} D^2, & i \in \sigma, \\ S^1, & i \in [m] \setminus \sigma. \end{cases} \tag{10}$$

Then the moment-angle complex $\mathcal{Z}_K$ on K is defined to be a subspace of $(D^2)^m$, as follows:

$$\mathcal{Z}_K := \bigcup_{\sigma \in K} B_\sigma \left(D^2, S^1\right) \subset \left(D^2\right)^m. \tag{11}$$

When $K = 2^{[m]}$, it is easy to see that $\mathcal{Z}_K = (D^2)^m$. On the other hand, when $K = 2^{[m]} \setminus \{[m]\}$, where $2^{[m]}$ denotes the power set of $[m]$, it can be easily shown that $\mathcal{Z}_K = S^{2m-1}$ (refer to, e.g., [7], Example 2.4). Since $(D^2)^m$ as a subspace of $\mathbb{C}^m$ is invariant under the standard action of T^m on $\mathbb{C}^m$ given by

$$\begin{aligned} &((g_1, g_2, \ldots, g_m), (z_1, z_2, \ldots, z_m)) \\ &\quad \longmapsto (g_1 z_1, g_2 z_2, \ldots, g_m z_m) \end{aligned} \tag{12}$$

$(D^2)^m$ inherits a natural T^m-action whose orbit space is the unit cube $I^m := [0,1]^m \subset \mathbb{R}^m_{\ge 0}$. This T^m-action on $(D^2)^m$ then induces a canonical T^m-action on the moment-angle complex $\mathcal{Z}_K$. Refer to [1], Chapters 6 and 7 for more details on a moment-angle complex $\mathcal{Z}_K$.

3. Proof of Theorem 3

The aim of this section is to give a proof of Theorem 3. Before doing it, we should remark that there seems to exist some confusing point at the end of [4], Section 8, where the authors erroneously claim the proof that C-rigidity implies B-rigidity for simple convex polytopes supporting quasitoric manifolds.

We begin with the following lemma.

Lemma 4. *Let P be a simple convex polytope supporting a quasitoric manifold M, and let $K = (\partial P)^*$ be the dual of the boundary of P. If K is B-rigid, then P is C-rigid.*

Proof. To prove it, suppose that K is B-rigid. Let P' be another simple convex polytope supporting a quasitoric manifold M' such that

$$H^* (M; \mathbf{k}) \cong H^* \left(M'; \mathbf{k}\right). \tag{13}$$

As before, let $K' = (\partial P')^*$ be the simplicial complex which is the dual of the boundary of the simple convex polytope P'. Then it follows from [4], Lemma 3.7, and Proposition 3.8 that as a ring

$$\mathrm{Tor}_{\mathbf{k}[\mathbf{v}]} (\mathbf{k}(K), \mathbf{k}) \cong \mathrm{Tor}_{\mathbf{k}[\mathbf{v}]} \left(\mathbf{k}\left(K'\right), \mathbf{k}\right). \tag{14}$$

Since we have a ring isomorphism $H^*(\mathcal{Z}_K; \mathbf{k}) \cong \mathrm{Tor}(\mathbf{k}(K), \mathbf{k})$ for any simplicial complex K by Buchstaber and Panov ([1], Theorem 7.6, and [8], Theorem 4.7), it follows from (14) that we have a ring isomorphism between $H^*(\mathcal{Z}_K; \mathbf{k})$ and $H^*(\mathcal{Z}_{K'}; \mathbf{k})$. Thus K is combinatorially equivalent to K' by the assumption that K is B-rigid. This implies that P is also combinatorially equivalent to P', which proves that P is C-rigid.

Next, we show the following lemma.

Lemma 5. *Let P (resp. P') be a simple convex polytope supporting a quasitoric manifold M (resp., M'), and let $K = (\partial P)^*$ (resp., $K' = (\partial P')^*$) be the dual of the boundary of P (resp., P'). Assume that as a ring*

$$H^* \left(\mathcal{Z}_K; \mathbf{k}\right) \cong H^* \left(\mathcal{Z}_{K'}; \mathbf{k}\right). \tag{15}$$

Then two Stanley-Reisner face rings $\mathbf{k}(K)$ and $\mathbf{k}(K')$ are isomorphic to each other, as rings.

Proof. To prove it, note first that as a ring

$$\mathrm{Tor}_{\mathbf{k}[\mathbf{v}]} (\mathbf{k}(K), \mathbf{k}) \cong \mathrm{Tor}_{\mathbf{k}[\mathbf{v}]} \left(\mathbf{k}\left(K'\right), \mathbf{k}\right). \tag{16}$$

In particular, we have

$$\beta_{i,2j}(K) = \beta_{i,2j}\left(K'\right) \tag{17}$$

for all i and j.

Next, we claim that the Stanley-Reisner face rings $\mathbf{k}(K)$ and $\mathbf{k}(K')$ are isomorphic to each other as rings. To prove it, let m (resp. m') denote the number of facets of P (resp., P'). Since $\beta_{1,2}(K)$ (resp., $\beta_{1,2}(K')$) is equal to the number of facets of P (resp. P') by the nice formula for bigraded Betti numbers by Hochster ([9], Theorem 5.2 or [4], Theorem 3.3), it follows from (17) that m should be equal to m'. Then, as in Section 2, consider an $\mathbb{N}^m$-graded minimal free resolution for $\mathbf{k}(K)$ as follows:

$$0 \longleftarrow \mathbf{k}(K) \xleftarrow{\phi_0} F_0 \xleftarrow{\phi_1} F_1 \xleftarrow{\phi_2} \cdots \xleftarrow{\phi_{h-1}} F_{h-1} \xleftarrow{\phi_h} F_h \longleftarrow 0, \tag{18}$$

where each homomorphism ϕ_i is an $\mathbb{N}^m$-graded degree-preserving homomorphism. Similarly, let

$$\begin{aligned} &0 \longleftarrow \mathbf{k}\left(K'\right) \xleftarrow{\phi'_0} F'_0 \xleftarrow{\phi'_1} F'_1 \\ &\xleftarrow{\phi'_2} \cdots \xleftarrow{\phi'_{h'-1}} F'_{h'-1} \xleftarrow{\phi'_{h'}} F'_{h'} \longleftarrow 0 \end{aligned} \tag{19}$$

be an $\mathbb{N}^m$-graded minimal resolution of $\mathbf{k}(K')$. Since the free resolution of $\mathbf{k}(K)$ is minimal, for each $i = 1, 2, \ldots, h$ we have

$$\begin{aligned} \mathrm{rank}_{\mathbf{k}[\mathbf{v}]}(F_i) &= \sum_j \beta_{i,2j}(K) = \sum_j \beta_{i,2j}\left(K'\right) \\ &= \mathrm{rank}_{\mathbf{k}[\mathbf{v}]}\left(F'_i\right). \end{aligned} \tag{20}$$

In particular, this implies that h is equal to h'.

Now, we show that there are $\mathbb{N}^m$-graded degree-preserving isomorphisms ψ_i $(0 \le i \le h)$ from F_i to F'_i so that the diagram (21) below commutes as follows:

$$\begin{array}{ccccccccccccc} 0 & \longleftarrow & \mathbf{k}(K) & \overset{\phi_0}{\longleftarrow} & F_0 & \overset{\phi_1}{\longleftarrow} & F_1 & \overset{\phi_2}{\longleftarrow} \cdots \overset{\phi_{h-1}}{\longleftarrow} & F_{h-1} & \overset{\phi_h}{\longleftarrow} & F_h & \longleftarrow & 0 \\ \| & & \downarrow \psi_{-1} & & \downarrow \psi_0 & & \downarrow \psi_1 & \cdots & \downarrow \psi_{h-1} & & \downarrow \psi_h & & \| . \\ 0 & \longleftarrow & \mathbf{k}\left(K'\right) & \overset{\phi_0'}{\longleftarrow} & F_0' & \overset{\phi_1'}{\longleftarrow} & F_1' & \overset{\phi_2'}{\longleftarrow} \cdots \overset{\phi_{h-1}'}{\longleftarrow} & F_{h-1'} & \overset{\phi_h'}{\longleftarrow} & F_h' & \longleftarrow & 0 \end{array} \quad (21)$$

To do so, recall first that by (16) there is a ring isomorphism κ_i from $\mathrm{Tor}^i_{\mathbf{k}[\mathbf{v}]}(\mathbf{k}(K),\mathbf{k})$ to $\mathrm{Tor}^i_{\mathbf{k}[\mathbf{v}]}(\mathbf{k}(K'),\mathbf{k})$ for each i. Hence, in particular, κ_i induces an isomorphism from $\mathrm{Tor}^{i,2}_{\mathbf{k}[\mathbf{v}]}(\mathbf{k}(K),\mathbf{k})$ to $\mathrm{Tor}^{i,2}_{\mathbf{k}[\mathbf{v}]}(\mathbf{k}(K'),\mathbf{k})$ for each i. On the other hand, since κ_i is a ring isomorphism and $\beta_{i,2j}(K)$ is equal to $\beta_{i,2j}(K')$ for all i and j, this implies that actually κ_i should be an $\mathbb{N}^m$-graded degree-preserving isomorphism from $\mathrm{Tor}^i_{\mathbf{k}[\mathbf{v}]}(\mathbf{k}(K),\mathbf{k})$ to $\mathrm{Tor}^i_{\mathbf{k}[\mathbf{v}]}(\mathbf{k}(K'),\mathbf{k})$ for each i. As noted in Section 2, recall also that there are $\mathbb{N}^m$-graded degree-preserving isomorphisms η_i and η_i' such that

$$\begin{aligned} \eta_i &: \mathrm{Tor}^i_{\mathbf{k}[\mathbf{v}]}\left(\mathbf{k}(K),\mathbf{k}\right) \longrightarrow F_i \otimes_{\mathbf{k}[\mathbf{v}]} \mathbf{k} \longrightarrow F_i, \\ \eta_i' &: \mathrm{Tor}^i_{\mathbf{k}[\mathbf{v}]}\left(\mathbf{k}\left(K'\right),\mathbf{k}\right) \longrightarrow F_i' \otimes_{\mathbf{k}[\mathbf{v}]} \mathbf{k} \longrightarrow F_i'. \end{aligned} \quad (22)$$

So it is now easy to see that $\psi_i = \eta_i' \circ \kappa_i \circ \eta_i^{-1}$ is an $\mathbb{N}^m$-graded degree-preserving isomorphism from F_i to F_i' which automatically makes the diagram (21) commute, as desired.

Moreover, by the standard argument using the diagram-chasing we can also construct an $\mathbb{N}^m$-graded degree-preserving homomorphism ψ_{-1} between $\mathbf{k}(K)$ and $\mathbf{k}(K')$ in such a way that the diagram (21) commutes. To be more precise, let $x \in \mathbf{k}(K)$. Then there is an element $y \in F_0$ such that $\phi_0(y) = x$. We then define $\psi_{-1} : \mathbf{k}(K) \to \mathbf{k}(K')$ by $\psi_{-1}(x) = \phi_0' \circ \psi_0(y)$. Then it is well defined; that is, this definition is independent of the choice of y. Indeed, let $y' \in F_0$ such that $\phi_0(y') = x$. Then, since $\phi_0(y-y') = 0$, there is an element $z \in F_1$ such that $y - y' = \phi_1(z)$. Thus we have

$$\begin{aligned} \phi_0' \circ \psi_0(y) - \phi_0' \circ \psi_0\left(y'\right) &= \phi_0' \circ \psi_0\left(y - y'\right) \\ &= \left(\phi_0' \circ \phi_1'\right) \circ \psi_1(z) = 0, \end{aligned} \quad (23)$$

as desired.

Finally, it is easy to see that by using the five-lemma ([10], p. 169) the ring homomorphism ψ_{-1} is also an $\mathbb{N}^m$-graded degree-preserving ring isomorphism between $\mathbf{k}(K)$ and $\mathbf{k}(K')$. This, in particular, completes the proof of Lemma 5.

Recall now that the Stanley-Reisner face ring $\mathbf{k}(K)$ is given by $\mathbf{k}[\mathbf{v}]/I_K$, where I_K denotes the Stanley-Reisner ideal defined as $\langle \mathbf{v}^\tau \mid \tau \notin K\rangle$. Let J be the ideal of $\mathbf{k}(K)$ generated by a regular sequence of homogeneous system of parameters of degree 2 elements. Then the cohomology ring $H^*(M;\mathbf{k})$ is isomorphic to

$$\frac{\mathbf{k}(K)}{J}. \quad (24)$$

Then we need the following lemma ([1], Lemma 3.35, or [4], Lemma 3.6).

Lemma 6. *Let J be an ideal generated by a regular sequence of $\mathbf{k}(K)$. Then we have the following algebra isomorphism:*

$$\mathrm{Tor}_{\mathbf{k}[\mathbf{v}]}\left(\mathbf{k}(K),\mathbf{k}\right) \cong \mathrm{Tor}_{\mathbf{k}[\mathbf{v}]/J}\left(\frac{\mathbf{k}(K)}{J},\mathbf{k}\right). \quad (25)$$

The following lemma will also play an important role in the proof of Theorem 3.

Lemma 7. *Under the same assumptions as in Lemma 5, two rings $\mathbf{k}(K)/J$ and $\mathbf{k}(K')/J'$ are isomorphic to each other.*

Proof. By Lemmas 5 and 6, we have an algebra isomorphism as follows:

$$\mathrm{Tor}_{\mathbf{k}[\mathbf{v}]/J}\left(\frac{\mathbf{k}(K)}{J},\mathbf{k}\right) \cong \mathrm{Tor}_{\mathbf{k}[\mathbf{v}]/J}\left(\frac{\mathbf{k}\left(K'\right)}{J'},\mathbf{k}\right). \quad (26)$$

By applying the same arguments as in the proof of Lemma 5 to $\mathbb{N}^m$-graded minimal free resolutions of $\mathbf{k}(K)/J$ and $\mathbf{k}(K')/J'$ derived from the isomorphism (26) as in (16), it is now straightforward to see that two rings $\mathbf{k}(K)/J$ and $\mathbf{k}(K')/J'$ are isomorphic to each other, as desired.

Finally, we are ready to prove our main theorem of this section, as follows.

Theorem 8. *Let P be a simple convex polytope supporting a quasitoric manifold M, and let $K = (\partial P)^*$ be the dual of the boundary of P. If P is C-rigid, then K is B-rigid.*

Proof. To prove it, as before let P (resp., P') be a simple convex polytope supporting a quasitoric manifold M (resp. M'), and let $K = (\partial P)^*$ (resp., $K' = (\partial P')^*$) be the dual of the boundary of P (resp., P'). Assume further that as a ring

$$H^*\left(\mathcal{Z}_K;\mathbf{k}\right) \cong H^*\left(\mathcal{Z}_{K'};\mathbf{k}\right). \quad (27)$$

By Lemma 7, we then have a ring isomorphism

$$H^*\left(M;\mathbf{k}\right) \cong \frac{\mathbf{k}(K)}{J} \cong \frac{\mathbf{k}\left(K'\right)}{J'} \cong H^*\left(M';\mathbf{k}\right). \quad (28)$$

Since M (resp., M') is a quasitoric manifold over a simple convex polytope (resp., P'), it follows from the assumption of P being C-rigid that P is combinatorially equivalent to P'. So K should be also combinatorial equivalent to K', completing the proof of Theorem 8.

Proof of Theorem 3. The proof of Theorem 3 now follows immediately from Lemma 4 and Theorem 8.

Conflict of Interests

The author declares that there is no conflict of interests regarding the publication of this paper.

Acknowledgments

The author is grateful to the anonymous reader for valuable comments on this paper. This work was supported by the National Research Foundation of Korea (NRF) Grant funded by the Korea Government (MSIP) (no. 2014001824).

References

[1] V. M. Buchstaber and T. E. Panov, *Torus Actions and Their Applications in Topology and Combinatorics*, vol. 24 of *University Lecture Series*, American Mathematical Society, Providence, RI, USA, 2002.

[2] M. W. Davis and T. Januszkiewicz, "Convex polytopes, Coxeter orbifolds and torus actions," *Duke Mathematical Journal*, vol. 62, no. 2, pp. 417–451, 1991.

[3] V. Buchstaber, "Lectures on toric topology, lecture note based on lectures at KAIST," *Trends in Mathematics*, vol. 10, pp. 1–64, 2008.

[4] S. Choi, T. Panov, and D. Y. Suh, "Toric cohomological rigidity of simple convex polytopes," *Journal of the London Mathematical Society*, vol. 82, no. 2, pp. 343–360, 2010.

[5] X. Cao and Z. Lü, "Möbius transform, moment-angle complexes and Halperin-Carlsson conjecture," *Journal of Algebraic Combinatorics*, vol. 35, no. 1, pp. 121–140, 2012.

[6] E. Miller and B. Sturmfels, *Combinatorial Commutative Algebra*, vol. 227 of *Graduate Texts in Mathematics*, Springer, Berlin, Germany, 2005.

[7] A. Bahri, M. Bendersky, F. R. Cohen, and S. Gitler, "The polyhedral product functor: a method of decomposition for moment-angle complexes, arrangements and related spaces," *Advances in Mathematics*, vol. 225, no. 3, pp. 1634–1668, 2010.

[8] T. Panov, "Cohomology of face rings, and torus actions," in *Surveys in Contemporary Mathematics*, vol. 347 of *London Mathematical Society Lecture Note Series*, pp. 165–201, Cambridge University Press, Cambridge, UK, 2008.

[9] M. Hochster, "Cohen-Macaulay rings, combinatorics, and simplicial complexes," in *Proceedings of the 2nd Oklahoma Conference*, pp. 171–223, Dekker, New York, NY, USA, 1977.

[10] S. Lang, *Algebra*, Addison-Wesley, New York, NY, USA, 3rd edition, 1999.

Some New Transformation Properties of the Nielsen Generalized Polylogarithm

Nina Shang, Qinghua Feng, and Huizeng Qin

Institute of Applied Mathematics, Shandong University of Technology, Zibo, Shandong 255049, China

Correspondence should be addressed to Huizeng Qin; qinhz_000@163.com

Academic Editor: Biren N. Mandal

Many of the properties of Nielsen generalized polylogarithm $S_{n,p}(z)$, for example, the special value and the transformation formulas, play important roles in the computation of higher order radiative corrections in quantum electrodynamics. In this paper, some transformation formulas of $z \to p(z)$, $p(z) = 1 - z, 1/z, 1/(1 - z), z/(z - 1)$, and $(1 - z)/z$ are obtained. In particular, the last three transformation formulas are new results so far in the literature. By use of these transformation formulas presented, new fast algorithms for Nielsen generalized polylogarithm $S_{n,p}(z)$ can be designed. For $s_{n,p} = S_{n,p}(1)$, a new recurrence formula is also given. The identities and the calculation of $\sigma_{n,p}$ and $a_{n,p}$ are also investigated.

1. Introduction

The Nielsen generalized polylogarithm, introduced by the Danish mathematician Niels Nielsen, reads as follows:

$$S_{n,p}(z) = \frac{(-1)^{n+p-1}}{(n-1)!p!} \int_0^1 \frac{\ln^{n-1} t \ln^p (1 - zt)}{t} dt, \qquad (1)$$

in which z is a complex and n and p are positive integers. It is a generalization of the polylogarithm function $Li_n(z)$. In fact, $Li_{n+1}(z) = S_{n,1}(z)$. In particular, the values of $S_{n,p}(x)$ for $x = 1, x = -1$, and $x = 1/2$ are usually used in the theory of the generalized polylogarithms. According to Nielsen's notation as follows,

$$s_{n,p} = S_{n,p}(1), \qquad \sigma_{n,p} = (-1)^p S_{n,p}(-1), \qquad a_{n,p} = S_{n,p}\left(\frac{1}{2}\right), \qquad (2)$$

the following relationship exists:

$$s_{n,p} = s_{p,n}, \qquad s_n = s_{n-1,1} = \zeta(n), \qquad \sigma_n = \sigma_{n-1,1} = \eta(n) = \left(1 - 2^{1-n}\right)\zeta(n), \qquad (3)$$

where $\zeta(n)$ is Riemann zeta function and $\eta(n)$ is the Dirichlet eta function.

The Nielsen generalized polylogarithm plays an important role in quantum electrodynamics (see [1, 2]). The properties of it can also find applications in group theory and geometry [3]. For the Nielsen generalized polylogarithm $S_{n,p}(z)$, Kölbig has done much in-depth and systematic study in [4, 5]. For example, the transformations $z \to 1 - z$ and $z \to 1/z$ are given as follows:

$$S_{n,p}(z) + \sum_{j=1}^{n} \frac{\ln^{n-j} z}{(n-j)!} \times \sum_{k=1}^{p} \frac{(-1)^{p-k} \ln^{p-k}(1-z)}{(p-k)!} S_{k,j}(1-z) = \sum_{j=1}^{n} \frac{\ln^{n-j} z}{(n-j)!} s_{j,p} + \frac{(-1)^p \ln^p (1-z) \ln^n z}{p!n!}, \qquad (4)$$

$$S_{n,p}(z) = (-1)^n \sum_{m=0}^{p-1} \frac{\ln^m(-z)}{m!} \sum_{k=m}^{p-1} (-1)^k C_{n+k-m-1}^{k-m} S_{n+k-m,p-k}\left(\frac{1}{z}\right) + (-1)^p \sum_{m=0}^{n-1} \frac{\ln^m(-z)}{m!} C_{n-m,p} + (-1)^p \ln_{n+p}(-z), \qquad (5)$$

where

$$C_{m,p} = \sum_{k=1}^{p-1} (-1)^{m+p-k-1} C_{m+k-1}^{k} \times \sum_{j=0}^{[(m+k-1)/2]} \frac{(-1)^j \pi^{2j}}{(2j)!} s_{m+k-2j,p-k} + (-1)^{m+p-1} (1-(-1)^m) \sum_{j=0}^{[(m-1)/2]} \frac{(-1)^j \pi^{2j}}{(2j)!} s_{m-2j,p} + (-1)^{[(m+p)/2]+n} \frac{\varepsilon_{m+p} \pi^{m+p}}{(m+p)(m-1)!p!} \tag{6}$$

(see (5.3) and (5.12) in [5]). In [5], the author also noted that by repeated use of (4)–(6) it is possible to find formulae for

$$S_{n,p}\left(\frac{1}{1-z}\right), \quad S_{n,p}\left(\frac{z-1}{z}\right), \quad S_{n,p}\left(\frac{z}{z-1}\right), \tag{7}$$

which only contains generalized polylogarithms apart from logarithms and known constants. These formulae soon become very complicated and a more systematic investigation needs to be done later.

In this paper, we use different methods from that in [5] to give the properties of the transformations $z \to p(z)$, $p(z) = 1-z, 1/(1-z), z/(z-1)$, and $(1-z)/z$ for the Nielsen generalized polylogarithm $S_{n,p}(z)$. For $s_{n,p} = S_{n,p}(1)$, we also give a new recurrence formula. Finally, we consider the identities and the calculation of $\sigma_{n,p}, a_{n,p}$ and high accuracy and fast calculation of $S_{n,p}(z)$.

2. The Properties of the Transformations $z \to 1-z$, $1/(1-z)$, $z/(z-1)$, and $(1-z)/z$

For the Nielsen generalized polylogarithm, in [5], by use of the following results (see [6, (2.15), (2.16)]),

$$S_{n,p}(z) = \sum_{j=0}^{n-1} \frac{\ln^j z}{j!} L_{n-j,p}(z), \qquad L_{n,p}(z) = \sum_{j=0}^{n-1} \frac{(-1)^j \ln^j z}{j!} S_{n-j,p}(z), \tag{8}$$

where

$$L_{n,p}(z) = \frac{(-1)^{n+p-1}}{(n-1)!p!} \int_0^z \frac{\ln^{n-1} t \ln^p (1-t)}{t} dt, \tag{9}$$

and some identities, the transformation properties (4)–(6) were obtained.

In this section, we use different methods to give the properties of the transformations $z \to 1-z, 1/z, 1/(1-z), z/(z-1)$, and $(1-z)/z$. First we introduce the following function:

$$S_{n,p}(z;x,y) = \frac{(-1)^{n+p-1}}{(n-1)!p!} \times \int_0^1 t^{x-1} (1-zt)^{y-1} \ln^{n-1} t \ln^p (1-zt)\, dt, \qquad s_{n,p}(x,y) = S_{n,p}(1,x,y). \tag{10}$$

It is easy to get the following relation:

$$\begin{aligned}
S_{n,p}(z) &= S_{n,p}(z;0,1), \\
s_{n,p} &= s_{n,p}(0,1), \\
B(z;x,y) &= z^x S_{1,0}(z;x,y), \\
B(x,y) &= S_{1,0}(1;x,y), \\
\frac{\partial^{n+p-1}}{\partial x^{n-1} \partial y^p} B(x,y) &= (-1)^{n+p-1} (n-1)!p! s_{n,p}(x,y), \\
S_{n,p}(z;x,y) &= \frac{(-1)^{n+p-1}}{(n-1)!p!} \frac{\partial^{n+p-1}}{\partial x^{n-1} \partial y^p} S_{1,0}(z;x,y) \\
&= \frac{(-1)^{n+p-1}}{(n-1)!p!} \frac{\partial^{n+p-1}}{\partial x^{n-1} \partial y^p} \left[z^{-x} B(z;x,y)\right], \\
\frac{\partial^{n+p-1}}{\partial x^{n-1} \partial y^p} B(z;x,y) &= (-1)^{n+p-1} (n-1)!p! z^x \times \sum_{j=1}^{n} \frac{(-1)^{n-j} \ln^{n-j} z}{(n-j)!} S_{j,p}(z;x,y),
\end{aligned} \tag{11}$$

where $B(x,y)$ and $B(z;x,y)$ are the Beta function and the incomplete Beta function, respectively.

Lemma 1. *When $x, y > 0$, for $S_{1,0}(z,x,y)$, there are the following properties:*

(1) $z \neq r$ $(r < 0$ *or* $r > 1)$:

$$S_{1,0}(z,x,y) = z^{-x} S_{1,0}(1;x,y) - z^{-x} (1-z)^y S_{1,0}(1-z,y,x). \tag{12}$$

(2) $z \neq r$ $(r > 0)$:

$$S_{1,0}(z,x,y) = z^{-x} S_{1,0}(1;x,y) - (-1)^{H(z)y} z^{-x} S_{1,0}(1;1-x-y,y) - (-z)^{y-1} S_{1,0}\left(\frac{1}{z};1-x-y,y\right), \tag{13}$$

where

$$H(z) = \begin{cases} -1, & \operatorname{Im} z > 0 \text{ or } z < 0 \\ 1, & \operatorname{Im} z < 0 \text{ or } z > 1; \end{cases} \tag{14}$$

(3) $z \neq r$ $(r \geq 1$ or $r < 0)$:

$$S_{1,0}(z,x,y) = z^{-x}S_{1,0}(1;x,y) - z^{-x}\left(\frac{1-z}{z}\right)^{y} \times S_{1,0}\left(\frac{z-1}{z};y,1-x-y\right): \quad (15)$$

(4) $z \neq r$ $(r \geq 1)$:

$$S_{1,0}(z,x,y) = (1-z)^{-x}S_{1,0}\left(\frac{z}{z-1};x,1-x-y\right) \quad (16)$$

(5) $z \neq r$ $(r > 0)$:

$$S_{1,0}(z,x,y) = (-z)^{-x}S_{1,0}(1;1-x-y,x) - \left(\frac{z-1}{z}\right)^{x}(1-z)^{y-1} \times S_{1,0}\left(\frac{1}{1-z};1-x-y,x\right). \quad (17)$$

Proof. (1) By (10) we have

$$\begin{aligned} S_{1,0}(z;x,y) &= \int_0^1 t^{x-1}(1-zt)^{y-1}dt \\ &= z^{-x}\int_0^z t^{x-1}(1-t)^{y-1}dt \\ &= z^{-x}\left(s_{1,0}(x,y) + \int_1^z t^{x-1}(1-t)^{y-1}dt\right) \\ &= z^{-x}\left(s_{1,0}(x,y) - \int_0^{1-z} t^{y-1}(1-t)^{x-1}dt\right) \\ &= z^{-x}\left(s_{1,0}(x,y) - B(1-z,y,x)\right). \end{aligned} \quad (18)$$

So (12) holds.

(2) For $B(z;x,y)$ we have

$$B(z;x,y) = \begin{cases} B(i;x,y) + \int_i^z t^{x-1}(1-t)^{y-1}dt, & \operatorname{Im} z > 0 \\ B(-i;x,y) + \int_{-i}^z t^{x-1}(1-t)^{y-1}dt, & \operatorname{Im} z < 0 \text{ or } z > 1, \end{cases} \quad (19)$$

where the second integral path does not go through the real axes. By using the variable substitution, we have

$$\begin{aligned} \int_i^z t^{x-1}(1-t)^{y-1}dt &= (-1)^{-y}\int_{-i}^{1/z} t^{-x-y}(1-t)^{y-1}dt, \quad \operatorname{Im} z > 0, \\ \int_{-i}^z t^{x-1}(1-t)^{y-1}dt &= (-1)^{y}\int_{i}^{1/z} t^{-x-y}(1-t)^{y-1}dt, \quad \operatorname{Im} z \leqslant 0. \end{aligned} \quad (20)$$

Substituting (20) into (19) we get that

$$B(z;x,y) = \begin{cases} B(i,x,y) - (-1)^{-y}B(-i,1-x-y,y) + (-1)^{-y}B\left(\frac{1}{z};1-x-y,y\right), & \operatorname{Im} z > 0 \\ B(-i,x,y) - (-1)^{y}B(i,1-x-y,y) + (-1)^{y}B\left(\frac{1}{z};1-x-y,y\right), & \operatorname{Im} z < 0 \text{ or } z > 1. \end{cases} \quad (21)$$

Letting $z \to 1$, we obtain that

$$\begin{aligned} B(x,y) - (-1)^{-y}B(1-x-y,y) &= B(i,x,y) - (-1)^{-y}B(-i,1-x-y,y), \\ B(x,y) - (-1)^{y}B(1-x-y,y) &= B(-i,x,y) - (-1)^{y}B(i,1-x-y,y). \end{aligned} \quad (22)$$

By (22) and (21) we get that

$$B(z;x,y) = B(x,y) + (-1)^{H(z)y} \times \left(B(1-x-y,y) - B\left(\frac{1}{z};1-x-y,y\right)\right). \quad (23)$$

Then by (11) one can see that (13) holds. By (13) and (12) we deduce that (15) holds.

Fully in accordance with the discussion of (2), we can also get that (15)–(17) hold.

By means of the application of Hadamard integral (the Hadamard finite part integral or neutrix calculus; see [6–8]) and integration by parts we have

$$\begin{aligned} \frac{\partial^{n-1}}{\partial x^{n-1}}B(z,0,1) &= \int_0^z t^{-1}\ln^{n-1}t\,dt = \frac{\ln^n z}{n}, \\ \frac{\partial^{n-1}}{\partial x^{n-1}}B(0,1) &= 0, \\ S_{n,0}(z) = \frac{(-1)^{n-1}}{(n-1)!}\frac{\partial^{n-1}}{\partial x^{n-1}}\left[z^{-x}B(z;x,y)\right]\Big|_{x=0,y=1} &= \frac{\ln^n z}{n!}, \\ s_{n,0} &= 0, \end{aligned} \quad (24)$$

$$zS_{n+1,p-1}(z,1,0)=S_{n,p}(z),$$
$$zS_{1,p-1}(z,1,0)=\frac{(-1)^p\ln^p(1-z)}{p!},$$
$$S_{n,p}(z,0,0)=S_{n,p}(z)+S_{n-1,p+1}(z),$$
$$s_{n,p}(0,0)=s_{n,p}+s_{n-1,p+1},\quad n>1,\ p>0,$$
$$S_{1,p}(z,0,0)=S_{1,p}(z)+\frac{(-1)^{p+1}\ln^{p+1}(1-z)}{(p+1)!},\tag{25}$$
$$s_{1,p}(0,0)=s_{1,p},\quad p>0$$
$$S_{1,0}(z,0,0)=-\ln(1-z),\quad s_{1,0}(0,0)=0,$$
$$S_{n,0}(z,0,0)=S_{n-1,1}(z),\quad s_{n,0}(0,0)=s_{n-1,1},\quad n>1.$$

By Lemma 1, (24), and (25), we will consider the transformation properties of $S_{n,p}(z)$ on z. First we prove (4) in another approach.

Theorem 2. *When the integers $n,p\geq 1$, (4) for the transformation $z\rightarrow 1-z$ holds for all complexes z.*

Proof. By (11) and (12), we have

$$\begin{aligned}S_{n,p}(z,x,y)&=z^{-x}\sum_{j=1}^{n}\frac{\ln^{n-j}z}{(n-j)!}s_{j,p}(x,y)\\&\quad-z^{-x}(1-z)^y\sum_{j=1}^{n}\frac{\ln^{n-j}z}{(n-j)!}\\&\quad\times\sum_{k=0}^{p}\frac{(-1)^{p-k}\ln^{p-k}(1-z)}{(p-k)!}\\&\quad\times S_{k+1,j-1}(1-z,y,x).\end{aligned}\tag{26}$$

In (26), letting $y=1,x=0$ and using (25), we have

$$\begin{aligned}S_{n,p}(z)&=\sum_{j=1}^{n}\frac{\ln^{n-j}z}{(n-j)!}s_{j,p}-(-1)^p\\&\quad\times\sum_{j=1}^{n}\frac{(-1)^j}{(n-j)!j!}\frac{\ln^n z\ln^p(1-z)}{p!}\\&\quad-\sum_{j=1}^{n}\frac{\ln^{n-j}z}{(n-j)!}\sum_{k=1}^{p}\frac{(-1)^{p-k}\ln^{p-k}(1-z)}{(p-k)!}\\&\quad\times S_{k,j}(1-z),\end{aligned}\tag{27}$$

where z can be extended to the whole complex plane. Then by (27) and $\sum_{j=1}^{n}((-1)^j/(n-j)!j!)=-1/n!$, we deduce that (4) holds.

Theorem 3. *When the integers $n,p\geq 1$, for the transformation $z\rightarrow 1/z$, there is*

$$\begin{aligned}S_{n,p}(z)&=\sum_{j=1}^{n}\frac{\ln^{n-j}z}{(n-j)!}s_{j,p}\\&\quad+\frac{1}{(n-1)!}\sum_{k=0}^{p}\frac{(-1)^{p-k}\ln^{p-k}(-z)\ln^{n+k}z}{k!(p-k)!(n+k)}\\&\quad+(-1)^n\sum_{k=1}^{p}\frac{(-1)^{p-k}\ln^{p-k}(-z)}{(p-k)!}\\&\quad\times\sum_{v=0}^{k-1}(-1)^vC_{n+v-1}^{v}S_{n+v,k-v}\left(\frac{1}{z}\right)\\&\quad-\sum_{j=1}^{n}\frac{(-1)^j\ln^{n-j}z}{(n-j)!}\sum_{k=1}^{p}\frac{(-\pi iH(z))^{p-k}}{(p-k)!}\\&\quad\times\sum_{v=0}^{k-1}(-1)^vC_{j+v-1}^{v}s_{j+v,k-v},\end{aligned}\tag{28}$$

where $z\neq 1$, $i=\sqrt{-1}$.

Proof. By (11) and (13) and noting $\ln(-1)=\pi i$, we have

$$\begin{aligned}S_{n,p}(z,x,y)&=\frac{1}{z^x}\sum_{j=1}^{n}\frac{\ln^{n-j}z}{(n-j)!}s_{j,p}(x,y)\\&\quad+\frac{(-1)^{n+p}}{(-z)^{1-y}}\sum_{k=0}^{p}\frac{\ln^{p-k}(-z)}{(p-k)!}\\&\quad\times C_{n+k-1}^{k}S_{n+k,0}\left(\frac{1}{z};1-x-y,y\right)\\&\quad+\frac{(-1)^n}{(-z)^{1-y}}\sum_{k=1}^{p}\frac{(-1)^{p-k}\ln^{p-k}(-z)}{(p-k)!}\\&\quad\times\sum_{v=0}^{k-1}(-1)^vC_{n+v-1}^{v}S_{n+v,k-v}\left(\frac{1}{z};1-x-y,y\right)\\&\quad+\frac{(-1)^{H(z)y}}{z^x}\sum_{j=1}^{n}\frac{(-1)^j\ln^{n-j}z}{(n-j)!}\sum_{k=0}^{p}\frac{(-\pi iH(z))^{p-k}}{(p-k)!}\\&\quad\times\sum_{v=0}^{k}(-1)^vC_{j+v-1}^{v}s_{j+v,k-v}(1-x-y,y).\end{aligned}\tag{29}$$

In (29), letting $x=0$, $y=1$ and using (24), we get that (28) holds.

Theorem 4. *When the integers $n, p \geq 1$, for the transformation $z \to (z-1)/z$, there is*

$$S_{n,p}(z) + (-1)^n \sum_{j=1}^{n} \frac{(-\ln z)^{n-j}}{(n-j)!} \sum_{k=1}^{p} \frac{(-\ln(1-z)/z)^{p-k}}{(p-k)!} \times \sum_{v=0}^{k-1} (-1)^v C_{j+v-1}^{v} S_{k-v,j+v}\left(\frac{z-1}{z}\right) = \sum_{j=1}^{n} \frac{\ln^{n-j} z}{(n-j)!} s_{j,p} + (-1)^n \times \sum_{k=0}^{p} \frac{(-\ln(1-z)/z)^{p-k}(-\ln z)^{n+k}}{(p-k)!(k+n)!}, \tag{30}$$

where $z \neq r$ $(r<0)$ and $z \neq 1$.

Proof. By (11) and (15) we have

$$S_{n,p}(z,x,y) = z^{-x} \sum_{j=1}^{n} \frac{\ln^{n-j} z}{(n-j)!} s_{j,p}(x,y) - \frac{((1-z)/z)^y}{z^x} \sum_{j=1}^{n} \frac{(-\ln z)^{n-j}}{(n-j)!} \sum_{k=0}^{p} \frac{(-\ln(1-z)/z)^{p-k}}{(p-k)!} \times \sum_{v=0}^{k} (-1)^v C_{j+v-1}^{v} S_{k+1-v,j+v-1}\left(\frac{z-1}{z}; y, 1-x-y\right). \tag{31}$$

In (31), letting $y=1, x=0$ and by (25), we have

$$S_{n,p}(z) = \sum_{j=1}^{n} \frac{\ln^{n-j} z}{(n-j)!} s_{j,p} - (-1)^n \times \sum_{k=0}^{p} \frac{(-1)^{p-k}\ln^{p-k}((1-z)/z)(-1)^{n+k}\ln^{n+k} z}{(p-k)!k!} \times \sum_{j=1}^{n} \frac{(-1)^j}{(n-j)!(j-1)!(j+k)} - (-1)^n \sum_{j=1}^{n} \frac{(-1)^{n-j}\ln^{n-j} z}{(n-j)!} \times \sum_{k=1}^{p} \frac{(-1)^{p-k}\ln^{p-k}(1-z)/z}{(p-k)!} \times \sum_{v=0}^{k-1} (-1)^v C_{j+v-1}^{v} S_{k-v,j+v}\left(\frac{z-1}{z}\right). \tag{32}$$

Noting $\sum_{j=1}^{n}((-1)^j/(n-j)!(j-1)!(j+k)) = -k!/(k+n)!$, one can see that (30) holds.

Theorem 5. *When the integers $n, p \geq 1$, for the transformation $z \to z/(z-1)$ and $z \to 1/(1-z)$, there is*

$$S_{n,p}(z) - (-1)^p \times \sum_{j=1}^{n} \frac{\ln^{n-j}(1-z)}{(n-j)!} \sum_{u=0}^{j-1} C_{p+u-1}^{u} (-1)^u S_{j-u,p+u}\left(\frac{z}{z-1}\right) = \frac{(-1)^p \ln^{n+p}(1-z)}{(n+p)!}, \tag{33}$$

where $z \neq r$ $(r \geq 1)$, and

$$S_{n,p}(z) + (-1)^p \sum_{k=1}^{p} \frac{\ln^{p-k}(1-z)}{(p-k)!} \times \sum_{j=1}^{n} \frac{(-1)^{n-j}\ln^{n-j}(z-1)/z}{(n-j)!} \times \sum_{u=0}^{j-1} C_{k+u-1}^{u} (-1)^u S_{k+u,j-u}\left(\frac{1}{1-z}\right) = (-1)^p \sum_{j=1}^{n} \frac{\ln^{n-j}(-z)}{(n-j)!} \times \sum_{u=0}^{j-1} C_{p+u}^{u} (-1)^u s_{p+u,j-u} + (-1)^p \times \sum_{j=0}^{n} \frac{(-1)^{n-j}\ln^{n-j}((z-1)/z)\ln^{p+j}(1-z)}{(n-j)!(p+j)!}, \tag{34}$$

where $z \neq r$ $(0 \leqslant r \leqslant 1)$.

Proof. By (11), (16), and (17) we obtain that

$$S_{n,p}(z,x,y) = -\frac{(-1)^p}{(1-z)^x} \sum_{j=1}^{n} \frac{(-1)^j \ln^{n-j}(1-z)}{(n-j)!} \times \sum_{u=0}^{j-1} C_{p+j-u}^{j-u} (-1)^u S_{1+u,p+j-1-u}\left(\frac{z}{z-1}; x, 1-x-y\right),$$

$$S_{n,p}(z,x,y) = \frac{(-1)^p}{(-z)^x} \sum_{j=1}^{n} \frac{\ln^{n-j}(-z)}{(n-j)!} \times \sum_{u=0}^{j-1} C_{p+u}^{u} (-1)^u s_{p+u+1,j-1-u}(1-x-y,x) - \frac{(-1)^p((z-1)/z)^x}{(1-z)^{1-y}} \sum_{k=0}^{p} \frac{\ln^{p-k}(1-z)}{(p-k)!}$$

$$\times \sum_{j=1}^{n} \frac{(-\ln (z-1)/z)^{n-j}}{(n-j)!} \times \sum_{u=0}^{j-1} C_{k+u}^{u} (-1)^{u} S_{k+u+1,j-1-u} \left(\frac{1}{1-z}; 1-x-y, x \right). \quad (35)$$

In (35), letting $y = 1, x = 0$ and using (25), we can deduce that (33) and (34) hold.

3. The Recursive Formula on $s_{n,p} = S_{n,p}(1)$

For $s_{n,p} = S_{n,p}(1)$ in Theorems 2–5, we can give the following recursive formula.

Theorem 6. *For the positive integers n, p ≥ 1, there is*

$$s_{n,p} = \frac{C_{n+p}^{n} \zeta(n+p)}{n+p} - \frac{1}{p} \sum_{i=1}^{n-1} \sum_{j=1}^{p-1} C_{n+p-i-j-1}^{n-i} s_{i,j} \zeta(n+p-i-j). \quad (36)$$

Proof. By calculating p-order partial derivatives on b for $B(a,b)$, we have

$$\begin{aligned} \frac{\partial^{p}}{\partial b^{p}} B(a,b) &= \frac{\partial^{p-1}}{\partial b^{p-1}} \left(B(a,b) \left(\psi(b) - \psi(a+b) \right) \right) \\ &= \sum_{j=0}^{p-1} C_{p-1}^{j} \frac{\partial^{j}}{\partial b^{j}} B(a,b) \left(\psi^{(p-1-j)}(b) - \psi^{(p-1-j)}(a+b) \right), \end{aligned} \quad (37)$$

$$\left. \frac{\partial^{p}}{\partial b^{p}} B(a,b) \right|_{b=1} = \sum_{j=0}^{p-1} C_{p-1}^{j} \left. \frac{\partial^{j}}{\partial b^{j}} B(a,b) \right|_{b=1} \times \left(\psi^{(p-1-j)}(1) - \psi^{(p-1-j)}(a+1) \right), \quad (38)$$

where $\psi(x)$ is the digamma function defined as

$$\psi(x) = \frac{d}{dx} \ln \Gamma(x) = -\gamma - \frac{1}{x} + \sum_{m=1}^{\infty} \left(\frac{1}{m} - \frac{1}{m+x} \right). \quad (39)$$

By calculating $n-1$-order partial derivatives on a in (38), we have

$$\begin{aligned} \left. \frac{\partial^{n+p-1}}{\partial a^{n-1} \partial b^{p}} B(a,b) \right|_{b=1} &= \sum_{j=1}^{p-1} C_{p-1}^{j} \left(\psi^{(p-1-j)}(1) - \psi^{(p-1-j)}(a+1) \right) \times \left. \frac{\partial^{n+j-1}}{\partial a^{n-1} \partial b^{j}} B(a,b) \right|_{b=1} \\ &+ \frac{\partial^{n-1}}{\partial a^{n-1}} \left[\left(\psi^{(p-1)}(1) - \psi^{(p-1)}(a+1) \right) B(a,b) \Big|_{b=1} \right] \\ &- \sum_{i=0}^{n-2} \sum_{j=1}^{p-1} C_{n-1}^{i} C_{p-1}^{j} \psi^{(n+p-2-i-j)}(a+1) \times \left. \frac{\partial^{i+j}}{\partial a^{i} \partial b^{j}} B(a,b) \right|_{b=1}. \end{aligned} \quad (40)$$

Using $\psi^{(k)}(1) = k!(-1)^{k+1} \zeta(k+1)$, we have

$$\begin{aligned} \left. \left(\psi^{(n+p-2-i-j)}(a+1) \frac{\partial^{i+j}}{\partial a^{i} \partial b^{j}} B(a,b) \right) \right|_{a=0,b=1} &= (n+p-2-i-j)! (-1)^{n+p-1} i! j! s_{i+1,j} \\ &\times \zeta(n+p-1-i-j), \\ & i \geq 0, \quad j \geq 1. \end{aligned} \quad (41)$$

By the property of the Beta function, it holds that

$$\begin{aligned} \left. \frac{\partial^{n-1}}{\partial a^{n-1}} \left(B(a,b) \left(\psi^{(p-1)}(1) - \psi^{(p-1)}(a+1) \right) \right) \right|_{a=0,b=1} &= \left. \frac{\partial^{n-1}}{\partial a^{n-1}} \left(\frac{\psi^{(p-1)}(1) - \psi^{(p-1)}(a+1)}{a} \right) \right|_{a=0} \\ &= -\frac{1}{n} \psi^{(n+p-1)}(1) \\ &= \frac{(-1)^{n+p-1}(n+p-1)!}{n} \zeta(n+p). \end{aligned} \quad (42)$$

Combining (10), (11), and (38)–(42), we obtain that (36) holds.

Remark 7. In [5], there is the following result:

$$s_{n,p} = (-1)^{n+p-1} \sum_{i=1}^{n-1} b_{n-i-1} \sum_{j=1}^{p-1} C_{i+j+1}^{j} b_{p-j} a_{i+j+1}, \quad (43)$$

where $a_0 = b_0 = 1$, and

$$a_k = \frac{1}{k}\sum_{m=1}^{k}(-1)^{m+1}\zeta(m)\,a_{k-m},$$
$$b_k = -\frac{1}{k}\sum_{m=1}^{k}(-1)^{m+1}\zeta(m)\,b_{k-m}, \quad k>0, \tag{44}$$

where $\zeta(1) = \gamma, \gamma$ is Euler's constant. From the perspective of recurrence relation, one can see that (36) is simpler and clearer than (43).

4. On the Identities of $\sigma_{n,p}$ and $a_{n,p}$ and Their Calculations

In (28), letting $z=-1$ and using $\ln(-1)=\pi i$, we have

$$2\sigma_{2n-1,p} + \sum_{v=1}^{p-1}C_{2n+v-2}^{v}\sigma_{2n+v-1,p-v} = (-1)^{p}c_{2n-1,p},$$
$$\sum_{v=1}^{p-1}C_{2n+v-1}^{v}\sigma_{2n+v,p-v} = (-1)^{p-1}c_{2n,p}, \tag{45}$$

where

$$c_{n,p} = \sum_{j=1}^{n}\frac{(\pi i)^{n-j}}{(n-j)!}s_{j,p} + \frac{(\pi i)^{n+p}}{(n-1)!\,p!\,(n+p)} - \sum_{j=1}^{n}\frac{(-1)^{j}}{(n-j)!}\sum_{k=1}^{p}\frac{(\pi i)^{n+p-j-k}}{(p-k)!}\sum_{v=0}^{k-1}(-1)^{v}C_{j+v-1}^{v}s_{j+v,k-v}. \tag{46}$$

Using (3) and (36), from (45) and (46), we have

$$\sigma_{2n+1,2} = \left(\frac{2n+1}{2^{2n+3}} - n\right)\zeta(2n+3) + \frac{1}{2}\sum_{j=1}^{n}\frac{(2n+1-2j)(-1)^{n-j}\pi^{2n+2-2j}}{(2n+2-2j)!}\zeta(2j+1) - \sum_{j=1}^{n}\frac{(-1)^{n-j}\pi^{2n-2j}}{(2n-2j)!}\sum_{l=1}^{j}\zeta(2l+1)\,\zeta(2j+2-2l), \tag{47}$$

$$\sum_{j=1}^{n}\frac{(-1)^{j-1}\zeta(2j)}{(2n+1-2j)!\,\pi^{2j}} = \frac{n}{(2n+1)!}. \tag{48}$$

Here (48) is a new formula.

In (34) and (4) letting $z=2$ we have

$$S_{n,p}(2) + (-1)^{p} \times \sum_{k=1}^{p}\frac{\ln^{p-k}(-1)}{(p-k)!}\sum_{j=1}^{n}\frac{(-1)^{j}\ln^{n-j}2}{(n-j)!}\sum_{u=0}^{j-1}C_{k+u-1}^{u}\sigma_{k+u,j-u}$$
$$= (-1)^{p}\sum_{j=1}^{n}\frac{\ln^{n-j}(-2)}{(n-j)!}\sum_{u=0}^{j-1}C_{p+u-1}^{u}(-1)^{u}s_{p+u,j-u} + (-1)^{p} \times \sum_{j=0}^{n}\frac{\ln^{n-j}2\,\ln^{p+j}(-1)}{(n-j)!\,(p+j)!},$$

$$S_{n,p}(2) + \sum_{k=1}^{p}\frac{(-1)^{p-k}\ln^{p-k}(-1)}{(p-k)!}\sum_{j=1}^{n}\frac{(-1)^{j}\ln^{n-j}2}{(n-j)!}\sigma_{k,j} = \sum_{j=1}^{n}\frac{\ln^{n-j}2}{(n-j)!}s_{j,p} + \frac{(-1)^{p}\ln^{p}(-1)\ln^{n}2}{p!\,n!}. \tag{49}$$

By (49) we get

$$\sum_{j=1}^{n}\frac{\ln^{n-j}2}{(n-j)!}\left(\sum_{k=1}^{p}\frac{(-1)^{j}(-1)^{k}\ln^{p-k}(-1)}{(p-k)!}\sigma_{k,j} - \sum_{k=1}^{p}\frac{(-1)^{j}\ln^{p-k}(-1)}{(p-k)!}\sum_{u=0}^{j-1}C_{k+u-1}^{u}\sigma_{k+u,j-u}\right)$$
$$= \sum_{j=1}^{n}\frac{\ln^{n-j}2}{(n-j)!}\left((-1)^{p}s_{j,p} - \frac{\ln^{p+j}(-1)}{(p+j)!} - \sum_{k=1}^{j}\frac{\ln^{j-k}(-1)}{(j-k)!}\sum_{u=0}^{k-1}C_{p+u-1}^{u}(-1)^{u}s_{p+u,k-u}\right). \tag{50}$$

Noting the bracket part independent of n in (50), it follows that

$$\sum_{k=1}^{[(p+1)/2]}\frac{\ln^{p+1-2k}(-1)}{(p+1-2k)!}\sigma_{2k-1,n} + \sum_{k=1}^{p}\frac{\ln^{p-k}(-1)}{(p-k)!}\sum_{j=1}^{n-1}C_{k+j-1}^{j}\sigma_{k+j,n-j}$$
$$= \frac{(-1)^{n}\ln^{p+n}(-1)}{(p+n)!} + (-1)^{n}\sum_{j=1}^{n}\frac{\ln^{n-j}(-1)}{(n-j)!} \times \sum_{u=0}^{j-1}C_{p+u-1}^{u}(-1)^{u}s_{p+u,j-u} - (-1)^{p+n}s_{n,p}. \tag{51}$$

Taking into account the real and imaginary parts, we have

$$\begin{aligned}
&\sum_{k=1}^{p}\frac{(-1)^{p-k}\pi^{2p-2k}}{(2p+1-2k)!}\left(2\sigma_{2k-1,n}+\sum_{j=1}^{n-1}C_{2k+j-2}^{j}\sigma_{2k-1+j,n-j}\right)\\
&=\delta_{n-1,2[(n-1)/2]}\frac{(-1)^{p+(n+1)/2}\pi^{2p+n-1}}{(2p+n)!}\\
&\quad-(-1)^{n}\sum_{j=1}^{[n/2]}\frac{(-1)^{j}\pi^{2j-2}}{(2j-1)!}\\
&\quad\times\sum_{u=0}^{n-2j}C_{2p+u-1}^{u}(-1)^{u}s_{2p+u,n+1-2j-u},\\
&\sum_{k=1}^{p}\frac{(-1)^{p-k}\pi^{2p-2k}}{(2p+1-2k)!}\sum_{j=1}^{n-1}C_{2k+j-1}^{j}\sigma_{2k+j,n-j}\\
&=\delta_{n,2[n/2]}\frac{(-1)^{p+n/2}\pi^{2p+n}}{(2p+n+1)!}-(-1)^{n}\sum_{j=1}^{[n/2]}\frac{(-1)^{j}\pi^{2j-2}}{(2j-1)!}\\
&\quad\times\sum_{u=0}^{n-2j}C_{2p+u}^{u}(-1)^{u}s_{2p+u+1,n+1-2j-u},\\
&\sum_{k=1}^{p}\frac{(-1)^{p-k}\pi^{2p-2k}}{(2p-2k)!}\sum_{j=1}^{n-1}C_{2k+j-1}^{j}\sigma_{2k+j,n-j}\\
&=\delta_{n,2[n/2]}\frac{(-1)^{p+n/2}\pi^{2p+n}}{(2p+n)!}-(-1)^{n}s_{n,2p}\\
&\quad+(-1)^{n}\sum_{j=0}^{[(n-1)/2]}\frac{(-1)^{j}\pi^{2j}}{(2j)!}\\
&\quad\times\sum_{u=0}^{n-2j-1}C_{2p+u-1}^{u}(-1)^{u}s_{2p+u,n-2j-u},\\
&\sum_{k=1}^{p}\frac{(-1)^{p-k}\pi^{2p-2k}}{(2p-2k)!}\left(2\sigma_{2k-1,n}\right.\\
&\qquad\left.+\sum_{j=1}^{n-1}C_{2k+j-2}^{j}\sigma_{2k+j-1,n-j}\right)\\
&=(-1)^{n}s_{n,2p-1}+\delta_{n-1,2[(n-1)/2]}\frac{(-1)^{p+(n+1)/2}\pi^{2p+n-1}}{(2p+n-1)!}\\
&\quad+(-1)^{n}\sum_{j=0}^{[(n-1)/2]}\frac{(-1)^{j}\pi^{2j}}{(2j)!}\\
&\quad\times\sum_{u=0}^{n-2j-1}C_{2p+u-2}^{u}(-1)^{u}s_{2p+u-1,n-2j-u}.
\end{aligned}\tag{52}$$

In (52), letting $n=2$, we get that

$$\begin{aligned}
\sigma_{2p-1,2}&=\left(1-p-(2p-1)2^{-2p-1}\right)\zeta(2p+1)\\
&\quad-\frac{1}{2}\sum_{k=1}^{p-1}\frac{(2k-1)\left(1-2^{-2k}\right)(-1)^{p-k}\pi^{2p-2k}\zeta(2k+1)}{(2p+1-2k)!}\\
&\quad+\sum_{k=1}^{p-1}\frac{(-1)^{p-k}\pi^{2p-2k}}{(2p+1-2k)!}\sigma_{2k-1,2},\\
\sigma_{2p-1,2}&=\left(1-p+(2p-1)2^{-2p-1}\right)\zeta(2p+1)\\
&\quad+\sum_{j=1}^{p-1}\zeta(1+j)\zeta(2p-j)\\
&\quad-\frac{1}{2}\sum_{k=1}^{p-1}\frac{(2k-1)\left(1-2^{-2k}\right)(-1)^{p-k}\pi^{2p-2k}\zeta(2k+1)}{(2p-2k)!}\\
&\quad-\sum_{k=1}^{p-1}\frac{(-1)^{p-k}\pi^{2p-2k}}{(2p-2k)!}\sigma_{2k-1,2}.
\end{aligned}\tag{53}$$

Thus we have established two recursive formulas for $\sigma_{2n-1,2}$.

In (30), letting $z=1/2$, we have

$$\begin{aligned}
&a_{n,p}+\sum_{j=1}^{n}\frac{(-1)^{n-j}\ln^{n-j}2}{(n-j)!}\sum_{v=0}^{p-1}C_{j+v-1}^{v}\sigma_{p-v,j+v}\\
&=\sum_{j=1}^{n}\frac{(-1)^{n-j}\ln^{n-j}2}{(n-j)!}s_{j,p}+\frac{(-1)^{n}\ln^{n+p}2}{(n+p)!}.
\end{aligned}\tag{54}$$

In (33), letting $z=1/2$, we have

$$\begin{aligned}
&a_{n,p}-\sum_{j=1}^{n}\frac{(-1)^{n-j}\ln^{n-j}2}{(n-j)!}\\
&\quad\times\sum_{u=1}^{j}C_{p+j-1-u}^{j-u}\sigma_{u,p+j-u}=\frac{(-1)^{n}\ln^{n+p}2}{(n+p)!}.
\end{aligned}\tag{55}$$

Equation (55) is just consistent with (9.11) in [6]. By (54) and (55), we have

$$\begin{aligned}
&\sum_{j=1}^{n}\frac{(-1)^{n-j}\ln^{n-j}2}{(n-j)!}\left(\sum_{u=0}^{j-1}C_{p+u-1}^{u}\sigma_{j-u,p+u}\right.\\
&\qquad\left.+\sum_{v=0}^{p-1}C_{j+v-1}^{v}\sigma_{p-v,j+v}-s_{j,p}\right)=0.
\end{aligned}\tag{56}$$

Similarly, noting bracket part independent of n in (56), we have

$$\sum_{u=0}^{n-1}C_{p+u-1}^{u}\sigma_{n-u,p+u}+\sum_{v=0}^{p-1}C_{n+v-1}^{v}\sigma_{p-v,n+v}=s_{n,p},\tag{57}$$

which is equivalent to (9.8) in [5].

TABLE 1: Time-consuming comparison of the algorithms (59), (4), (28), (30), (33), and (34) with PolyLog for different precistions ($n = 4$, $p = 5$, and $m = 16$).

	z0	r	$T1_{32}, T2_{32}, T1_{32}/T2_{32}$	err1, err2	$T1_{64}, T2_{64}, T1_{64}/T2_{64}$	err1, err2
(59)	0	1/8	0.0654, 0.0009, 67.0	$0. * 10^{-32}, 0. * 10^{-31}$	0.1728, 0.0019, 88.5	$0. * 10^{-64}, 0. * 10^{-64}$
(59)	0	1/4	0.0673, 0.0019, 34.5	$0. * 10^{-32}, 0. * 10^{-32}$	0.1748, 0.0039, 44.7	$0. * 10^{-64}, 0. * 10^{-64}$
(4)	1	1/8	0.5136, 0.0107, 47.8	$0. * 10^{-26}, 0. * 10^{-30}$	1.7880, 0.0195, 91.5	$0. * 10^{-58}, 0. * 10^{-62}$
(4)	1	1/4	0.4628, 0.0156, 29.6	$0. * 10^{-25}, 0. * 10^{-29}$	1.5361, 0.0292, 52.4	$0. * 10^{-57}, 0. * 10^{-61}$
(28)	10	1	0.4326, 0.0185, 23.3	$0. * 10^{-30}, 0. * 10^{-29}$	1.2070, 0.0380, 31.6	$0. * 10^{-62}, 0. * 10^{-61}$
(28)	100	10	0.4472, 0.0195, 22.8	$0. * 10^{-31}, 0. * 10^{-29}$	1.1201, 0.0380, 29.4	$0. * 10^{-63}, 0. * 10^{-61}$
(30)	1	1/8	0.5117, 0.0195, 26.1	$0. * 10^{-26}, 0. * 10^{-30}$	1.7988, 0.0380, 47.2	$0. * 10^{-58}, 0. * 10^{-62}$
(30)	1	1/4	0.4658, 0.0292, 15.8	$0. * 10^{-25}, 0. * 10^{-29}$	1.5361, 0.0595, 25.7	$0. * 10^{-57}, 0. * 10^{-61}$
(33)	−1	1/8	0.4091, 0.0449, 9.1	$0. * 10^{-25}, 0. * 10^{-29}$	1.7275, 0.0957, 18.0	$6. * 10^{-57}, 0. * 10^{-61}$
(33)	−1	1/4	0.4111, 0.0468, 8.7	$0. * 10^{-25}, 0. * 10^{-29}$	1.6464, 0.1826, 9.0	$6. * 10^{-57}, 0. * 10^{-61}$
(34)	−10	1	0.4423, 0.0156, 28.3	$0. * 10^{-29}, 0. * 10^{-29}$	1.3437, 0.0302, 44.3	$0. * 10^{-61}, 0. * 10^{-61}$
(34)	−100	10	0.4443, 0.0156, 28.4	$0. * 10^{-30}, 0. * 10^{-30}$	1.0556, 0.0302, 34.8	$0. * 10^{-63}, 0. * 10^{-62}$

TABLE 2: Time-consuming comparison of the algorithms (59), (4), (28), (30), (33), and (34) with PolyLog for different precistions ($n = 4$, $p = 5$, and $m = 16$).

	z0	r	$T1_{96}, T2_{96}, T1_{96}/T2_{96}$	err1, err2	$T1_{128}, T2_{128}, T1_{128}/T2_{128}$	err1, err2
(59)	0	1/8	0.2236, 0.0039, 57.2	$0. * 10^{-96}, 0. * 10^{-96}$	0.5664, 0.0048, 116.0	$0. * 10^{-128}, 0. * 10^{-128}$
(59)	0	1/4	0.2275, 0.0058, 38.8	$0. * 10^{-96}, 0. * 10^{-96}$	0.5664, 0.0078, 72.5	$0. * 10^{-128}, 0. * 10^{-128}$
(4)	1	1/8	5.8251, 0.0302, 192.4	$0. * 10^{-89}, 0. * 10^{-94}$	8.7773, 0.0410, 214.	$0. * 10^{-118}, 0. * 10^{-126}$
(4)	1	1/4	3.2138, 0.0458, 70.0	$0. * 10^{-89}, 0. * 10^{-93}$	5.9726, 0.0634, 94.0	$0. * 10^{-121}, 0. * 10^{-125}$
(28)	10	1	2.2666, 0.0585, 38.6	$0. * 10^{-94}, 0. * 10^{-93}$	6.1044, 0.0791, 77.1	$0. * 10^{-126}, 0. * 10^{-125}$
(28)	100	10	2.2167, 0.0576, 38.4	$0. * 10^{-95}, 0. * 10^{-93}$	4.1611, 0.0781, 53.2	$0. * 10^{-127}, 0. * 10^{-125}$
(30)	1	1/8	3.7890, 0.0595, 63.6	$0. * 10^{-89}, 0. * 10^{-94}$	6.6630, 0.0820, 81.2	$0. * 10^{-118}, 0. * 10^{-126}$
(30)	1	1/4	3.1523, 0.0917, 34.3	$0. * 10^{-89}, 0. * 10^{-93}$	26.0390, 6.3847, 4.0	$0. * 10^{-121}, 0. * 10^{-125}$
(33)	−1	1/8	3.3681, 5.0224, 0.67	$0. * 10^{-89}, 0. * 10^{-93}$	5.9814, 5.0312, 1.18	$0. * 10^{-121}, 0. * 10^{-125}$
(33)	−1	1/4	1.4921, 4.8085, 0.31	$0. * 10^{-89}, 0. * 10^{-93}$	5.5576, 4.8798, 1.1	$0. * 10^{-121}, 0. * 10^{-125}$
(34)	−10	1	2.3085, 0.0478, 48.2	$0. * 10^{-93}, 0. * 10^{-93}$	5.6611, 0.0664, 85.2	$0. * 10^{-125}, 0. * 10^{-125}$
(34)	−100	10	2.2587, 0.0488, 46.2	$0. * 10^{-95}, 0. * 10^{-94}$	4.0634, 0.0664, 61.1	$0. * 10^{-127}, 0. * 10^{-126}$

5. High Accuracy and Fast Calculation for the Nielsen Generalized Polylogarithm $S_{n,p}(z)$

In this section, we apply the transformation formulas established previously to design algorithms of high accuracy and fast calculation for the Nielsen generalized polylogarithm $S_{n,p}(z)$. In Mathematica, there is an internal command, PolyLog[n, p, z], designed to calculate the Nielsen generalized polylogarithm $S_{n,p}(z)$. When $|z| < 1$, and the integers $n, p \geq 1$ in the Nielsen generalized polylogarithm $S_{n,p}(z)$, using (1) and

$$\ln^{p}(1-t) = (-1)^{p} p! \sum_{m=p}^{\infty} \frac{t^{m} H_{m-1}^{(p-1)}}{m}, \tag{58}$$

where

$$H_{m}^{(k+1)} = \sum_{j=k+1}^{m} \frac{1}{j} H_{j-1}^{(k)},$$
$$H_{m}^{(1)} = H_{m}, \tag{59}$$
$$H_{m}^{(0)} = 1,$$

satisfying the following recursive formula,

$$H_{n}^{(k)} = H_{n-1}^{(k)} + \frac{1}{n} H_{n-1}^{(k-1)}, \tag{60}$$

we have

$$S_{n,p}(z) = \sum_{m=p}^{\infty} \frac{z^{m} H_{m-1}^{(p-1)}}{m^{n+1}}. \tag{61}$$

Concerning the computational speed and accuracy, we make a comparison between PolyLog and (59), (4), (28), (30), (33), and (34). We will make the comparison in m well-distributed points from a circle with the center z_0 and the radius r. Numerical results are given in Tables 1 and 2.

Here $T1_P$, err 1 are the average running time and the relative error of PolyLog[n,p,N[z,Prec]], while $T2_P$, err 2 are the average running time and the relative error of (59), (4), (28), (30), (33), and (34) in precision P, respectively. We always use second as the unit of time.

Seen from Table 1, when the precision $P \leqslant 64$, (59), (4), (28), (30), (33), and (34) are faster than the PolyLog. Seen from Table 2, in addition to (33), with improved precision, it is more obvious that the calculation speed of (59), (4), (28), (30), and (34) is faster than PolyLog.

Remark 8. When $z_1 = 1/2 + \sqrt{3}i/2$ or $z_2 = 1/2 - \sqrt{3}i/2$, since $|1-z_i| = |1/(1-z_i)| = |z_i/(z_i-1)| = 1, i = 1,2$, is calculated as a value, (59), (4), (28), (30), (33), and (34) cannot be used. But so far we did not find such a transformation $|(a+z_i)/(b+z_i)| < 1, i = 1,2$. Although in the literature [6] the author gives some transformation $z \to (a+z)/(b+z)$, it essentially still belongs to the above type of transformation. For example, in [6], the author gives the following results:

$$L_{n,p}\left(\frac{a+z}{b+z}\right) = s_{n,p} - \sum_{k=0}^{p} C_{n+p-k-1}^{n-1} M_{k+1,n+p-k-1}\left(\frac{b-a}{b+z}\right) + L_{p+1,n-1}\left(\frac{b-a}{b+z}\right), \quad (62)$$

where $M_{n,p}(z) = ((-1)^{n+p-1}/(n-1)!p!)\int_0^z(\ln^{n-1}t\ln^p(1+t)/t)dt$. If the $(b-a)/(b+z)$into z, (62) becomes

$$L_{n,p}(1-z) = s_{n,p} - \sum_{k=0}^{p} C_{n+p-k-1}^{n-1} M_{k+1,n+p-k-1}(z) + L_{p+1,n-1}(z), \quad (63)$$

this is clearly the type of (4).

6. Conclusions

In this paper, for the Nielsen generalized polylogarithm $S_{n,p}(z)$, the formulas for the transformations $z \to p(z)$, $p(z) = 1-z, 1/z, 1/(1-z), z/(z-1)$, and $(1-z)/z$ have been established, and the last three transformation formulas are new. It is worthy to note that the transformation formula (28) is faster than (1.5) in [5] after taking into account numerical calculation. By use of these transformation formulas established above, new fast algorithms for Nielsen generalized polylogarithm $S_{n,p}(z)$ can be designed. Numerical results show that when $|z - z_i] > 2/5$, where $z_1 = 1/2 + i\sqrt{3}/2, z_2 = 1/2 - i\sqrt{3}/2$, the new algorithm is superior to Mathematica's internal functions PolyLog tens of times faster or even a hundred times. For $s_{n,p} = S_{n,p}(1)$, we have also given a new recursive formula (36). For $a_{n,p} = S_{n,p}(1/2)$ and $\sigma_{n,p} = S_{n,p}(-1)$, we have got some new identities, some of which are different from those in [5].

Conflict of Interests

The authors declare that there is no conflict of interests regarding the publication of this paper.

Acknowledgment

This work is supported by National Natural Science Foundation of China (61379009).

References

[1] I. S. Gradshteyn and I. M. Ryzhik, *Table of Integrals, Series and Products*, Academic Press, New York, NY, USA, 1980.

[2] H. Cheng and T. Wu, *Expanding Protons: Scattering at High Energies*, The MIT Press, Cambridge, Mass, USA, 1987.

[3] W. Maier and H. Kiesewetter, *Funktionalgleichungen Mit Analytischen Lösungen*, Deutscher Verlag der Wissenschaften, Berlin, Germany, 1971.

[4] K. S. Kölbig, J. A. Mignaco, and E. Remiddi, "On Nielsen's generalized polylogarithms and their numerical calculation," *BIT*, vol. 10, no. 1, pp. 38–73, 1970.

[5] K. S. Kölbig, "Nielsen's generalized polylogarithms," *The SIAM Journal on Mathematical Analysis*, vol. 17, pp. 1232–1258, 1986.

[6] H. Jacques, *Lectures on Cauchy's Problem in Linear Partial Differential Equations*, Dover Phoenix, New York, NY, USA, 1923.

[7] D. F. Paget, "The numerical evaluation of Hadamard finite-part integrals," *Numerische Mathematik*, vol. 36, no. 4, pp. 447–453, 1981.

[8] B. Bialecki, "A sinc quadrature rule for Hadamard finite-part integrals," *Numerische Mathematik*, vol. 57, no. 1, pp. 263–269, 1990.

Somewhat Slightly Generalized Double Fuzzy Semicontinuous Functions

Fatimah M. Mohammed,[1,2] M. S. M. Noorani,[1] and A. Ghareeb[3,4]

[1] *School of Mathematical Sciences, Universiti Kebangsaan Malaysia, 43600 UKM Bangi, Selangor, Malaysia*
[2] *College of Education, Tikrit University, Iraq*
[3] *Department of Mathematics, College of Science in Al-Zulfi, Majmaah University, Al-Zulfi, Saudi Arabia*
[4] *Department of Mathematics, Faculty of Science, South Valley University, Qena, Egypt*

Correspondence should be addressed to A. Ghareeb; nasserfuzt@hotmail.com

Academic Editor: Nawab Hussain

The aim of this paper is to introduce the concepts of somewhat slightly generalized double fuzzy semicontinuous functions and somewhat slightly generalized double fuzzy semiopen functions in double fuzzy topological spaces. Some interesting properties and characterizations of these functions are introduced and discussed. Furthermore, the relationships among the new concepts are discussed with some necessary examples.

1. Introduction

In 1968, Chang [1] was the first to introduce the concept of fuzzy topological spaces. These spaces and their generalization are later developed by Goguen [2], who replaced the closed interval $[0, 1]$ by more general lattice L. On the other hand, by the independent and parallel generalization of Kubiak and Šostak's [3, 4], made topology itself fuzzy besides their dependence on fuzzy set in 1985.

Various generalizations of the concept of fuzzy set have been done by many authors. In [5–10], Atanassove introduced the notion of intuitionistic fuzzy sets. Later Çoker [11] defined intuitionistic fuzzy topology in Chang's sense. Then, Mondal and Samanta [12] introduced the intuitionistic gradation of openness of fuzzy sets. Gutiérrez García and Rodabaugh [13], in 2005, replaced the term "intuitionistic" and concluded that the most appropriate work is under the name "double."

In 1980, Jain [14] introduced the notion of slightly continuous functions. On the other hand, Nour [15] defined slightly semicontinuous functions as a weak form of slight continuity and investigated their properties. In [16], Noiri introduced the concept of slightly β-continuous functions. Sudha et al. [17] introduced slightly fuzzy ω-continuous functions. Also in 2004, Ekici and Caldas [18] introduced the notion of slight γ-continuity (slight b-continuity).

In this paper, the concepts of somewhat slightly generalized double fuzzy semicontinuous functions and somewhat slightly generalized double fuzzy semiopen functions are introduced. Several interesting properties and characterizations are introduced and discussed. Furthermore, the relationships among the concepts are obtained and established with some interesting counter examples.

2. Preliminaries

Throughout this paper, let X be a nonempty set, I the unit interval $[0, 1]$, $I_0 = (0, 1]$, and $I_1 = [0, 1)$. The family of all fuzzy sets in X is denoted by I^X. $P_t(X)$ is the family of all fuzzy points in X. By $\underline{0}$ and $\underline{1}$ we denote the smallest and the greatest fuzzy sets on X. For a fuzzy set $\lambda \in I^X$, $\underline{1} - \lambda$ denotes its complement. Given a function $f : X \to Y$, $f(\lambda)$ and $f^{-1}(\lambda)$ defined the direct image and the inverse image of f, defined by $f(\lambda)(y) = \bigvee_{f(x)=y} \lambda(x)$ and $f^{-1}(\mu)(x) = \mu(f(x))$ for each $\lambda \in I^X$, $\mu \in I^Y$, and $x \in X$, respectively. All other notations are standard notations of fuzzy set theory.

Definition 1 (see [12, 13]). A double fuzzy topology (τ, τ^*) on X is a pair of maps $\tau, \tau^* : I^X \to I$, which satisfies the following properties:

(O1) $\tau(\lambda) \leq \underline{1} - \tau^*(\lambda)$ for each $\lambda \in I^X$;

(O2) $\tau(\lambda_1 \wedge \lambda_2) \geq \tau(\lambda_1) \wedge \tau(\lambda_2)$ and $\tau^*(\lambda_1 \wedge \lambda_2) \leq \tau^*(\lambda_1) \vee \tau^*(\lambda_2)$ for each $\lambda_1, \lambda_2 \in I^X$;

(O3) $\tau(\bigvee_{i\in\Gamma}\lambda_i) \geq \bigwedge_{i\in\Gamma}\tau(\lambda_i)$ and $\tau^*(\bigvee_{i\in\Gamma}\lambda_i) \leq \bigvee_{i\in\Gamma}\tau^*(\lambda_i)$ for each $\lambda_i \in I^X, i \in \Gamma$.

The triplet (X, τ, τ^*) is called double fuzzy topological spaces (dfts, for short). A fuzzy set λ is called an (r, s)-fuzzy open ((r, s)-fo, for short) if $\tau(\lambda) \geq r$ and $\tau^*(\lambda) \leq s$, λ is called an (r, s)-fuzzy closed ((r, s)-fc, for short) if and only if $\underline{1} - \lambda$ is an (r, s)-fo set, and λ is called (r, s)-fuzzy clopen ((r, s)-fco, for short) if and only if λ is (r, s)-fo set and (r, s)-fc set. Let (X, τ_1, τ_1^*) and (Y, τ_2, τ_2^*) be two dfts's. A function $f : X \to Y$ is said to be a double fuzzy continuous if and only if $\tau_1(f^{-1}(\nu)) \geq \tau_2(\nu)$ and $\tau_1^*(f^{-1}(\nu)) \leq \tau_2^*(\nu)$ for each $\nu \in I^Y$.

Before starting to present our results, there are two questions that we must ask ourselves. First, what is the difference between classical topology and double fuzzy topology? Secondly, where we can apply our results?

To answer the first question, we should know that double fuzzy sets and hence double fuzzy topological spaces deal with obscurities. In addition to that, we observed that the concept of double fuzzy topological spaces is a generalization of fuzzy topological spaces and classical topology. For example, when the first condition in Definition 1 does not hold, we get the definition of fuzzy topological spaces in Kubiak-Šostak's sense [3, 4]. Also, in the same definition, when we replace 2^X with I^X, we will get results in double gradation fuzzifying topological spaces [19]. Appropriate changes can be made to get results in the classical topological spaces.

With regard to applications, since double fuzzy topology forms an extension of fuzzy topology and general topology, we think that our results can be applied in the fuzzy mathematics, which has many applications in different branches of engineering and ICT. For example, recently double fuzzy topological spaces have been applied to study sensor bias [20] and there exist well-established applications of fuzzy topological spaces in the areas of digital topology [21], image processing [22], and geographic information systems (GIS) problems [23].

Theorem 2 (see [24, 25]). *Let (X, τ, τ^*) be a dfts. Then for each $r \in I_0$, $s \in I_1$, and $\lambda \in I^X$, one defines an operator C_{τ,τ^*} : $I^X \times I_0 \times I_1 \to I^X$ as follows:*

$$C_{\tau,\tau^*}(\lambda, r, s) = \bigwedge\left\{\mu \in I^X \mid \lambda \leq \mu, \tau(\underline{1} - \mu) \geq r, \tau^*(\underline{1} - \mu) \leq s\right\}. \quad (1)$$

For $\lambda, \mu \in I^X$, $r, r_1, r_2 \in I_0$ and $s, s_1, s_2 \in I_1$, the operator C_{τ,τ^*} satisfies the following statements:

(C1) $C_{\tau,\tau^*}(\underline{0}, r, s) = \underline{0}$;

(C2) $\lambda \leq C_{\tau,\tau^*}(\lambda, r, s)$;

(C3) $C_{\tau,\tau^*}(\lambda, r, s) \vee C_{\tau,\tau^*}(\mu, r, s) = C_{\tau,\tau^*}(\lambda \vee \mu, r, s)$;

(C4) $C_{\tau,\tau^*}(\lambda, r_1, s_1) \leq C_{\tau,\tau^*}(\lambda, r_2, s_2)$ *if* $r_1 \leq r_2$ *and* $s_1 \geq s_2$;

(C5) $C_{\tau,\tau^*}(C_{\tau,\tau^*}(\lambda, r, s), r, s) = C_{\tau,\tau^*}(\lambda, r, s)$.

Theorem 3 (see [24, 25]). *Let (X, τ, τ^*) be a dfts. Then for each $r \in I_0$, $s \in I_1$, and $\lambda \in I^X$, one defines an operator I_{τ,τ^*} : $I^X \times I_0 \times I_1 \to I^X$ as follows:*

$$I_{\tau,\tau^*}(\lambda, r, s) = \bigvee\left\{\mu \in I^X \mid \mu \leq \lambda, \tau(\mu) \geq r, \tau^*(\mu) \leq s\right\}. \quad (2)$$

For $\lambda, \mu \in I^X$, $r, r_1, r_2 \in I_0$ and $s, s_1, s_2 \in I_1$, the operator I_{τ,τ^*} satisfies the following statements:

(I1) $I_{\tau,\tau^*}(\underline{1} - \lambda, r, s) = \underline{1} - C_{\tau,\tau^*}(\lambda, r, s)$;

(I2) $I_{\tau,\tau^*}(\underline{1}, r, s) = \underline{1}$;

(I3) $I_{\tau,\tau^*}(\lambda, r, s) \leq \lambda$;

(I4) $I_{\tau,\tau^*}(\lambda, r, s) \wedge I_{\tau,\tau^*}(\mu, r, s) = I_{\tau,\tau^*}(\lambda \wedge \mu, r, s)$;

(I5) $I_{\tau,\tau^*}(\lambda, r_1, s_1) \geq I_{\tau,\tau^*}(\lambda, r_2, s_2)$ *if* $r_1 \leq r_2$ *and* $s_1 \geq s_2$;

(I6) $I_{\tau,\tau^*}(I_{\tau,\tau^*}(\lambda, r, s), r, s) = I_{\tau,\tau^*}(\lambda, r, s)$;

(I7) *If* $I_{\tau,\tau^*}(C_{\tau,\tau^*}(\lambda, r, s), r, s) = \lambda$, *then* $C_{\tau,\tau^*}(I_{\tau,\tau^*}(\underline{1} - \lambda, r, s), r, s) = \underline{1} - \lambda$.

Definition 4 (see [26]). Let (X, τ, τ^*) be a dfts. For each $\lambda, \mu \in I^X$, $r \in I_0$, and $s \in I_1$.

(1) A fuzzy set λ is called (r, s)-fuzzy semiclosed (briefly, (r, s)-fsc) if $I_{\tau,\tau^*}(C_{\tau,\tau^*}(\lambda, r, s), r, s) \leq \lambda$. λ is called (r, s)-fuzzy semiopen (briefly, (r, s)-fso) if and only if $\underline{1} - \lambda$ is an (r, s)-fuzzy semiclosed set.

(2) An (r, s)-fuzzy semiclosure of λ is defined by $SC_{\tau,\tau^*}(\lambda, r, s) = \bigwedge\{\mu \in I^X \mid \lambda \leq \mu$ and μ is (r, s)-fsc$\}$.

Definition 5 (see [26]). Let (X, τ, τ^*) be a dfts. For each $\lambda, \mu \in I^X$, $r \in I_0$ and $s \in I_1$.

(1) A fuzzy set λ is called (r, s)-generalized fuzzy semiclosed (briefly, (r, s)-gfsc) if $SC_{\tau,\tau^*}(\lambda, r, s) \leq \mu$, $\lambda \leq \mu$ and $\tau(\mu) \geq r$, $\tau^*(\mu) \leq s$. λ is called (r, s)-generalized fuzzy semiopen (briefly, (r, s)-gfso) if and only if $\underline{1} - \lambda$ is (r, s)-gfsc set.

(2) An (r, s)-fuzzy generalized semiclosure of λ is defined by $GSC_{\tau,\tau^*}(\lambda, r, s) = \bigwedge\{\mu \in I^X \mid \lambda \leq \mu$ and μ is (r, s)-gfsc$\}$.

(3) An (r, s)-fuzzy generalized semi-interior of λ is defined by $GSI_{\tau,\tau^*}(\lambda, r, s) = \bigvee\{\mu \in I^X \mid \mu \leq \lambda$ and μ is (r, s)-gfso$\}$.

Definition 6 (see [27]). Let (X, τ_1, τ_1^*) and (Y, τ_2, τ_2^*) be dfts's. A function $f : (X, \tau_1, \tau_1^*) \to (Y, \tau_2, \tau_2^*)$ is called

(1) slightly double fuzzy continuous (briefly, sdfc) if for every $\lambda \in I^X$, $\mu \in I^Y$, $r \in I_0$, and $s \in I_1$ such that μ is (r, s)-fco set and $f(\lambda) \leq \mu$, there exists $\nu \in I^X$ such that $\tau_1(\nu) \geq r$, $\tau_1^*(\nu) \leq s$, $\lambda \leq \nu$, and

$$f(\nu) \leq \mu, \quad (3)$$

(2) slightly generalized double fuzzy semicontinuous (briefly, sgdfsc) if for each $\lambda \in I^X$, $\mu \in I^Y$, $r \in I_0$, and $s \in I_1$ such that μ is (r,s)-fco set and $f(\lambda) \le \mu$, there exists an (r,s)-gfso set $\nu \in I^X$ such that $\lambda \le \nu$ and

$$f(\nu) \le \mu. \quad (4)$$

3. Somewhat Slightly Generalized Double Fuzzy Semicontinuous Functions

Definition 7. Let (X, τ_1, τ_1^*) and (Y, τ_2, τ_2^*) be dfts's. A function $f : (X, \tau_1, \tau_1^*) \to (Y, \tau_2, \tau_2^*)$ is called somewhat slightly generalized double fuzzy semicontinuous (briefly, swsgdfsc) if for each fuzzy set $\lambda \in I^X$, $\mu \in I^Y$, $r \in I_0$, and $s \in I_1$ such that $f^{-1}(\mu) \ne \underline{0}$ and $f(\lambda) \le \mu$, there exists an (r,s)-gfso set $\underline{0} \ne \nu \in I^X$ such that $\lambda \le \nu$ and

$$\nu \le f^{-1}(\mu). \quad (5)$$

Definition 8. A fuzzy set λ in a dfts (X, τ, τ^*) is called (r,s)-generalized fuzzy semidense (resp., (r,s)-fuzzy-*dense**) set if there exists no (r,s)-gfsc (resp., (r,s)-fco) set $\mu \in I^X$, $r \in I_0$, and $s \in I_1$ such that

$$\lambda < \mu < \underline{1}. \quad (6)$$

Example 9. (1) Let $X = \{a, b\}$. Define λ_1 and λ_2 as follows:

$$\lambda_1(a) = 0.1, \quad \lambda_1(b) = 0.2,$$
$$\lambda_2(a) = 0.8, \quad \lambda_2(b) = 0.7. \quad (7)$$

And define $\tau(\lambda)$ and $\tau^*(\lambda)$ as follows:

$$\tau(\lambda) = \begin{cases} 1, & \text{if } \lambda = \underline{0} \text{ or } \underline{1}; \\ \frac{1}{3}, & \text{if } \lambda = \lambda_1; \\ \frac{2}{3}, & \text{if } \lambda = \lambda_2; \\ 0, & \text{otherwise;} \end{cases}$$
$$\tau^*(\lambda) = \begin{cases} 0, & \text{if } \lambda = \underline{0} \text{ or } \underline{1}; \\ \frac{2}{3}, & \text{if } \lambda = \lambda_1; \\ \frac{1}{3}, & \text{if } \lambda = \lambda_2; \\ 1, & \text{otherwise.} \end{cases} \quad (8)$$

So, if $\lambda(a) = 0.9$, $\lambda(b) = 0.8$, then there exists no $(1/3, 2/3)$-gfsc set μ in I^X such that $\lambda < \mu < \underline{1}$. Therefore, λ is $(1/3, 2/3)$-generalized fuzzy semidense set in I^X.

(2) In (1), let λ_1 and λ_2 be defined as follows:

$$\lambda_1(a) = 0.1, \quad \lambda_2(b) = 0.2,$$
$$\lambda_2(a) = 0.9, \quad \lambda_2(b) = 0.8. \quad (9)$$

So, if $\lambda(a) = 0.8$, $\lambda(b) = 0.9$, then there exists no $(1/3, 2/3)$-fco set μ in I^X such that $\lambda < \mu < \underline{1}$. Therefore, λ is $(1/3, 2/3)$-fuzzy-*dense** set in I^X.

Definition 10. Let (X, τ, τ^*) be a dfts. For a fuzzy set $\lambda \in I^X$, $r \in I_0$, and $s \in I_1$, I^*_{τ,τ^*} and C^*_{τ,τ^*} are defined as follows:

(1) $I^*_{\tau,\tau^*}(\lambda, r, s) = \bigvee\{\mu \in I^X \mid \mu \le \lambda \text{ and } \mu \text{ is } (r,s)\text{-fco}\}$;

(2) $C^*_{\tau,\tau^*}(\lambda, r, s) = \bigwedge\{\mu \in I^X \mid \lambda \le \mu \text{ and } \mu \text{ is } (r,s)\text{-fco}\}$.

Proposition 11. *Let (X, τ_1, τ_1^*) and (Y, τ_2, τ_2^*) be dfts's, and let $f : (X, \tau_1, \tau_1^*) \to (Y, \tau_2, \tau_2^*)$ be any function. Then the following are equivalent.*

(1) *f is swsgdfsc function.*

(2) *If λ is an (r,s)-fco set such that $f^{-1}(\lambda) \ne \underline{1}$ and $\lambda \le f(\underline{1} - \nu)$, for each $\nu \in I^X$, $r \in I_0$, and $s \in I_1$, then there exists an (r,s)-gfsc set $\mu \le \underline{1} - \nu \in I^X$ such that $\mu \ge f^{-1}(\lambda)$.*

(3) *If λ is (r,s)-gfs-dense set in I^X, then $f(\lambda)$ is (r,s)-fuzzy-dense* set in I^Y such that every (r,s)-fco set $\mu \le f(\underline{1} - \nu)$, for each $\nu \in I^X$, $r \in I_0$, and $s \in I_1$.*

Proof. (1) ⇒ (2) Suppose f is swsgdfsc function, and let λ be any (r,s)-fco set in I^Y such that $f^{-1}(\lambda) \ne \underline{1}$ and $\lambda \le f(\underline{1} - \nu)$, for each $\nu \in I^X$, $r \in I_0$, and $s \in I_1$. Then, $\underline{1} - \lambda$ is (r,s)-fco in I^Y such that $f^{-1}(\underline{1} - \lambda) \ne \underline{0}$ and $f(\nu) \le \underline{1} - \lambda$. Then by the hypothesis, there exists an (r,s)-gfso set $\underline{0} \ne \alpha \in I^X$, $r \in I_0$, and $s \in I_1$ such that $\nu \le \alpha$ and $\alpha \le f^{-1}(\underline{1} - \lambda)$. That is, $\underline{1} - \alpha$ is an (r,s)-gfsc set and

$$\underline{1} - \alpha \ge \underline{1} - f^{-1}(\underline{1} - \lambda) = f^{-1}(\lambda). \quad (10)$$

Put $\underline{1} - \alpha = \mu$. Then μ is an (r,s)-gfsc set in I^X such that $\mu \ge f^{-1}(\lambda)$.

(2) ⇒ (3) Let λ be an (r,s)-gfs-dense set in I^X, and suppose that $f(\lambda)$ is not a fuzzy-*dense** set in I^Y, such that each (r,s)-fco set $\mu \le f(\underline{1} - \nu)$, for each $\nu \in I^X$, $r \in I_0$, and $s \in I_1$. Then, there exists an (r,s)-fco set $\alpha \in I^Y$ such that

$$f(\lambda) < \alpha < \underline{1}, \quad (11)$$

since

$$\alpha < \underline{1}, \quad f^{-1}(\alpha) \ne \underline{1}. \quad (12)$$

Now, α is an (r,s)-fco set such that $f^{-1}(\alpha) \ne \underline{1}$ and $f(\underline{1} - \nu) \ge \alpha$, for each $\nu \in I^X$, $r \in I_0$, and $s \in I_1$. Then by the hypothesis, there exists an (r,s)-gfsc set $\gamma \le \underline{1} - \nu \in I^X$ such that $\gamma \ge f^{-1}(\alpha)$.

But

$$f^{-1}(\alpha) > f^{-1}(f(\lambda)) = \lambda. \quad (13)$$

That is, $\gamma \ge \lambda$. Therefore, there exists an (r,s)-gfsc set $\gamma \in I^X$, $r \in I_0$, and $s \in I_1$ such that $\gamma \ge \lambda$, which is a contradiction. Therefore, $f(\lambda)$ is an (r,s)-fuzzy *dense** set in I^Y such that

$$\gamma \le f(\underline{1} - \nu), \quad (14)$$

for each $\nu \in I^X$ and (r,s)-fco set $\gamma \in I^Y$.

(3) $\Rightarrow$ (1) Let λ be an (r,s)-fco set such that $f^{-1}(\lambda)\neq\underline{0}$ and $f(\nu)\leq\lambda$, for each $\nu\in I^X$, $r\in I_0$, and $s\in I_1$. Then, $\lambda\neq\underline{0}$. Now, suppose that $\nu\leq\alpha$ and $\mathrm{GSI}_{\tau,\tau^*}(f^{-1}(\lambda),r,s)=\underline{0}\in I^X$. Then,

$$\mathrm{GSC}_{\tau,\tau^*}\left(\underline{1}-f^{-1}(\lambda),r,s\right)=\underline{1}\in I^X. \tag{15}$$

That is, $\underline{1}-f^{-1}(\lambda)$ is an (r,s)-gfs-dense in I^X. Then by (3), $f(\underline{1}-f^{-1}(\lambda))$ is an (r,s)-fuzzy *dense** set such that there exists an (r,s)-fco set $\mu\leq f(\underline{1}-\nu)$, for each $\nu\in I^X$, $r\in I_0$, and $s\in I_1$.

But

$$f\left(\underline{1}-f^{-1}(\lambda)\right)=f\left(f^{-1}\left(\underline{1}-\lambda\right)\right)\leq\underline{1}-\lambda<\underline{1}, \tag{16}$$

since $\underline{1}-\lambda$ is an (r,s)-fco and

$$\begin{gathered}f\left(\underline{1}-f^{-1}(\lambda)\right)\leq\underline{1}-\lambda,\\ C^*_{\tau,\tau^*}\left(f\left(\underline{1}-f^{-1}(\lambda)\right),r,s\right)\leq\underline{1}-\lambda.\end{gathered} \tag{17}$$

That is,

$$\underline{1}-\lambda\geq\underline{1}\Longrightarrow\lambda=\underline{0}, \tag{18}$$

which is a contradiction, since $\lambda\neq\underline{0}$. Therefore, $\nu\leq\alpha$ and $\mathrm{GSI}_{\tau,\tau^*}(f^{-1}(\lambda),r,s)\neq\underline{0}$. So f is swsgdfsc. □

4. Somewhat Slightly Generalized Double Fuzzy Semiopen Functions

Definition 12. Let (X,τ_1,τ_1^*) and (Y,τ_2,τ_2^*) be dfts's. A function $f:(X,\tau_1,\tau_1^*)\to(Y,\tau_2,\tau_2^*)$ is called

(1) generalized double fuzzy semiopen (briefly, gdfso) if for each (r,s)-gfso set $\lambda\in I^X$, $r\in I_0$, and $s\in I_1$, $f(\lambda)$ is an (r,s)-gfso in I^Y;

(2) slightly generalized double fuzzy semiopen (briefly, sgdfso) if for each (r,s)-gfso set $\lambda\in I^X$ and each $\mu\in I^X$, $r\in I_0$, and $s\in I_1$ such that $\lambda\leq\mu$, $f(\lambda)$ is an (r,s)-fco set in I^Y and

$$f(\lambda)\leq f(\mu); \tag{19}$$

(3) somewhat generalized double fuzzy semiopen (briefly, swgdfso) if for each (r,s)-gfso set $\underline{0}\neq\lambda\in I^X$, $r\in I_0$, and $s\in I_1$, there exists an (r,s)-gfso set $\underline{0}\neq\mu\in I^Y$ such that

$$f(\lambda)\geq\mu; \tag{20}$$

(4) somewhat slightly generalized double fuzzy semiopen (briefly, swsgdfso) if for each (r,s)-gfso set $\underline{0}\neq\lambda\in I^X$ such that $\lambda\leq\nu$ and for each $\nu\in I^X$, $r\in I_0$, and $s\in I_1$, there exists an (r,s)-fco set $\underline{0}\neq\mu\in I^Y$, $\mu\leq f(\nu)$ such that

$$f(\lambda)\geq\mu. \tag{21}$$

That is, $I^*_{\tau,\tau^*}(f(\lambda),r,s)\neq\underline{0}$, and there exists an (r,s)-fco set μ such that $f(\nu)\geq\mu$ and $\lambda\leq\nu$, for each $\nu\in I^X$, $r\in I_0$, and $s\in I_1$.

Proposition 13. *Let (X,τ_1,τ_1^*), (Y,τ_2,τ_2^*), and (Z,τ_3,τ_3^*) be dfts's. If $f:(X,\tau_1,\tau_1^*)\to(Y,\tau_2,\tau_2^*)$ and $g:(Y,\tau_2,\tau_2^*)\to(Z,\tau_3,\tau_3^*)$ are swsgdfso functions, then $g\circ f:(X,\tau_1,\tau_1^*)\to(Z,\tau_3,\tau_3^*)$ is a swsgdfso function.*

Proof. Let $\underline{0}\neq\lambda\in I^X$ be an (r,s)-gfso set $r\in I_0$ and $s\in I_1$ such that $\lambda\leq\mu$, for each fuzzy set $\mu\in I^X$, $r\in I_0$, and $s\in I_1$. Since f is swsgdfso, then there exists an (r,s)-fco set $\underline{0}\neq\nu\in I^Y$, and $f(\mu)\geq\nu$ such that $f(\lambda)\geq\nu$.

Now, $\mathrm{GSI}_{\tau,\tau^*}(f(\lambda),r,s))$ is an (r,s)-gfso in I^Y such that

$$\begin{gathered}\mathrm{GSI}_{\tau,\tau^*}(f(\lambda),r,s)\neq\underline{0},\\ \mathrm{GSI}_{\tau,\tau^*}(f(\lambda),r,s)\leq f(\mu),\end{gathered} \tag{22}$$

for each $f(\mu)\in I^Y$.

Since g is swsgdfso, then there exists an (r,s)-fco set $\underline{0}\neq\gamma\in I^Z$ and $\gamma\leq g(f(\mu))$ such that

$$\gamma\leq g\left(\mathrm{GSI}_{\tau,\tau^*}(f(\lambda),r,s)\right). \tag{23}$$

But

$$g\left(\mathrm{GSI}_{\tau,\tau^*}(f(\lambda),r,s)\right)\leq g(f(\lambda)). \tag{24}$$

Thus, there exists an (r,s)-fco set $\underline{0}\neq\gamma\in I^Z$ and

$$(g\circ f)(\mu)\geq\gamma, \tag{25}$$

such that

$$(g\circ f)(\lambda)\geq\gamma. \tag{26}$$

Therefore, $g\circ f$ is swsgdfso. □

Proposition 14. *Let (X,τ_1,τ_1^*) and (Y,τ_2,τ_2^*) be dfts's, and let $f:(X,\tau_1,\tau_1^*)\to(Y,\tau_2,\tau_2^*)$ be a bijective function. Then the following are equivalent.*

(1) *f is swsgdfso function.*

(2) *If λ is an (r,s)-gfsc set in I^X such that $f(\lambda)\neq\underline{1}$ and $\lambda\geq\nu$ for each $\nu\in I^X$, then there exists an (r,s)-fco set $\mu\in I^Y$, $\mu\neq\underline{1}$, and $f(\nu)\leq\mu$ such that $f(\lambda)\leq\mu$.*

Proof. (1) $\Rightarrow$ (2) Let λ be an (r,s)-gfsc set in I^X such that $f(\lambda)\neq\underline{1}$ and $\lambda\geq\nu$, for each $\nu\in I^X$, $r\in I_0$, and $s\in I_1$. Then, $\underline{1}-\lambda$ is an (r,s)-gfso set in I^X such that $f(\underline{1}-\lambda)\neq\underline{0}$ and $\underline{1}-\lambda\leq\underline{1}-\nu$, for each $\nu\in I^X$. So

$$\underline{1}-\lambda\neq\underline{0}. \tag{27}$$

Since f is a swsgdfso, then there exists an (r,s)-fco set $\underline{0}\neq\delta\in I^Y$ and $f(\underline{1}-\nu)\geq\delta$ such that

$$f(\underline{1}-\lambda)\geq\delta. \tag{28}$$

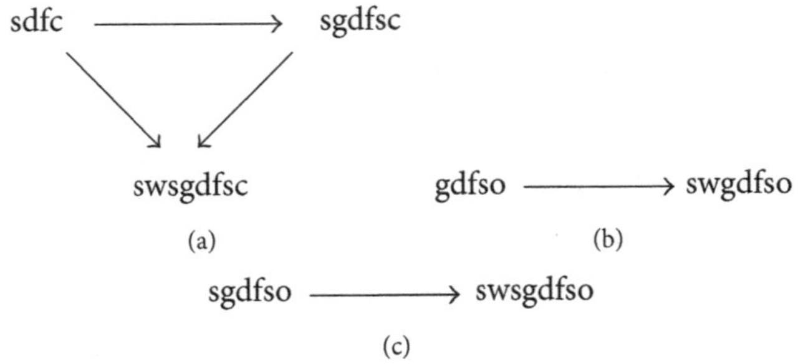

FIGURE 1

Now, $\underline{1}-\delta$ is an (r,s)-fco set in I^Y such that $\underline{1}-\delta\neq\underline{1}$ and $\underline{1}-\delta\geq f(\nu)$ such that

$$\underline{1}-\delta\geq f(\lambda). \tag{29}$$

Take

$$\underline{1}-\delta=\mu, \tag{30}$$

so (2) is proved.

(2) $\Rightarrow$ (1) Let $\lambda\neq\underline{0}$ be any (r,s)-gfso set in I^X such that $\lambda\leq\nu$, for each $\nu\in I^X$. Then, $\underline{1}-\lambda$ is an (r,s)-gfsc set in I^X such that $\underline{1}-\lambda\neq\underline{1}$ and $\underline{1}-\lambda\geq\underline{1}-\nu$ for each $\nu\in I^X r\in I_0$ and $s\in I_1$. Now,

$$f(\underline{1}-\lambda)=\underline{1}-f(\lambda)\neq\underline{1}. \tag{31}$$

For, if $\underline{1}-f(\lambda)=\underline{1}$, then

$$f(\lambda)=\underline{0}\Longrightarrow\lambda=\underline{0}. \tag{32}$$

Hence by the hypothesis, there exists an (r,s)-fco set $\mu\in I^Y$, $\underline{1}\neq\mu\geq f(\underline{1}-\nu)$, such that

$$f(\underline{1}-\lambda)\leq\mu. \tag{33}$$

That is,

$$\underline{0}\neq\underline{1}-\mu\leq f(\nu), \tag{34}$$

such that

$$\underline{1}-\mu\leq f(\lambda). \tag{35}$$

Let $\underline{1}-\mu=\gamma$. Then, $\gamma\neq\underline{0}$ is an (r,s)-fco set in I^Y such that $f(\nu)\geq\gamma$ and $f(\lambda)\geq\gamma$. Therefore, f is swsgdfso function. □

5. Interrelations

The following implication illustrates the relationships between different functions in Figure 1.

None of these implications is reversible where $A\rightarrow B$ represents A implies B, as shown by the following examples.

Example 15. Let $X=\{a,b\}$.

(1) Let $f:(X,\tau_1,\tau_1^*)\rightarrow(X,\tau_2,\tau_2^*)$ be the identity function. Define $\lambda_1,\lambda_2,\mu_1$, and μ_2 as follows:

$$\begin{aligned}\lambda_1(a)&=0.1, & \lambda_1(b)&=0.2,\\ \lambda_2(a)&=0.9, & \lambda_2(b)&=0.8,\\ \mu_1(a)&=0.5, & \mu_1(b)&=0.5,\\ \mu_2(a)&=1.0, & \mu_2(b)&=0.7.\end{aligned} \tag{36}$$

And define (τ_1,τ_1^*) and (τ_2,τ_2^*) as follows:

$$\begin{gathered}\tau_1(\lambda)=\begin{cases}1, & \text{if } \lambda\in\{\underline{0},\underline{1}\},\\ \frac{1}{4}, & \text{if } \lambda=\lambda_1,\\ \frac{1}{8}, & \text{if } \lambda=\lambda_2,\\ 0, & \text{otherwise},\end{cases}\\ \tau_1^*(\lambda)=\begin{cases}0, & \text{if } \lambda\in\{\underline{0},\underline{1}\},\\ \frac{1}{8}, & \text{if } \lambda=\lambda_1,\\ \frac{1}{4}, & \text{if } \lambda=\lambda_2,\\ 1, & \text{otherwise},\end{cases}\\ \tau_2(\lambda)=\begin{cases}1, & \text{if } \lambda\in\{\underline{0},\underline{1}\},\\ \frac{1}{4}, & \text{if } \lambda=\mu_1,\\ \frac{1}{8}, & \text{if } \lambda=\mu_2,\\ 0, & \text{otherwise},\end{cases}\\ \tau_2^*(\lambda)=\begin{cases}0, & \text{if } \lambda\in\{\underline{0},\underline{1}\},\\ \frac{1}{8}, & \text{if } \lambda=\mu_1,\\ \frac{1}{4}, & \text{if } \lambda=\mu_2,\\ 1, & \text{otherwise}.\end{cases}\end{gathered} \tag{37}$$

Then, f is sgdfsc function but not sdfc.

(2) In (1), f is swsgdfsc function but not sdfc.

(3) Let $f:(X,\tau_1,\tau_1^*)\rightarrow(X,\tau_2,\tau_2^*)$ be a function defined by

$$f(a)=a,\qquad f(b)=a. \tag{38}$$

Define λ_1,μ_1, and μ_2 as follows:

$$\begin{aligned}\lambda_1(a)&=1.0, & \lambda_1(b)&=0.9,\\ \mu_1(a)&=0.0, & \mu_1(b)&=0.2,\\ \mu_2(a)&=1.0, & \mu_2(b)&=0.8,\end{aligned} \tag{39}$$

and define (τ_1, τ_1^*) and (τ_2, τ_2^*) as follows:

$$\tau_1(\lambda) = \begin{cases} 1, & \text{if } \lambda \in \{\underline{0}, \underline{1}\}, \\ \frac{1}{2}, & \text{if } \lambda = \lambda_1, \\ 0, & \text{otherwise}, \end{cases}$$

$$\tau_1^*(\lambda) = \begin{cases} 0, & \text{if } \lambda \in \{\underline{0}, \underline{1}\}, \\ \frac{1}{2}, & \text{if } \lambda = \lambda_1, \\ 1, & \text{otherwise}, \end{cases}$$

$$\tau_2(\lambda) = \begin{cases} 1, & \text{if } \lambda \in \{\underline{0}, \underline{1}\}, \\ \frac{1}{3}, & \text{if } \lambda = \mu_1, \\ \frac{2}{3}, & \text{if } \lambda = \mu_2, \\ 0, & \text{otherwise}, \end{cases} \quad (40)$$

$$\tau_2^*(\lambda) = \begin{cases} 0, & \text{if } \lambda \in \{\underline{0}, \underline{1}\}, \\ \frac{2}{3}, & \text{if } \lambda = \mu_1, \\ \frac{1}{3}, & \text{if } \lambda = \mu_2, \\ 1, & \text{otherwise}. \end{cases}$$

Then, f is swsgdfsc function but not sgdfsc.

(4) Let $f : (X, \tau_1, \tau_1^*) \to (X, \tau_2, \tau_2^*)$ be the identity function. Define λ_1, λ_2, and μ as follows:

$$\begin{aligned} \lambda_1(a) &= 0.0, & \lambda_1(b) &= 1.0, \\ \lambda_2(a) &= 1.0, & \lambda_2(b) &= 0.0, \\ \mu(a) &= 0.0, & \mu(b) &= 0.3. \end{aligned} \quad (41)$$

And define (τ_1, τ_1^*) and (τ_2, τ_2^*) as follows:

$$\tau_1(\lambda) = \begin{cases} 1, & \text{if } \lambda \in \{\underline{0}, \underline{1}\}, \\ \frac{1}{3}, & \text{if } \lambda = \lambda_1, \\ \frac{2}{3}, & \text{if } \lambda = \lambda_2, \\ 0, & \text{otherwise}, \end{cases}$$

$$\tau_1^*(\lambda) = \begin{cases} 0, & \text{if } \lambda \in \{\underline{0}, \underline{1}\}, \\ \frac{2}{3}, & \text{if } \lambda = \lambda_1, \\ \frac{1}{3}, & \text{if } \lambda = \lambda_2, \\ 1, & \text{otherwise}, \end{cases}$$

$$\tau_2(\lambda) = \begin{cases} 1, & \text{if } \lambda \in \{\underline{0}, \underline{1}\}, \\ \frac{1}{3}, & \text{if } \lambda = \mu, \\ 0, & \text{otherwise}, \end{cases}$$

$$\tau_2^*(\lambda) = \begin{cases} 0, & \text{if } \lambda \in \{\underline{0}, \underline{1}\} \\ \frac{2}{3}, & \text{if } \lambda = \mu, \\ 1, & \text{otherwise}. \end{cases} \quad (42)$$

Then, f is swgdfso function but not gdfso.

(5) Let $f : (X, \tau_1, \tau_1^*) \to (X, \tau_2, \tau_2^*)$ be the identity function. Define λ_1, μ_1, and μ_2 as follows:

$$\begin{aligned} \lambda_1(a) &= 0.1, & \lambda_1(b) &= 0.1, \\ \mu_1(a) &= 0.05, & \mu_1(b) &= 0.02, \\ \mu_2(a) &= 0.95, & \mu_2(b) &= 0.98. \end{aligned} \quad (43)$$

And define (τ_1, τ_1^*) and (τ_2, τ_2^*) as follows:

$$\tau_1(\lambda) = \begin{cases} 1, & \text{if } \lambda \in \{\underline{0}, \underline{1}\}, \\ \frac{1}{3}, & \text{if } \lambda = \lambda_1, \\ 0, & \text{otherwise}, \end{cases}$$

$$\tau_1^*(\lambda) = \begin{cases} 0, & \text{if } \lambda \in \{\underline{0}, \underline{1}\}, \\ \frac{2}{3}, & \text{if } \lambda = \lambda_1, \\ 1, & \text{otherwise}, \end{cases}$$

$$\tau_2(\lambda) = \begin{cases} 1, & \text{if } \lambda \in \{\underline{0}, \underline{1}\}, \\ \frac{1}{3}, & \text{if } \lambda = \mu_1, \\ \frac{2}{3}, & \text{if } \lambda = \mu_2, \\ 0, & \text{otherwise}, \end{cases} \quad (44)$$

$$\tau_2^*(\lambda) = \begin{cases} 0, & \text{if } \lambda \in \{\underline{0}, \underline{1}\}, \\ \frac{2}{3}, & \text{if } \lambda = \mu_1, \\ \frac{1}{3}, & \text{if } \lambda = \mu_2, \\ 1, & \text{otherwise}. \end{cases}$$

Then, f is swsgdfso function but not sgdfso.

Conflict of Interests

The authors declare that there is no conflict of interests regarding this paper.

Acknowledgments

The authors would like to acknowledge the financial support received from Universiti Kebangsaan Malaysia under the research Grant GUP-2013-040. The authors also wish to gratefully acknowledge all those who have generously given their time to referee their paper.

References

[1] C. L. Chang, "Fuzzy topological spaces," *Journal of Mathematical Analysis and Applications*, vol. 24, no. 1, pp. 182–190, 1968.

[2] J. A. Goguen, "L-fuzzy sets," *Journal of Mathematical Analysis and Applications*, vol. 18, no. 1, pp. 145–174, 1967.

[3] T. Kubiak, *On fuzzy topologies [Ph.D. thesis]*, A. Mickiewicz, Poznan, Poland, 1985.

[4] A. P. Šostak, "On a fuzzy topological structure," in *Proceedings of the 13th Winter School on Abstract Analysis. Section of Topology*, pp. 89–103, Circolo Matematico di Palermo, Palermo, Italy, 1985.

[5] K. Atanassov and S. Stoeva, "Intuitionistic fuzzy sets," in *Proceedings of the Polish Symposium on Interval and Fuzzy Mathematics*, pp. 23–26, Poznan, Poland, August 1983.

[6] K. T. Atanassov, "Intuitionistic fuzzy sets," *Fuzzy Sets and Systems*, vol. 20, no. 1, pp. 87–96, 1986.

[7] K. T. Atanassov, "New operations defined over the intuitionistic fuzzy sets," *Fuzzy Sets and Systems*, vol. 61, no. 2, pp. 137–142, 1994.

[8] K. T. Atanassov, "More on intuitionistic fuzzy sets," *Fuzzy Sets and Systems*, vol. 33, no. 1, pp. 37–45, 1989.

[9] K. T. Atanassov, "Remarks on the intuitionistic fuzzy sets-III," *Fuzzy Sets and Systems*, vol. 75, no. 3, pp. 401–402, 1995.

[10] K. Atanassov, *Intuitionistic Fuzzy Sets*, Physica, New York, NY, USA, 1999.

[11] D. Çoker, "An introduction to intuitionistic fuzzy topological spaces," *Fuzzy Sets and Systems*, vol. 88, no. 1, pp. 81–89, 1997.

[12] T. K. Mondal and S. K. Samanta, "On intuitionistic gradation of openness," *Fuzzy Sets and Systems*, vol. 131, no. 3, pp. 323–336, 2002.

[13] J. Gutiérrez García and S. E. Rodabaugh, "Order-theoretic, topological, categorical redundancies of interval-valued sets, grey sets, vague sets, interval-valued "intuitionistic" sets, "intuitionistic" fuzzy sets and topologies," *Fuzzy Sets and Systems*, vol. 156, no. 3, pp. 445–484, 2005.

[14] R. C. Jain, *The role of regularly open sets in general topology [Ph.D. thesis]*, Meerut University, Institute of Advenced Studied, Meerut-India, 1990.

[15] T. M. Nour, "Slightly semi-continuous functions," *Bulletin of Calcutta Mathematical Society*, vol. 87, pp. 187–195, 1995.

[16] T. Noiri, "Slightly β-continuous functions," *International Journal of Mathematics and Mathematical Sciences*, vol. 28, no. 8, pp. 469–478, 2001.

[17] M. Sudha, E. Roja, and M. K. Uma, "Slightly fuzzy ω-continuous mappings," *International Journal of Mathematical Analysis*, vol. 5, no. 13–16, pp. 779–787, 2011.

[18] E. Ekici and M. Caldas, "Slightly γ-continuous functions," *Boletim da Sociedade Paranaense de Matematica*, vol. 22, no. 2, pp. 63–74, 2004.

[19] R. N. Bhaumik and S. E. Abbas, "(L, M)-intuitionistic fuzzy topological spaces," *Bulletin of Kerala Mathematics Association*, vol. 5, pp. 25–47, 2008.

[20] K. Dong, H. Wang, and Y. Liu, "Formation track association algorithm in the presence of sensor bias based on double fuzzy topology," *Applied Mechanics and Materials*, vol. 536-537, pp. 329–332, 2014.

[21] P. K. Saha, "Fuzzy digital topology and geometry and their applications to medical imaging," in *Pattern Recognition and Machine Intelligence*, vol. 8251, pp. 13–29, Springer, Berlin, Germany, 2013.

[22] I. Bloch, "On fuzzy distances and their use in image processing under imprecision," *Pattern Recognition*, vol. 32, no. 11, pp. 1873–1895, 1999.

[23] K. Liu and W. Shi, "Quantitative fuzzy topological relations of spatial objects by induced fuzzy topology," *International Journal of Applied Earth Observation and Geoinformation*, vol. 11, no. 1, pp. 38–45, 2009.

[24] M. Demirci and D. Çoker, "An introduction to intuitionistic fuzzy topological spaces in Šostak's sense," *Busefal*, vol. 67, pp. 67–76, 1996.

[25] Y. C. Kim and S. E. Abbas, "Several types of fuzzy regular spaces," *Indian Journal of Pure and Applied Mathematics*, vol. 35, no. 4, pp. 481–500, 2004.

[26] S. E. Abbas, "Several types of double fuzzy semiclosed sets," *The Journal of Fuzzy Mathematics*, vol. 20, pp. 89–102, 2012.

[27] F. M. Mohammed, M. S. M. Noorani, and A. Ghareeb, "Slightly double fuzzy continuous functions," *Journal of the Egyptian Mathematical Society*, 2014.

Permissions

The contributors of this book come from diverse backgrounds, making this book a truly international effort. This book will bring forth new frontiers with its revolutionizing research information and detailed analysis of the nascent developments around the world.

We would like to thank all the contributing authors for lending their expertise to make the book truly unique. They have played a crucial role in the development of this book. Without their invaluable contributions this book wouldn't have been possible. They have made vital efforts to compile up to date information on the varied aspects of this subject to make this book a valuable addition to the collection of many professionals and students.

This book was conceptualized with the vision of imparting up-to-date information and advanced data in this field. To ensure the same, a matchless editorial board was set up. Every individual on the board went through rigorous rounds of assessment to prove their worth. After which they invested a large part of their time researching and compiling the most relevant data for our readers. Conferences and sessions were held from time to time between the editorial board and the contributing authors to present the data in the most comprehensible form. The editorial team has worked tirelessly to provide valuable and valid information to help people across the globe.

Every chapter published in this book has been scrutinized by our experts. Their significance has been extensively debated. The topics covered herein carry significant findings which will fuel the growth of the discipline. They may even be implemented as practical applications or may be referred to as a beginning point for another development.

The editorial board has been involved in producing this book since its inception. They have spent rigorous hours researching and exploring the diverse topics which have resulted in the successful publishing of this book. They have passed on their knowledge of decades through this book. To expedite this challenging task, the publisher supported the team at every step. A small team of assistant editors was also appointed to further simplify the editing procedure and attain best results for the readers.

Our editorial team has been hand-picked from every corner of the world. Their multi-ethnicity adds dynamic inputs to the discussions which result in innovative outcomes. These outcomes are then further discussed with the researchers and contributors who give their valuable feedback and opinion regarding the same. The feedback is then collaborated with the researches and they are edited in a comprehensive manner to aid the understanding of the subject.

Apart from the editorial board, the designing team has also invested a significant amount of their time in understanding the subject and creating the most relevant covers. They scrutinized every image to scout for the most suitable representation of the subject and create an appropriate cover for the book.

The publishing team has been involved in this book since its early stages. They were actively engaged in every process, be it collecting the data, connecting with the contributors or procuring relevant information. The team has been an ardent support to the editorial, designing and production team. Their endless efforts to recruit the best for this project, has resulted in the accomplishment of this book. They are a veteran in the field of academics and their pool of knowledge is as vast as their experience in printing. Their expertise and guidance has proved useful at every step. Their uncompromising quality standards have made this book an exceptional effort. Their encouragement from time to time has been an inspiration for everyone.

The publisher and the editorial board hope that this book will prove to be a valuable piece of knowledge for researchers, students, practitioners and scholars across the globe.

List of Contributors

Hiroyoshi Oda and Makoto Tsukada
Department of Information Sciences, Toho University, Funabashi, Chiba 274-8501, Japan

Takeshi Miura
Yamagata University, Yonezawa, Yamagata 273-0866, Japan

Yuji Kobayashi
Toho University, Funabashi, Chiba 274-8501, Japan

Sin-Ei Takahasi
Toho University, Yamagata University, Funabashi, Chiba 273-0866, Japan

Asma Ali and Phool Miyan
Department of Mathematics, Aligarh Muslim University, Aligarh 202002, India

Howard E. Bell
Department of Mathematics, Brock University, St. Catharines, ON, Canada

Byung Keun Sohn
Department of Mathematics, Inje University, Kimhae 621-749, Republic of Korea

Hyang Sook Kim
Department of Applied Mathematics, Institute of Basic Science, Inje University, Gimhae 621-749, Republic of Korea

Jin Suk Pak
Kyungpook National University, Daegu 702-701, Republic of Korea

G. Muhiuddin and Abdullah M. Al-roqi
Department of Mathematics, University of Tabuk, Tabuk 71491, Saudi Arabia

Young Bae Jun
Department of Mathematics Education (and RINS), Gyeongsang National University, Jinju 660-701, Republic of Korea

Yilmaz Ceven
Department of Mathematics, Faculty of Science, Suleyman Demirel University, 32260 Isparta, Turkey

Slavko Simic
Mathematical Institute SANU, Kneza Mihaila 36, 11000 Belgrade, Serbia

Young Jae Sim and Oh Sang Kwon
Department of Mathematics, Kyungsung University, Busan 608-736, Republic of Korea

Abdolamir Karbalaie, Hamed Hamid Muhammed and Bjorn-Erik Erlandsson
Division of Informatics, Logistics and Management, School of Technology and Health (STH), Royal Institute of Technology (KTH), 100 44 Stockholm, Sweden

Gun Sunyeekhan
50 Department of Mathematics, Faculty of Science, Kasetsart University, Bangkok 10900, Thailand

Paul Bracken
Department of Mathematics, University of Texas, Edinburg, TX 78541-2999, USA

Joshua Sack
Institute of Logic, Language and Computation, Universiteit van Amsterdam, P.O. Box 94242, 1090 GE Amsterdam, The Netherlands

Saleem Watson
Department of Mathematics, California State University, Long Beach, CA 90840, USA

Samit Bhattacharyya and Suma Ghosh
Departments of Mathematics and Biology, University of Utah, Salt Lake City, UT 84112, USA

Esra Betul Koc Ozturk and Ufuk Ozturk
Department of Mathematics, Faculty of Sciences, University of Cankiri Karatekin, Cankiri 18100, Turkey
School of Mathematics & Statistical Sciences, Arizona State University, Room PSA442, Tempe, AZ 85287-1804, USA

Kazim Ilarslan
Department of Mathematics, Faculty of Sciences and Art, University of Kırıkkale, Kırıkkale 71450, Turkey

Emilija Nešovic
University of Kragujevac, Faculty of Science, Department of Mathematics and Informatics, Kragujevac 34000, Serbia

Md. Azizul Baten
Department of Decision Science, School of Quantitative Sciences, Universiti Utara Malaysia (UUM), Sintok, 06010 Kedah, Malaysia

Anton Abdulbasah Kamil
Mathematics Program, School of Distance Education, Universiti Sains Malaysia (USM), 11800 Penang, Malaysia

Mowaffaq Hajja
Department of Mathematics, Yarmouk University, Irbid, Jordan

Armands Gritsans and Felix Sadyrbaev
Daugavpils University, Department of Mathematics, Parades Street 1, 5400 Daugavpils, Latvia

Daniel Simson and Katarzyna Zajac
Faculty of Mathematics and Computer Science, Nicolaus Copernicus University, Ulica Chopina 12/18, 87-100 Torun, Polan

E. Karapinar
Department of Mathematics, Atilim University, Incek, 06836 Ankara, Turkey

D. K. Patel and D. Gopal
Department of Mathematics and Humanities, S. V. National Institute of Technology, Surat, Gujarat 395 007, India

M. Imdad
Department of Mathematics, Aligarh Muslim University, Aligarh 202 002, India

Huda Aldweby and Maslina Darus
School of Mathematical Sciences, Faculty of Science and Technology, Universiti Kebangsaan Malaysia, 43600 Bangi, Selangor, Malaysia

M. Y. Waziri
Department of Mathematics, Faculty of Science, Universiti Putra Malaysia, 43400 Serdang, Selangor, Malaysia
Department of Mathematical Sciences, Faculty of Science, Bayero University Kano, Kano, Nigeria

Z. A. Majid
Department of Mathematics, Faculty of Science, Universiti Putra Malaysia, 43400 Serdang, Selangor, Malaysia
Institute for Mathematical Research, Universiti Putra Malaysia, 43400 Serdang, Selangor, Malaysia

Biljana Jolevska-Tuneska
Faculty of Electrical Engineering and Informational Technologies, Saints Cyril and Methodius University of Skopje, Karpos II bb, 1000 Skopje, Macedonia

Tatjana Atanasova-Pacemska
Faculty of Computer Sciences, Goce Delcev University of Stip, Krste Misirkov, Br 10A, 2000 Stip, Macedonia

Aliasghar Alikhani-Koopaei
Department of Mathematics, Berks College, Pennsylvania State University, Tulpehocken Road, P.O. Box 7009, Reading, PA 19610-6009, USA

Pentti Haukkanen, Mika Mattila and Jorma K. Merikoski
School of Information Sciences, University of Tampere, 33014, Finland

Timo Tossavainen
School of Applied Educational Science and Teacher Education, University of Eastern Finland, P.O. Box 86, 57101 Savonlinna, Finland

Jin Hong Kim
Department of Mathematics Education, Chosun University, 309 Pilmundaero, Dong-gu, Gwangju 501-759, Republic of Korea

Nina Shang, Qinghua Feng and Huizeng Qin
Institute of Applied Mathematics, Shandong University of Technology, Zibo, Shandong 255049, China

M. S. M. Noorani
School of Mathematical Sciences, Universiti Kebangsaan Malaysia, 43600 UKM Bangi, Selangor, Malaysia

Fatimah M. Mohammed
School of Mathematical Sciences, Universiti Kebangsaan Malaysia, 43600 UKM Bangi, Selangor, Malaysia
College of Education, Tikrit University, Iraq

A. Ghareeb
Department of Mathematics, College of Science in Al-Zulfi, Majmaah University, Al-Zulfi, Saudi Arabia
Department of Mathematics, Faculty of Science, South Valley University, Qena, Egypt